Environmental Radiochemical Analysis II

Environmental Radiochemical Analysis II

Edited by

Peter Warwick

Department of Chemistry, Loughborough University, Leicestershire

RS•C

advancing the chemical sciences

The proceedings of the 9th International Symposium on Environmental Radiochemical Analysis held on 18–20 September 2002 in Maidstone, Kent, UK.

Special Publication No. 291

ISBN 0-85404-618-6

A catalogue record for this book is available from the British Library

Published by The Royal Society of Chemistry,
Thomas Graham House, Science Park, Milton Road,
Cambridge CB4 0WF, UK

Registered Charity Number 207890

For further information see our web site at www.rsc.org

Printed by Athenaeum Press Ltd, Gateshead, Tyne & Wear, UK

Preface

Environmental Radiochemical Analysis II is a collection of refereed papers presented at the 9th International Symposium on Environmental Radiochemical Analysis held in Maidstone, Kent, UK in September 2002. The Symposium is part of the Radiochemical Methods Group's programme and is held every four years. The papers provide the latest information on new methods of radionuclide analyses, developments and improvements in existing methods, method comparisons, environmental studies, method uncertainties, underground laboratories and software development. The papers were presented as posters or oral presentations and were delivered by experts from many different parts of the world. The book therefore presents the latest information concerning environmental analysis.

Particular thanks go to members of the Organising Committee: Prof Susan Parry (Imperial College), Dr Anthony Ware, Mr Peter Hodson (Veternary Laboratories Agency), Dr Belinda Colston (De Monfort University) and Dr Kinson Leonard (CEFAS) for their help in making the Symposium a success.

Peter Warwick

Contents

DEVELOPMENT OF AN IN-VITRO PROCEDURE TO MEASURE THE AVAILABILITY OF SOIL-ASSOCIATED RADIONUCLIDES FOR UPTAKE AFTER INADVERTENT INGESTION BY HUMANS

S. Shaw and N. Green

National Radiological Protection Board, Chilton, Didcot, Oxon OX11 0RQ, UK

1 INTRODUCTION

The ingestion of soil by humans is relatively common and can be inadvertent or purposeful. Inadvertent ingestion of soil generally occurs through primitive living conditions or professions that have close and continual contact with soil. For example, dust may adhere to hands or plants and be subsequently transferred to the mouth and ingested. Inadvertent ingestion is of particular importance to children because they may be less meticulous about hygiene than adults. Purposeful ingestion of soil can occur as a result of pica (an eating disorder manifested by a craving to ingest any material not fit for food) or geophagy, a special case of pica where substances such as chalk, clay or earth are ingested.

Radiological assessments of the impact of routine and accidental releases of radionuclides into the environment generally take into account the dose from inadvertent ingestion of radionuclides associated with soil or sediment. However, a lack of relevant data has meant that radionuclides associated with soil or sediment are assumed to have the same availability for gut uptake as those incorporated into food. Studies investigating the availability of soil-associated radionuclides after ingestion have been mainly conducted on ruminant animals.[1-5] The ruminant digestive tract is totally different to that of man, and so data from animal experiments may not be valid in the case of humans. The importance of inadvertent ingestion in radiological assessments can be illustrated using a recent study on the potential incursion of marine sediment inland in Northwest England.[6] The results indicated that inadvertent ingestion of deposited material could account for a substantial fraction of the estimated dose, especially for children.

Studies to investigate the availability and the possible absorption of soil-associated radionuclides in humans have to be carried out *in vitro*. The digestive system in humans is difficult to simulate because of the complex processes involved. Many experiments designed to simulate human digestion have been carried out to determine the availability/absorption of trace elements and heavy metals in nutritional studies.[7-15] These studies either used expensive complex equipment or were simple acid extractions carried out on a small scale.

This paper summarises the development of a relatively simple inexpensive *in vitro* digestion procedure (enzymolysis) that would be able to simulate *in vivo* conditions as closely as possible. The enzymolysis procedure was then used to determine the availability of specific radionuclides, for potential absorption across the gut, in a range of soil types. A full report is available at NRPB's web site.[16]

2 MATERIALS

2.1 Soil Types and Radionuclides of Interest

Loam, peat and sand soils were chosen for this study to cover a broad spectrum of cultivated soils found in the UK. The soils were collected from NRPB's established lysimeter facility, set up in 1983 and containing elevated levels of ^{137}Cs, ^{90}Sr, ^{239}Pu and ^{241}Am. Collected soil was stored in a field moist condition at 4 °C until required; sub-samples of soil were air-dried before use.

2.2 Reagents

The porcine enzymes pepsin (catalogue number P-7000), pancreatin (catalogue number P-1750), α-amylase (catalogue number A-1376) and bile salts (catalogue number B-8756) were purchased from Sigma Chemicals. The enzymes are available from the manufacturer in different strengths and therefore the catalogue numbers have been given here for clarification. All other reagents used were Analytical Reagent grade or equivalent.

2.3 Equipment

The vessel used to carry out the digestion was a 2 litre magnetic culture vessel. The vessel had two access ports and contained a magnetic stirrer and baffles to aid mixing. A magnetic hotplate stirrer connected to an electronic contact thermometer was used to keep the solution well mixed and at a constant temperature of 37 °C (body temperature).

3 ANALYTICAL METHODS

3.1 Sample Preparation

The liquid and solid phases of a sample were separated by centrifugation immediately after completion of the enzymolysis procedure described in Section 4. The liquid phase was then decanted through a series of cellulose nitrate membrane filters (5.0, 0.65 and 0.45μm). The membrane filters were dissolved in acetone and evaporated to dryness for analysis with the solid phase.

3.1.1 Liquid Phase. The sample volume was reduced, the solution was acidified with 8 M nitric acid and poured into a suitable container for ^{137}Cs determination, if required. After ^{137}Cs measurement the solution was evaporated to dryness and ashed at

500°C. The ash was dissolved in the minimum volume of 8 M nitric acid. Any undissolved residue was separated from solution and treated with small amounts of a 3:1 hydrofluoric acid: nitric acid mixture, to ensure total dissolution, before re-dissolving in 8 M nitric acid and combining with the bulk sample.

3.1.2 Solid Phase. After separation from the liquid phase, the residue was evaporated to dryness and then dried at 105°C. The dried residue was lightly ground and packed into a suitable container for ^{137}Cs determination, if required. After ^{137}Cs determination the sample was then combined with the membrane filters and carefully ashed at 500°C. Dissolution of the ash was carried out as described in Section 3.1.1. If after the hydrofluoric acid: nitric treatment solids still remained, then a caustic fusion was carried out to obtain total dissolution. The caustic fusion procedure was a modification of the method originally described by Bock,[17] the full method is given in NRPB W-17.[16]

3.2 Sample Analysis

Caesium-137 was measured by gamma-ray spectrometry using hyperpure Ge detectors, housed in a purpose built facility and appropriately calibrated. The separation and purification of ^{90}Sr was carried out using Sr resin (EIChroM Industries, Inc.) and based on a method developed by Shaw and Green.[18] After a suitable period of ingrowth followed by separation, ^{90}Sr was determined by beta counting of its ^{90}Y daughter on a low background gas flow beta counter. The analytical methods used for ^{239}Pu and ^{241}Am were based on those described by Krey and Beck[19] and Ham[20] respectively. All of the analytical methods employed are in regular operational use at NRPB's laboratories and are carried out within formal agreement with the United Kingdom Accreditation Service (UKAS).

4 ENZYMOLYSIS PROCEDURE DEVELOPMENT

After evaluation of several published methods,[7, 8, 11, 13-15, 21-24] a two phase enzymolysis procedure was designed based on a method developed by McKay and Memmott.[24] The first phase of the procedure simulated the conditions found in the stomach and the second phase those found in the small intestines. The full detailed enzymolysis procedure is given in NRPB W-17.[16] Briefly, the stomach phase consisted of pepsin dissolved in a saline hydrochloric solution which was heated to 37°C and then air-dried soil added. The stomach phase was then incubated for 1 h. After 1 h, the simulated intestinal fluid (intestinal enzymes dissolved in a sodium bicarbonate solution) heated to 37°C was added to the stomach phase and the pH adjusted to 7.5 with sodium hydroxide. The intestinal phase was then incubated for 2 h. When the digestion was complete the solid and liquid phases were separated by centrifugation.

The experimental protocol was developed by carrying out various studies to ensure that parameters which could possibly affect the extraction of radionuclides from the soil were optimised. The incubation times were chosen to reflect retention times of food in the stomach and small intestines and were investigated to see if longer incubation times increased the amount of activity extracted from the soil. It was found that longer incubation times had little effect on the extraction of the soil-associated radionuclides. The extraction of the radionuclides in the different phases was also studied. These experiments showed that a two phase procedure was necessary to simulate *in vivo* conditions.

Whilst evaluating the enzymolysis procedure, problems with soil ageing were also discovered. It is common practice to collect a bulk sample of soil and store in a refrigerator when sub-samples are required for a series of experiments. The storage of soil was found to have an indirect effect on the amount of activity extracted from the soil, especially for ^{239}Pu. Therefore, subsequent experiments used freshly sampled soil is used for experiments. The detailed enzymolysis procedure development is given in NRPB W-17.[16]

5 AVAILABILITY OF SPECIFIC RADIONUCLIDES FOR UPTAKE

The enzymolysis procedure was used to determine the availability of ^{137}Cs, ^{239}Pu, ^{241}Am and ^{90}Sr in loam, peat and sand soils. Each experiment was carried out in quadruplicate to ensure reproducibility of the results. The results are given in Table 1.

Table 1 *Effect of soil type on the percentage of activity extracted using the enzymolysis procedure*

Soil type	*^{137}Cs*	*^{239}Pu*	*^{241}Am*	*^{90}Sr*
Loam	1.8 ± 0.4	6.1 ± 0.5	2.7 ± 0.5	51 ± 5
	2.1 ± 0.6	5.6 ± 0.4	2.7 ± 0.5	Lost sample
	1.1 ± 0.5	8.2 ± 0.7	2.2 ± 0.4	55 ± 5
	1.8 ± 0.4	6.1 ± 0.5	2.9 ± 0.5	53 ± 5
Mean	1.7 sd 0.42	6.5 sd 1.2	2.6 sd 0.30	53 sd 2
Peat	2.7 ± 0.5	7.1 ± 0.6	1.2 ± 0.2	20 ± 2
	2.9 ± 0.5	5.7 ± 0.5	1.0 ± 0.2	18 ± 2
	3.3 ± 0.6	5.6 ± 0.4	1 ± 0.1	18 ± 2
	3.1 ± 0.6	6.2 ± 0.5	1 ± 0.1	20 ± 2
Mean	3.0 sd 0.26	6.2 sd 0.69	1.1 sd 0.10	19 sd 1.2
Sand	3.3 ± 0.8	10.4 ± 0.8	2.3 ± 0.3	Lost sample
	3.2 ± 0.8	10.3 ± 0.8	2.5 ± 0.3	46 ± 5
	2.9 ± 0.8	11.5 ± 0.9	1.8 ± 0.2	49 ± 5
	3.3 ± 0.9	13.1 ± 1	3.6 ± 0.4	49 ± 5
Mean	3.2 sd 0.19	11.3 sd 1.2	2.5 sd 0.76	48 sd 1.7

The soil type did not have a marked effect on the amount of ^{137}Cs and ^{241}Am that was extracted, the range being between 1% and 3%. However, the amount of ^{239}Pu and ^{90}Sr extracted was dependent on soil type. For ^{239}Pu, approximately 6% was extracted from loam and peat soils, whereas 11 % was extracted from sand soil. The effect of soil type was even greater for ^{90}Sr, approximately 20% being extracted from peat soil compared to around 50% extracted from loam and sand soil. If the soil type is unknown it is recommended that the generalised availability factors 3%, 3%, 10% and 50% for ^{137}Cs,

^{241}Am, ^{239}Pu and ^{90}Sr respectively are used. There was little variation between the results of the quadruplicate samples indicating that the procedure gave reproducible data.

6 DISCUSSION

The fraction of ingested activity that is transferred across the gut is usually represented by the gut uptake factor f_1. However, a radionuclide that is ingested in an available form need not necessarily have an f_1 value of unity even when administered in solution.[3, 25] Studies carried out using ruminant animals introduced the concept of an availability factor, A_f.[3] This was a measure of the proportion of ingested activity that could be solubilised by the rumen and therefore be considered potentially available for absorption across the gut. A simple multiplication of the A_f value by the f_1 value could then be used to estimate the transfer of soil-associated radionuclides to animal products.[26]

It would be reasonable to assume that the A_f values derived from this study for humans could be used in a similar way. It should be borne in mind however, that activity not absorbed across the gut leads to exposure of the gut wall; this is reflected in the calculations of appropriate dose coefficients. Dose coefficients for adults were re-calculated by the computer program Pleiades using the estimated A_f value (A.Phipps, personal communication). Table 2 shows the recalculated values and compares them with the simpler calculation of the published dose coefficient for an adult[28] multiplied by the A_f value.

Table 2 *Comparison of dose coefficients*

Soil	*^{137}Cs*		*^{239}Pu*		*^{241}Am*		*^{90}Sr*	
type	*Pleiades*	*ICRP**	*Pleiades*	*ICRP**	*Pleiades*	*ICRP**	*Pleiades*	*ICRP**
Loam	1.5 E-9	2.2 E-10	2.0 E-8	1.6 E-8	9.9 E-9	5.2 E-9	1.5 E-8	1.5 E-8
Peat	1.7 E-9	3.9 E-10	1.9 E-8	1.6 E-8	6.7 E-9	2.2 E-9	6.8 E-9	5.3 E-9
Sand	1.7 E-9	4.2 E-10	3.2 E-8	2.8 E-8	97 E-9	5.0 E-9	1.4 E-8	1.3E-8

* The product of the determined A_f value (Table 1) and the ICRP recommended dose coefficient.

For the radionuclides considered here, the simple multiplication of the published ICRP dose coefficients[27] by the derived A_f value underestimates the dose coefficents as calculated by Pleiades. The use of the simple multiplication technique would not necessarily be justified for assessments of effective dose; where possible dose coefficients should be recalculated.

Generalised Derived Limits (GDLs) have been published by NRPB as a means of making broad assessments of the radiological impact of activity released into the environment.[28] Inadvertent ingestion of well-mixed soil and sediment was not found to be an important contributor to the overall dose for the radionuclides considered in this study. However, the dominant exposure pathway for GDLs for ^{210}Pb and ^{210}Po in marine and fresh water sediment is inadvertent ingestion.[29] The dose coefficients used in the calculation of GDLs were those applicable to general foodstuffs, and so if the availability factor was very much less than unity then the numerical value of the GDL could be affected significantly. Therefore, the determination of availability factors for ^{210}Pb and ^{210}Po in marine and fresh water sediment is important for future work.

Food would normally be present in the stomach when ingestion of soil occurs, which may affect the availability of the soil-associated radionuclides. Different food types have already been shown to increase or decrease the absorption of plutonium across the gut.[30] In addition, other studies on heavy metal contamination have shown that soil can enhance or inhibit bioavailability.[15] Therefore, interactions between food and soil-associated radionuclides merit further investigation.

7 CONCLUSIONS

A study has been carried out to develop a simple enzymolysis procedure to simulate human digestion and to determine availability factors of soil-associated radionuclides using that procedure.

- Given practical constraints, the procedure adopted closely simulates human digestion, is relatively simple to carry out, has minimal costs and gives reproducible results for a range of radionuclides.
- Soil storage can affect the results from the enzymolysis procedure, and so freshly collected soil should be used.
- The availability factors A_f for soil-associated ^{137}Cs, ^{239}Pu, ^{241}Am and ^{90}Sr for uptake by the human gut have been determined for loam, peat and sand soils. In some cases the A_f values were dependent upon soil type.
- When the soil type is unknown it is recommended that generalised availability factors of 3%, 3%, 10% and 50% for ^{137}Cs, ^{241}Am, ^{239}Pu and ^{90}Sr respectively be used for radiological assessments.
- The use of a simple multiplication of published dose coefficients by the availability factor should be carried out with caution; where possible dose coefficients should be recalculated.
- Future work is required to determine the availability factors for ^{210}Pb and ^{210}Po in marine and fresh water sediment and to investigate the interactions between food and soil-associated radionuclides.

References

1 M. Belli, M. Blasi, E. Capra, A. Drigo, S. Menegon, E. Piasentier and U. Sansone, *Sci. Total Environ.*, 1993, **136**, 243.
2 N.A. Beresford, R.W. Mayes, B.J. Howard, H.F. Eayres, C.S. Lamb, C.L. Barnett and M.G. Segal, *Radiation Protection Dosimetry*, 1992, **41**, 87.
3 A.I. Cooke, *PhD Thesis, University of Newcastle*, 1995.
4 A.I. Cooke, N. Green, D.L. Rimmer, T.E.C. Weekes and B.T. Wilkins, *J. Environ. Radioactivity*, 1995, **28**, 191.
5 A.I. Cooke, N. Green, D.L. Rimmer, T.E.C. Weekes, B.T. Wilkins, N.A. Beresford and J.D. Fenwick, *Sci. Total Environ.*, 1996, **192**, 21.
6 D.A. Huntley, K.R. Dyer, D. Cavrot, M.J. Tooley, I.K. Haslam, N. Green and B.T. Wilkins, *NRPB-R284*, London HMSO, 1996.

7 H.M. Crews, J.A. Burrell and D.J. Mc Weeny, *J. Sci. and Food Agric.*, 1983, **34**, 997.
8 R.A Culp and A.B. Rawitch, *J. of Paint Tech.*, 1973, **45**, 38.
9 A. Jacobs and D.A. Greenman, *Brit. Med. J.*, (1969), **1**, 673.
10 S. Lock and A.E. Bender, *Brit. J. Nutrition*, 1980, **43**, 413.
11 D.D. Miller, B.R. Schricker, R.R. Rasmussen and D. Van Campen, *The Am. J. Clin.Nut.*, 1981, **34**, 2248.
12 B.S. Narasinga Rao and T. Prabhavathi, *Am. J. Clin. Nut.*, 1978, **31**, 169.
13 R.R. Rodriguez and N.T. Basta, *Environ. Sci. Technol.*, 1999, **33**, 642.
14 M.V. Ruby, A. Davis, R. Schoof, S. Eberle and C.M. Sellstone, *Environ. Sci. Technol.*, 1996, **30**, 422.
15 S.C. Sheppard, W.G. Everden and W.J. Schwartz, *J. Environ. Qual.*, 1995, **24**, 498.
16 S. Shaw and N. Green, *The availability of soil-associated radionuclides for uptake after inadvertent ingestion*, W-17, 2002, from URL nrpb.org/publications/reports/w-17.
17 R. Bock, *A handbook of decomposition methods*, Analytical Chemistry's Glasgow International Textbook Company, 1979.
18 S. Shaw and N. Green, *Determination of ^{90}Sr in calcium rich environmental samples using extraction chromatography*, National Radiological Protection Board, Chilton, Memorandum NRPB-M1284, 2001.
19 P.W. Krey and H.L. Beck, *Environmental measurements laboratory procedures manual*, US Dept of Energy, HASL-300, New York, 1990.
20 G.J. Ham, *Sci. Total, Environ.*, 1995, **173/174**, 19.
21 A.M. Aura, H. Härkönen, M. Fabritius and K. Poutanen, *J. Cereal Sci.*, 1999, **29**, 139.
22 D.A.Garrett, M.L. Failla and R.J. Sarama, *J. Agric. Food Chem.*, 1999, **47**, 4301.
23 M. Lucarini, R. Canali, G. Di Lullo and G. Lombardi-Boccia, *Food Chem.*, 1999, **64**, 519.
24 W.A. McKay and S.D. Memmott, *Food Additives and Contam.*, 1991, **8**, 781.
25 J.D. Harrison, *Sci. Total Environ.*, 1991, **100**, 43.
26 B.T. Wilkins, A.I. Cooke, N. Green, D.L. Rimmer and T.E.C. Weekes, *Rad. Protection Dos.*, 1997, **69**, 111.
27 International Commission on Radiological Protection, *Age-dependent doses to members of the public from intake of radionuclides: part 5 compilation of ingestion and inhalation dose coefficients*, ICRP Publication 72, Annals of ICRP, **26**(1), 1996.
28 National Radiological Protection Board, *Revised generalised derived limits for radioisotopes of strontium, ruthenium, iodine, caesium, plutonium, americium and curium*, Doc. NRPB, London HMSO, 1998, **9**(1).
29 National Radiological Protection Board, *Generalised derived limits for radioisotopes of polonium, lead, radium and uranium*, Doc. NRPB, London HMSO, 2000, **11, 43**.
30 D.M. Taylor, J.R.Duffield and S.A. Proctor, *Speciation of Fission and Activation Products in the Environment*, ed. R.A. Bulman and J.R. Cooper, Elsevier, New York, 1986, p. 208.

SEPARATION OF IRON-55/59 FROM FISSION AND ACTIVATION PRODUCTS USING DI ISOBUTYLKETONE-BASED EXTRACTION CHROMATOGRAPHIC MATERIALS

P E Warwick & I W Croudace

Southampton Oceanography Centre, Southampton, SO14 3ZH

1 INTRODUCTION

Iron-55 is produced by neutron activation of stable ^{54}Fe (5.9% natural abundance). Iron-55 is therefore widely found, along with other common activation products, in wastes arising from nuclear power generation. The ^{55}Fe activity in such wastes has been determined, following limited chemical purification, via its X-ray emission although the technique is not particularly sensitive. Low-level determination of ^{55}Fe requires more rigorous chemical separation followed by liquid scintillation analysis of the ^{55}Fe activity. However, chemical separation is usually complicated by the large amounts of stable iron often present in the sample.

Purification of Fe has been performed using anion exchange chromatography[1]. Fe is retained as the $FeCl_4^-$ species in HCl concentrations greater than 6M and is either eluted using dilute HCl or more rapidly using HNO_3. However, a wide range of other elements will follow Fe, including Co, Zn, Ga, Sn, Sb, Po and U. Improved isolation of Fe is readily achieved from most other radionuclides by extracting the iron chloro- complex from >6M HCl into a solvent such as di isopropyl ether, methyl *iso*butylketone or ethyl acetate.

Little has been published on the extraction chromatographic purification of ^{55}Fe. One of the main limitations in the use of this technique is the requirement for large loading capacities necessary to accommodate the levels of stable Fe routinely found in samples. Konig *et al*[2] used Chelex 100 resin to preconcentrate Fe from steel and concrete samples. The technique was not Fe-specific and solvent extraction was still required to purify the ^{55}Fe prior to liquid scintillation counting. Testa *et al*[3] employed a tri-octylamine-based extraction chromatographic material to isolate ^{55}Fe from 3M HNO_3. TRU resin has also been employed[4] for isolation of ^{55}Fe. Fe is retained from 8M HNO_3 and subsequently eluted with 2M HNO_3. However, the maximum Fe loading for the procedure is limited to 3mg for a standard column containing 0.7g of resin. Silica-immobilised formylsalicylic acid was found to efficiently extract Fe[5]. Loading capacities of approximately 54 mg Fe per gram of resin were reported. However, the extractant was not specific to Fe and the column was therefore most useful as a preconcentrator.

The incorporation of an organic extractant such as ethyl acetate or di *iso*butyl ketone on an inert support would produce a chromatographic material exhibiting the specificity for Fe observed in traditional solvent extraction-based techniques whilst incorporating the superior separation and handling characteristics achieved by column chromatography.

2 METHODOLOGY

2.1 Reagents and equipment

Di *iso*butylketone (GPR grade) was supplied by Merck Ltd, Poole, Dorset and used without any further purification. Amberlite XAD-7 was supplied by Aldrich chemicals, Gillingham, UK. The XAD-7 supplied was too coarse for the current application. The material was therefore ground and sieved to produce a material with a particle size of 125 – 250 μm. Amberchrom CG-71 (an inert polyacrylamide resin with a particle size of 100-150μm) was supplied by Eichrom Industries, Darrien, USA.

All stable element standards were supplied by BDH chemicals, Poole, UK. Chromium-51, ^{54}Mn, ^{59}Fe, ^{60}Co and ^{65}Zn were produced by neutron irradiation of the corresponding stable element at the Imperial College Reactor, Silwood Park, UK. Plutonium-242 and ^{209}Po were supplied by AEA Technology, Harwell, UK.

All other reagents were supplied by Fisher Scientific Ltd, Loughborough, UK. Milli-Q grade water was used throughout. Unless otherwise stated all reagents were of analytical grade or better.

2.2 Choice of extractant and preparation of chromatographic material

Initial attempts to load ethyl acetate onto an inert XAD-7 support were unsuccessful. Approximately 1 g of XAD-7 was slurried with 5ml of ethyl acetate and approximately 50ml of 6M HCl was mixed into the slurry. The ethyl acetate-loaded XAD-7 was allowed to settle and the excess HCl was decanted off. The resulting mixture was used to prepare a 3 x 0.5 cm column. On passing 1mg of Fe in 5ml 6M HCl through the column a distinct yellow band was visible at the top of the column. However, this band of Fe passed rapidly through the column and breakthrough of Fe was noted after only 2ml of 6M HCl washings. The rapid elution of Fe from the column was explained by the relatively high solubility of ethyl acetate in 6M HCl and the hydrolytic degradation of the ethyl acetate. Both factors would result in a rapid decrease in the amount of ethyl acetate on the column and subsequent removal of the $FeCl_4^-$ complex from the column.

Di *iso*butyl ketone (DIBK) loaded onto an inert XAD-7 support was also investigated. The solvent exhibited higher extraction coefficients for Fe compared to the more commonly used di *iso*propyl ether (DIPE). Although DIBK is more soluble in HCl than DIPE, the improved extraction coefficient makes the use of DIBK preferable to DIPE. 2g of Amberlite XAD-7 resin were slurried with 3.5ml of DIBK. The DIBK was added dropwise to the XAD-7 with constant stirring to produce a damp white solid. This solid was then slurried with 6M HCl again with dropwise addition of the HCl and constant stirring of the mixture. Rapid addition of HCl with insufficient stirring resulted in a hydrophobic solid that was unsuitable for column preparation. 2.5g of extraction material were sufficient to prepare one column of 5 x 0.9 cm (3.4ml bed volume).

Amberlite XAD-7 was initially chosen as the inert support. As Amberlite XAD-7 was only available in a coarse mesh size, the resin was ground using an agate pestle and mortar. The ground resin was sieved and the 125 - 250μm fraction was subsequently used. In later experiments, Amberchrom CG1 was used as the inert support. This material is available commercially in 100-150μm particle size and was used without any further treatment.

Table 1 *Measured distribution coefficients and loading capacities of Fe for various solvents*

Solvent	Partition coefficient for Fe from 6M HCl	Effective loading capacity mg Fe/ ml solvent
Ethyl acetate	65	95
Di *iso*butyl ketone	38	49
Di *iso*propyl ether	4	59

Table 2 : *Physical properties of Amberchrom CG-71*

Particle size	76% 100-150μm < 10% < 100μm
Surface area	> 475 m^2/g dry resin
Porosity	0.50 – 0.70 cm^3/cm^3

Data supplied by Eichrom Indiustries

3 RESULTS

3.1 Physical characterisation of the DIBK/XAD-7 column

Unless otherwise stated, all evaluations were performed using a 5 x 0.9cm column. The column dead volume and bed volume were determined as 0.17ml and 3.39ml respectively. Elution profiles of non-adsorbed ^{137}Cs showed complete elution of the element after only 4ml of eluent had passed through the column.

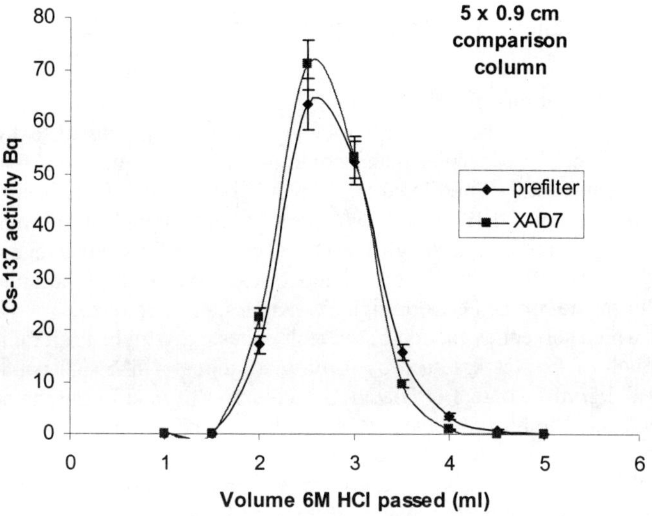

Figure 1 *Elution of non-adsorbed* ^{137}*Cs from the DIBK column.*

3.2 Uptake of Fe on the DIBK/XAD-7 column

A standard was prepared containing 1mg Fe in 2ml of 6M HCl. The sample was loaded onto the column with dimensions of 5 x 0.9 cms and the column repeatedly washed with 2ml aliquots of 6M HCl. A total of 60ml of 6M HCl was passed through the column (equivalent to approximately 30 column volumes). The Fe was eluted with 3 x 2ml of deionised water. Each eluent fraction was diluted to 10ml with 2% nitric acid. The concentration of Fe was then determined using atomic absorbance spectrometry.

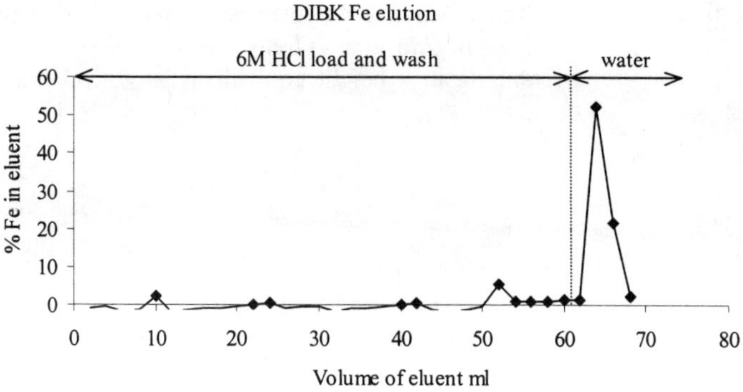

Figure 2 *Retention and elution of Fe on a DIBK / XAD-7 column 5 x 0.9cm*

3.3 Maximum loading of Fe on the DIBK/XAD-7 column

One of the main limitations of any technique for the separation of Fe is the maximum loading of Fe that the technique can tolerate. In general solvent extraction-based separations have higher Fe capacities than a column-based technique. A solvent extraction-based technique using 20ml of undiluted DIBK would be capable of extracting 980 mg Fe. The loading of Fe on DIBK-based extraction chromatographic column would be significantly lower. Therefore, although the DIBK column offers advantages in terms of improved purification of Fe from other elements excessively large columns would be required for the separation of Fe from high Fe samples, such as ferrous metals, soils and sediments, for which solvent extraction-based techniques may still be preferable.

Breakthrough of Fe was assessed by loading solutions of 6M HCl containing either 0.1mg/ml or 1.0 mg/ml Fe onto a DIBK/XAD-7 column and monitoring the concentration of Fe in the eluents. The breakthrough of Fe will be dependent on both the amount of Fe loaded onto the column as well as the number of theoretical plates present on the column. For a 0.1mg/ml Fe solution, breakthrough of Fe was observed after 80ml of load solution had passed through the column (equivalent to 8mg of Fe loaded onto the column). When the Fe concentration in the load solution was increased to 1.0 mg/ml, breakthrough was observed after only 30ml of the load solution had passed through the column (equivalent to 30mg of Fe on the column). The shape and position of the breakthrough curves may be used to calculate the plate height of the column using the following equation[6] :

$$N = \frac{V_{mb} \times V'}{(V_{mb} - V')^2}$$

where

N = the number of plates
V_{mb} = the volume of eluent passed through the column to obtain $C/C_0 = 0.5$
V' = the volume of eluent passed through the column to obtain $C/C_0 = 0.157$.

For 0.1mg/ml Fe loading, V_{mb} = 84.67ml and V' = 80.86ml whilst for the 1.0g/ml loading V_{mb} = 31.79ml and V' = 30.33ml. The number of theoretical plates was calculated as 470 and 450 for the 0.1mg/ml and 1.0 mg/ml load solutions respectively. In both cases, the column height was 50mm, resulting in a height equivalent theoretical plate (HETP) of 0.11mm.

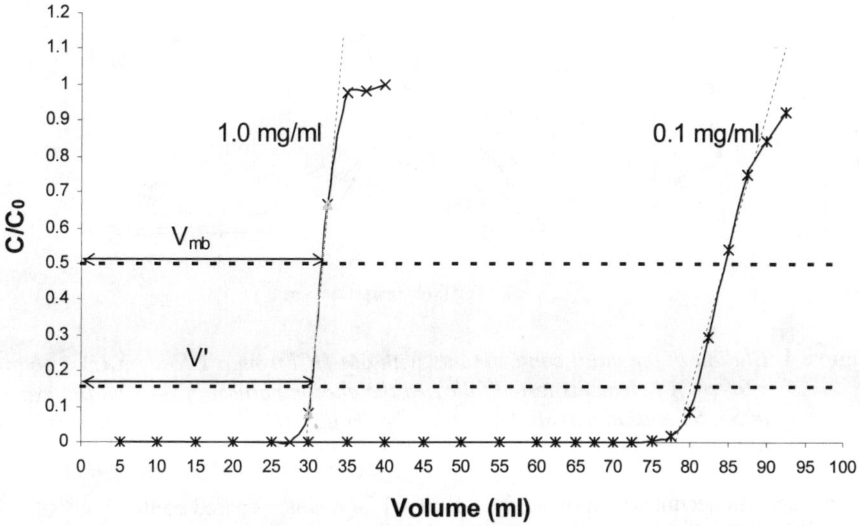

Figure 3 *Breakthrough curves for 0.1 mg/ml and 1.0 mg/ml Fe in 6M HCl. 5 x 0.9cm column containing 1g DIBK/g XAD7*

The HETP was also determined for the DIBK/CG-71 column. 0.1mg/ml Fe was loaded onto 2 x 0.9cm column containing 1.7g DIBK / g CG-71. A HETP of 0.18mm was determined using the approach described above.

3.4 Separation of Fe from other elements

As ^{55}Fe is normally associated with a wide range of other activation and fission products in nuclear wastes it is vital that any separation chemistry effectively isolates the ^{55}Fe from these other radionuclides. Elution profiles for a range of elements were therefore determined.

A standard was prepared containing 0.01mg rare earth elements, Ba, Co, Ni, U, Zr, Y, Th and Sn in 2ml of 6M HCl. The sample was loaded onto the column with dimensions of 5 x 0.9 cms and the column repeatedly washed with 2ml aliquots of 6M HCl. A total of 60ml of 6M HCl was passed through the column (equivalent to approximately 30 column volumes). The column was eluted with 3 x 2ml of deionised water. Each eluent fraction was diluted to 10ml with 2% nitric acid. 100 μl of each eluent fraction was diluted to 10ml with 2% nitric acid and the concentration of all analytes was determined by ICP-MS.

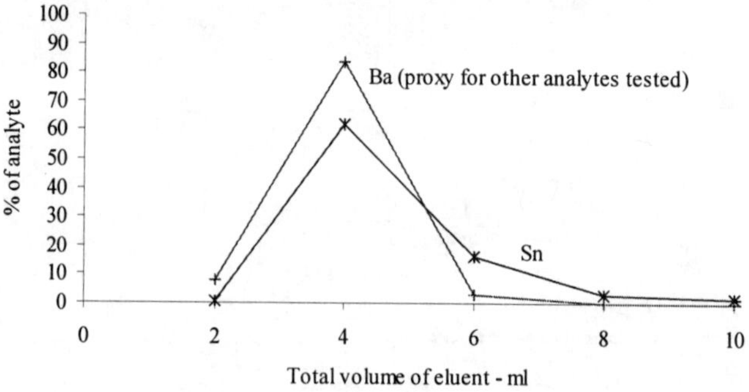

Figure 4: *Elution of potential contaminants with 6M HCl from a DIBK / XAD-7 column 5 x 0.9cm All elements tested followed the elution pattern of Ba with the exception of Sn. Sn elution was slightly delayed although 100% of Sn was eluted in 10ml of 6M HCl.*

In a second decontamination study, a load solution was prepared containing ^{51}Cr, ^{54}Mn, ^{59}Fe, ^{60}Co and ^{65}Zn in 5ml 6M HCl. This combination of radionuclides was chosen to be representative of the activation products commonly associated with ^{55}Fe in nuclear wastes. This solution was loaded onto a 5 x 0.9cm DIBK/XAD-7 resin prepared and conditioned as described previously. The column was then washed with 2ml aliquots of 6M HCl to a total volume of 25ml. The column was then eluted with 4 x 2ml of water. Each fraction was collected separately and counted on a Canberra well-type HPGe gamma spectrometry system.

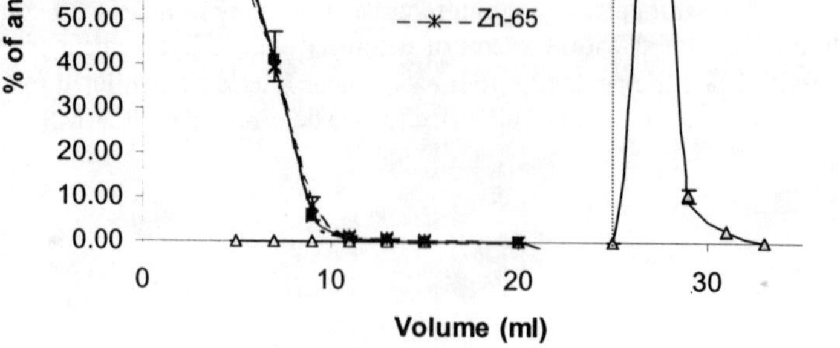

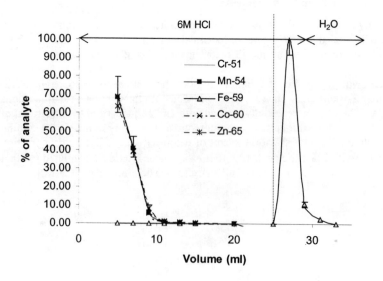

Figure 5 *Elution of activation products from a 5 x 0.9cm DIBK/XAD-7 column*

In a third decontamination study, the separation of Pu and Po from Fe were investigated. Both Pu and Po will form potentially-extractable anionic complexes in 6M HCl. One isotope of Pu, ^{241}Pu, is a low energy beta-emitting radionuclide (E_{max} = 21 keV) that would interfere with the liquid scintillation measurement of ^{55}Fe. Polonium will not directly interfere with the measurement of ^{55}Fe in nuclear wastes. However, ^{210}Po has been detected in purified ^{55}Fe fractions from sediments where conventional solvent extraction has been employed[7]. Approximately 8Bq of ^{209}Po and 7Bq of Pu(alpha) isotopes in 5ml of 6M HCl were loaded onto two separate 5 x 0.9cm DIBK-CG-71 columns. The columns were washed with 5ml aliquots of 6M HCl followed with 10ml 0.3M HNO_3 (to simulate the elution of ^{55}Fe). 1ml aliquots of the 6M HCl solutions and 5ml aliquots of the 0.3M HNO_3 solutions were transferred to 22ml polythene scintillation vials and mixed with Gold Star scintillation cocktail. All fractions were then counted on a Wallac 1220 Quantulus liquid scintillation counter to determine the total alpha activity.

Pu was rapidly eluted from the DIBK column, with 99% of the alpha activity being detected in the load and first 5ml wash solution. Po elution was delayed compared with the Pu, but > 95% of the Po activity was eluted with 20ml of 6M HCl.

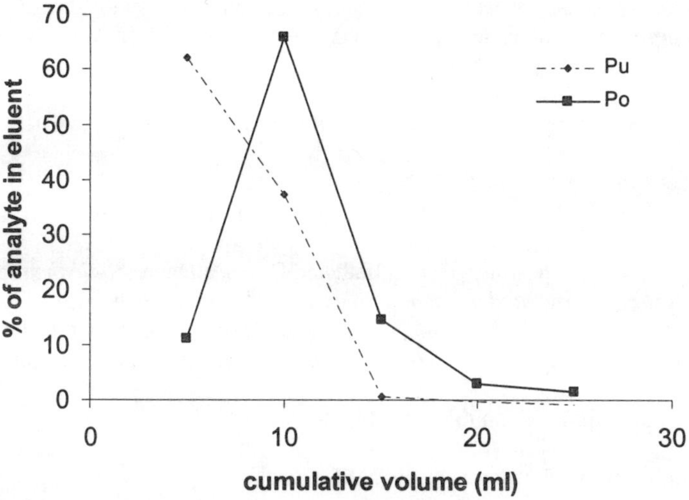

Figure 6 *Elution of Po and Pu from a DIBK column in 6M HCl*

4 CONCLUSIONS

Di *iso*butylketone (DIBK) coated onto an inert support has been successfully used for the isolation of ^{55}Fe. Up to 1.7g of DIBK can be loaded onto 1g of resin giving a final loading of 63% w/w of final material. Iron is effectively retained on the resin column from > 6M HCl. The loading capacity of the column is typically 20mg maximum but the effective loading capacity of the column is determined both by the absolute capacity of the resin and the gradual elution of Fe through the column. Potential interferents, including ^{51}Cr, ^{54}Mn, ^{60}Co, ^{65}Zn, Pu and Th are rapidly eluted from the column. Sn and Po exhibit limited retention from the column but may be effectively eluted using 5 c.v. 6M HCl. Fe is quantitatively recovered from the column by eluting with water. The operation of the columns is less time consuming than conventional solvent extraction-based techniques and may be more readily scaled up to large sample batches whilst reducing the volumes of organic solvents used. However, the chromatographic-based technique still retains the selectivity of the solvent extraction techniques whilst achieving a greater degree of purification of the final Fe fraction.

Acknowledgements

This study was supported by the Geosciences Advisory Unit, Southampton Oceanography Centre and AEA Technology, Harwell.

References

1 C. W. Baker, G. A. Sutton and J. W. R. Dutton, *Studies of liquid radioactive effluent discharged to the aquatic environment from CEGB nuclear power stations; the determination of the major beta-emitting isotopes*. Symposium on the determination of radionuclides in environmental and biological materials, CEGB, Sudbury House, 2-3 April 1973.

2 W. König, R. Schupfner and H Schüttelkopf, *Radioanal. Nucl. Chem.,* Articles, 1995, **193,** 119.

3 C. Testa, D Desideri, M. A. Meli and C Roselli, *Radioact. Radiochem*, 1991, **2**, 46.

4 A. Bohnstedt, M. Langer-Lüer, H. Stuhlfath and D. C. Aumann, *Radiochemical characterisation of low- and medium-activity waste by destructive assay of long-lived alpha and beta emitting nuclides*, In Odoj R., Baier J, Brennecke P. and Kühn K. (eds). Radioactive waste products 1997, Forschungszentrum Jülich GmbH, Germany (1998).

5 Mahmoud and Soliman, *Talanta,* 1997, **44**, 15.

6 E. Glueckauf, *Trans. Faraday Soc., 1955,* **51**, 34.

7 P. E. Warwick, *The determination of pure beta emitters and their behaviour in a saltmarsh environment.* PhD Thesis Unpubl., University of Southampton, UK, 1999.

SYNTHESIS OF A NOVEL ION EXCHANGER, ZIRCONIUM VANADATE, FOR IMMOBILISING $^{134,137}Cs$ RADIONUCLIDES

Kamalika Roy[1], D. K. Pal[1], S. Basu[1], Dalia Nayak[2], A. De[2] and Susanta Lahiri[2*]

[1]Department of Chemistry, The University of Burdwan, Burdwan 713 104, INDIA
[2*]Chemical Sciences Division, Saha Institute of nuclear Physics, 1/AF Bidhannagar, Kolkata 700 064, INDIA

1 INTRODUCTION

Radioactive waste management is one of the most challenging areas of present day research. ^{134}Cs (2.06 y) and ^{137}Cs (30 y) are long-lived radionuclides and persistent in the environment. Even today, the extent of contamination near Chernobyl is defined as zones of ^{137}Cs. Thus from safety consideration of radioactive wastes, it is necessary to develop new materials which will immobilise long-lived Cs radionuclides. The separation and immobilisation of ^{134}Cs (2.02y) and ^{137}Cs (30.02y) from high level waste is essential as it reduces the volume of disposed wastes. Safety consideration for a repository of radioactive wastes located in a salt formation includes the scenario of water ingression into the storage area and waste may contact with saturated salt brine.[1] In a thermal breeder reactor mixed oxide fuel is first oxidised, then dissolved in a molten chloride salt mixture (NaCl-KCl). In all of these processes ^{137}Cs has a high probability of being housed in the bulk of NaCl, since it is in the same group. Thus, it is also important to immobilise Cs radionuclides in the presence of bulk sodium chloride.

Inorganic ion exchangers having high thermal, radiation and chemical stabilities may serve as a tool in solving the problem. Acidic salts of polyvalent metals like zirconium, thorium, titanium, cerium, etc., have been described in the literature.[2] Zirconium phosphate is the widely studied inorganic ion exchanger and its role in separating a huge number of cations from one another has also been studied.[3] The general formula of this type of ion exchanger is $M^{IV}(HX^{V}O_4)_2.\ nH_2O$ (where M =tetravalent metal and X may be P, As, Mo, W, Sb, V, etc.). A radiochemical separation scheme for Cs has also been reported by Mishra et al., (2001) using zirconium phosphate ion exchanger.[4] Preparations of both poorly crystalline and amorphous compounds have been reported earlier, amongst which zirconium phosphate is the most widely studied.[5] Furthermore, compounds like antimonate, molybdate, tungstate and silicate salts of zirconium, titanium and thorium have also been reported.[6] These compounds are mostly high temperature resistant, and have chemical and radiation stability, exchangeable hydrogen ions and hence cation exchange properties.

Recently we have synthesised an amorphous variety of zirconium vanadate which has a higher exchange capacity for Na^+ than the amorphous variety of zirconium phosphate.[7] A simple column chromatographic operation allowed separation and immobilisation of both ^{134}Cs and ^{137}Cs in this matrix. The immobilisation of caesium radionuclides has also been checked in the presence of a large amount of sodium chloride.

2 EXPERIMENTAL

The new inorganic ion exchanger was synthesised and characterised. A 0.1 M sodium vanadate solution (100 mL) was added dropwise into a solution of 0.1 M zirconium oxychloride (50 mL) in 2 M HCl with constant stirring. After the addition was complete, a yellow precipitate appeared, the reaction mixture was diluted to 1 L and allowed to settle for 24 hours. The precipitate was washed several times with deionised water, filtered and again washed with hot water until chloride-free. The pH of the final solution was slightly acidic (around 4). The washed solid was dried by gentle heating and preserved in a desiccator.[7]

The exchanger was found to contain 26.18% Zr and 29.93% V, which approximates to the ratio Zr:V = 1:2. The thermal and radiation stabilities of the exchanger were satisfactory. The exchanger was found to be stable upto a temperature of 350^0C and even after absorbing a total dose of 7500 Gy. The exchanger is moderately stable in different chemical environments. The ion exchange capacity, IEC, for Na^+ ion was found to be 4.08 meq/g.

The immobilization of caesium was observed through three different sets of experiments.

2.1 Immobilization in the presence of mineral acids

The zirconium vanadate exchanger was preconditioned with 0.01 M HCl solution and a column of 5 cm length was packed with this material. ^{134}Cs ($3.5x10^4$ Bq) was loaded on the top of the column and the column was eluted with the help of 0.01 M HCl solution. The flow rate was maintained constant at 1 drop/minute and fractions were collected after every 5 minutes for counting. Gamma spectrometry was carried out using a HPGe detector with a resolution of 2.0 keV at 1.33MeV. The same experiment was repeated with 0.01 M HNO_3.

2.2 Immobilisation in the presence of a large amount of NaCl

The exchanger, preconditioned with a 0.01 M HCl solution was packed into a column of 5 cm. 1mL of 0.1 M NaCl solution spiked with ^{22}Na ($1.0x10^5$ Bq) and ^{134}Cs ($3.5x10^4$ Bq) was slowly poured on the top of the column. The column was then eluted with 0.01 M HCl solution. A flow rate of 1 drop/min was maintained and counts were taken after every 5 minutes.

2.3 Immobilisation after heating at very high temperature

In the third set of experiments, the zirconium vanadate exchanger, preconditioned with 0.01 M HCl was allowed to absorb the ^{134}Cs ($3.5x10^4$ Bq) radionuclide and the material was subjected to heating at 900^0C for 15 hours. The hot molten mass was then quenched to room temperature and the liquid solidified into a hard solid. This solid was then vigorously shaken with different aqueous phases (e.g. HCl, HNO_3 and alkaline solution).

3 RESULTS AND DISCUSSIONS

When the column was eluted with a 0.1 M HCl solution, no trace of ^{134}Cs in any fraction was found. This implied that the exchange site of the exchanger is perfect for the unipositive alkali metals and the added caesium was absorbed completely.

On repeating the same experiment with 0.01 M HNO_3 it was observed that the added caesium was again completely retained by the column. But on passing an excess of HNO_3, a small fraction of Cs started to elute. This may have been due to leaching of the column material.

The exchanger was previously observed to exchange alkali metals highly. To establish the immobilisation of ^{134}Cs in the exchanger, the isotope was allowed to absorb in the presence of an excess of sodium chloride solution. The observation after elution with 0.01 M HCl solution is that ^{134}Cs was absorbed even in the presence of an excess NaCl solution. In HNO_3 medium ^{134}Cs absorption also took place, but after eluting with a large volume of 0.01 M HNO_3, the absorbed caesium started to elute in small fractions.

When the exchanger, in a caesium-absorbed condition, was heated to 900°C, and then solidified, the caesium atoms get firmly embedded inside the solid matrix and the immobilisation occurs to such an extent that even strong eluting agents like 6 M HCl solution can leach out only 1.5% of the total caesium absorbed. The results of leaching with different aqueous phases have been tabulated in Table 1.

Table 1. *Desorption of ^{134}Cs from solidified zirconium vanadate*

Solvents	Desorption of caesium (%)
0.1 M HCl	0.62
1 M HCl	1.12
6 M HCl	1.58
0.1 M HNO_3	0.19
1 M HNO_3	0.28
6 M HNO_3	1.09
1 M NaOH	0.42

The exchangeable hydrogen atom of the exchanger is possibly replaced by the caesium and sodium ions until all the hydrogen atoms are exhausted in the first set of experiments. When the exchanger, in a caesium-absorbed condition, is subjected to heating at very high temperature (900^{0}C), the solid melts and when it solidifies the caesium atoms get completely merged in the matrix of the exchanger material. Leaching of caesium from the matrix using different eluting agents under natural conditions now becomes very difficult and caesium is thus immobilised.

4 CONCLUSION

The zirconium vanadate ion exchanger is highly selective for alkali metals. Its utility towards immobilisation and hence confinement of Cs radionuclides may thus make it an important agent in the field of environmental protection.

5 REFERENCES

1. Y. H. Wen, S. Lahiri, Z. Quin, X. L. Wu and W. S. Liu, *J. Radioanal. Nucl. Chem.*, 2002, **253**, (*in press*).
2. B. Sarkar and S. Basu, *Asian J. Chem.*, 1992, **2**, 313.
3. V. Lobo and Z. R. Turel, *J. Radioanal. Nucl. Chem.*, 2001, **247**, 221.
4. S. P. Mishra and D. Tiwari, *J. Nucl. Radiochem. Sci.*,2001, **2**, 200.
5. A. Clearfield and H. Hagiwara, *J. Inorg. Nucl. Chem.*,1978, **40**, 907.
6. A. I. Bortun, L. Bortun, and A. Clearfield, *Solv. Extract. Ion Exch.*, 1996, **14**, 341.
7. K. Roy, D. K. Pal, S. Basu, D. Nayak and S. Lahiri, *Appl. Radiat. Isot.*, 2002, **57,** 471.

INVESTIGATION OF A METHOD FOR MEASURING RADON IN IRISH DOMESTIC GROUNDWATER SUPPLIES

S.Sequeira, L. McKittrick, T.P. Ryan and P.A. Colgan

Radiological Protection Institute of Ireland, 3 Clonskeagh Square, Clonskeagh Road, Dublin 14, Ireland.

1 INTRODUCTION

Radon-222 is a member of the uranium decay series and its presence in the environment is associated mainly with the trace amounts of its immediate parent, radium-226, in rocks and soil. Because of its gaseous nature, radon can move freely through porous media such as soil or fragmented rock. Where pores are saturated with water, like soil and rock under the water table, radon dissolves in the water and is transported with it. Studies, such as those carried out in Sweden,[1] have shown that, in general, the highest radon concentrations in water are found in wells drilled in crystalline rocks, usually associated with high uranium concentrations in bedrock. Surface waters are generally least affected by radon.

Radon in domestic water supplies can cause human exposure to a radiation dose both through inhalation and ingestion. Radon is easily released from water into the atmosphere by agitation or heating. Many domestic uses of water result in such a release and can contribute to the total indoor airborne radon concentration. It has been estimated that 1000 Bq/l of radon in water will, on average, increase the indoor radon concentration by 100 Bq/m^3. The long-term risk from the inhalation of high concentrations of radon is lung cancer. The greatest risk associated with the ingestion of water containing radon and radon progeny is considered to be stomach-colon cancer.[2]

Recent attention has been focused on the issue of radon in drinking water by a European Commission Recommendation[3] proposing that surveys should be undertaken in Member States to determine the scale and nature of exposures caused by radon in domestic drinking water supplies. The Commission recommends 1000 Bq/l as the radon concentration in private drinking water supplies above which remedial action to reduce the concentration should be taken. Similar remedial action should be taken when public supplies are found to be in excess of 100 Bq/l.

The use of liquid scintillation techniques to measure radon in water is well established.[4,5] However there have been very few systematic studies made to estimate the loss of radon in the period between sampling and measurement. In this investigation particular emphasis was placed on the method of sampling and the losses that are likely to occur during sampling. This was done as part of a pilot study of radon in drinking water.

County Wicklow was selected for the study on the basis that high radon concentrations in air have been predicted in a significant number of dwellings in the county [6], the underlying geology is predominantly granite with an anticipated elevated uranium content and there is a high usage of drinking water from private supplies. The findings of the pilot study are presented in this report.

2 METHOD AND RESULTS

2.1 Measurement of Radon in Water using Liquid Scintillation

The rationale for the method is that radon in water is selectively extracted into an organic scintillant which is immiscible with water. The extracted radon reaches secular equilibrium with its short lived daughters after 3 hours resulting in beta and alpha emissions. The alpha activity is detected and measured in the alpha channel of a liquid scintillation counter which uses pulse shape analysis to separate the alphas from the betas.

The method used was based on that described by Cantaloub[7] using a Packard TriCarb-2770 TR/SL. Alpha/beta separation on this counter is set by the pulse decay discriminator (PDD). The optimum PDD setting is largely dependent on the chemical nature of the scintillant. The PDD setting for this measurement was established using an aqueous radium-226 standard with a number of proprietary scintillants over a range of PDD settings. The best alpha efficiency/low background combination was obtained with UltimaGold™ F at a PDD setting of 175. The counter was calibrated for radon-222 measurements with a certified radium-226 standard as follows: an accurately measured amount of radium-226 aqueous standard was added to a scintillation vial and the volume made up to 11 ml with purified water that had been previously boiled to eliminate any dissolved radon and then cooled; 11 ml of UltimaGold™ F were added to the vial after which the vial was tightly capped and stored in a cool place for 27 days to allow secular equilibrium with radon-222 to be established (at least 99% ingrowth); the vial was then counted using the conditions set out in Table 1.

2.1.1 Sample Preparation Optimisation. Two methods of transferring the radon from the water into the scintillant were investigated. In the first method radon is allowed to passively diffuse into the scintillant. In the second method extraction by mixing is used to accomplish the transfer. An 11 ml aliquot of the radium-226 standard, which was in secular equilibrium with radon-222, was added to each of 2 vials containing scintillant. One vial (passive diffusion method) was placed directly in the counter while the other (extraction method) was mixed by continuous inversion for 1 minute before being placed in the counter. Both vials were counted at regular intervals for 3 days. The results are shown in Figures 1 and 2.

Table 1 *Measurement details for measuring radon in aqueous samples*

Parameter	Details
Instrument	Packard TriCarb TR/SL 2270
Counting Vial	Glass, 24 ml
Scintillant	Ultima Gold™ F
SampleVolume	11 ml
Scintillant Volume	11 ml
Count Time	30 minutes
Count Mode	Low level
PDD Setting	175
Window	300 to 1000 keV
Temperature mode	Cooling unit "on"
Radon-222 Counting "Efficiency"	260%
Typical Blank Count Rate	1.5 cpm
Minimum Detectable Activity Concentration, (30 minute count)	0.5 Bq/l
Uncertainity at 2SD	Typically 10%

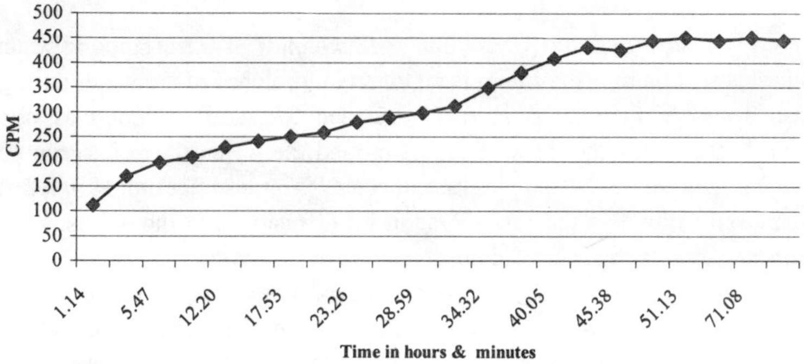

Figure 1 *The alpha count rate from a "passive diffusion" radon vial with respect to time*

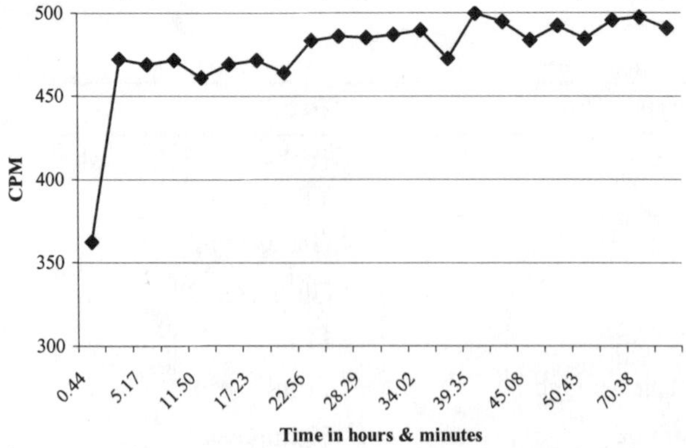

Figure 2 *The alpha count rate from an "extracted" radon vial with respect to time*

The results indicate that, with passive diffusion, equilibrium is attained after 2 days or more, whereas with extraction equilibrium is reached after 3 hours. Extraction by mixing was therefore used as the standard method for sample preparation.

2.2 Sampling and Measurement Evaluation

In a pilot study of domestic groundwater supplies in County Wicklow, kitchen-tap water samples were collected in duplicate from 166 households. The study was designed to elucidate some of the scientific issues surrounding radon in drinking water in an Irish context. County Wicklow, in the east of Ireland, was selected for the study on the basis that high radon concentrations in air have been predicted in a significant number of dwellings in the County, the underlying geology is predominantly granite with an anticipated elevated uranium content and there is a high usage of drinking water from private supplies.

2.2.1 Sampling and Measurement Techniques. When sampling water for radon it is important that there is minimum aeration and turbulence of the water during sampling. It is also desirable that the sample be analysed for radon as soon as possible after collection. In this study the tap was turned on and the water allowed to run gently for at least five minutes before the sample was collected. Samples were stored in a fridge prior to measurement to minimise the loss of radon by diffusion into the atmosphere. Care was taken to ensure that the samples did not freeze when in storage.

Water samples were collected using one or both of two methods: the "Routine Method" and the "Direct Method". In the "Routine Method" the sample was collected and stored in an air-tight container which was completely filled. The container chosen for this study was a 40 ml Teflon-rubber disc capped vial which has the property of completely excluding air from the sample. On arrival in the laboratory the sample was stored in the fridge for at least 30 minutes prior to preparation. The container was then uncapped and the water gently poured into a vial containing 11 ml of UltimaGold™ F until the scintillant level was just above the neck of the vial. The weight of vial and scintillant had been tared to zero on a three-figure balance so that the mass of water added could be determined. The vial was then tightly capped and mixed by continuous inversion for 1 minute. It was then placed in the counter and allowed to equilibrate for at least three hours before it was counted. In the "Direct Method" the water from the tap was sampled directly into a vial containing 11 ml of UltimaGold F until the scintillant level was just above the neck of the vial. The vial was then tightly capped and mixed by continuous inversion for one minute. The capped vial and scintillant had been pre-weighed on a three-figure balance so that the mass of water added could be determined when the sample arrived in the laboratory. It was then placed in the counter and allowed to equilibrate for at least three hours before it was counted. A water blank was also prepared using purified water which had been boiled (to eliminate any dissolved radon) and then cooled.

To calculate the radon-222 activity concentration in the sample the following variables were used: the mass of sample (=volume of sample, assuming the density of the sample to be unity); the time interval between the time of sampling and the mid-point of counting, to decay correct all results to the time of sampling; the net count rate in the alpha channel; a counting "efficiency" value of 260% for radon-222.

2.2.2 Assay Characteristics. To estimate radon loss between the time of sampling and the time of vial preparation using the "Routine Method", 15 samples were measured using both the "Routine Method" and the "Direct Method" simultaneously. The "Direct Method" samples were counted approximately 4 hours after sampling, and the "Routine Method" samples, approximately 24 hours after sampling. All results were decay corrected to the time of sampling. These are shown in Figure 3.

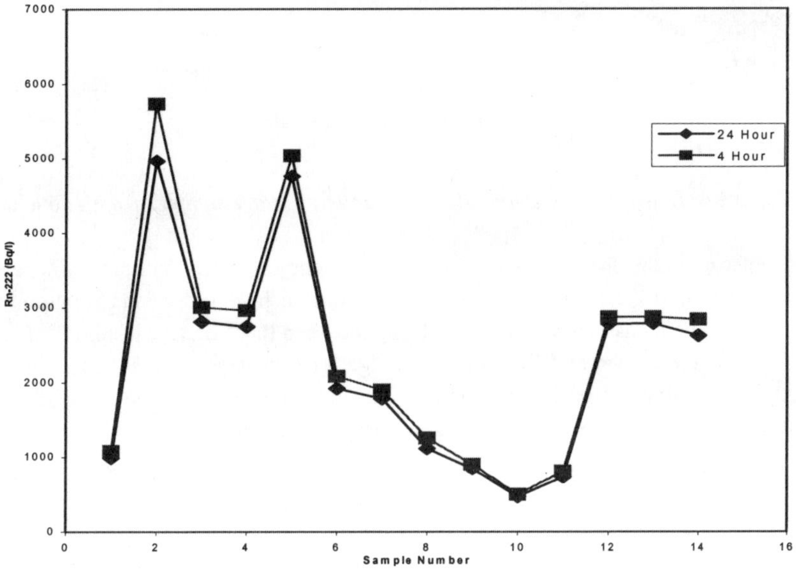

Figure 3 *Radon measurements using the "Routine Method" (24 hours) and the "Direct Method" (4 hours)*

The results indicate that there is a small loss of approximately 6% when using the "Routine Method". This is typically within the uncertainty of the measurement.
Agreement between duplicates on measurement of duplicate samples was used as an index of reliability of the sampling technique. The relationship between the duplicates of the samples taken from 166 supplies is shown in Figure 4.

It was concluded from the excellent agreement observed that the sampling and measurement techniques employed to measure radon are reliable and can be used with confidence in surveys of radon in drinking water.

To estimate the loss of radon from a routine sample container over time, four water samples were collected using the "Routine Method" from the same location at the same time. The samples were transported back to the laboratory and were stored in a fridge. The water from each container was analysed for radon once, at increasing time intervals between measurements. All results were decay corrected to the time of sampling. The results are shown in Table 2. These indicate that, under optimum conditions, there is no loss of radon from the sampling container over a period of 7 days.

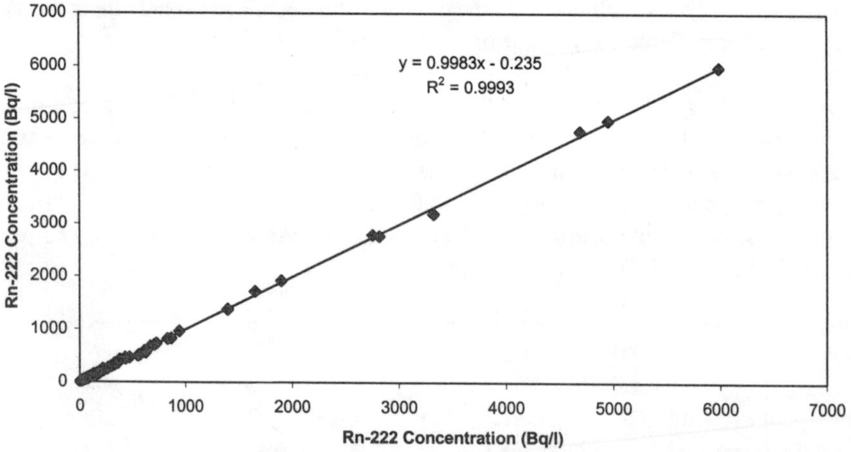

Figure 4 *Linear regression analysis of duplicate water samples from 166 houses*

To determine the loss of radon from a bulk sample taken without using a strict sampling protocol, a 2.5 l sample of water was collected from one of the locations along with samples to be measured by the "Direct Method" and the "Routine Method". The bulk sample was tightly capped and stored in the fridge for four days. It was then analysed for radon and the result compared to the "Direct Method" and the "Routine Method" results. All results were decay corrected to the time of sampling and are shown in Table 3.

Table 2 *Radon measurements of 4 samples taken from the same location and measured on different days*

Time of measurement	Radon-222 concentration (Bq/l)
Day 1	1070±107
Day 5	1198±120
Day 7	1068±110
Day 17	1083±117

Table 3 *Radon measurements of 3 samples taken from the same location using different sampling conditions*

Description	Radon-222 concentration (Bq/l)
Bulk Sample	2965
"Direct Method"	5736
"Routine Method"	4971

The results suggest that, under optimum conditions, bulk samples lose between 40% and 50% of dissolved radon over a period of four days.

2.3 Results of the County Wicklow Survey

Drinking water from 166 houses that use water from boreholes as their primary supply in Co. Wicklow were sampled and analysed in duplicate for radon. Of the supplies sampled, four had concentrations of radon which exceeded 1000 Bq/l, (representing 2.4% of the total). These concentrations were 1396 Bq/l, 1720 Bq/l, 3316 Bq/l and 5736 Bq/l. Fifteen houses had concentrations between 500 and 1000 Bq/l (9.5%), 51 had concentrations between 100 and 500 Bq/l (31.4%) and 96 had concentrations less than 100 Bq/l (56.8%).

To confirm that the activity detected in the samples is radon, a range of samples with differing radon levels were boiled, allowed to cool and then measured for radon. Radon was not detected in any of the boiled samples. Since the process of boiling eliminates radon from the water, it was reasonably assumed that the measured activity in the original samples was radon. These samples were in a fridge for 30 days, then removed and counted again for radon to determine whether the radon detected is supported or unsupported. In all cases no radon was detected. This indicates that the radon originally detected in the samples was unsupported.

One supply, with radon concentrations greater than 1000 Bq/l, provided samples at regular intervals (mainly fortnightly) to study temporal trends. The results of this analysis are presented in Figure 5

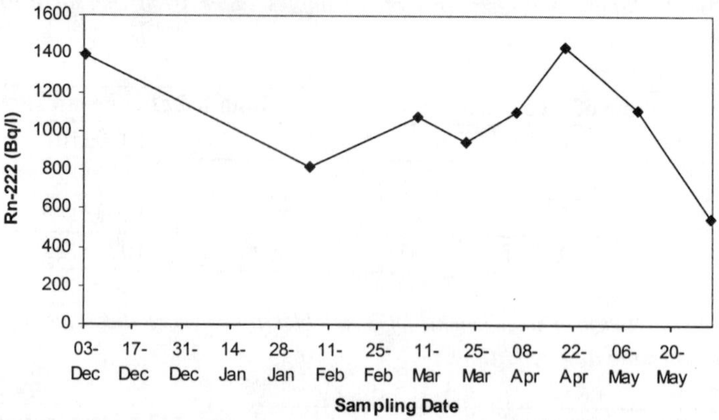

Figure 5 *Variation of radon levels with time at a location in Co. Wicklow*

Significant variation in radon concentrations were found - in this case the radon concentration was found to vary between 550 Bq/l in May 2002 and 1437 Bq/l in April 2002.

Additional analyses were carried out on the supplies whose radon levels exceeded 1000 Bq/l. These analyses included total alpha measurements, polonium-210 and radium-226 using radiochemical techniques. The Analytical and Regional Geochemistry Group of the British Geological Survey in Nottingham, England measured uranium and thorium mass concentrations using ICP-MS. The results of these analyses are shown in Table 4.

Table 4 *Radionuclide measurements in water from supplies with radon levels exceeding 1000 Bq/l.*

Rn-222 (Bq/l)	Thorium (μg/l)	Uraniu m (μg/l)	Uranium (Bq/l)	Gross alpha (Bq/l)	Ra-226 (Bq/l)	Po-210 (Bq/l)
5736	<0.02	4.61	0.12	0.391	< 0.5	0.194
2216	<0.02	2.16	0.06	0.404	< 0.5	0.020
1396	<0.01	27.1	0.70	0.768	< 0.5	0.208
1720	<0.02	1.14	0.03	0.049	< 0.5	0.009

Three of the four houses had total alpha activities above the screening level of 0.1 Bq/l generally applied in monitoring programmes. In this case individual radionuclide concentrations were determined. Two of the houses had polonium-210 concentrations of 0.194 Bq/l and 0.208 Bq/l respectively which leads to an indicative dose of greater than 0.1 mSv per annum. Thorium was not detectable in any of the water samples from the four houses. Uranium concentrations in the waters ranged between 1.14 μg/l and 27.0 μg/l (Table 4). The WHO recommends that uranium concentrations in drinking water should be below 2 μg/l, while the equivalent recommendation in the United States is 20 μg/l. In all sixteen water supplies were measured for their uranium content and a poor correlation was found between uranium concentration (R^2=0.06) in drinking water and radon in drinking water (Figure 6). However this correlation is greatly improved with the elimination of two outliers (R^2=0.83) and clearly more data would be valuable in determining the strength of the relationship.

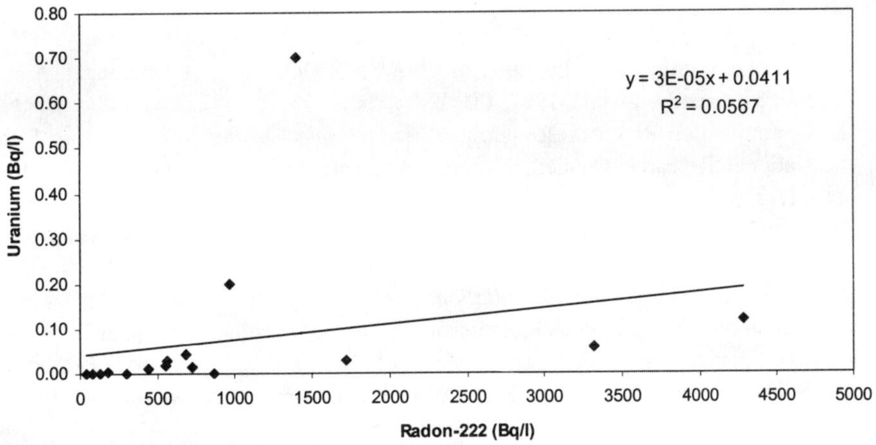

Figure 6 *Linear regression analysis between uranium and radon concentrations in Co. Wicklow water supplies.*

3 CONCLUSIONS

This study confirms the findings of other workers that the measurement of radon in water by liquid scintillation has several benefits: it is convenient and non-technically demanding; it is a high sensitivity technique with low detection levels; advances in liquid scintillation technology have made alpha/beta separation possible, thus improving specificity; automated counting makes it eminently suitable for batch counting and data handling. However, in this investigation the importance of validating and optimising a reliable sampling method prior to measurement has been identified as a key factor in the accurate assessment of radon levels in water. It is recommended that any publication on the subject should include validation data of the sampling procedures used. The "Direct Method" described in this paper is, arguably, the most accurate means of determining radon levels at source.

The findings of the Wicklow survey are comparable to a study carried out recently in Devon in the UK, a known High Radon Area. [8] The results of this study showed that 8% of the 100 water supplies examined had radon concentrations above 1000 Bq/l. The highest concentration found was 5340 Bq/l. A study carried out in Sweden[1] estimated that nationally, 5% of drilled wells have radon concentrations in excess of 1000 Bq/l. The maximum concentration observed in Sweden was 57 000 Bq/l.

Acknowledgements

The study was undertaken in close co-operation with Wicklow County Council, whose staff were responsible for the collection of water samples. This contribution is gratefully acknowledged.

References

1. G. Akerblom and J. Lindgren in *IAEA-TECDOC--980,* International Atomic Energy Agency, Vienna, 1997, **461**, 237.
2. Committee on Risk Assessment, in *Risk Assessment of Radon in Drinking Water,* National Research Council, National Academy Press, Washington, D.C., 1999, ch1-4, p.1
3. The Commission of the European Communities, *Official Journal of the European Communities*, 2001, **L344**, 85.
4. L. Salonen in *Liquid Scintillation Spectrometry 1992,* ed. J.E. Noakes, F. Schonhöfer and H.E. Polack, Radiocarbon Tucson, Arizona, 1993, p. 361.
5. J.D. Spaulding and J.E. Noakes in *Liquid Scintillation Spectrometry 1992,* ed. J.E. Noakes, F. Schonhöfer and H.E. Polack, Radiocarbon Tucson, Arizona,1993, p. 373
6. S.G. Fennell, G.M. Mackin, J.S. Madden, A.T. McGarry, J.T. Duffy, M. O'Colmáin, P.A. Colgan and D. Pollard in *Radon In Dwellings. The Irish National Radon Survey,* The Radiological Protection Institute of Ireland, Dublin, 2002, RPII-02/1, p. 4.
7. M.G. Cantaloub in *Practical Aspects of Environmental LSC, Bioassay Conference (BAER '99),* Packard Instruuments Company, Gaithersburg, Maryland, 1999.
8. D.J. Talbot, J.R. Davis, and M.P. Rainey, *DETR Report – DETR/RAS/00.010,* Department of the Environment, Transport and Regions, UK, 2002.

A COMPARISON OF THREE METHODOLOGIES FOR THE DETERMINATION OF GROSS ALPHA ACTIVITY IN WATER SAMPLES

Chongyang Zhou, Rengang Zhou, Shouqun Lin, Ziwei Cheng, Shujuan Feng, Kang Zhao

Northwest Institute of Nuclear Technology, Xi'an, Shannxi Province, 710024, P.R. China

1 INTRODUCTION

Gross alpha and beta counting is mainly used for the screening of radionuclides in water. Usually, there are two methods used to prepare the source for gross alpha and beta activities analysis: co-precipitation and evaporation. For the special requirements of gross alpha and beta determination there are no fixed standards for calibration. For this reason, many different methods have been developed to determine the activity by different calibration techniques. Particularly for gross alpha analysis, several standards have been used to calibrate methods, such as ^{241}Am, which is employed by the USEPA, ^{239}Pu, ^{226}Ra and natural uranium. Even if the same radionuclide is adopted to calibrate a method, it may have different physical forms, for example, a ^{239}Pu source may be in solution or already deposited on a planchet; uranium oxide (U_3O_8) may be also in solution or in the form of a fine powder. All of these different standards may lead to different results for the same sample. The results obtained by different laboratories using these different methods are the source of much controversy.

Up-to date methods, using liquid scintillation counting system for the determination of gross alpha and beta, provide an efficiency of nearly one hundred percent. For example, Eichrom Inc has developed a new method for determination of gross alpha and beta in water, where the alpha emitting radionuclides are scavenged by Eichrom's Actinide Resin and counted by LSC.

In China, there are two National Standard Methods for the analysis of gross alpha: saturation thickness and precipitation by ferric oxyhydroxide. The method of relative comparison, which is similar to the USEPA's standard method, is adopted widely with a calibration standard of fine powder uranium oxide. This paper describes a comparison of the three methods, as they have been applied to water samples in China.

2 STANDARDS AND INSTRUMENT

The alpha standard was ^{241}Am powder source, mixed in calcium sulphate, provided by China Institute of Metrology with an activity of 13.4 Bq g^{-1}. The beta standard source was KCl reagent, and its activity is 14.7 Bq g^{-1}.The 5 cm diameter stainless steel planchet ^{239}Pu standard source (2.28×10^4 dpm/2π) and (Ra, Ba)CO_3 powder standard source were purchased from China Academy of Atomic Energy.

An ALPHA-BETA MULTIDETECTOR MIN20 (EURISYS MESURES, France) was used to count alpha and beta activity. This instrument was a proportional gas-flow counter with four chambers and the counting gas was P-10 gas, mixed by 90% argon and 10% methane. The counting time was 3600 s. In this experiment, the counting error was not more than 30% and the detection limits for alpha and beta were 0.042Bq L^{-1} and 0.037Bq L^{-1}, respectively. The samples were placed in a 5 cm diameter shallow stainless steel tray for counting.

A LMU-3 laser fluorescence analyser (China Institute of Geology) was used for the determination of uranium. The determination range was 0.20~20 μg L^{-1}.

3 METHODS

3.1 Sample

Plastic containers (5L volume) were rinsed more than three times with water prior to collection. All samples were stored in the containers and acidified to pH 2~3 with HNO_3 for subsequent preparation. There was no visible precipitation in the samples.

3.2 Determination of alpha activity

3.2.1 Evaporation 5mL of 9mol L^{-1} sulfuric acid (H_2SO_4) was added to 1L of water, which was then evaporated on a hotplate to about 30mL. It was then transferred to a weighed ceramic crucible, again evaporated to nearly dryness, ashed in a muffle furnace under 500°C, weighed and finely ground. The solid residue, weighing 0.22 g, was put in a steel tray for counting.

3.2.2 Saturation thickness [1] Because alpha particles can travel only short distances through the solid residue, the counts from the alpha particles will not vary with the source thickness once it has reached a saturation thickness. According to this principle, the alpha source was prepared a bit thicker than the saturation thickness to determine the gross alpha activity.

3L of water was evaporated, the same procedure as the sample above, after adding 10mL of 9mol L^{-1} sulfuric acid (H_2SO_4) and 6mL of uranium solution (1000ppm mL^{-1}). A series of solid residue masses were put in steel trays for counting alpha activity. Two straight lines were obtained by plotting the alpha counting against the mass thickness of solid residue (mg cm^{-2}). It can be seen that the slope initially increases and then is nearly parallel to the horizontal axis. A change in the gradient indicates the saturation thickness of the source. From Figure 1, the saturation thickness of water solid residue is 11.2mg cm^{-2}.

The gross alpha activity in water is given by the following formula:

$$A = \frac{6.67 \times 10^{-2} W(n_c - n_b)}{S\eta V\delta Y} \qquad (2)$$

Where

A: gross alpha activity in water, Bq L^{-1};
W: weight of solid residue in water, mg;
n_b: background count rate, cpm;
n_c: total count rate of background and sample, cpm;

Y: recovery, %; in this experiment Y=100%;
η: efficiency of instrument by counting ^{239}Pu planchet source, %;
V: volume of water, L;
δ: saturation thickness, mg cm^{-2};
S: area of sample in the tray, cm^{2};
6.67×10^{-2}: constant to convert alpha activity to sample solid specific activity, Bq to mg dpm^{-1})$^{-1}$.

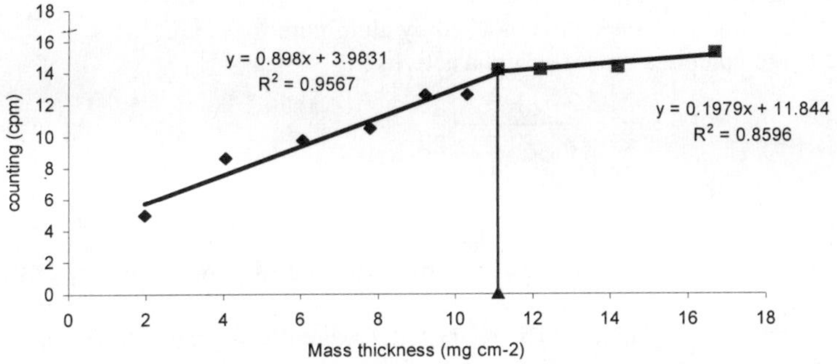

Figure 1 *Alpha saturation thickness of water solid residue*

3.2.3 Relative comparison A sample of ^{241}Am powder source, weighing approximately the same mass as the samples, was transferred to a steel tray and counted for alpha activity. The instrumental efficiency was determined, and the gross alpha activity in water calculated by this formula:

$$A = \frac{(n_c - n_b)\times W}{K\times E\times W_a\times V} \quad (2)$$

Where

E: efficiency by counting ^{241}Am powder source, %;
K=60;
W_a: the mass of water solid residue used for counting, g.

3.2.4 Precipitation by ferric oxyhydroxide[1] 2L of water were treated with 6 mol L^{-1} hydrochloric acid , added drop-wise, to remove the carbon dioxide in the water, with agitation. Then 16mL of 50mg mL^{-1} ferric (III) chloride was added, followed by 14 mol mL^{-1} of ammonium hydroxide, added drop-wise, to adjust the solution to pH 8,, and then it was allowed to stand for more than 4 h. The precipitation was filtered, transferred to a ceramic crucible and ashed in a muffle furnace for 30min at a temperature of under 500°C. After cooling to room temperature the the ferric oxyhydroxide was ground to a fine powder, weighing 0.22g, which was counted in a steel tray.

The planchet ^{239}Pu source was counted to determine the efficiency of the instrument.

An aluminum foil of the same diameter (5 cm) was weighed and the mass thickness (mg cm^{-2}) calculated. It was then placed on the top of the planchet ^{239}Pu source, and counted in the same position. The equation (3) below was used to calculate the absorbed thickness of aluminum foil (δ_{Al}):

$$\delta_{Al} = d_{Al} \times \frac{n_0}{n_0 - n_x} \qquad (3)$$

where

δ_{Al}: absorbed thickness of aluminum foil, mg cm^{-2};
d_{Al}: mass thickness of aluminum foil, mg cm^{-2}, here it is 8.03 mg cm^{-2};
n_0: the counting of ^{239}Pu source, cpm;
n_x: the counting of ^{239}Pu source blocked by aluminum foil, cpm.

Use formula (4) to calculate the gross alpha activity:

$$A = \frac{33.3 \times n \times W}{S \times \eta \times \delta_{Al} \times 1.23 \times 2} \qquad (4)$$

Here: δ_{Al}=8.05 mg cm^{-2}.

3.3 The determination of gross beta, uranium, radium and γ spectrometry of the solid residue

3.3.1 Gross beta determination The KCl (A.R.) reagent was ground to fine powder, which is used for the beta standard source to calibrate the evaporation residue. Equation (2) was used to calculate the gross beta activity.

3.3.2 Uranium determination The method of liquid laser fluorescence was employed to determine uranium by adding a reagent of fluorescence strength in the water to form a simple complex with uranyl ions. The wavelength was 337nm. This is the Chinese National standard method to determine uranium in water [2].

3.3.3 Radium determination Radium was co-precipitated with Ba(Ra,Pb)SO_4, by adding H_2SO_4 and $BaCl_2$ to10L of water. After filtration, the disodium ethylenedinitriloacetate (Na_2EDTA) solution was introduced to separate radium from lead by drop addition of glacial acetic acid to form a precipitate of (Ba,Ra)SO_4, whereas lead remained in the aqueous phase. A very thin precipitate is used to prepare source for counting in low-level α/β background counter. This method is the Chinese National Standard to determine the radium in water [3].

3.3.4 γ spectrometry determination LOAX γ spectrometry is used for scanning the solid sample (6.22g) in order to determine the existence some other gamma-ray emitting radionuclides. The counting time was 250000 s. The results are shown in Figure 2.

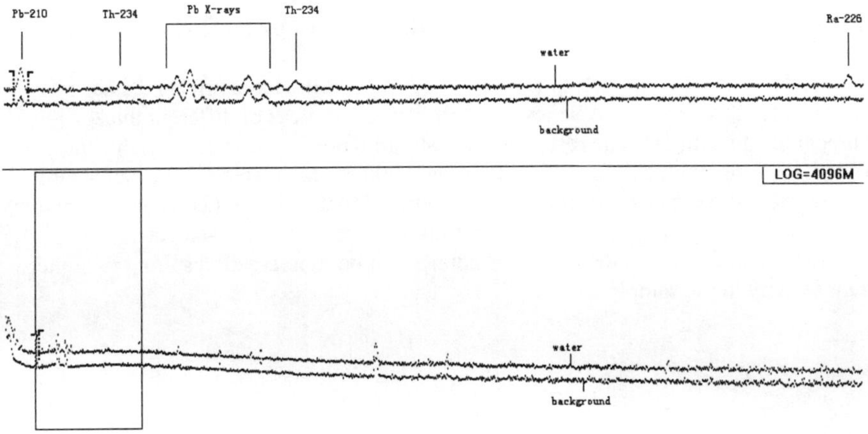

Figure 2 *The scanning result of the solid residue and background by γ spectrometry*

4 RESULTS AND DISCUSSIONS

4.1 Gross alpha, beta, uranium and radium

Table 1 shows the results of the determination of gross alpha, gross beta, uranium and radium in the water.

Table 1 *The results of gross alpha, beta, uranium and radium in the water*

Number	Solid residue ($g\ L^{-1}$)	Gross alpha ($Bq\ L^{-1}$)			Gross beta ($Bq\ L^{-1}$)	Uranium ($\mu g\ L^{-1}$)	Radium ($Bq\ L^{-1}$)
		Evaporation		Precipitation			
		Saturation thickness	Relative comparison				
1	0.7289	0.15	0.82		0.33±0.14	19.8	
2	0.7236	0.11	0.61		0.32±0.08	20.2	0.0029
3	0.7524	0.11	0.60		0.30±0.01	22.6	0.0042
4	0.6992	0.15	0.67		0.39±0.13	21.0	0.0052
5	0.8019	0.18	0.78		0.37±0.18	20.9	
6	0.772±0.037	0.17±0.04	0.91		0.32±0.07	29.4	
7	0.9631	0.20	1.11	0.060±0.01	0.49±0.26	29.7	0.0064
8	0.8930	0.16	0.85		0.62±0.30	27.9	0.0044
9	0.9238	0.14	0.60		0.38±0.16	23.4	0.0064
Mean		0.15±0.03	0.77±0.17		0.39±0.10	23.9±4.0	0.0049±0.0014
WHO Guideline values			0.1		1.0	50	1.1

Clearly there are three significantly different results for gross alpha using the different methods. The result of the relative comparison using ^{241}Am powder source gives the highest results of the three methods.

4.2 The cause of the difference in alpha activities obtained by the different methods

It is possible that the saturation thickness of ^{241}Am powder source affects the accuracy of the results for that method. A series of ^{241}Am powder sources of different thicknesses were prepared and counted. The results are shown in Figure 3. The saturation thickness is shown to be 3.2 mg cm^{-2}, although the saturation thickness calculated for the water sample is 11.2 mg cm^{-2} , three times that of the ^{241}Am source. From equation (2) it can be seen that the activity is inversely proportional to saturation thickness and if the standard is thicker than the saturation thickness, the measured counts will be suppressed, resulting in higher apparent activity in the sample.

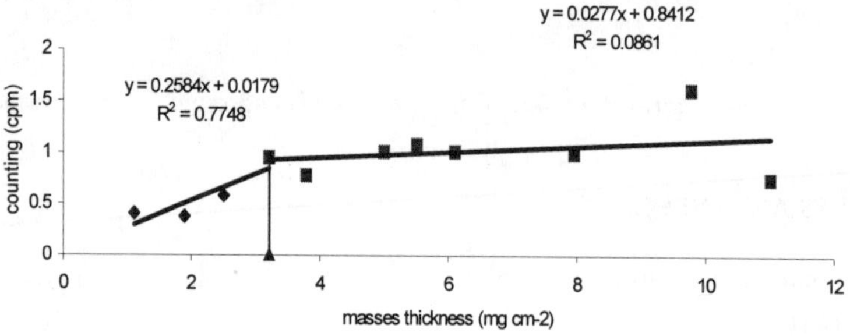

Figure 3 *Saturation thickness of ^{241}Am powder source*

4.3 The result of γ Spectrometry

From Figure 2, the result of gamma-ray spectrometry demonstrates the absence of other gamma-ray emitting radionuclides, such as ^{137}Cs , which could contribute to the activity of the sample.

4.4 Evaluation of the water sample

From the results showed in table 1, it can be seen that gross alpha by evaporation gives higher values than the WHO Guideline Values. Therefore the samples were analysed for uranium and radium, as required by WHO if the gross alpha value is higher than 0.1Bq L^{-1}. The activities of these two radionuclides are lower than the WHO Guideline Values and from these data, it is concluded that the water is suitable for drinking.

5 CONCLUSIONS

The results of this work indicate that it is not easy to get consistent results in the determination of gross alpha and gross beta by different methods. The saturation thickness of the solid water residue is more than three times that of ^{241}Am powder source and this is probably the reason for obtaining different results with the two methods. The activity of gross alpha in this water is mainly due to natural uranium. According to the result of gross alpha, gross beta, uranium, radium and γ spectrometry, the water is suitable for drinking.

References

1 The determination method of gross alpha in natural drinking ore water. GB 8538.56—87 (UDC 663.6 : 543.06) (in Chinese).
2 The analysis method of trace uranium in water. GB 6768—86 (UDC 628.54:543.06) (in Chinese).
3 The determination for alpha – radionuclides of radium in water. GB11218—89. (in Chinese)

EFFICIENCY AND RESOLUTION OF GERMANIUM DETECTORS AS A FUNCTION OF ENERGY AND INCIDENT GEOMETRY

R.M. Keyser

New Product Development, ORTEC, 801 South Illinois Avenue, Oak Ridge, TN 37831

1 INTRODUCTION

The use of germanium detectors for the identification and quantification of radionuclides in unknown and non-standard counting geometries has increased in the recent past with the need to clean up and verify sites used for radionuclide processing, smuggling detection and other purposes. The calculation of the amount of radionuclide present requires a knowledge of the efficiency of the detector in the counting geometry. Several methods of determining the efficiency in these unusual geometries have been developed over the years. These methods are used in waste measurement, field (in situ) measurements and material verification. One method currently being used is the modeling of the detector response using Monte Carlo simulation programs, especially Monte Carlo N-Particle Code (MCNP). The input to MCNP is a detailed knowledge of the detector construction details, detector crystal details and the source being measured. This paper addresses the difficulty in obtaining the required detector crystal details, even when the common specifications are well known.

In addition, the Institute of Electrical and Electronic Engineers (IEEE) Germanium Test Standard (325-1996) is widely used for the specification of High Purity Germanium (HPGe) detectors. The two important specifications are the efficiency and the resolution of the detector. However, the IEEE 325-1996 standard only specifies the Full Width at Half Maximum (FWHM) measurement at one geometry and two energies. Thus, this standard does not apply in most counting geometries used today, especially for environmental samples.

Modeling programs, such as MCNP, use the physical dimensions of the detector crystal to predict the response of HPGe detectors on the assumption that the detector response can be related to the physical dimensions and that the detector crystal response is independent of the position of the interaction in the crystal. Other investigators[1], have shown that the peak resolution (FWHM, FW.1M and FW.02M) change with position of the incident gamma ray on the front of the detector. Such variability has possible implications for the accuracy of peak shape and area determination, since the calibration is potentially a function of angle of incidence.

To demonstrate that the efficiency and resolution vary as a function of energy and gamma-ray point of incidence, measurements have been made on several coaxial detectors of various crystal types and sizes for different energies. The full-energy peaks from 60 keV to 2.6 MeV were used.

2 EXPERIMENTAL SETUP

Several detectors were chosen as being representative of the majority of detectors in use today. Large detectors were selected as it was expected that these would show the most variation. However, it is shown that the variations are most significant in older detectors. The detectors were placed in a low-background shield to reduce any contribution from external sources. None of the detectors tested was a low-background type. The detector dimensions are shown in Table 1.

Table 1. Detectors Studied

Detector Number	Diameter (mm)	Length (mm)
N21240A	60.5	60.4
N31626B	53.0	47.8
P41075A	84.5	100.5
P41182A	86.1	112.4

The sources used were ^{241}Am and ^{60}Co point sources. The ^{241}Am source was collimated to a 1 mm diameter beam by a 3 cm long by 17 mm diameter lead collimator. The ^{60}Co source was collimated to a 2 mm diameter beam by a 8 cm long by 8 cm diameter tungsten collimator. In addition, ^{60}Co data were collected without a collimator. Complete details of the collimators were presented earlier[2].

The measurements were made on the top or front of the detector in East-West and North-South modes and on the side of the detector in Up-Down mode as shown in Fig. 1. The actual positions were selected at random with respect to the crystal orientation.

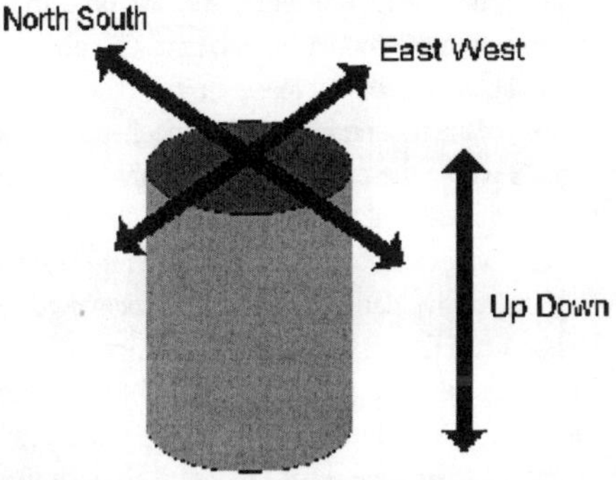

Figure 1. *Definition of Scan Directions*

The spectra were collected with a Digital Signal Processing MCA (DSPEC Plus), with 12 μs rise time (corresponding to an analog shaping time of about 6 μs), 1 μs flattop and a cusp of 1. The deadtime was always less than 10%. The number of channels was 16k, giving at least 6 channels in the FWHM of the narrowest 59 keV peak.

The peak areas and widths were calculated using the methods described in IEEE 325-1996 (equivalent to International Electrotechnical Commission standard IEC 61976-2000). The position numbers are arbitrary scale readings and are related to the detector crystal position by the count rate data.

3 RESULTS

3.1 Efficiency at 59 keV

The first detector studied was an ORTEC GMX type (n-type) with dimensions of 60 mm diameter and 60 mm length. The ^{241}Am scan on the front is shown in Fig. 2. The crystal diameter is shown by the black bar. Note the uniform sensitivity across the front of the detector, indicating that the crystal has uniform dead layer on the front.

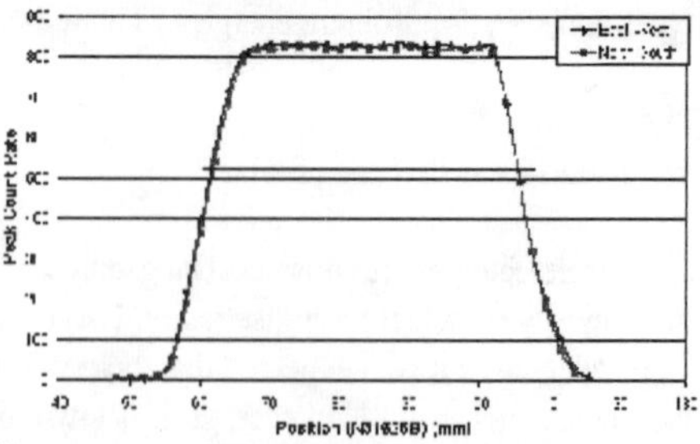

Figure 2. *Peak Area vs Position on Front for 59 keV in NS Direction*

The scan down the side of a similar detector is shown in Fig. 3. This shows the 59 keV peak area for the Up-Down scans for 0 degrees and 90 degrees. The intensity of the peak shows the three thick regions in the mounting cup of the detector. More importantly, the two scans are nearly the same. The slight variation between the front of the crystal (left side) and the bottom of the crystal in the reduction in count with position is attributed to the increase in material at the bottom of the mounting cup. The peak area in the regions where the cup is thin, show that the dead layer is uniform along the length of the crystal. Other GMX detectors showed the similar results.

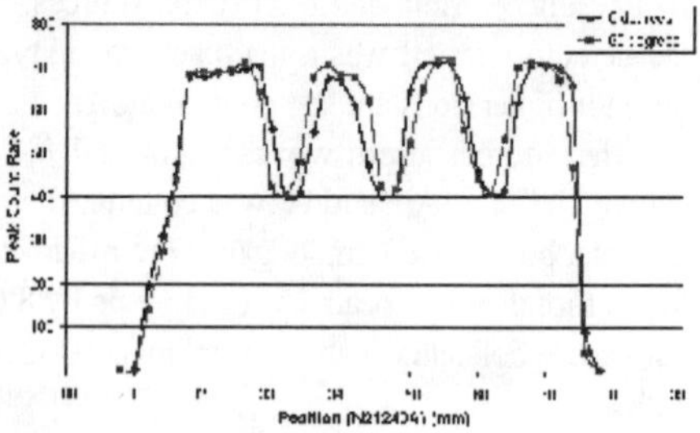

Figure 3. *Peak Count Rate at 59 keV vs Position on Side of Detector*

The relative dead layer was also measured for larger ORTEC GEM type detectors. Fig. 4 shows the relative intensity of the 59 keV peak for a scan along the length (UD) of the crystal. For comparison, the results of Ref. 3 for a similar, but not identical, detector are shown. Note that the first thick band is visible, but the second band is obscured by the thickening dead layer. In both of these detectors, the dead layer appears to be constant for a portion of the detector (starting at the closed end or front) and then increasing in thickness from some point to the bottom of the crystal.

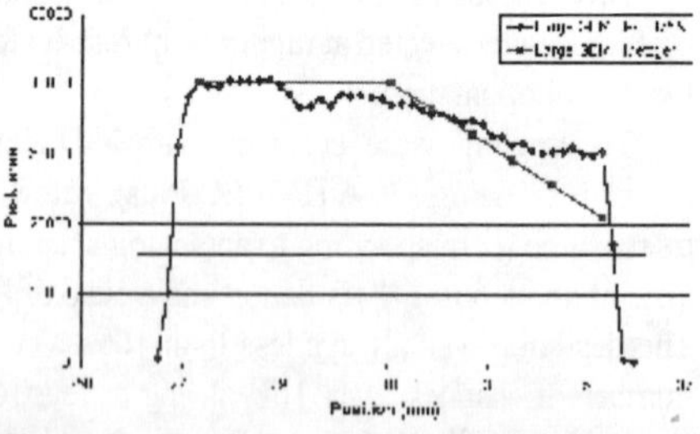

Figure 4. *Peak Count Rate at 59 keV vs Position on Side of Detector P41075A*

In contrast to these two detectors, Fig. 5 shows a 59 keV scan of another GEM detector, again of similar length and diameter. In this scan, both the thick bands of the cup can be seen and the dead layer is otherwise nearly uniform from front to rear of the crystal.

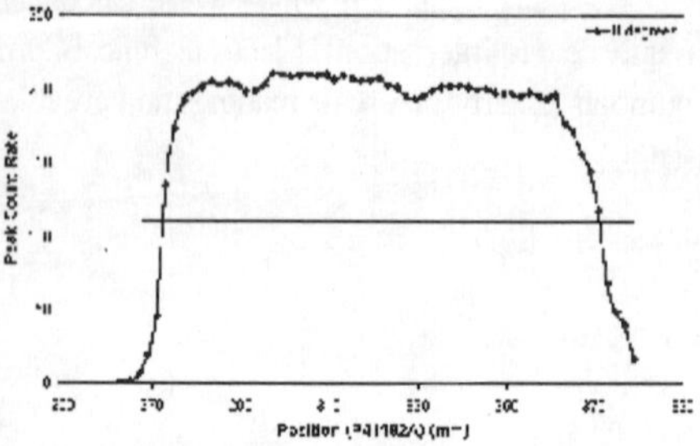

Figure 5. *Peak Count Rate at 59 keV vs Position on Side of Detector for P41182A*

3.2 Resolution at 59 keV

The peak shape of ^{241}Am on the front of a GMX (N31626B) is shown in Fig. 6. The crystal diameter is shown by the black bar. Another detector is shown in Fig. 7. Note the uniformity of the peak shape over the front, with the exception of the increase for a region in Fig. 7 and the general broadening near the edges of the crystal.

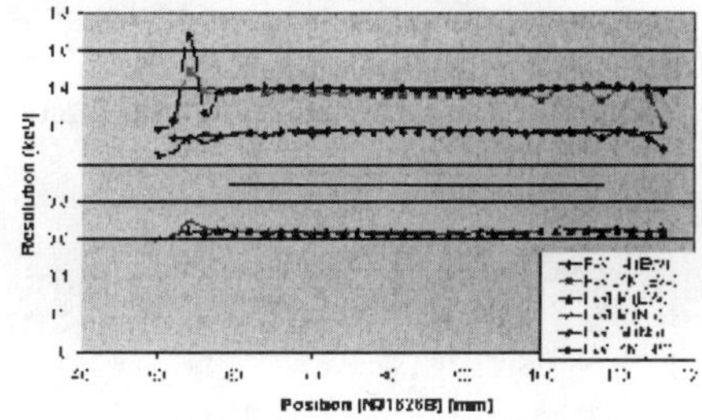

Figure 6. *Resolution at 59 keV Across Front of GMX (N31626B).*

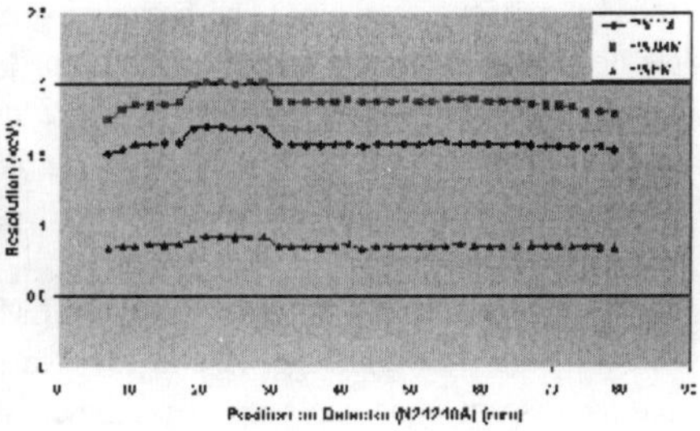

Figure 7. *Resolution at 59 keV Across Front of GMX (N21240A).*

The peak FWHM for the scans down the length of the crystal are shown Figs. 8, 9, and 10. These are again uniform, except for small variations at the ends of the crystals and at some positions down the length. With the exception of Fig. 7, it is expected that the detectors would exhibit this behavior because the low-energy gamma rays interact near the surface.

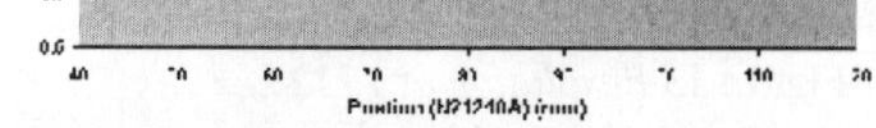

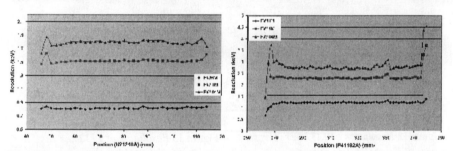

Figure 8. *Resolution at 59 keV the Side of GMX (N20140A).*

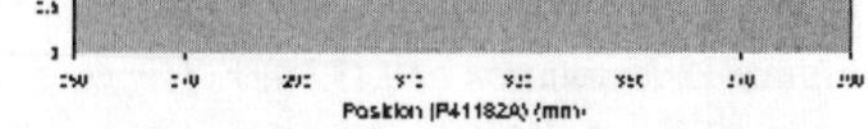

Figure 9. *Resolution at 59 keV on the Side of GEM (P41182A).*

3.3 Resolution at 1.1 and 1.3 MeV

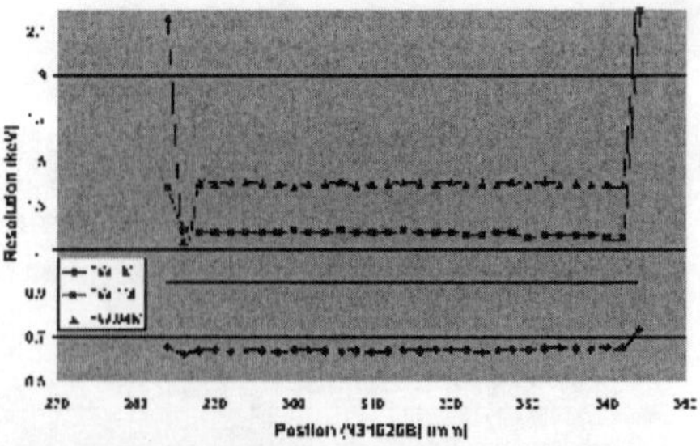

The resolution for the ^{60}Co peaks for scans across the front of a GMX detector is shown in Fig. 12. The GMX detectors do not show the increase in width at the center of the crystal, as seen by previous workers. The broadening of the peak near the center of the crystal is due to the center hole of the detector. This hole extends to about 9 mm of the front surface, giving the possibility of loss of energy into the hole.

Figure 10. *Resolution at 59 keV on the Side of GMX (N31626B).*

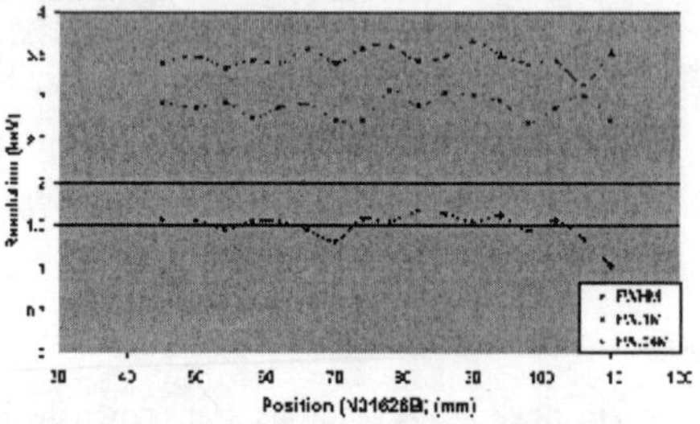

The ^{60}Co scans across the front for the GEM detectors are shown in Figs. 12 and 13. Note that these scans show the same dependance of the FW.1M and FW.04M with front position as seen before. The 1173 and 1332 keV peaks show similar results. This is a general increase in peak width, but is more pronounced at the base of the peak.

Figure 11 *Resolution at 1332 keV Across Front of GMX (N31626B).*

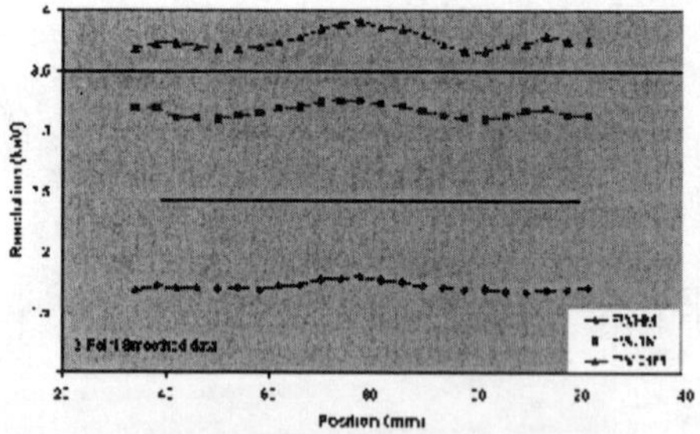

Figure 12 *Resolution at 1332 keV Across Front of GEM (P41075A).*

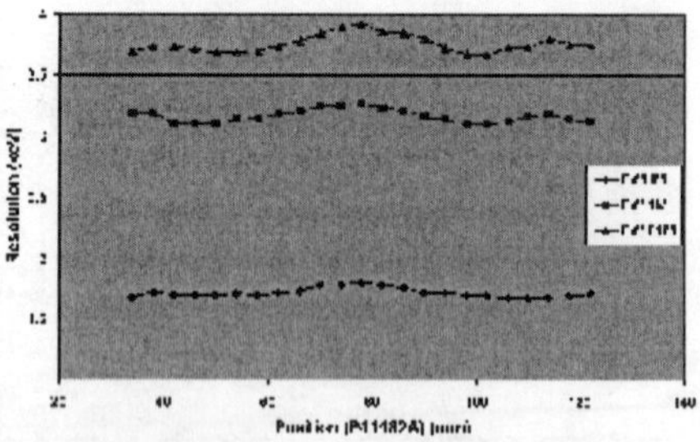

Figure 13 *Resolution at 1332 keV Across Front of GEM (P41182A).*

The scan on the side for GEM detectors is shown in Figs. 14 and 15. The shape is very nearly constant over the length except at the bottom of the crystal for Fig. 14, but begins to significantly increase about 75% of the distance from the top in Fig. 15.

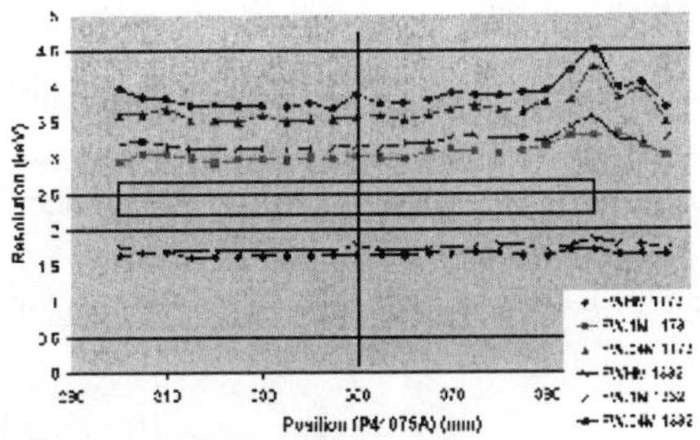

Figure 14. *Resolution at 1332 on Side of GEM (P41075A).*

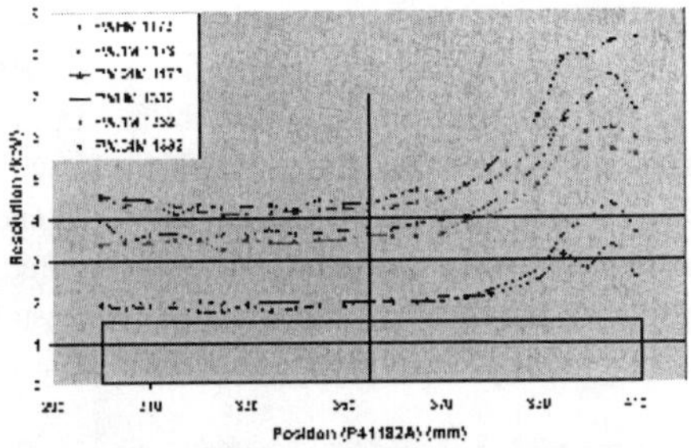

Figure 15. *Resolution at 1332 on Side of GEM (P41182A).*

3.4 Efficiency at 1.1 and 1.3 MeV

The sensitivity scan UD of a GEM detector for both 1173 and 1332 keV is shown in Fig. 16. This is the same detector and position as the detector in Fig. 14. The 1173 and 1332 keV UD scan has

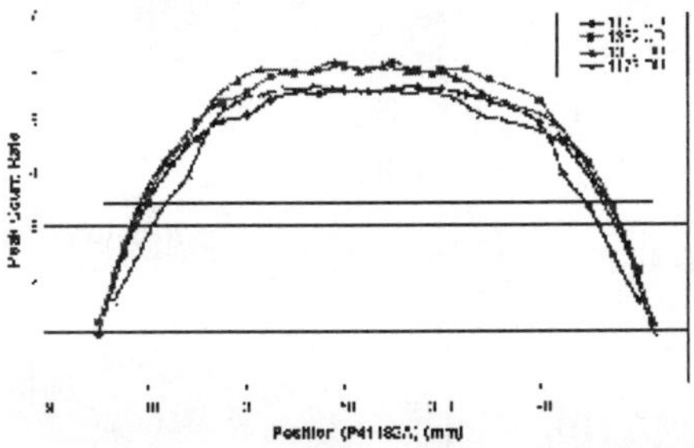

Figure 16. *Peak Count Rate at 1173 and 1332 keV vs Position on Side of Detector P41182A*

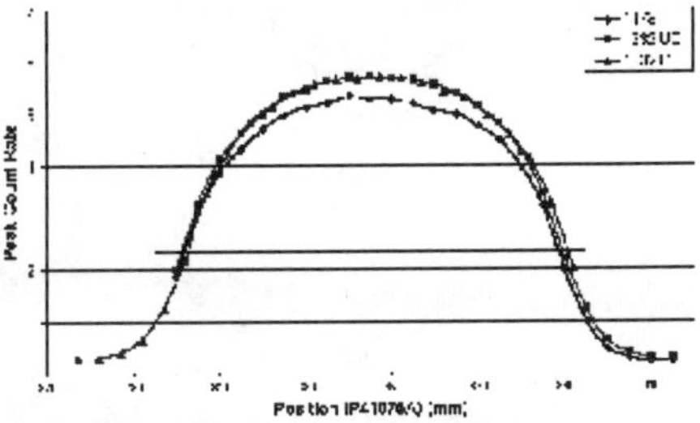

***Figure** 17. Peak Count Rate at 1173 and 1332 keV vs Position on Side of Detector P41075A*

been plotted UD and DU to show the similarity between the front and rear of the crystal. This plot shows the sensitivity is uniform away from the ends of the crystal.

Fig. 17 shows the same plot for the GEM detector in Fig. 15. The 1173 and 1332 keV UD scan has been plotted UD and DU. Note that the detector is very symmetric top-to-bottom, which is not seen in the 59 keV scan (Fig. 4).

4 DISCUSSION and CONCLUSION

Previous workers[1,3] measured several detectors in preparation for use in MCNP and other programs. These scans showed significant variation in the sensitivity (efficiency) and peak shape with the incident position of the gamma-ray beam. Detectors from several manufacturers were used. In one work, the actual thickness of the dead layer as a function of the position on the detector crystal surface was very important to the measurement geometry.

These variations could have a significant impact on the efficacy of the calculations. Thus it is important to measure the variation for more detectors.

The peak shape and sensitivity for the detectors studied did not show the wide variations with incident beam position as seen by others. However, similar detectors, in terms of easily measured physical properties (length and diameter), do show this variation. Thus, any model of a detector performance will, of necessity, require a very precise characterization of the exact detector. Future work on other detectors will extend these measurements and compare the MCNP predictions for different assumptions with experimental results.

REFERENCES

1. R. J. Gehrke, R. P. Keegan, and P. J. Taylor, "Specifications for Today's Coaxial HPGe Detectors," 2001 ANS Annual Meeting, Milwaukee, WI

2. R. M. Keyser, "Resolution and Sensitivity as a Function of Energy and Incident Geometry for Germanium Detectors", 2002 IRRMA Meeting, June 2002, Bologna, Italy

3. R. L. Metzger, private communication, see also: R. L. Metzger, K. A. Van Riper, and K. J. Kearfott, "Radionuclide Depth Distribution by Collimated Spectroscopy," 2002 ANS Topical Meeting, Santa Fe, NM

NEW METHOD FOR DETERMINATION OF TRANSPLUTONIUM ISOTOPES BY USING A PM-143 TRACER

A. Ermakov[1], T. Kuprishova[1], L. Velichko[1], Yu. Maslov[1], S. Malinovsky[1], A. Sobolev[1], A. Novgorodov[2]

[1]Scientific&Industrial Association "Radon", 7-th Rostovsky lane, 2/14, Moscow, 119121, Russian Federation
[2]Laboratory of Nuclear Problems, Joint Institute of Nuclear Research, Dubna, The Moscow district, 141980, Russian Federation

1 INTRODUCTION

Radionuclide survey of exhausts and discharges produced by nuclear installations and plants is one of the most topical and at the same time, complicated problems. Information on radionuclide pollution is important from the point of view of examining non-proliferation of nuclear materials and weapons, as well as for an impartial estimate of the operational state of nuclear installations and their protective constructions. For a solution to all the above-mentioned problems, the determination of americium and curium contents in environmental samples is rather important. Very often it requires the analysis of not only the total content, but also the precise isotopic ratio, as it is necessary to discriminate between the discharged radionuclides over background contamination.

Usually, for assaying long-lived americium and curium isotopes in environmental objects, alpha-spectrometry based on semiconductor detectors (for ^{241}Am, ^{243}Am, ^{242}Cm, ^{243}Cm, ^{244}Cm) or LS-spectrometry for registration of beta- and electron emission (for ^{242}Am, ^{242m}Am) are applied. Before measurement of their characteristic radiation, it is necessary to use complex and time-consuming procedures to remove the bulk of matrix material and also overlapping emitters during the preparation of the counting source. At the final stages of purification, chromotographic procedures are applied in a traditional manner. For preconcentration of TPu and REE, materials in which the stationary phase is either HDEHP[1] or a mixture CMPO with TBP (TRU-resin)[2] are utilized.

Differences in the stability of their thiocyanate complexes helps in separation of TPU and REE in chromotographic procedures. The most frequently applied exchangers are either the anion exchanger, Dowex 1x4,[3] or materials with aliphatic quaternary amine as the active component (TEVA-resin).[4] Sometimes extraction chromatography with a HDEHP-supported column[5] or with TRU-resin is applied.[6]

Radiochemical separation of Am and Cm is routinely performed with ^{243}Am or ^{244}Cm as a tracer, which does not allow the detailed determination of TPu isotopic ratio without parallel analysis. In this work, as a key procedure for transplutonium elements separation, cation-exchange chromatography with Dowex 50x8 column, using the complexing agent Ammonium 2-Hydroxy-2-methylbutirate (α-HMBA) has been applied. ^{143}Pm tracer, which is eluted between the Curium and Americium fractions, may be collected into a single fraction. For determining the chemical yield, gamma-spectrometry has been applied.

2 METHOD AND RESULTS

2.1 Reagents and tracers

All reagents were analytical grade. Dowex® 1x4 resin, (100-200 mesh, analytical grade) and Dowex® 50Wx8 resin, (200-400 mesh, analytical grade), were purchased from Serva Feinbiochemika GmbH&Co., Heidelberg, Germany. The 40-50 μ fraction of Dowex® 50Wx8 resin, obtained by elutriation, was used in this study. TRU*Spec® resin, 100-150 μ, was purchased from EIChroM Inc., Darien, USA.

Radioisotope tracers used were ^{143}Pm, ^{153}Gd, ^{159}Dy, ^{172}Lu obtained from Laboratory of Nuclear Problems, Joint Institute of Nuclear Research, Dubna. ^{144}Ce/^{144}Pr, ^{152}Eu, ^{241}Am, ^{244}Cm were obtained from All-Russian Scientific Research Institute of Metrology, St-Petersburg.

2.2 Instrumentations

A TRICARB 2550 TR/AB (PACKARD Inc.) LSC was used in this study. The "RadSpectraDec" spectrum-processing programme[7] was used for identification of radionulides and for calculation of their activities. DIN-based ULTIMA GOLD AB emulsion type LSC-cocktail and 20 ml (Packard Cd37,60081117) polyethylene counting vials were used throughout.

Gamma-spectrometer using a GC10023 (CANBERRA Inc.) 100% HPGe coaxial detector was applied for the determination of gamma-emitting radionuclides.

Alpha spectrometer, consisting of four units model 7401 (CANBERRA Inc.) equipped with PIPS detectors of effective surface 600 mm^{2}, was used for the precise determination of alpha-emitters activity ratios in samples.

A high performance microwave digestion unit (MLS 1200 mega (Milestone) with HPR1000/6 rotor and high-pressure TFM vessels) was used for sample leaching and for digestion procedures.

2.3 Design of analytical procedure

To avoid spiking of the samples with transplutonium element isotopes we approximated their behaviour by using rare-earth analogs. A classical cation-exchange elution technique was used on their derivatives of butyric acids. At a concentration of complexing agent of about 0.25M and pH 4.5 it is possible to separate Americium and Curium fractions, using Promethium to determine their elution positions. Thus rare earths, the content of which in the earth crust is maximal, i.e. cerium and lanthanum, are retained in the column. Eluted fractions of curium and americium should contain only impurities of samarium and neodymium.

A general procedure for isolation and measurement of Tpu is:

- conversion of analytes into the acid-dissolved state, isotope exchange with
- chemical yield tracers;
- preconcentration and purification of TPu, removing the bulk of matrix material and also overlapping alpha-emitters;
- final purification of TPu,
- preparation of a counting source by electrodeposition or microprecipitation.

The detailed analytical procedure includes the following basic stages **(Figure 1)**

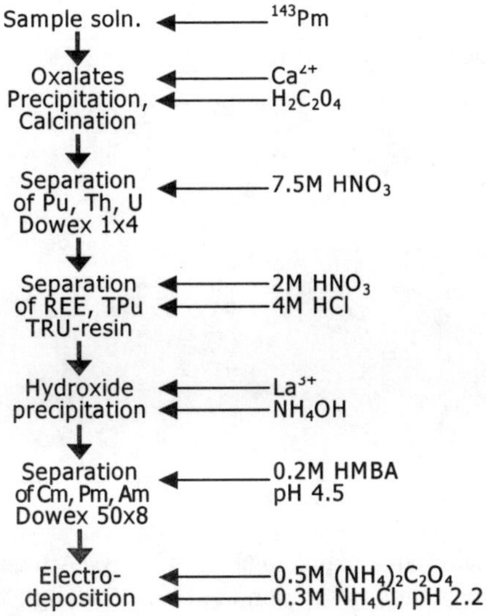

Figure 1 *The block-diagram of analytical procedure*

During the investigations of each process shown in Figure 1, composite radionuclide solutions were used and Pm behaviour was compared with that of Am and Cm.

2.3.1 Leaching/digestion of samples This procedure is determined by the speciations of TPu present in samples. Thus, the IAEA-135 sample, which contained sorbed forms of Am and Cm, was treated by boiling in solutions of mineral acids. RM IAEA-367 which contains highly refractory TPu forms, was dissolved in nitric acid with subsequent complete decomposition of a solid residue in a microwave system in a mixture of concentrated HNO_3, HF and HCl. Sample "Zab-1" from the West of the Bryansk area, in which Pu and TPu are included in dispersed fine fuel particles, was processed in a microwave system in concentrated HNO_3 with the addition of HF.

2.3.2 Coprecipitation with calcium oxalate This stage is frequently applied in analytical procedures as it allows initial concentration of analytes and also separation of them from the abundant elements and some other radionuclides. It is important that such elements as iron, titanium and zirconium, which considerably impede the extraction chromatography utilizing phosphorus-organic compounds, are removed.

Coprecipitation of trace amounts of REE and TPu with calcium oxalate depends on many factors, among which acidity of medium and concentration of the major ions are the most important. It is known that TPu are precipitated quantitatively at pH greater than 1.5.[3] During model experiments we determined an optimum Ca^{2+}-ion concentration to provide complete precipitation of Pm, Cm and Am and, to avoid overloading the TRU-resin and Sr-resin columns with Ca when subsequently separating TPu and strontium **(Table 1).**

Table 1 *Efficiency of coprecipitation with calcium oxalate for* **REE** *and* **Am** *Precipitation conditions:* $[Ca^{2+}]$ *– 200, 600 and 2000 mg/l,* $[La^{3+}]$ *– 2 mg/l,* $[C_2O_4^{2-}]$ *– 0.2M, pH – 3.0.*

Radionuclide	Ca^{2+}*-ion concentration, mg/l*		
	200	600	2000
^{144}**Ce**	87[a]	92	>98
	(4)	(2)	
^{241}**Am**	91	>98	>98
	(4)		
^{143}**Pm**	96	97	>98
	(4)	(3)	
^{152}**Eu**	89	94	>98
	(4)	(3)	
^{153}**Gd**	82	91	95
	(4)	(3)	(3)
^{159}**Dy**	87	92	96
	(4)	(3)	(3)

[a] Quota of precipitated activity, %
[b] Combined uncertainty

Table 1 shows that for quantitative coprecipitation of REE, Am and Cm, it is necessary to provide a concentration of calcium in solution of at least 500 mg/l.

2.3.3 Preconcentrating of REE and TPu with TRU-resin column This procedure was investigated by analyzing the elution behaviour of different REE cations, and Am (**Figure 2**). A standard 2 ml chromatographic column from Eichrom Inc. (calculated free column volume fcv - 1.36 ml) was used in these experiments.

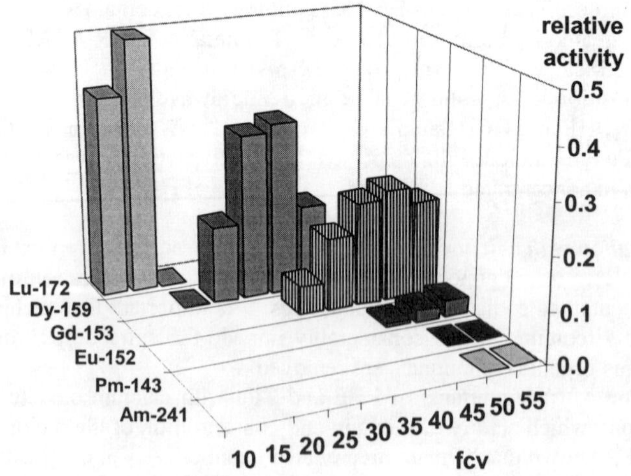

Figure 2 *Elution of REE and TPu on TRU*Spec column. Feed solution:* $^{144}Ce/^{144}Pr$, ^{143}Pm, ^{152}Eu, ^{153}Gd, ^{159}Dy, ^{172}Lu, ^{241}Am, ^{244}Cm; Ca^{2+} *- 1 g,* La^{2+} *- 2 mg in 15 ml of 2 M* HNO_3*. Washing with 2 M* HNO_3*.*

Differences in the elution behaviour of Pm, Am and Cm in the experiments simulating their extraction-chromotographic purification using TRU-resin column were not seen. When analysing actual samples, a purification procedure was used which included washing of a column with 30 ml (22 fcv) of 2 M HNO_3 after loading of the sample solution.

2.3.4 Separation of Cm, Pm and Am by elutive cation chromatography Coprecipitation of REE hydroxides and optimization of separation of REEs by elutive cationic chromatography using α-HIBA, including the choice of flow rate, temperature, and a column bed height are reported elsewhere.[8] In simulating experiments for the separation of REE, (Am and Cm) we utilized a column (3 mm × 40-50 mm) containing cation exchange resin Dowex® 50Wx8, 40-50 μ. Elements were eluted from the column with a buffer solution of Ammonium 2-Hydroxy-2-methylbutirate (α-HMBA). The concentration of complexing agent varied from 0.2 M to 0.3 M, and the pH value varied from 4.0 to 4.7. Effluent was collected drop by drop, each four drops were combined into a single fraction. One of the elution diagrams is shown below **(Figure 3)**.

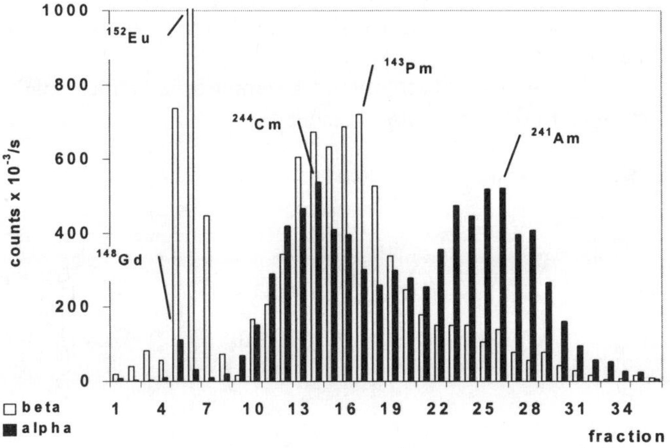

Figure 3 *Elution of REE and TPu on Dowex50x8 column (3 mm × 46 mm). Feed solution: tracers $^{144}Ce/^{144}Pr$, ^{143}Pm, ^{152}Eu, ^{153}Gd, ^{159}Dy, ^{172}Lu, ^{241}Am, La^{3+} - 2 mg in 3 ml of 0.4 M NH_4Cl. Elution with 0.2M α-HMBA, pH – 4.5, flow rate – 0.75 ml/min*

For analyses of RM and real samples we used the following conditions for the separation: column dimensions – 3 mm × 46 mm, elution solution 0.24M α-HMBA, pH 4.5

2.3.5 Preparation of alpha sources After completely destroying the organic compounds in the eluate by evaporation with concentrated HNO_3 and 30% H_2O_2, alpha sources were prepared by electrodeposition onto stainless steel disks. The cell was made of Teflon and the effective area of electrodeposition was 8 cm^2 (∅32 mm). As an anode, a platinum disk of 32 mm dia was used. Electroplating solution used in this study consisted of 0.5M $(NH_4)_2C_2O_4$ and 0.3M NH_4Cl at pH 2.2, volume of solution was 25 ml. Current value during electrodeposition was 2A (250 mA/cm^2), duration of the process – 1 hour.

The electroplating stage is the most crucial in determining the total chemical yield for the analytical procedure. Losses of the analytes on this stage ranged from 5% to 40% and therefore the electroplating procedure requires further improvements.

2.4 Calculation of activities and evaluation of uncertainty components

The specific activities of alpha-emitting radionuclides using ^{143}Pm gamma-tracer may be calculated from:

$$a_i = \frac{A_i}{m_{smp}} = \left(\frac{N_i}{T_i} - \frac{N_{Bkgi}}{T_{Bkg}}\right) \cdot \frac{1}{\varepsilon_\alpha} \cdot \frac{(A_{Pm\,ref})}{(A_{Pm\,smp})} \cdot \frac{V_{Pm\,smp}}{V_{Pm\,ref}} \cdot \frac{1}{m_{smp}} \quad (1)$$

where: N_i – peak areas of measured radionuclide in the alpha-spectrum of the analysed sample;

N_{Bkgi} – number of events in ROI of radionuclide in the background alpha-spectrum;
T_i and T_{Bkg} – measurement duration of sample and background (s);
$A_{Pm\text{-}143ref}$ – measured tracer activity in reference source, corrected to $A_{Pm\text{-}143\,smp}$ Measurement date
$A_{Pm\text{-}143smp}$ – measured tracer activity in the sample after radiochemical separations,
ε_α – alpha-spectrometry counting efficiency.

Combined standard uncertainty:

$$\frac{u(A_i)}{A_i} = \sqrt{\frac{\frac{N_i}{T_i^2} + \frac{N_{Bkgi}}{T_{Bkgi}^2}}{\left(\frac{N_i}{T_i} - \frac{N_{Bkgi}}{T_{Bkgi}}\right)^2} + \left(\frac{u(\varepsilon_\alpha)}{\varepsilon_\alpha}\right)^2 + \left(\frac{u(A_{Pmsmp})}{A_{Pmsmp}}\right)^2 + \left(\frac{u(A_{Pmref})}{A_{Pmref}}\right)^2 + 2\cdot\left(\frac{u(V_{Pmref})}{V_{Pmref}}\right)^2 + \left(\frac{u(m_{smp})}{m_{smp}}\right)^2} \quad (2)$$

2.5 Analyses of RM and real samples

The described method was used to analyse RM materials with a known content of radionuclides and by comparative parallel analyses of sample containing ^{243}Am tracer. TPu were separated from REE on the anion exchanger Dowex1x4 column in a composite medium consisting of ammonium thiocyanate, hydrochloric acid and ethanol. The results obtained are shown in **Table 2.**

Table 2 *Results of analyses of RM and real samples*

Concentration	^{241}**Am**, *Bq/kg*		^{243,244}**Cm**, *Bq/kg*		*This method*	
Sample	*Target value*	*This method*	*Target value*	*This method*	*Chemical yield, %*	*Resolution, keV*
IAEA-135	318 (8)[a]	330 (19)	-	2.7 (0.19)	37.4 (0.7)	28
Zab-1	4.74 (0.21)	4.38 (0.26)	0.38 (0.04)	0.53 (0.07)	39.6 (0.8)	36
D-16-12	38 (4)	46 (5)	7.8 (1.0)	8.9 (1.1)	38.1 (0.8)	32

[a] Combined standard uncertainty

Examples of alpha-spectra obtained as the result of the analyses of RM and the real samples are shown in **Figure 4**.

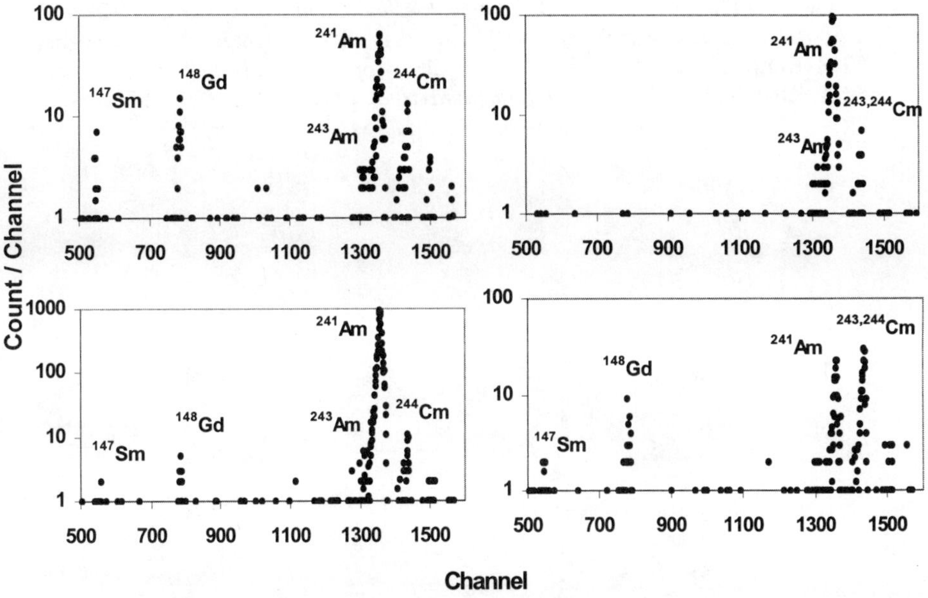

Figure 4 *Alpha spectra of TRu fractions, recovered from solid samples when using* ^{143}Pm *gamma-tracer. Samples: Zab-1 (upper), IAEA-135 (lower) - left side, D-16-12 Am fraction (upper), D-16-12 Cm fraction (lower) - right side*

Twenty one analyses of TPu content in soil and bottom sediment have been carried out with the proposed method. The masses of the samples varied from 5 up to 20 grammes. The mean chemical yield value was 37% (ranging from 32% to 45%). Alpha-peak resolution of ^{241}Am line 5485 keV ranged from 26 to 48 keV.

3 CONCLUSION

The proposed method allows the determination of the total content and isotopic ratio of alpha-emitting TPU and REE in environmental samples using ^{143}Pm as a gamma-tracer. This method is useful when analysing a series of samples with a varying TPu isotope ratio. If the TPu ratio in samples is similar the method may be applied for the determination of TPu content by LSC-method and thus the electrodeposition stage is excluded. Also it is possible to determine the content of beta-emitting ^{147}Pm in samples by the LCS-method by analysis of a parallel subsample without ^{143}Pm tracer.

References

1 D. Desideri, M.A. Meli, C. Roselli and C .Testa, *J. Radioanal. Nucl. Chem.*, 2002, **251**, 37.

2 S.J. Goldstein, J. Steven; C.A. Hensley, C.E. Armenta, R.J. Peters, *Anal. Chem.,* 1997, **69**, 809.
3 A.Yamato, *J. Radioanal. Nucl. Chem.*, 1982, **75**, 265.
4 M. Pimpl, R. H. Higgy, *J. Radioanal. Nucl. Chem.*, 2001, **248**, 537.
5 D.B. Martin, D.G. Pope, *Anal.Chem.* 1982, **54**, 2552.
6 F. Goutelard, M. Morello and D. Calmet, *J. Alloys Compounds*, 1998, **271-273**, 25.
7 I.A. Kashirin, A.I. Ermakov, S.V. Malinovskiy, S.V. Belanov, Yu.A. Sapozhnikov, K.M. Efimov, V.A. Tikhomirov, A.I. Sobolev, *Appl. Radiat. Isot.*, 2000, **53**, 303.
8 V.A. Khalkin, N.A. Lebedev, *J. Radioanal. Nucl. Chem.*, 1985, **88/1**, 153.

ULTRA LOW-BACKGROUND GAMMA SPECTROMETRY FOR THE MONITORING OF ENVIRONMENTAL NEUTRONS

K. Komura

Low Level Radioactivity Laboratory, Institute of Nature and Environmental Technology, Kanazawa University, Wake, Tatsuokuchi, Ishikawa 923-1224, JAPAN

1 INTRODUCTION

In order to measure extremely low-levels of natural and artificial radionuclides, the Ogoya Underground Laboratory (OUL) was constructed in the tunnel (depth = 270 meters water equivalent) of former Ogoya copper mine,[1] located about 20 km from the Low Level Radioactivity Laboratory (LLRL) of Kanazawa University in Tatsunokuchi, Ishikawa Pref., Japan[1] (Figure 1). By the use of Ge detectors specially designed for ultra low-background gamma spectrometry, it became possible to measure extremely low levels of radioactivity as low as 1 cpd (count per day). Various kinds of measurements have been performed in OUL since 1995, such as assessment of the environmental impact of the JCO criticality accident,[2] which occurred in 1999 and investigations of "natural" radionuclides induced by environmental neutrons. As of September 2002, eight ultra low-background Ge detectors (1 coaxial Ge, 4 well type Ge and 3 planar type Ge) are set in OUL.

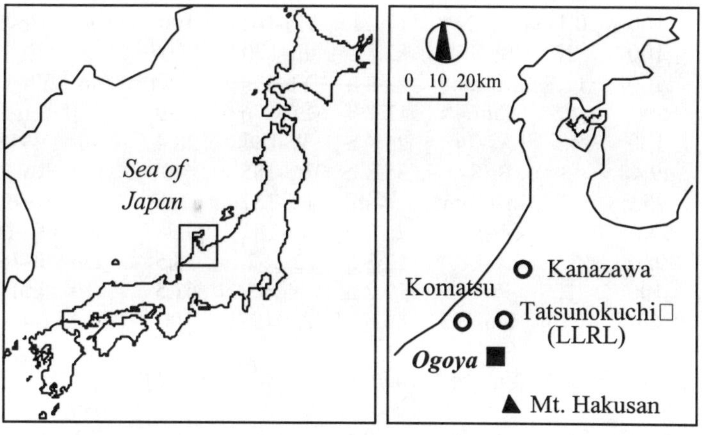

Figure 1 *Location of Ogoya Underground Laboratory.*

"Natural" ^{60}Co and ^{152,154,155}Eu induced by environmental neutrons were first detected in old cobalt reagents produced prior to the World War-II and in europium oxide.[2] A study aimed at explaining the discrepancy between observed and calculated ^{60}Co and ^{152}Eu in atomic-bomb exposed samples was undertaken (observed >> theoretical values in the samples collected from > 1 km). However, natural productions of ^{60}Co and ^{152}Eu were found to be too low to explain the discrepancy.

The project "Survey of natural radionuclides induced by environmental neutrons" has been conducted since 1998. More than 20 nuclides induced by environmental neutrons have been identified by ground-level experiments, mountain-climbing experiments and high altitude experiments by domestic and/or intercontinental flights.[2] This paper summarizes the experiments in order to detect neutron-induced radionuclides in the reagents and/or environmental samples and some applications made for the monitoring of environmental neutrons.

2 METHOD AND RESULTS

2.1 Candidates of Target Material

It is known that the flux of environmental neutrons is n x 10^{-3} cm^{-2} s^{-1} (n = 5-10) at sea level[3] and it increases with altitude by mean free path of about 200 g cm^{-2}. Activity of neutron induced nuclides can be estimated by the following equation.

$$A = n\,\phi\,\sigma[1 - \exp(-0.693t/T)]$$

Where A, n, ϕ and σ are activity of product nuclide, number of target atoms, neutron flux and neutron capture cross section, respectively, and t and T show exposure time and half-life of product nuclide. In this calculation, the energy spectrum of environmental neutrons and the excitation function of neutron capture reactions must be taken in account.

Table 1 *Candidates of target-product pair.*

Target nuclide	Abund. (%)	σ (b)	Product nuclide	Half-life	Target nuclide	Abund. (%)	σ (b)	Product nuclide	Half-life
Na-23	100	0.53	Na-24*	15.0 h	Tb-159	100	46	Tb-160	72.3 d
Al-27	100	0.11	Na-24*	15.0 h	Ho-165	100	64	Ho-166	26.9 h
Sc-45	100	27	Sc-46*	83.8 d	Er-170	27.07	9	Er-171	7.52 h
Mn-55	100	13.3	Mn-56*	2.58 h	Yb-174	31.84	46	Yb-175*	4.20 d
Cu-63	69.1	4.5	Cu-64*	12.7 h	Lu-176	2.59	2100	Lu-177	6.71 d
As-75	100	4.5	As-76*	26.4 h	W-186	28.4	40	W-187*	23.7 h
Br-81	49.48	3	Br-82*	35.3 h	Re-185	37.07	110	Re-186*	3.72 d
In-115	95.7	73	In-116m*	54 m	Re-187	62.93	70	Re-188*	16.7 h
Sb-121	57.25	6	Sb-122*	2.70 d	Ta-181	99.988	21	Ta-182*	114.4 d
La-139	99.911	8.9	La-140*	1.68 d	Ir-191	38.5	750	Ir-192*	73.83 d
Pr-141	100	12	Pr-142	19.2 h	Ir-193	61.5	110	Ir-194*	17.40 h
Sm-152	26.63	210	Sm-153	46.8 h	Au-197	100	98.8	Au-198*	2.695 d
Sm-154	22.53	5	Sm-155	0.4 h					
			Eu-155*	4.76 y		*Half-life*	*>1 year*		
Eu-151	47.77	2800	Eu-152m*	9.30 h	Co-59	100	20	Co-60*	5.27 y
Gd-158	24.9	3.4	Gd-159	18.5 h	Cs-133	100	28	Cs-134*	2.06 y
Dy-164	28.18	800	Dy-165	2.35 h	Eu-151	47.77	5300	Eu-152*	13.33 y
					Eu-153	52.23	320	Eu-154*	8.80 y

In order to obtain high *A*, it is obvious that n and σ must be large and half-life of product nuclide should be appropriately long from the constraint of measurement (longer than ~1 h and shorter than ~1 y). Candidates of target-product pairs selected accordingly are listed in Table 1. Radionuclides marked by asterisk were detected already by the experiments given below.

2.2 Ground-Level Experiments

Target materials used in the ground-level experiments are Sc-oxide (5g), metallic As (50~200 g), Cs-chloride (50 g), metallic Ta (100 g), Ir-oxide (5~20 g), and various gold items (0.7 mm thick Au-plate, 1-2mm Au-grain, necklace, ring and coin). Most of these targets were purchased from the market except some of gold items. Measurements of long-lived nuclides such as ^{46}Sc ^{134}Cs, ^{182}Ta were made soon after purchase by assuming that production of these nuclides were at equilibrium state before purchase. Other short-lived nuclides were measured experimentally after appropriate exposure time (> 5 half-lives). All measurements were performed using Ge detectors in OUL coupled with 4K or 8K channel pulse height analyzer.

Gold experiments were most frequently made not only because production rate of ^{198}Au is sufficiently high but also because half-life of ^{198}Au (2.695 d) is conveniently long considering the practical constraints of exposure and counting time. More than 100 experiments have been performed for gold under various conditions from ground level to an altitude of 12 km, in the water, soil and paraffin.

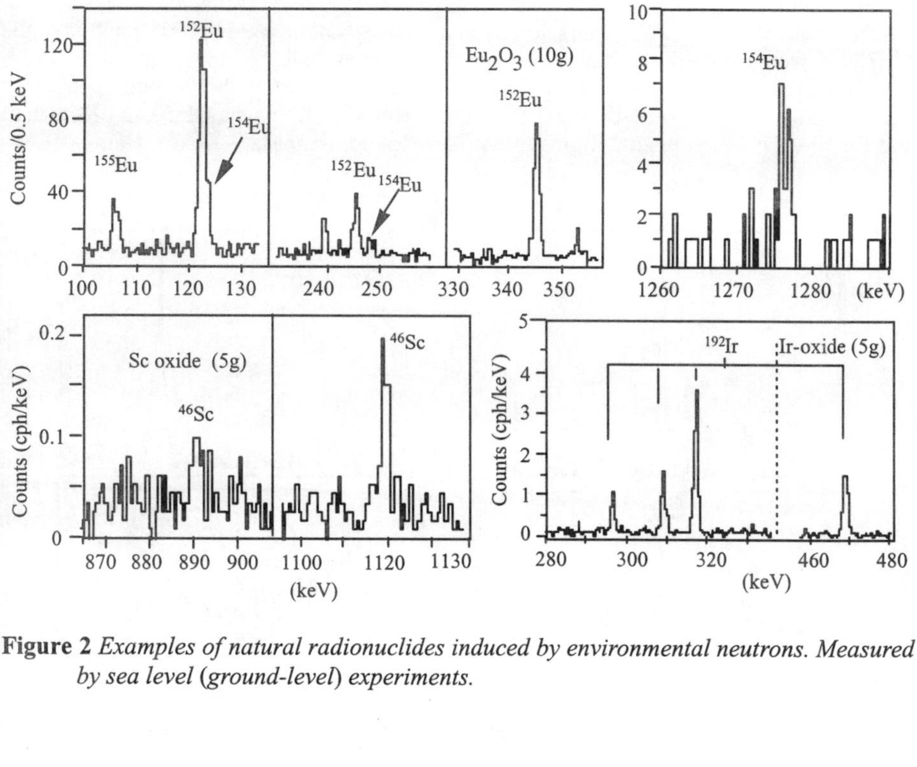

Figure 2 *Examples of natural radionuclides induced by environmental neutrons. Measured by sea level (ground-level) experiments.*

2.3 Mountain-Climbing Experiments

Some short-lived radionuclides such as ^{56}Mn (2.58h), ^{188}Re (16.7h) and ^{194}Ir (17.4 h) were expected to be detectable when sufficient amount of targets were exposed at high altitude of at least 2-3 km, where neutron flux is 3-5 times higher than at sea level. Mountain-climbing experiments were thus performed in August of 2000 and 2002 at Mt. Hakusan (2702 m) located about 50 km from LLRL (Figure 1). In the first experiment, 100 g of metallic Mn was used as target. The Mn target was sealed in a polyethylene bag and carried in back-pack together with gold monitors. After 3 h stay in the hut at 2100 m above sea level, the Mn target was transported quickly to OUL and gamma ray measurement was initiated 3 h after climbing down. Gold targets set at various altitudes between 950~2100 m, were collected 20 days after the ^{56}Mn experiment and measured to investigate the altitude dependence of thermal neutron flux. In the second experiment (August 2002), Re powder (20 g), Ir-oxide (20 g) and Sc-oxide (20 g) were chosen as target materials. These targets were exposed to neutrons for 12 h (1 night stay) in mountain hut at 2400 m. Gamma ray measurements of Re and Ir were initiated 6 h after climbing down. Gamma-ray spectra of ^{56}Mn and ^{188}Re are shown in Figure 3. Scandium and gold targets will be recovered in mid October of 2002.

2.4 Flight Experiments

Flight experiments can be made even for short-lived nuclide with a half-life less than 1 h since OUL is located only 25 km from Komatsu Airport. The first experiment was performed for ^{116m}In (54 min) and ^{56}Mn by Tokyo-Komatsu flight (schedule time is 55 min, cruising time at 8 km was 20 min). In this experiment, indium oxide (100 g) and metallic Mn (80 g) were exposed to cosmic-ray neutrons at high altitude. Gamma ray measurements were initiated 60 min after the landing at Komatsu Airport. Other experiments were performed by 2-3 h of domestic flights (Misawa - Haneda - Komatsu, Okinawa - Komatsu) and 8-12 h of intercontinental flights (New York-Tokyo, Denpasar - Osaka, Cairo - Osaka etc.)

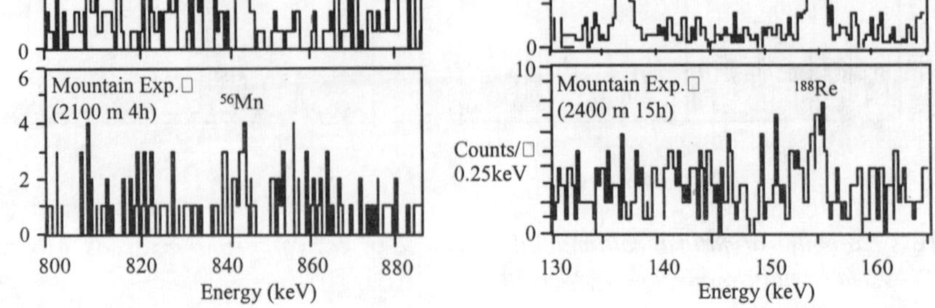

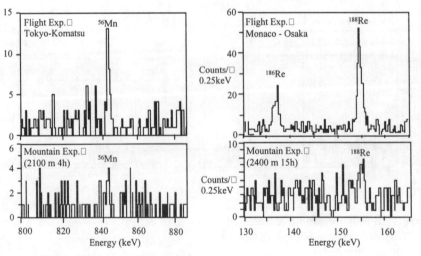

Figure 3 *^{56}Mn, ^{188}Re and ^{186}Re induced by flight experiments (upper spectra) and mount-climbing experiment (lower spectra).*

Target materials used in flight experiments were cooking foil (Al), table salt (NaCl), copper coins, NaBr, As-metal, La-oxide, Eu-oxide, Yb-oxide, W-metal, Re-metal and Ir-oxide. These targets were simply carried in a suitcase for cosmic-ray exposure. Examples of Mn and Re spectra obtained by flight experiments are shown in Figure 3 together with those obtained by mountain-climbing experiments. As seen in Figure 3, ^{186}Re could be detected only by flight experiment because neutron flux is insufficient for the production of detectable amount of ^{186}Re at lower altyitudes.

2.5 Evaluation of Neutron Fluence of JCO Criticality Accident

Neutron monitoring is not commonly undertaken except for special facilities such as nuclear reactors or nuclear fuel reprocessing plants. In the JCO criticality accident, which occurred in 1999, lack of neutron monitoring system in the JCO campus gave a long delay in notification about the criticality. Within 1 week after the accident, a research group composed mainly of radiochemists and a few physicists was organized to assess the environmental impacts of the accident. More than 400 environmental samples were collected in the JCO campus, Tokai-mura and Naka-machi to evaluate neutron and gamma ray doses.

These samples were measured at more than 10 laboratories including HADES in Belgium.[5] The ^{198}Au induced in gold items such as necklaces, finger rings and coins, which were borrowed from inhabitants in the vicinity of JCO, was measured using Ge detectors in OUL. Influence of the JCO accident was observed for ^{198}Au up to 1400 m from the JCO campus.[6] Measured ^{198}Au activity was well reproduced by theoretical calculation based on the DOT-3.5 code. Neutron induced ^{51}Cr in stainless steel was detected up to 400 m by nondestructive gamma spectrometry.[7,8] Fast neutron product ^{32}P in table salts and sulfur containing reagents was radiochemically separated and measured by ultra low-background beta ray counter in OUL. Large directional dependence was observed for ^{32}P collected within 200 m range due to the shielding by buildings in the JCO campus.[9] Fast neutron induced ^{54}Mn by $^{54}Fe(n,p)^{54}Mn$ reaction could successfully be detected in OUL by applying radiochemical separation.[10]

2.6 Neutron Monitoring by Gold Activation

Among the nuclides listed in Table 1, gold is considered to be most useful for neutron monitoring, because it has a large thermal neutron cross section and the product nuclide, ^{198}Au, has a suitable half-life. Furthermore, gold is not so expensive (~10 USD g^{-1}) compared to many neutron counters. It has also high chemical stability, which makes it possible to set a gold target in extremely severe environment, i.e. wide ranges of temperature from 0 K to melting point) and humidity (0-100%), even in soil and water. Practical neutron monitoring has been performed by 20 g of gold grains packed in a polyethylene bag of 20 mm x 20 mm. The target was exposed to environmental neutrons for about one month to attain saturation activity of ^{198}Au. Then the 412 keV gamma-ray from ^{198}Au was measured using ultra low-background Ge detectors in OUL. The production rate of ^{198}Au was measured to be ~50 atoms g^{-1} (20% of uncertainty) at sea level in Japan (23 degree of geomagnetic latitude) when 1-2 mmϕ of gold grains are used as target. The detection limit of neutron flux is ~10^{-3} n cm^{-2}-s^{-1} when 20 g of gold is used as target.

Neutron monitoring has been conducted since 1999 at various locations under a wide variety of environmental conditions. As an example, the result of neutron monitoring at Rokkasho-Mura, Aomori Prefecture, Japan, where nuclear conglomerate (^{235}U enrichment plant, nuclear fuel processing plant and nuclear waste disposal plant) are under construction, is shown in Figure 4. Due to the transportation problem from Rokkasho to LLRL (2-3 days), statistical uncertainty was rather large, however, observed values were found to be within the range of ground level values in other locations.

3 CONCLUSION

(1) Gold is considered to be most useful for the monitoring of neutrons in variety of environments not only because production rate is rather high but also because half-life and gamma-ray of ^{198}Au are adequate for ultra low-background gamma spectrometry.
(2) If longer exposure (several months) is available, iridium is considered to be more sensitive than gold not only because saturation activity of ^{192}Ir is 3 times higher but also because longer half-life (20 times) allows longer measuring time of over 10 days. As a result, a neutron flux of the order of 10^{-4} cm^{-2} s^{-1} can be measured by using 20g of iridium target.
(3) In order to monitor high altitude neutrons, ^{187}W, ^{186}Re, ^{188}Re, ^{194}Ir are recommended because of their high production rate in a short exposure time.
(4) Neutron monitoring by the activation method can successfully be applied without influenced by complex radiation environment such as high altitude of commercial light and/or space shuttle, where contributions of cosmic ray derived protons, alphas and heavy ionizing particles are extremely high and complex.

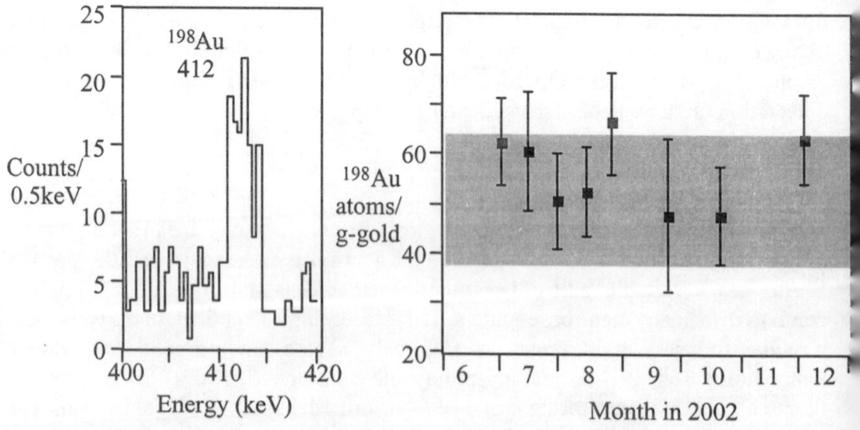

Figure 4. *Neutron monitoring at Rokkasho-Mura by gold activation method. Example of gamma ray spectrum (left) and ^{198}Au atoms per g-gold (right). Hatched area shows normal level obtained at other area.*

References

1 K. Komura, *Proc. Int. Symp. on Environmental Radiation*, Ed. T. Tsujimoto and H. Ogawa, Tsuruga, Fukui, Japan, Oct. 20, 1997, p. 56.

2 K. Komura, A. M. Yousef, *Proc. Int. Workshop on Distribution and Speciation of Radionuclides in the Environment*, Rokkasho, Aomori, Japan, Oct. 11-13, 2000, ed by J. Inaba, S. Hisamatsu, Y. Otsuka, 2000, p. 210.
3 J. E. Hewitt, L. Hughes et al, *Natural Radiation Environment III*, **Vol. 2**, USDOE, 1980, p. 855.
4 K. Komura et al., *J. Environmental Radioactivity*, 2000, **50 (1,2)**, 3.
5 K. Komura, *J. Radiation Research,* 2001, Suppl. **42**, S17.
6 K. Komura, A. M. Yousef, Y. Murata, T. Mitsugashira, R. Seki, T. Imanaka, *J. Environmental Radioactivity*, 2000, **50 (1,2)**, 77.
7 S. Endo, N. Tosaki, K. Shizuma, M. Ishikawa, J. Takada, R. Suga, K. Kitagawa, M. Hoshi, Journal of Environmental Radioactivity. 2000, **50 (1,2)**, 83.
8 M. Hult, M. J. Martínez Canet, N. Peter, P. N. Johnston, K. Komura, *J. Environmental Radioactivity*, 2002, **60**, 307.
9 H. Kofuji, K. Komura, Y. Yamada, M. Yamamoto, *J. Environmental Radioactivity*. (2000) **50 (1,2),** 49.
10 Y. Murata, T. Muroyama, H. Kofuji, M. Yamamoto, K. Komura, *J. Radiation Research,* 2001, Suppl. **42**, S69.

ESTIMATION OF UNCERTAINTY IN RADIOCHEMICAL ANALYSIS FROM SUB-SAMPLING SOLID MATRICES

G. J. Ham and M. J. Youngman

National Radiological Protection Board, Chilton, Didcot, OXON, OX11 0RQ, UK

1 INTRODUCTION

With the introduction of ISO/IEC 17025 and its adoption by the United Kingdom Accreditation Service (UKAS), accredited laboratories are facing increasing pressure to estimate all causes of uncertainty in their measurements. One potentially large source of uncertainty in radiochemical analyses of environmental samples is the sub-sampling of solid matrices. In such analyses, it is often necessary to take separate sub-samples for the determination of different radionuclides. Practical considerations mean that these aliquots are often taken from dried milled material, rather than a solution. Liggett et al in 1984[1] published measurements on the homogeneity of three National Bureau Standards reference materials. The conclusions were that despite the vigorous grinding and mixing of such materials the concentrations in sub-samples were not homogeneous and that the degree of in-homogeneity varied between the three materials. Time and cost constraints are likely to preclude such vigorous sample homogenisation on routine environmental samples, resulting in a larger degree of in-homogeneity in such samples than those reported by Liggett et al.

The estimation of the uncertainty caused by sub-sampling is very time consuming, particularly as the degree of in-homogeneity is likely to be both sample and analyte dependent. It requires the collection and preparation of a large amount of the environmental material and the analysis of many replicate aliquots. The low concentration of many radionuclides in environmental materials presents a further complication to this type of study. The effort and cost involved in this process make it impractical to analyse the required number of replicates for every sample. Therefore, the approach used by this laboratory was to undertake the estimation of the likely uncertainty incurred by sub-sampling for a restricted range of materials and analytes. The data were then used to estimate likely uncertainties for the sub-sampling performed under operational conditions in the laboratory.

As part of the evaluation, two vegetation samples were dried, ground, sub-divided, and used for an intercomparison exercise. This exercise was run by the Analysts Informal Working Group (AIWG), which includes most of the organisations involved in low-level radionuclide analysis in the UK. The results were used to validate the conclusions drawn from the in house study.

2 MATERIALS AND METHODS

2.1 Samples

2.1.1 Sample selection. With the exception of ^{210}Po, concentrations of all of the radionuclides of interest in typical environmental samples would be so low that the measurement uncertainties would obscure any sampling variability. Hence, most samples were taken from non-typical locations (either lysimeters where the soils had been deliberately contaminated or environments where activity concentrations were known to be elevated). In all cases however, the locations selected had been stable for some years and so from the analytical point of view the radionuclides were likely to behave similarly to those in samples from situations that are more typical.

The main environmental materials analysed in this laboratory are vegetation and soil. The types of vegetation selected for this study were grass and cabbage. Grass is the most commonly analysed type of vegetation, for which one important complicating factor for sub-sampling is the likely presence of soil. Many of the radionuclides of interest have a low soil to plant transfer factor, and so the concentrations of radionuclides in soil will be higher than the corresponding values in grass. Soil may in fact be the main contributor to the total activity in the sample.

The first sample of grass was collected from lysimeters that had been artificially contaminated with ^{137}Cs, $^{239+240}$Pu, ^{241}Am and ^{90}Sr in 1983. A second sample was collected from open land in Southern England. A ball headed variety of cabbage was chosen to represent a vegetable where the edible portion is protected from surface contamination by inedible parts. Soil should not therefore be present in the edible portion, which is the part normally analysed for radiological protection purposes. The sample of cabbage was collected from land that had been reclaimed from the sea in the late 1970's on the Lancashire coast[2]. The radionuclides of interest in this sample were ^{137}Cs and ^{40}K for the main study and ^{137}Cs, ^{40}K, $^{239+240}$Pu, ^{241}Am and ^{90}Sr for the intercomparison exercise.

Sediment from the Ravenglass estuary in north-west England was selected to represent soil. Radionuclide concentrations in this area are elevated as a result of authorised discharges to sea from the nearby Sellafield nuclear fuel reprocessing plant. Concentrations of a range of radionuclides are therefore higher than in inland soils across the UK.

2.1.2 Sample collection and preparation. Both grass samples were cut by hand using shears; care was taken to minimise contamination by soil. Each sample was then oven dried at 105^{0}C in a fan-assisted oven. In accordance with standard practice in this laboratory, the sample was not washed. The cabbages were cut by hand. The outer leaves were removed and the inner ball rinsed with water before being cut into slices and oven dried in the same way as the grass. The water content of the cabbage (94%) was much higher than the grass (65%). All of the dried vegetation samples were passed through a knife mill fitted with a 3mm mesh. The sediment sample was collected by hand from the inter-tidal zone, oven dried at 105^{0}C and milled in a disc mill to a fine powder.

All milled samples were collected in plastic bags, sealed and shaken by hand to mix. This procedure follows that normally carried out for such samples in this laboratory.

2.1.3 Sub-sampling. The effect of using a sample divider was investigated by the following procedure. About 500g of dried, milled grass was divided in two using a riffling box. From one portion, 10 aliquots of 20g were taken for subsequent analysis using a hand scoop. The other portion was further sub divided into 10 aliquots of 20g using a rotary sample divider. Alternate aliquots were used individually in the analytical scheme described below. To confirm the reproducibility of the analytical method the remaining aliquots were bulked and dissolved. Five aliquots were then taken from the resultant solution. All of these samples were analysed for ^{210}Po.

For the remainder of the study, all sub-samples were taken by scooping the appropriate mass of material from the bagged bulk sample.

2.2 Analytical methods

2.2.1 Gamma ray emitting radionuclides. Three sample geometries were used, half litre marinelli beakers, one litre marinelli beakers and 90mm diameter petri dishes. All samples were packed so as to fill the appropriate container completely. Gamma-ray emitting radionuclides were determined using hyperpure germanium detectors housed in a purpose built facility and appropriately calibrated. The aim of the experiment was to investigate reproducibility between similar samples and not to give absolute measurements of their radionuclide content. Consequently, no correction for sample density was applied to the measurements. The reproducibility of the spectrometry system was investigated by measuring the radionuclide concentrations in one sample of grass contained in a petri dish six times. The sample was removed from the instrument after each measurement so that any uncertainty produced by differences in sample positioning would be included in the results.

2.2.2 Alpha particle emitting radionuclides. Plutonium isotopes and ^{210}Po were determined using validated in house methods. Briefly, the samples for plutonium analysis were spiked with ^{242}Pu, dry ashed, dissolved and the plutonium separated by ion-exchange. Sources were prepared by electroplating on to stainless steel discs. Samples for ^{210}Po analysis were spiked with ^{208}Po, wet ashed with nitric acid and the polonium separated by deposition on to silver discs. All radionuclides were determined by alpha spectrometry using planar implanted silicon detectors operating *in vacuo*.

2.2.3 Quality assurance. The laboratory holds UKAS accreditation for all these analyses and the analyses were performed as described in the laboratory's quality manuals. However due to the experimental nature of this work some sample preparation did not follow the accredited procedures and as already noted no correction for density was made on the gamma-ray spectrometry results. Therefore, the results reported fall outside the scope of the accreditation.

2.3 Intercomparison exercise

To investigate the effect of sub-sampling further, two of the vegetation samples were used for the intercomparison exercise run by the AIWG. Each participant received an aliquot of grass taken from the lysimeter (130g dry mass) and the cabbage (100g dry mass). These had been taken by hand from the bulk sample using a scoop. Each participating laboratory took sub-samples and analysed them for gamma-ray emitting radionuclides, plutonium, ^{241}Am and ^{90}Sr using their standard procedures. Results were reported by the participants along with their estimate of the uncertainty of the measurement.

3 RESULTS AND DISCUSSION

3.1 Effect of sample dividers

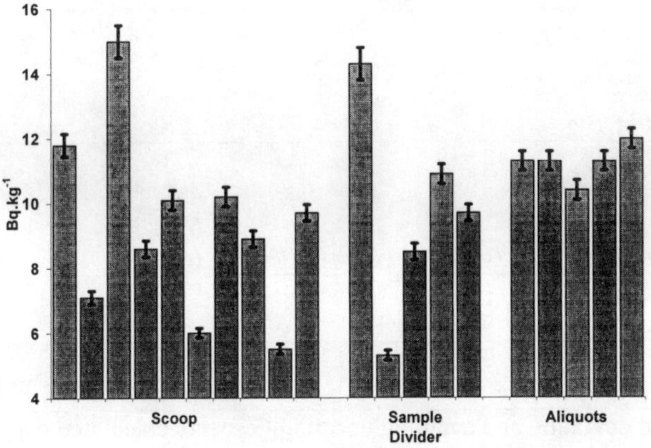

Figure 1 ^{210}*Po concentration in sub-samples of grass. (The displayed result uncertainties are from counting statistics only at k=1)*

The preparation procedure gave three sets of results. There were 5 individual results from the aliquots taken from the solution, 5 from the samples produced using the rotary divider and 10 from the samples produced using the scoop. These are displayed graphically in Figure 1.

The reproducibility from the aliquots taken from the bulk solution was good. The standard deviation on the results was 5% and the u-test score[3] confirmed that none of the individual results differed significantly from the mean. This confirmed the reproducibility of the analytical method. The variability of the results from the two sets of individual solid aliquots was much larger, standard deviations being 37% for those taken by scoop and 43% for those taken by sample divider. These results suggest that there is no improvement in reproducibility of aliquots produced by a rotary divider when compared with those taken by hand. This conclusion led to the decision that all further aliquots used in this study would be taken by hand using a scoop.

3.2 Vegetation

3.2.1 Gamma-ray emitting radionuclides in grass. Three sets of measurements were made on dried grass samples taken from the lysimeter. One set used approximately 250g aliquots in one-litre marinelli beakers, one set used approximately 20g aliquots in petri dishes and one set utilised a single petri dish sample, which was counted repeatedly. Both ^{137}Cs and ^{40}K were measurable in all cases although the ^{40}K result for one count was rejected due to poor peak shape.

The results are summarised in Table 1. The stability and reproducibility of the counting equipment was confirmed by the results for the repeated counting of a single sample. The variability in the results for ^{137}Cs for the other two sets of data were much larger than would be expected solely from the counting uncertainty for both sample sizes.

Table 1 *Measurements on grass samples.*

Sample mass and approximate activity concentration	*Number of samples or measurements*	*Standard deviation[a]*	*Counting uncertainty[a]*
^{137}Cs (≈400 Bq.kg^{-1})			
250g	5	8%	0.4%
20g	8	16%	1%
Repeat count of 20g	6	1.1%	1.1%
^{40}K (≈300 Bq.kg^{-1})			
250g	4	4%	2%
20g	8	6%	7%
Repeat count of 20g	6	6%	8%
^{210}Po (≈7 Bq.kg^{-1})			
20g	15	30%	3%
^{239}Pu (≈0.002 Bq.kg^{-1})			
400g	7	9.5%	8%

[a]Both standard deviation and counting uncertainties were calculated using a coverage factor (k) of 1.

However, the variability on the smaller sample size was twice that for the larger. In contrast, the variability in the results for ^{40}K was comparable with the counting uncertainties for both sample sizes. This may be because ^{40}K concentrations in both grass and soil should be comparable, while the ^{137}Cs concentration in soil would be at least an order of magnitude higher than that in dried grass[4]. Therefore, the contribution from a small amount of soil to the total activity in the grass would be less significant for ^{40}K than for ^{137}Cs.

3.2.2 *Alpha particle emitting radionuclides in grass.* The results for ^{210}Po analysis in grass are shown in Figure 1; all measurements on grass are summarised in Table 1. The standard deviation from all the individual measurements on 20g-samples was about 10 times the counting uncertainties. In contrast, the analysis of seven 400g samples for $^{239+240}Pu$ gave similar values for the standard deviation and counting uncertainties. This follows the pattern of the ^{137}Cs results where the variability was inversely proportional to the aliquot mass. Unfortunately, the low concentration of plutonium in grass and the limited availability of sample precluded the investigation of other sample sizes.

3.2.3 *Gamma-ray emitting radionuclides in cabbage.* This part of the study made use of five petri dishes containing approximately 50g of dried cabbage, and one marinelli beaker containing 750 g. The results for the measurements on the 50g aliquots are summarised in Table 2. The concentration of ^{137}Cs and ^{40}K measured in the 750g aliquot was not significantly different from those from the 50g aliquot. For both radionuclides, the standard deviation from all six measurements was the same or less than the uncertainty from the counting statistics. The much better reproducibility of the measurements on cabbage samples when compared with grass could be because the edible part of the cabbage was protected from external contamination by the inedible outer leaves.

Therefore, the radionuclide content of the sample is likely to be more inherently homogeneous.

Table 2 *Measurements on cabbage samples.*

Sample mass and approximate activity concentration	*Number of samples*	*Standard deviation*[a]	*Counting uncertainty*[a]
^{137}Cs (≈18.9 Bq.kg^{-1})			
50g	5	0.4%	7.7%
^{40}K (≈1.2 Bq.kg^{-1})			
50g	5	1.8%	5.3%

[a]Both standard deviation and counting uncertainties were calculated using a coverage factor (k) of 1.

3.3 Sediment samples

Two sets of measurements were made on sediment, one using approximately 750g in half litre marinelli beakers and one using 80g aliquots in petri dishes. Caesium-137 and ^{241}Am were both measurable with sufficient precision to give meaningful comparisons. The results are summarised in Table 3. For both radionuclides, the standard deviation was about ten times higher than the counting uncertainties and the reproducibility was no better in the larger aliquots than in the smaller ones.

Table 3 *Measurements on sediment samples.*

Sample mass and approximate activity concentration	*Number of samples*	*Standard deviation*[a]	*Counting uncertainty*[a]
^{137}Cs (≈2,500 Bq.kg^{-1})			
750g	3	8.5%	0.4%
80g	8	7.8%	0.2%
^{241}Am (≈4,000 Bq.kg^{-1})			
750g	3	7.8%	0.8%
20g	8	4.8%	0.3%

[a]Both standard deviation and counting uncertainties were calculated using a coverage factor (k) of 1.

3.4 Intercomparison exercise.

Nine laboratories participated in the exercise and submitted results for ^{137}Cs, ^{40}K, $^{239+240}$Pu, ^{238}Pu, ^{241}Am and ^{90}Sr. The results are summarised in Table 4. Some laboratories did not report results for all radionuclides while some performed more than one measurement. The number of measurements for each radionuclide is listed in Table 4. As each participant used their own established operational conditions, the aliquot sizes used for the individual measurements varied. The average aliquot mass used is shown in Table 4.

Generally, the variability in the results reported for cabbage was lower than the corresponding value for grass, which was consistent with the results of the main study. The one exception was for the measurements of ^{90}Sr, a radionuclide that had not been investigated in the in house study. However, ^{90}Sr has a relatively high soil to plant transfer factor[5]. This means that activity from soil contamination of the vegetation should not be significant. Also most participants reported measurement uncertainties at the higher end of the reported range. These relatively high uncertainties when compared with the inter-aliquot variability make it difficult to draw any conclusions from the data. The high standard deviation on the results for ^{40}K in grass was due to the results from two laboratories, which were dramatically different from the data supplied by the other participants.

Table 4 *Results of the intercomparison exercise.*

	Average aliquot mass, g	*Median Concentration, Bq.kg^{-1}*	*Range, Bq.kg^{-1}*	*Standard deviation^a, % (n)*	*Reported range of uncertainties^a, %*
Dried Grass sample					
^{137}Cs	34	380	326 – 452	9 (11)	1 – 15
^{40}K	34	297	240 – 598	40 (11)	2 – 42
$^{239+240}Pu$	20	1.85	0.61 – 2.44	28 (12)	1 – 12
^{241}Am	20	1.94	1.07 – 2.44	18 (11)	2 – 16
^{90}Sr	20	1140	1080 - 1250	5 (10)	1 – 12
Dried Cabbage sample					
^{137}Cs	54	19	18 – 23	9 (11)	2 – 15
^{40}K	54	1010	923 – 1200	9 (11)	2 – 14
$^{239+240}Pu$	34	0.043	0.036 – 0.046	9 (8)	3 – 21
^{241}Am	34	0.061	0.040 – 0.076	16 (8)	2 – 16
^{90}Sr	34	4.3	3.4 – 6.0	20 (7)	4 – 20

[a]Both standard deviation and uncertainties were calculated using a coverage factor (k) of 1.

Disregarding these two results lowered the standard deviation to 8%, similar to the values in Table 1. It was unclear whether the two anomalous results were due to sample variability or some error in the measurement.

3.5 Effect of sub-sample size.

Overall, the results reported above demonstrated that for grass the size of the sub-sample was one of the major factors affecting the variability. The relationship was most noticeable where the concentration of the analyte was significantly higher in soil than in the vegetation. As expected this led to the conclusion that the uncertainty from sub-sampling would depend on the material being analysed and the analyte as well as the size of the sub-sample taken. Therefore, the results of this investigation cannot be directly applied to other similar materials. However, they can be used to produce an estimate of the likely uncertainty from sub-sampling different sized aliquots (Table 5).

As the values from Table 5 are not directly applicable to all materials and analytes, it has been decided in this laboratory not to include them in the calculation of total uncertainty for individual measurements. Instead, reports state clearly that the reported uncertainty on individual results specifically excludes the uncertainty from sub-sampling. The likely uncertainties from the sub-sampling are then indicated separately using the values from Table 5.

Table 5 *Estimated uncertainty from sub-sampling.*

Aliquot size, g	*Uncertainty*[a], %
<=30	10
30.1 – 100	5
100.1 – 200	2.5
>200	0

[a]Calculated using a coverage factor (k) of 1.

4 CONCLUSIONS

The uncertainty introduced to an analysis by taking a sub-sample from a milled dried bulk was dependent on the analyte, the matrix and in some cases the size of the aliquot. Generally, the results of this study suggested that the largest variability would be found in samples of vegetation, particularly those that may contain external contamination, notably soil. This might be expected because of the relative concentrations of many radionuclides in soil and vegetation and the inherent in-homogeneity of any soil contamination.

Where it was possible to compare different aliquot sizes, the variability in results for grass samples was dependent on the aliquot size. The 400g aliquots used for the analysis of plutonium showed little variability, whereas the 20g aliquots used for the determination of caesium-137 and polonium-210 showed a variability of 16 and 30% respectively.

The measurements of polonium-210 on grass showed that, for this type of material, commercial sample dividers did not improve the homogeneity of the sub-samples significantly. This conclusion has important implications because the increased effort and cost associated with the use of a sample divider could only be justified if it improved the reproducibility of the sub-samples.

The results of this investigation can only give an indication of the likely variability of sub-samples from a particular type of material. However, effort and cost considerations preclude analysing several replicates of routine samples to determine their individual variability while ISO / IEC 17025 requires laboratories to estimate all sources of uncertainty in their reported results. The approach used in this laboratory has been to use the data from this study to estimate the likely uncertainty from sub-sampling different sized aliquots (Table 5). The reported uncertainty on individual results specifically excludes the uncertainty from sub-sampling; the values in Table 5 are used to report these separately.

5 ACKNOWLEDGMENTS

The authors wish to acknowledge the assistance of Mr L W Ewers in sample preparation and analyses of ^{210}Po. They would also like to thank the members of the AIWG for their analyses of the intercomparison samples.

References

1 W.S. Ligget, K.G.W. Inn and J.M.R. Hutchinson, *J. Environment International,* 1984, **10,** 143.
2 N. Green, B.T. Wilkins and D.J. Hammond, *J. Environ. Radioactivity,* 1994, **23**, 151.

3 C.J. Brooks, I.G. Betteley and S.M. Loxston, 'Significance tests' in *Fundamentals of mathematics and statistics,* Wiley, NewYork, 1979, pp 369-377.
4 Radioactivity in food and the environment, 2000, Food Standards Agency and Scottish Environment Protection Agency, London, RIFE – 6, September 2001.
5. A.F. Nisbet and S. Shaw, *J. Environ. Radioactivity,* 1994, **23**, 1.

A RAPID DETERMINATION OF RA-226 AND RA-224 USING EXTRACTION CHROMATOGRAPHY

A.H. Thakkar[1], M.J. Fern[1] and D. McCurdy[2]

[1]Eichrom Technologies, Darien, IL USA
[2]491 Howard street, Northboro, MA 01532

1 INTRODUCTION

A number of recent regulatory issues in the United States have created renewed interest in the development of more efficient methods for the analysis of radium isotopes in water samples. Radium has the lowest maximum permissible concentration in drinking waters among regulated radionuclides[1]. A new rule by the US Environmental Protection Agency (EPA)[2] has created a new set of sampling requirements designed to better monitor the US water supply. Because of these changes, the more than 50,000 individual drinking water supplies in the US will need to perform incremental analyses for radium isotopes over the next couple of years. This will create a large increase in the demand for radioanalytical services. A comparison of eight different methods for radium analysis was recently completed by Kohler, et al[3]. These available methods are lengthy, inconsistent, and time consuming. It is unlikely that the laboratory community will have sufficient capacity to provide the analyses required by the new US regulations, unless more efficient methods are available.

Additionally, in recent years, it has been discovered that a number of drinking water supplies in the eastern part of the US contain high levels of short-lived Ra-224. The currently approved methods for alpha emitting radium isotopes are all based on radon emanation. Due to the long ingrowth periods required by these methods, any Ra-224 in the sample decays away and goes undetected. A faster method for direct measurement of alpha emitting radium isotopes would allow for the detection and measurement of Ra-224.

For these reasons, we have proposed the following alpha spectrometric method for determination of Ra-223, Ra-224 and Ra-226. It can be readily coupled with a rapid method for determination of Ra-228 to provide an efficient means of measurement of alpha and beta emitting isotopes of radium. This combined method is currently under review for approval by ASTM, International technical committee D19.04, *Radioactivity in Water*. The ultimate goal of this committee is approval of the method by US EPA for use in drinking water analyses.

2 EXPERIMENTAL

2.1 Standards

National Institute of Standards and Technology (NIST) traceable radionuclides ^{133}Ba and ^{226}Ra were used to characterise the radionuclides of interest.

2.2 Reagents and Materials

All reagents utilised in the analysis were American Chemical Society reagent grade; 0.1M HNO_3; 8M HNO_3; 0.095M HNO_3; concentrated HNO_3; 40% Na_2SO_4; 70% Na_2SO_4; concentrated H_2SO_4; 0.75 mg/ml $BaCl_2$; and 1:1 acetic acid. Prepacked Ln Resin columns and cation exchange resin, 8% cross-linked, 100-200 mesh, hydrogen form were obtained from Eichrom Technologies, Inc. (Darien, IL, USA). Ln Resin is an extraction chromatographic resin which has di-(2-ethylhexyl) orthophosphoric acid as an extractant[4]. The resin has strong affinity for tri-valent species such as actinium and also separates several 'rare-earths'.

2.3 Sample Preparation and Separation Procedure using Cation Exchange Resin and Ln Resin

A 500 to 1000 ml aliquot of the sample was acidified with nitric acid to pH 2. Ten ml cation exchange resin was slurried in DI water and packed into 1.5 cm diameter columns. The column was preconditioned with 50 ml of 0.1M HNO_3. After the acidified sample passed through the cation exchange column, it was rinsed with 30 ml of 0.1M HNO_3. Ra and Ba were eluted with 100 ml of 8M HNO_3.

The eluted Ra/Ba fraction was then evaporated to dryness. The sample residue was re-dissolved in 10 ml of 0.095M HNO_3. A pre-packed 2 ml Ln Resin column was preconditioned with 5 ml 0.095M HNO_3. A cleaned, labeled beaker was placed under the column and the sample solution loaded on the column. Three 5 ml rinse of 0.095M HNO_3 were passed through the column and collected. Radium passed through this column while actinides and rare earth elements were retained. This ensured decontamination of the radium fraction from many alpha emitters that could interfere in the alpha spectrum. The combined load and rinse solution (containing radium) was evaporated to an approximate volume of 10 ml. The solution was then transferred to a 50 ml centrifuge tube.

Although this paper will discuss only Ra-226 and Ra-224 measurements, one can also measure Ra-228 via Ac-228 with this technique. Burnett, et al.[5] demonstrated the use of Ln Resin for Ra-228 measurement of water samples. After the elution of the Ra/Ba fraction from the Ln column, Ac-228 can be eluted with 10 ml of 0.35M HNO_3.

A micro-precipitation of barium was carried out on the evaporated Ra/Ba fraction. This technique is a modified version of the one described in Sill 1987[6]. Seventy-five μg of barium carrier was added to the centrifuge tube, followed by 3 ml of 40% Na_2SO_4 and a few drops of 1:1 acetic acid solution. The sample was gently stirred and then 0.2 ml of a seeding suspension added and mixed immediately. The sample was placed in a cold-water bath for at least 30 minutes. The solution was then filtered through a 0.1 micron, 25 mm diameter polypropylene filter. The filter was mounted on a planchet and counted for Ra-223, Ra-226 and Ra-224 by alpha spectrometry. A typical alpha spectrum for a sample containing Ra-224, Ra-226 and daughters is shown in figure 1. After the completion of alpha spectrometry counting, the filter is counted for Ba-133 by gamma spectrometry.

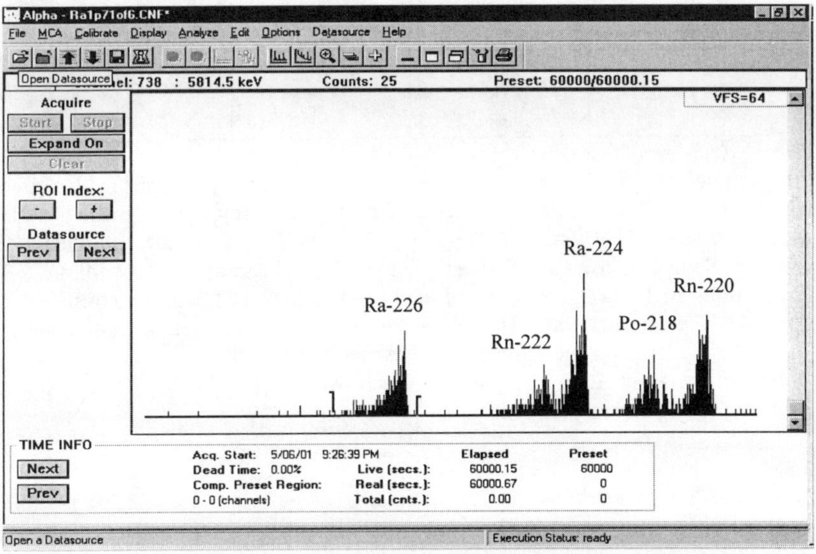

Figure 1. *Alpha Spectrum of sample containing Ra-226, Ra-224 and daughters*

3 RESULTS AND DISCUSSION

3.1 Water Samples

Table 1 summarises the recoveries of Ra-226 tracers achieved by the method on DI and tap water samples. The Ra-226 recoveries averaged 85% on 0.5L DI water samples, 84% for 0.5 l tap water and 93% on 1 l tap waters.

Table 1. *Ra-226 recoveries in water samples*

Water Type	Volume	Ra-226 Yield (%)	Number of samples
DI water	0.5 L	85 ± 8	14
Tap water	0.5 L	84 ± 13	4
Tap water	1.0 L	93 ± 8	5

Several samples that had been previously tested by other laboratories with different techniques for Ra-226 were analysed to validate the accuracy of this method. These included EPA cross check samples (September 18, 1998)[7], and samples supplied by the New Jersey State Department of Health and Georgia Institute of Technology. The results, reported in Table 2, compared to the known values well, with no consistent bias high or low. The expected value shown for the EPA cross check sample is the reference value reported by EPA for the samples. The expected values for the samples supplied by the New Jersey State Lab and Georgia Tech were determined by their in-house laboratories using gamma spectrometry.

Table 2. *Ra-226 recoveries in cross check samples*

Water Type	Ba-133 Yields (%)	Eichrom Value pCi/L (mBq/L)	Exp. Value PCi/L (mBq/L)
EPA Cross check	95.7 ± 6	1.41 ± 0.17 (52.2 ± 6.3)	1.7 ± 0.5 (62.9 ± 18)
New Jersey	91	8.64 ± 1 (320 ± 37)	9.1 ± 0.5 (337 ± 18)
Georgia Tech WS14776	88	4.14 ± 0.58 (153 ± 21)	3.3 ± 0.3 (122 ± 11)
Georgia Tech S8933	88	22.6 ± 3 (836 ± 111)	18.0 ± 2 (666 ± 74)

3.2 Matrix Effects

The first step in this procedure is a concentration of Ra and Ba on cation exchange resin, which does not have a high degree of selectivity for these elements over calcium and magnesium. Since Ca and Mg are often present in high concentrations in environmental waters, there is a potential for interference with the uptake of radium and barium by the cation exchange resin. Therefore, the effect of increasing amounts of these two elements on the uptake of radium and barium in this procedure was tested. Results, shown in Figure 2, indicated that, increasing concentrations of Mg did not have a significant impact on Ra recoveries even at levels as high as 1000ppm in a 500 ml sample. Calcium on the other hand, had a significant effect. Up to 200 ppm Ca, chemical recoveries of Ra were acceptable. Two samples with Ca content of 600 and 1000 ppm were analysed. With these two samples, there was too much precipitate in the final alpha source preparation to allow them to be counted.

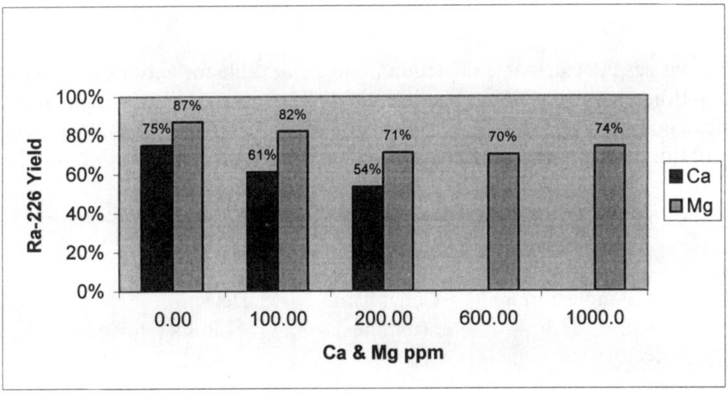

Figure 2. *Effect of Ca and Mg on Ra-226 Recoveries in 500 ml of sample*

In order to deal with samples with Ca levels higher than 200 ppm, we experimented with reducing the volume of the sample to 100 ml. Results of the experiment are shown in Figure 3. Ra-226 and Ba-133 spikes showed equal and acceptable chemical recoveries even up to 500 ppm Ca in a sample. These results show that reduction in sample volume can successfully address the calcium matrix issue. The comparable radium and barium recoveries also indicate that the Ba-133 is a suitable yield monitor for radium in this method.

Most drinking waters contain no more than 200 ppm Ca so the analysis of 500 ml samples is acceptable. In those situations where the calcium content is greater than 200 ppm, a 100 ml aliquot can be analysed. This will require longer counting times to achieve a given detection limit.

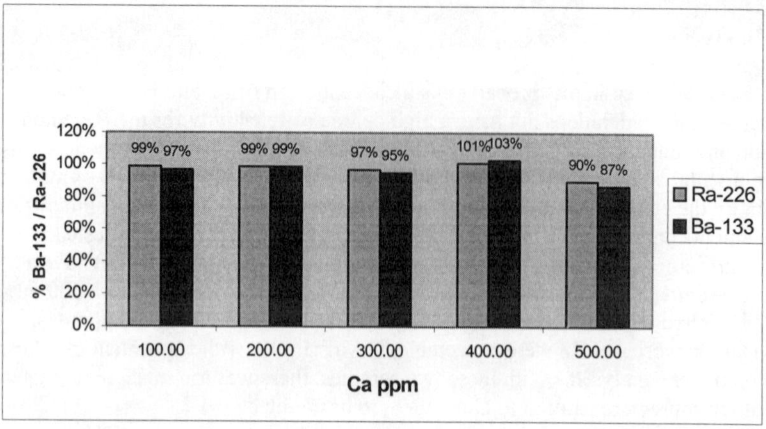

Figure 3. *Effect on Ca on Ba-133 and Ra-226 recoveries in 100 mL sample*

Table 3 summarises the calculated detection limits achievable for various sample sizes and counting times, assuming 90% chemical yield. The current US EPA regulations require a detection limit of 1 pCi/L (37 mBq/L). This is readily achievable with 100-minute count time and 500 ml sample or 480-minute count time and 100 ml sample.

Table 3. *Detection limits based on different volume and count time*

Sample Volume (L)	Count Time (minutes)	Detection Limits pCi/L
0.5 L	1000	0.09 (3.33 mBq/L)
0.5 L	240	0.3 (11.1 mBq/L)
0.5 L	100	0.8 (29.6 mBq/L)
0.1 Liter	480	0.9 (33.3 mBq/L)

4 CONCLUSION

The method described here shows promise as a more rapid alternative to existing methods for Ra-224 and Ra-226 measurements. Detection limits required by US EPA are achievable with reasonable sample sizes and short counting times. The applicability of this method to the European regulatory environment has not been determined. More demanding detection limits may require method improvements to accommodate larger sample sizes.

The most significant limitation to adoption of this method is regulatory approval. It is our hope that the efforts of ASTM D19.04 should remedy this.

References

1 Code of Federal Regulations, Title 10, part 20, revised April 30, 1975.
2 U.S. Environmental Protection Agency, "National Interim Primary Drinking Water Regulations", EPA-570/9-76-003 Washington, D.C.
3 M. Kohler, W.Presusse, B. Gleisberg, I. Schafer, T.Heinrich, and B.Knobus, *Applied Radiation and Isotopes,* 2002, **56**, 387-392.
4 E.P. Horwitz and C.A.A. Bloomquist, *J. Inorg. Nucl. Chem*, 1975, **17**, 425-434.
5 W. Burnett and P.Cable, *Radioactivity & Radiochemistry,* 1995, **6**, (3), 36-44.
6 C.W. Sill, *Nucl. Chem. Waste Manage.* 1987, **7**, 239-256.
7 Uranium-Radium in Water Performance Evaluation Study, A Statistical Evaluation of the Sept 18, 1998 Data, Environmental Protection Agency, National Exposure Research Laboratory, Environmental Sciences Division, Las Vegas, Nevada

Acknowledgements

The authors would like to thank Dr. B. Parsa, New Jersey State Department of Health and Dr. B. Kahn, Georgia Institute of Technology, for providing the samples from their laboratory for the validation of this method.

CONSTRUCTION OF AN UNCERTAINTY BUDGET FOR ALPHA SPECTROMETRY:- A CASE STUDY (THE DETERMINATION OF URANIUM ISOTOPIC ACTIVITY LEVELS IN VEGETATION SAMPLES USING EXTRACTION CHROMATOGRAPHY)

T. Gingell and M. Harwood

Dstl Radiation Protection Services, Institute of Naval Medicine, Crescent Road, Gosport, Hants. PO12 2DL

1 INTRODUCTION

At DRPS, the analysis method used to determine uranium activity in samples of vegetation by alpha spectrometry is UKAS (United Kingdom Accreditation Service) accredited. In November 1999 UKAS formally adopted ISO 17025[1] as its accreditation standard. This standard requires that accredited testing laboratories like ourselves produce estimates of uncertainty of measurement for all testing using accepted methods of analysis. The methods used to estimate the uncertainties must also be documented. Though this requirement existed in the previous UKAS accreditation standard M10*, much greater emphasis is given in ISO 17025 to the need for method validation and for the estimation of measurement of uncertainty. One of the ways to satisfy this requirement is to construct an uncertainty budget in accordance with the 'bottom up' approach set out in the Guide to the Expression of the Uncertainty of Measurement (GUM)[2]. This process involves listing all of the relevant components of uncertainty that affect the results of the analysis, along with a reasonable estimate of their magnitude. The total uncertainty to be applied to the final analysis result can then be determined by combining the principal (dominant) components in the uncertainty budget. Components of uncertainty whose magnitude is less than a third of the largest component may be excluded from the calculation of the total uncertainty. This approach has worked very well in the realm of physical measurements where uncertainty budgets can be accurately estimated and used in turn to better define and control the testing environment[3]. Application of the same approach to the field of radiochemical measurement is less straightforward. Normally a radiochemical measurement is a multi-stage analysis process that involves a combination of physical measurements, chemical separation of the nuclide of interest and quite often selection of the test portion from a larger portion by subsampling. An understanding of the chemistry of the separation process is vital before reliable results can be achieved. A lot of useful data is usually acquired during the early phases of method optimisation and development, some of which can be used to estimate the magnitude of certain of the uncertainty components. With particular reference to alpha spectrometry, optimisation of the chemical recovery of the analyte of interest is one of the main objectives of the method development process.

In the UK during the 1990s the number of radiochemistry laboratories gaining UKAS accreditation for their analysis methods gradually increased and there emerged an increasing awareness of the importance of setting up realistic uncertainty budgets. At first however there was little in the way of guidance on how to do this for radiochemical procedures. The publication of the GUM[2] in 1993 formally established general rules for evaluating and expressing uncertainty in measurement across a broad spectrum of measurements but it wasn't the easiest of documents to use for practical situations. Since then guidance more relevant to those making practical measurements has appeared in the literature[4,5,6] and workshops and training courses in uncertainty measurement in radiochemistry are now being provided. Analytical techniques, practices and procedures have generally improved and more consistent and 'realistic' assessments of uncertainty are being made. There still remains however the issue of how to deal with uncertainty due to sampling. A complete total uncertainty budget would include contributions from primary sampling in the field and from any secondary sampling in the laboratory that is done to obtain a test portion (for our lab this is the 1 gram of ash that is actually analysed) from the laboratory sample (this is the 200g of fresh grass that arrives for analysis). In common with many radiochemistry laboratories the DRPS laboratory is not responsible for sample procurement in the field and makes no attempt to quantify the uncertainty of this process. However the laboratory does undertake subsampling during the sample preparation stage in order to arrive at the test portion. It is known that sampling uncertainties can be the largest contributors to the total measurement uncertainty, possibly as high as 1000% of the result for field sampling and up to 50% for secondary sampling within the lab[7]. Even when the magnitudes of these uncertainties have been estimated, the best way to include them into the overall combined uncertainty is still under discussion[8].

This paper describes how the laboratory constructed its uncertainty budget for the analysis of uranium isotopic activity levels in vegetation. It describes briefly some of the work done to optimise the chemical recovery of the uranium isotopes. For presentational clarity, the uncertainty budget has been determined only for ^{238}U; it was assumed that the ^{234,235}U isotopes behave in similar fashion during the tests performed. Uncertainty budgets for these nuclides will be identical except for differences in the magnitude of the counting uncertainties. The title of this paper bears the qualifier "case study" as the uncertainty estimates presented here are specific to the particular method used in our laboratory. It can not be assumed that the uncertainties that occur in the analysis of other nuclides such as ^{239}Pu or ^{137}Cs in vegetation will be of the same magnitude. A laboratory must construct its own uncertainty budgets for those methods that it operates. However it is hoped that the data presented in this report will add to the growing database of information on uncertainty in radiochemical measurements.

2 IDENTIFICATION OF THE UNCERTAINTY COMPONENTS

The process of measurement uncertainty estimation can be considered to be a four step process[4]. The first step is to specify the measurand, the second step is to list the possible sources of uncertainty, the third step is to measure or estimate the size of the uncertainty component associated with each source of uncertainty, and the final step is to combine those uncertainty components that make significant contributions into a total uncertainty that can be applied to the analysis results. (An insignificant source of uncertainty is one whose estimated magnitude is less than a third that of the most significant uncertainty component). The first two steps in the construction of the uncertainty budget are described in this section.

3.1 Specifying the measurand

In the context of uncertainty estimation this means having a statement of what is being measured and a quantitative expression relating the value of the measurand to the parameters on which it depends. In our case the statement is "the measurement of uranium isotopic activity levels (in units of Bq/g dry sample) in samples of vegetation sent to the laboratory by the customer". The quantitative expression for the activity (*Act*) of each uranium isotope is:-

$$Act = \left(\frac{C_{gr}^{238} - C_{bgd}^{238}}{C_{gr}^{232} - C_{bgd}^{232}}\right) \times S \times \left(\frac{M_{tracer}^{232}}{T \times M_{sample\ ash}}\right) \times \left(\frac{A}{D}\right) \qquad (1)$$

where

$\left(C_{gr}^{238} - C_{bgd}^{238}\right)$ - the net ^{238}U count recorded in the alpha spectrum
$\left(C_{gr}^{232} - C_{bgd}^{232}\right)$ - the net ^{232}U count recorded in the alpha spectrum
S - the specific activity of the ^{232}U tracer (Bq/g)
M_{tracer}^{232} - the mass of ^{232}U tracer added to the ashed sample (g)
$M_{sample\ ash}$ - the mass of the ashed vegetation used in the analysis (g)
T - the sample count time (s)
$\left(\frac{A}{D}\right)$ - the ratio of the ashed sample weight (A) to the dried sample weight (W)

Equation 1 identifies the seven parameters whose variation would directly affect the value of the activity, so they are the first sources of uncertainty to be identified.

3.2 The analysis method

The next stage of the identification process involves consideration of the analysis method

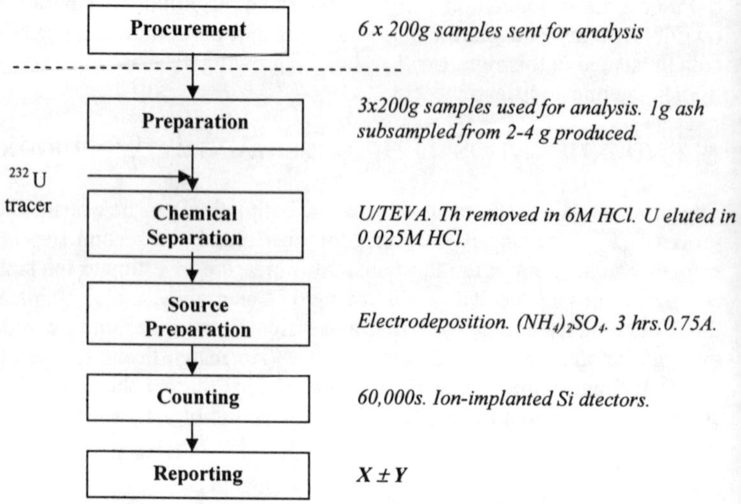

Figure 1 *Schematic of analysis method*

used. This is outlined in figure 1. As part of a programme of terrestrial monitoring six grass samples of approximate weight 200 g are procured by the customer at each of his sampling locations. A sample location is an area measuring no greater than 20 m x 20 m from which the six grass sampling locations are chosen at random. The top 10 cm of the grass is sampled.and each sample is taken over a measured area of 1 m^2. The grass at each sampling location has been left uncut for at least six weeks prior to cutting. All six samples are sent to the laboratory for analysis. The customer is interested in the spatial distribution of any uranium activity that is found in addition to the magnitude of the activity levels. Therefore no aggregation of the three grass samples chosen for analysis occurs in our method. Each of the three chosen samples is air-dried to constant weight and then ashed @ 550°C for 24 hours. The ashes are compacted and tumbled on a rolling mill for 16 hours. 1 g of the ash is taken for analysis and ^{232}U tracer is added to it. This is taken to dryness twice with concentrated nitric acid, filtered, and then loaded on to an Eichrom U/TEVA extraction column in 4M nitric acid. Thorium is eluted from the column with 6M hydrochloric acid. Uranium is eluted with 0.025M hydrochloric acid. This is evaporated in the presence of potassium hydrogen sulphate, then redissolved in 1M ammonium suphate. A thin source suitable for alpha counting is prepared by electrodeposition of the ammonium suphate solution after adjustment of its pH to 2.5 ± 0.1. The electrodeposition is carried out for a minimum of 3 hours at a current of 0.75 A. The source is counted for 60,000 seconds in an Ortec Octete using implanted Si detectors of area 450 mm^2. The resulting peaks in the uranium alpha spectra are integrated manually from which the estimated activity results are derived. The final stage is reporting the activity estimates to the customer in the form Act = X ± Y Bq/g dry sample weight.

Sources of uncertainty arising from consideration of the method may be discovered by asking questions such as "Is the ashing temperature of 550°C critical to the procedure?", "What effect would adding the ^{232}U tracer to the dried grass samples earlier in the procedure have upon the result?" and "Do you know if the 1 g of ash that constitutes the test portion is representative of the 2 to 4 grams of ash that is the result of the sample preparation process?"

For the first few years of the monitoring programme the customer sent very large quantities (1.5-2.5 kg fresh weight) of grass from some of the sampling locations. This practice necessitated the taking of smaller aliquots of the fresh material (200 g fresh weight) for drying and subsequent analysis. In addition to air-drying, freeze drying was used to supplement the smaller sample processing capacity that we had at the time. This was further confounded by having four muffle furnaces of varying capacity in which samples were ashed. The magnitude of the uncertainties due to the subsampling and variation in sample preparation techniques is discussed in section 4.

3.3 Insignificant sources of uncertainty

A list of the sources of uncertainty relevant to our analysis method is shown in Table 1. Those sources listed from number 9 to 13 inclusive are briefly discussed below then will not be considered further in the estimation of the total uncertainty budget as their contribution is considered to be insignificant compared with that of the other sources.

3.3.1 Weighing. The uncertainty $u\left(M_{tracer}^{232}\right)$ in the weighing out of the ^{232}U tracer is commonly observed to be of the order of 0.0005 g in a total weight of 0.2 g of tracer. Hence the relative standard uncertainty $\left(\frac{u\left(M_{tracer}^{232}\right)}{M_{tracer}^{232}}\right)$ is (0.0005/0.2000)_= 0.0025

Table 1. *List of relevant sources of uncertainty*

	Description		*Description*
1	Gross ^{238}U count $\left(C_{gr}^{238}\right)$	8	Method error $\left(X_{method}\right)$
2	Background ^{238}U count $\left(C_{bgd}^{238}\right)$	9	Operator error
3	Gross ^{232}U count $\left(C_{gr}^{232}\right)$	10	Weight of ^{232}U tracer added $\left(M_{tracer}^{232}\right)$
4	Background ^{232}U count $\left(C_{bgd}^{232}\right)$	11	Weight of ash used for analysis $\left(M_{sampleash}\right)$
5	^{232}U tracer standard (S)	12	Use of nuclear data (ND)
6	Subsampling $\left(X_{sub}\right)$	13	Measurement of count time (T)
7	Ash weight to dry weight ratio (A/D)		

The relative standard deviation associated with weighing of the 1 g of ash $M_{sampleash}$ is even smaller.

3.3.2 Nuclear data. The one nuclear decay parameter thought likely to affect the activity estimate was the half life of the ^{232}U tracer. This is given as 69.8 ± 0.5 years*. The reference date for our ^{232}U tracer is 05/07/94 which means that any variation in the half life could affect the calculation of the ^{232}U activity level tracer used today. The laboratory normally uses the value of 69.8 years to compute the decay correction factor. Substitution of a half life of 70.3 years was used to calculate a new value of the decay correction factor. There is a difference of 0.05% between the two decay correction factors which is insignificant as a source of uncertainty.

3.3.3 Count time. Uncertainty in the count time only becomes a factor in a high count rate situation. This is unlikely to occur during counting of uranium in vegetation by alpha spectrometry.

3.3.4 Operator error. This is an ever present source of uncertainty and one that is very difficult to quantify. It is often observed for instance that uranium chemical recoveries fall temporarily when a new operator commences routine alpha spectrometry analysis. This does not affect the accuracy of the activity estimate but will contribute to extra counting uncertainty. Also a new operator can make mistakes when integrating regions of interest in an alpha spectrum. Though these mistakes have a potentially large effect on the activity estimates the problem is usually circumvented by a combination of good on the job training and the timely checking of all analysis data by an approved countersignatory.

4 QUANTIFICATION OF THE UNCERTAINTY COMPONENTS.

4.1 Counting uncertainties; $u(C_{gr}^{238}), u(C_{bgd}^{238}), u(C_{gr}^{232}), u(C_{bgd}^{232})$

4.1.1 ^{238}U Gross Count [$u(C_{gr}^{238})$]. Under normal circumstances the biggest contribution to the uncertainty budget arises from the small number of counts acquired in the ^{238}U region of the spectrum. This is a direct consequence of the very low level of uranium activity found in vegetation grown in natural conditions within the UK[9]. Under the normal operating conditions of DRPS' analysis method, an average uranium activity level of 0.25 mBq.g^{-1} dry weight will give a gross count of less than 20 in the ^{238}U region of the alpha spectrum. The standard uncertainty $u(C_{gr}^{238})$ associated with this count is $\sqrt{20} = 4.47$, hence the relative standard uncertainty is $\frac{4.47}{20} = 0.2235$ (22.35%).

The ^{238}U region of interest corresponding to an alpha particle energy of ~ 4.2 Mev does not suffer from spectral overlap interference so no additional uncertainty is applied to $u(C_{gr}^{238})$. The majority of analyses performed over the last couple of years in our laboratory are producing ^{238}U counts of less than 20 and over half of the activity estimates are reported as less than the limit of detection of the method. For convenience of calculation the value of $u(C_{gr}^{238})$ is assigned a value of 0.2 (20%).

4.1.2 ^{232}U Gross Count [$u(C_{gr}^{232})$] The activity level of ^{232}U tracer added to the sample produces an average gross count of 800 counts. The relative standard uncertainty associated with this term $u(C_{gr}^{232})$ is therefore $\frac{\sqrt{800}}{800}$ ~ 0.035 (3.5%). The ^{232}U region of interest corresponding to an alpha particle energy of 5.32 Mev does suffer from spectral overlap interference due to the proximity of the ^{228}Th daughter (5.42 Mev). This would become a significant problem if the separation chemistry failed to remove any thorium present within the sample. However in this case there would be an accompanying ^{232}Th peak at ~ 4.0 Mev that would indicate the magnitude of this interference. No additional uncertainty is therefore added to $u(C_{gr}^{232})$.

4.1.3 Background Counting Terms [$u(C_{bgd}^{238})$, $u(C_{bgd}^{232})$] The background counting uncertainty described by these two terms has two sources. The first is that due to the presence of ^{238}U and ^{232}U counts in the reagent blank that is analysed with each batch of samples. The second is that due to build up of alpha recoil contamination on the detectors. This contamination will in turn contribute counts to the ^{238}U and ^{232}U regions of interest. For ^{238}U these two sources of uncertainty are normally insignificant. There are normally less than 5 counts in the ^{238}U region of the reagent blank. For grass analysis where the average gross ^{238}U count is often less than 20, the reagent blank can form a significant proportion of the ^{238}U gross count, but in practice the algorithm used for activity calculation will return a "less than limit of detection" result when the combined uncertainty of the gross ^{238}U count and the reagent blank ^{238}U count exceeds 30%.

The alpha recoil detector backgrounds are well characterised as backgrounds are acquired every two months over a 240,000 second count time. These backgrounds are very low for ^{238}U (< 2). It is more of a problem in the ^{232}U region as recoil contamination by its

^{228}Th daughter gradually builds up with usage of the detector. If the ^{232}U recoil detector background exceeds 50 counts in 60,000 seconds it is no longer used for uranium counting, as at this level it would become a significant proportion of the expected gross ^{232}U count. For these reasons the uncertainty terms $u\left(C_{bgd}^{238}\right)$ and $u\left(C_{bgd}^{232}\right)$ are not considered further in the construction of the total uncertainty budget.

It must be remembered that the total relative standard uncertainty due to counting $u\left(C_{gr}^{TOT}\right)$ is a combination of the two gross counting terms $u\left(C_{gr}^{238}\right)$ and $u\left(C_{gr}^{232}\right)$. However due to the dominance of the $u\left(C_{gr}^{238}\right)$ term in this expression it is taken to be numerically equal to the value of $u\left(C_{gr}^{TOT}\right)$. The term $u\left(C_{gr}^{232}\right)$ is not considered further in the construction of the total uncertainty budget.

4.1.4 Optimisation of chemical recovery. Due to the reasons mentioned above it is highly desirable to maximise the uranium chemical recovery. During the method development phase the effect on the ^{232}U recovery of varying the electrodeposition conditions and the variation in performance of the U/TEVA columns was investigated.

4.1.5 Effect of varying electrodeposition conditions. The most variable parameter in setting up the electrodeposition cell used in the laboratory is the position of the platinum anode with respect to the stainless steel disc cathode. The body of the cell is constructed from a polythene bottle which may be disposed of at the end of the electrodeposition process. The principal reason for having this type of system is to avoid cross contamination from previously analysed samples. The disadvantage is that the platinum anode – steel disc geometry is not as reproducible as it is in some of the more rigid cell designs that are in use. Therefore the effect of varying this geometry on the recovery of ^{232}U tracer solution added directly to the cell was investigated. The platinum anode was placed at several 'grid positions' relative to the steel disc. The grid position (20,-10) is illustrated. The results are shown in Table 2.

Table 2. *Variation of chemical recovery as Pt electrode-steel disc geometry is varied*

Grid Position	*Total electrode displacement (mm)*	*Recovery (%)*
(0,0)	0	85.2
(0,-10)	10.0	92.3
(0,10)	10.0	91.0
(10,0)	10.0	80.0
(10,-10)	14.1	96.8
(20,0)	20.0	94.0
(20,-10)	22.4	89.8
	Mean	91.0
	Relative Standard Deviation (%)	6.22

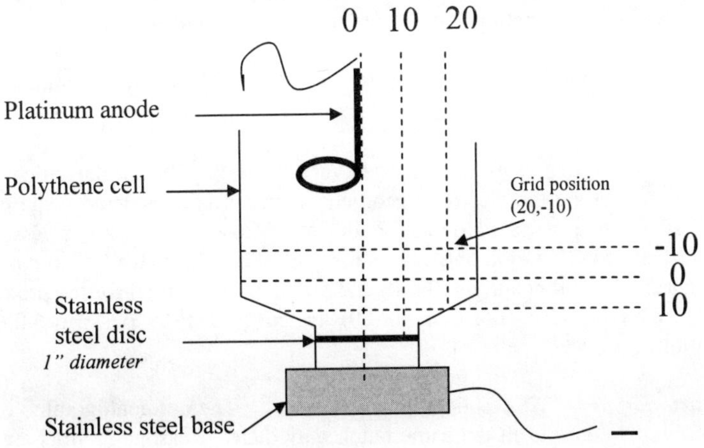

Figure 2 *Construction of electrodeposition cell showing grid line spacings (in mm) used to investigate the effect of varying the platinum anode to steel disc geometry on the recovery of ^{232}U.*

A second investigation was carried out to study the effect of keeping the platinum anode to steel disc geometry constant at grid position (0,0). The results are shown in Table 3 .

Table 3 *Variation of chemical recovery under conditions of fixed Pt electrode-steel disc geometry*

Grid Position	*Total electrode displacement (mm)*	*Recovery (%)*
(0,0)	0	84.4
(0,0)	0	91.8
(0,0)	0	93.3
(0,0)	0	89.7
(0,0)	0	89.9
	Mean	90.0
	Relative Standard Deviation (%)	3.75

Table 4 *Estimation of variation in count rate due to use of U/TEVA columns*

	Relative standard deviation (%)
Normalised count rate (cpm.g^{-1} eluant)	8.17
Mass of ^{232}U tracer added	0.60
Mass of eluant analysed by liquid scintillation	0.21
Liquid scintillation counting repeatability	0.54
^{232}U counting uncertainty	2.2

Allowing for a counting uncertainty for the deposited ^{232}U of 4.4% the relative standard uncertainty $\left(\frac{\sqrt{u(edep, var)}}{u(edep, var)}\right)$ due to variation of the electrode geometry is estimated to be ~ 4.4 %. i.e.

$(6.22)^2 = ((4.4)^2 + (u(edep,var))^2$, $\therefore$ u(edep,var) = 4.4 %. By similar calculation the uncertainty due to keeping the electrode geometry constant is considered to be negligible as the observed relative standard deviation of the recovery (3.75 %) is less than the counting uncertainty (4.4 %).This outcome is not particularly surprising. Some of the data in Table 1 suggest that grid positions (0,-10) and (10,-10) are optimal for the production of the highest recoveries. Grid position (0,-10) is now used to perform all uranium electrodepositions .

4.1.6 *Variation in U/TEVA column performance.* The chromatographic efficiency of different columns taken from the same batch may differ. Reasons for this are packing differences of the resin, storage conditions, age of the resin, and differences in the mass of the resin in each column. Variations in any of the above factors could contribute extra uncertainty to the gross ^{238}U and gross ^{232}U counts. To estimate these uncertainties a nominal 7 Bq of ^{232}U tracer was placed onto 15 randomly chosen columns and the usual chemical separation procedure was followed. 3g from the 15 g of eluted ^{232}U was counted by liquid scintillation. To evaluate the repeatability of the liquid scintillation counting fifteen replicate ^{232}U solutions were prepared and counted. The results are summarised in Table 4.

The normalised count rate is calculated by dividing the liquid scintillation count (cpm) by the weight of eluant being counted. The magnitude of the combined uncertainty due to variation in the terms listed in Table 4 does not explain the relative standard deviation of 8.17 %. The only other source of uncertainty in the normalised countrate must be variation in the performance of the U/TEVA resin as the ^{232}U passed through it. The relative standard deviation of this variation in performance can be calculated to be 7.7 %. Because this variance will affect ^{238}U and ^{232}U in like manner it will not affect the estimate of the uranium isotopic activity so it is not included in the final uncertainty budget.

More recent work involving checking of the laboratory's ^{238}U and ^{232}U standards by passage of replicate samples through U/TEVA columns has indicated a repeatability of about 5%. This suggests the variation in the performance of the U/TEVA resin is currently less than 5%.

4.2 ^{232}U tracer uncertainty; (u(S))

The calibration certificate that accompanies the ^{232}U tracer used in the laboratory states that the total uncertainty of the activity level is ± 1.7 % for a coverage factor (k = 1). The tracer is used for the analysis of intercomparison samples and for internal laboratory quality assurance where its stability is checked against a ^{238}U standard obtained from NPL. Table 5 shows some of the results obtained in these activities. When considering the laboratory's performance in the intercomparison a good performance is deemed to be had if the u-value is <1.96 and the deviation from the target value is < 10 %. If the u-value is < 1.64 the laboratory result is statistically indistinguishable from the target value.

Table 5 *DRPS' performance in past intercomparison exercises and the checking of ^{238}U standards*

Description of test	*Deviation from target (%)*	*u-value*
NPL Intercomp 1996	-7.24	2.62
NPL Intercomp 1998	-5.23	0.93
NPL ^{238}U std	-8.10	2.01
NPL ^{238}U std	-11.00	1.98
NPL ^{238}U std	-7.80	1.35
NPL ^{238}U std	-16.50	3.08
PROCORAD Reference U 1996	-6.60	0.60
1997	-10.30	1.71
1998	-4.88	0.91
1999	-3.30	0.73
2002	2.56	0.18
PROCORAD Solution A 1997	-5.40	0.77
1998	6.40	0.77
1999	-5.63	0.91
2002	1.23	0.09

Though it can be seen that there have been some good performances (NPL intercomparison 1998) the data in Table 5 strongly suggest that the ^{232}U tracer is negatively biased. This has always been the case since 1994 when we were advised by the NAMAS assessors to procure our ^{232}U tracer from a supplier with which they had a mutually reciprocal agreement. Particularly worrying is the data obtained when using the ^{232}U tracer to estimate the activity of the ^{238}U standard acquired from NPL. Based on this evidence the uncertainty in the ^{232}U tracer activity is taken to be 8%.

4.3 Subsampling uncertainty; u(X_{sub}^{TOT})

The method currently used within the laboratory no longer requires subsampling of the fresh samples sent by the customer so this contribution to the total uncertainty has decreased since the first years of the uranium monitoring programme. There still remains a component of uncertainty **u**(X_{sub}^{ash})due to subsampling of 1 g of the ash that is formed in the sample preparation process. A number of years ago the receipt of grass samples with a ^{238}U activity of about 11mBq/g provided an ideal opportunity to undertake the investigation of the magnitudes of uncertainties due to subsampling. This level of activity produced a ^{238}U count of 550 in 60,000 seconds (counting uncertainty of 4.24%) as opposed to the usual 20 counts in 60,000 seconds. It had frequently been observed that when one of the six fresh samples was found to have uranium activity levels of this magnitude, analysis of the other two accompanying replicate samples would reveal similarly elevated but significantly different activity levels to be present. This could have been due to real differences in the spatial distribution of uranium depositing itself across the sampling locations. Alternatively it could have been due to the subsampling of the fresh vegetation that was necessitated at that time. In cases where the uranium is not homogeneously distributed in the sample single aliquots taken from a large sample may not be representative. This inhomogeneity will contribute to uncertainty in the resulting estimate of the uranium activity in the sample. The magnitude of this uncertainty is also dependent upon the size of aliquot taken. Two of the grass samples of about 11 $mBq.g^{-1}$ ^{238}U activity were taken for analysis. In consideration of the 'average' uncertainty due to subsampling it was decided to establish 'best case' and 'worst case' scenarios, estimate the subsampling uncertainty for each case and then decide where on the scale between these two extremes our sample preparation scheme lay.

4.3.1 Creating a homogeneous sample ("best case"). One sample (A) was used to create as homogeneous a sample as reasonably possible. About 1 kg of fresh sample was removed from its bag, weighed and then air dried for 24 hours at 65°C, after which it was finely chopped by hand, mixed and reweighed. The mixed sample was then separated into five equal mass portions of 200 g weight, each portion in turn being placed into a beaker and then ashed for 24 hours at a temperature of 500°C . After ashing the five portions were recombined, mixed, weighed and placed into a container and tumbled for 24 hours. Five replicate samples (Aliquots 1-5) from it were then analysed for uranium using the normal analysis procedure. This situation is the "best case" scenario as there would not ordinarily be the time available to prepare each sample in such a meticulous way. The results of this analysis are given in Table 6. The reported uncertainty terms are expanded counting uncertainties where k = 2.

4.3.2 Creating two inhomogeneous samples ("worst case"). The second sample (B) was removed from its bag and divided into two equal 500 g portions (B1 and B2) by weight. Sample B1 was then further divided into five 100 g sub-portions (B1:Aliquots 1-5) by taking what came to hand as the fresh sample lay in its retaining tray. These were air-dried for 24 hours at 65°C. Portion B2 was similarly divided into five subportions (B2:Aliquots 1-5). These sub-portions were freeze-dried. The dry weights of each sub-portion were recorded individually. Each was then ashed for 24 hours at a temperature of 500°C . After ashing, the samples were weighed individually and then placed into labelled containers. The ten sub-portions (B1:Aliquots 1-5and B2:Aliquots 1-5) were then analysed for uranium using the standard analysis procedure. As any mixing that had occurred was uncontrolled samples were treated as heterogeneous. The sample ash:dry ratios were compared to see what effect the different sample drying techniques had on B1 (heating cabinet) and B2 (freeze dryer). This situation represented the "worst case" scenario as considerably more effort was made in the laboratory to form subportions by taking handfuls of fresh grass from randomly chosen parts of the grass as it lay in its retaining

Table 6 *Analysis of "best case" homogeneous grass sample*

Sample Descriptor	^{238}U estimated activity (Bq/g dry)
A:Aliquot 1	11.37 ± 0.91
A:Aliquot 2	9.95 ± 0.90
A:Aliquot 3	12.05 ± 1.02
A:Aliquot 4	11.38 ± 0.96
A:Aliquot 5	12.07 ± 1.02
Mean (Bq/g)	11.19
Standard Deviation (Bq/g)	0.88
Relative Standard Deviation (%)	7.90

Table 7 *Analysis of "worst case" inhomogeneous grass sample, air-dried*

Sample Descriptor	^{238}U estimated activity (Bq/g dry)
B1:Aliquot 1	11.36 ± 0.87
B1:Aliquot 2	12.52 ± 1.04
B1:Aliquot 3	14.12 ± 1.16
B1:Aliquot 4	9.29 ± 0.83
B1:Aliquot 5	12.15 ± 0.98
Mean (Bq/g)	12.09
Standard Deviation (Bq/g)	0.88
Relative Standard Deviation (%)	15.38

Table 8 *Analysis of "worst case" inhomogeneous grass sample, freeze-dried*

Sample Descriptor	^{238}U estimated activity (Bq/g dry)
B2:Aliquot 1	7.42 ± 0.61
B2:Aliquot 2	6.04 ± 0.57
B2:Aliquot 3	23.86 ± 1.78
B2:Aliquot 4	9.34 ± 0.81
B2:Aliquot 5	15.12 ± 0.95
Mean (Bq/g)	11.60
Standard Deviation (Bq/g)	0.88
Relative Standard Deviation (%)	61.54

tray. Also the resulting ash would be pulverised and well mixed in the laboratory before the 1 g of ash was removed for analysis. The results of these analyses are shown in Tables 7 and 8.

4.3.3 Calculation of subsampling uncertainty;u(X_{sub}^{TOT}) for the best case. The relative standard deviation of 7.9% in Table 6 is due to the combined principal uncertainty terms (i.e. the counting uncertainty, the ^{232}U tracer uncertainty, the inherent subsampling uncertainty, the uncertainty in the ash:dry ratio, and any uncertainty due to error in the method which remains unquantified for our particular analysis) that we have been considering in this report. This can be written as:-

$$(7.9)^2 = ((\boldsymbol{u}^2(C_{gr}^{238}) + (\boldsymbol{u}^2(C_{gr}^{232})) + (\boldsymbol{u}^2(S)) + (\boldsymbol{u}^2(X_{sub}^{TOT}) + \boldsymbol{u}^2(A/D) + \boldsymbol{u}^2(X_{method}))$$

Leaving out the unquantitfied component $\boldsymbol{u}(X_{method})$ the following numerical values can be substituted into the expression above:-

$(7.9)^2 = ((4.2)^2 + (3.5)^2 + ((8)^2/3) + (\boldsymbol{u}^2(X_{sub}^{TOT}) + (6)^2)$ (The value of $\boldsymbol{u}(A/D)$ is taken to be 6% and will be described in the next section 4.4.).

Therefore it can be seen that ($u(X_{sub}^{TOT})$ for the "best case" grass preparation scenario is insignificantly small as the other uncertainty terms ultimately combine to a greater value than that given by the value of the square of the measured relative standard deviation . This is not a surprise considering the amount of preparation effort expended on the sample.

4.3.4 Calculation of subsampling uncertainty; $u(X_{sub}^{TOT})$ for the "worst" case, air-dried. A larger value of the relative standard deviation for the analysis is now evident in Table 7 as would be expected. Interestingly the value of the mean ^{238}U activity is very similar to that derived for the best case scenario. Assuming the other components of uncertainty have the same value as used before the value of $u(X_{sub}^{TOT})$ for this scenario is ~ 12%

4.3.5 Calculation of subsampling uncertainty; $u(X_{sub}^{TOT})$ for the "worst" case, freeze-dried. In an entirely analogous way to calculating $u(X_{sub}^{TOT})$ for the previous two cases using the data in Table 8 and the uncertainty terms as before (apart from u(A/D) which was found to be 26% using freeze-drying), the value of $u(X_{sub}^{TOT})$ is estimated to be ~ 50%. If however the results of the analysis are recalculated in units of Bq/g dry weight, the uncertainty component u(A/D) should be excluded. Results are shown in (Table 9). In this case the value of $u(X_{sub}^{TOT})$ is estimated to be 38%.

Because the uncertainty due to method error has not been included the value of the subsampling uncertainty estimates could probably go down somewhat. However the exercise was useful in that it indicated quite vividly the large magnitude of the variance that can occur if solid samples are not properly prepared. The 'working' or standard value of $u(X_{sub}^{TOT})$ is somewhere between the extremes derived above. $u(X_{sub}^{TOT})$ contains

Table 9 *Analysis of "worst case" inhomogeneous grass sample, expressed in units of Bq/g ash.*

Sample Descriptor	^{238}U estimated activity (Bq/g ash)
B2:Aliquot 1	70.6 ± 5.8
B2:Aliquot 2	52.6 ± 5.0
B2:Aliquot 3	136.4 ± 10.2
B2:Aliquot 4	88.7 ± 7.6
B2:Aliquot 5	103.0 ± 9.0
Mean (Bq/g)	96.46
Standard Deviation (Bq/g)	0.88
Relative Standard Deviation (%)	38.9

contributions from the uncertainty in subsampling from the fresh sample $u(X_{sub}^{fresh})$ and from $u(X_{sub}^{ash})$ of the uncertainty in subsampling of the 1g ash test portion. It is instinctively felt that $u(X_{sub}^{fresh})$ is greater than $u(X_{sub}^{ash})$. Therefore the value of $u(X_{sub}^{fresh})$ is set to range from 10% to 40% and $u(X_{sub}^{ash})$.is estimated to have a value of 6%-7%.

4.4 Ash:Dry ratio uncertainty; u(A/D)

This has been estimated by observation of large numbers of these ratios over the six years that the monitoring programme has been in operation. The uncertainty is dependent on which method is used for the drying process. Freeze drying has only been used when sample loading has outstripped the capacity of the air-drying cabinets. The larger uncertainty term for freeze drying is probably due to the greater difficulty of establishing when a sample is dry. The efficiency of the process definitely drops when vacuum seals begin to break down. The average operator finds the process of air-drying a more conveniently straightforward procedure to use. All of the grass samples are ashed in a large capacity oven. One source of uncertainty here is the possibility of differential heating across the oven interior. It could also depend on the way that samples are packed within it. Improperly ashed samples will then cause problems with the ensuing separation chemistry.

The value of u(A/D) is often 25%-30% when data from the 19 sampling locations that form one of the customer's sites are examined. Much of this is undoubtedly due to the natural variation in species that are commonly referred to as grass. Each species will have a particular ashing characteristic. The relevant value of u(A/D) for our analyses is that for the variation within the six replicate grass samples that are taken at each location. This value can be below 5% if the samples are carefully prepared. However there are a disturbing number of grass samples that are found to have soil inclusions within them and this obviously increases the measured ash:dry ratios. The working value for u(A/D) that is put into the uncertainty budget is optimistically estimated at 6%-10%.

4.5 Uncertainty due to method error; u((X_{method}))

Normally this is estimated by analysis of a relevant certified reference material and also by participation in intercomparison exercises where the nuclides of interest are available in closely matched matrices. This is not available for uranium in grass. The best way forward

Table 10 *The principal sources of uncertainty*

Symbol	*Source*	*Type of distribution*	*Value (%)*
(C_{gr}^{238})	^{238}U counts	Normal	20
(u(S))	^{232}U tracer	Rectangular	8
u(A/D)	Ash:Dry ratio	Normal	6-10
X_{sub}^{ash}	Subsampling (Ash)	Normal	6-7
X_{sub}^{fresh}	Subsampling (Fresh)	Normal	(10-40)
	Method	?	?
u(TOT)	TOTAL	Normal	24+

probably lies in locally arranged intercomparisons that occur from time to time in some of the working groups.

5 COMBINATION OF THE UNCERTAINTY TERMS

The next stage is to estimate the value of u(TOT) the combined standard uncertainty for the analysis method and this is done by summing in quadrature the estimated values of the principal components of the uncertainty budget. Table 10 lists the principal components of the uncertainty budget for the uranium in grass analysis method. i.e.

$$U^2(TOT) = ((20)^2 + ((8)^2)/3) + (10)^2 + (7)^2)$$

$$U(TOT) = \sim 24\%$$

The value for X_{sub}^{fresh} is left in parentheses to denote that it would need to be factored into the total uncertainty should subsampling ever be used again in the method

6 EXPANDED UNCERTAINTY

The activity result *Act* (Bq/g dry sample) should be stated with the expanded uncertainty *U*, where *U* = u(TOT) multiplied by a coverage factor (k). For a level of confidence of approximately 95% , k = 2. In this case the activity estimate would be reported in the following form:-
Net ^{238}U activity estimate is $Act \pm 0.44{*}Act$ Bq/g dry sample.

7 CONCLUSION

The analysis method has been investigated and a list of uncertainty components were identified. Those components whose uncertainty contribution was less than a third of that of the largest uncertainty component were not considered further. Four principal components contribute to the overall uncertainty term. The uncertainty component due to errors in the analysis method remains unquantified.

The exercise has occasionally been very time consuming but generally worth the effort. There is still a feeling that some of the uncertainty estimates have been arrived at by a process of educated guesswork rather than one of scientific rigour. Obviously a lot more work remains to be done. At the end of the day all this work has been done in order to quote the uranium activities with more realistic uncertainty values, but most of our samples are reported as below the limit of detection. However the same principles used to construct the uncertainty budget described in this report can be applied to other radiochemical procedures . Sometimes this process makes more evident a source of uncertainty that had been overlooked, in our case this was the uncertainty in the ash:dry ratios. It also identifies areas where further investigative effort ought to be applied. For our laboratory this will involve procurement of a closely matched matrix reference material to assess the errors in the method.

References

1 ISO/IEC 17025 *General requirements for the competence of testing and calibration laboratories,* Geneva, Switzerland
2 BIPM, IEC, IFCC, ISO, IUPAC, IUPAP, OIML *Guide to the Expression of Uncertainty in Measurement,* ISO, Geneva, Switzerland. ISBN 92-67-10188-9, First Edition 1993.
3 J.L. Love. *Accreditation and Quality Assurance*, 2002, **7 (3)** 95-100
4 EURACHEM/CITAC *Quantifying uncertainty in analytical measurement,* Second Edition (2000)

5 K. Birch *An Intermediate Guide to Estimating and Reporting Uncertainty of measurement in Testing* Measurement Good Practice Guide No.36. BMTA March 2001 ISSN 1368-1550
6 UKAS Publication LAB 12, *Guide to the Expression of Uncertainties in Testing,* Edition 1 January 2001
7 P. Gy, in *Sampling for analytical purposes,* Wiley, Chichester, UK, 1998
8 P. de Zorzi, M. Belli, S. Barbizzi, S. Menegon, A. Deluisa, *Accred Qual Assur,* 2002,**7,** 182
9 D.W.K. Jenkins, F.J. Sandalls, and R.J. Hill, *Uranium and Thorium in British crops and arable grass,* 1989, AERE R 13442, HMSO

SPATIAL AND TEMPORAL VARIATION OF TRITIUM ACTIVITIES IN COASTAL MARINE SEDIMENTS OF THE SEVERN ESTUARY (UK)

P.E. Warwick[1], I.W. Croudace[1], A.G. Howard[2] A.B. Cundy[3], J Morris[1] & K Doucette[2]

[1]Southampton Oceanography Centre, European Way, Southampton, SO14 3ZH
[2]Dept. of Chemistry, University of Southampton, Highfield, Southampton, SO17 1BJ
[3]School of Chemistry, Physics and Environmental Science, University of Sussex, Brighton, BN1 9QJ,

1 INTRODUCTION

Tritium is a low energy pure beta emitting radionuclide (E_{max} = 18.6 keV) that is routinely released from certain nuclear power stations and reprocessing plants into the marine environment. Typically, ^{3}H is released as tritiated water, which is rapidly diluted in the water mass to which it is released, dispersing with prevailing currents and subsequently being lost via radioactive decay with a half-life of 12.3 years. However, discharges from the radiolabelled compound manufacturer, Amersham plc, located near Cardiff, UK are unusual in that they contain a proportion of ^{3}H present as organically-bound tritium. Since 1982, Amersham plc have discharged ^{3}H into a local sewage system producing a mixture of radiolabelled organic compounds and sewage wastes, which is subsequently discharged into the Severn Estuary. The exact chemical composition of the effluent from the site is unknown as it contains reaction intermediates and by-products as well as the final radiolabelled materials. It is estimated[1] that around 30% of the ^{3}H discharged into the sewage system up to May 1998 consisted of organically-bound tritium. In May 1998, certain tritiated methanolic wastes and tritiated water were withheld from the discharge resulting in a significant decline in total ^{3}H discharged. However, the proportion of organically-bound ^{3}H increased to *ca* 80% at this time.

The behaviour of organically-bound tritium in the marine environment will vary considerably compared with that of tritiated water and will be determined by the chemical properties of the organic molecule containing the ^{3}H. The situation is further complicated by any chemical or physical interactions of the original tritiated species when mixed with the sewage wastes. Whereas no bioaccumulation for tritiated water would be expected, significant uptake of ^{3}H in biota from the Severn Estuary has been observed. Flounder from this area are routinely analysed for ^{3}H and activity levels of ^{3}H of 50000 Bq/kg fresh weight have been reported[2] in 2000. However, following the reduction in ^{3}H discharge activities in May 1998, levels of ^{3}H in flounder have remained high suggesting that either the source of the ^{3}H is still present in the discharge or that there is now an environmental sink of ^{3}H providing a source of ^{3}H to the food chain. As elevated levels of ^{3}H have also been observed in lower benthos, such as burrowing worms, it has been suggested that sediment may play an important role in the transfer of ^{3}H to the foodchain[3]. This study aimed at investigating the spatial and temporal variability of ^{3}H in sediments in the Severn Estuary and the factors that may control the release of ^{3}H from the sediment as well as evaluating the importance of sediments in the retention of Amersham-derived ^{3}H in the Severn Estuary.

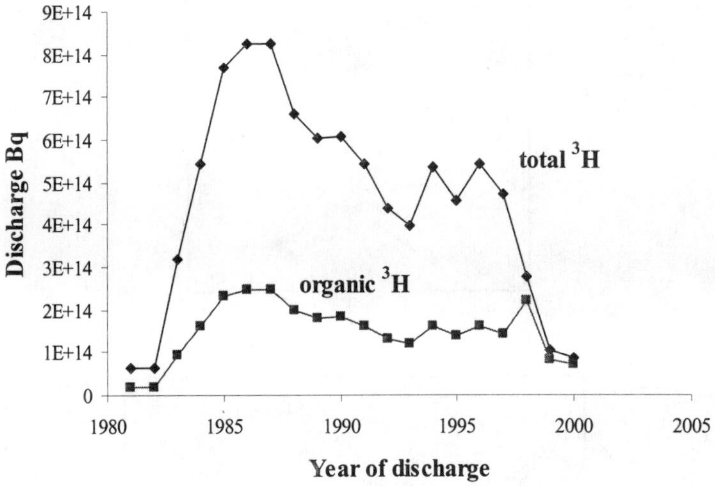

Figure 1 *Historical authorised discharges of tritium from Amersham plc, Cardiff, into the Severn Estuary. Values for organic ^{3}H are estimates only (J Williams, pers comm)*

2. METHODOLOGY

2.1 Sampling

Samples were collected from February 2000 to February 2002 on a monthly basis where possible from locations 1 to 7, shown in Figure 2. Surface sediment samples were collected over an area of 25 by 25 cm and to a depth of 1 cm. Due to the dynamic nature of the Severn Estuary and the severe weather conditions experienced towards the end of 2000, it was not always possible to obtain sufficient sediment at all the locations. Additional samples were collected in 2002 from points 0, 8, 9 and 10 to determine the levels of ^{3}H at distance from the discharge point. Surface sediment samples were also collected from the Cumbrian coastline, which is exposed to the marine discharges of tritiated water arising from the BNFL Sellafield site.

An aliquot of the sediment was removed and freeze-dried to determine the wet/dry ratio. The remaining fresh sample was stored, prior to analysis, frozen at -20°C in tightly sealed containers. When required for analysis the samples were thawed, homogenised and an aliquot removed for ^{3}H analysis.

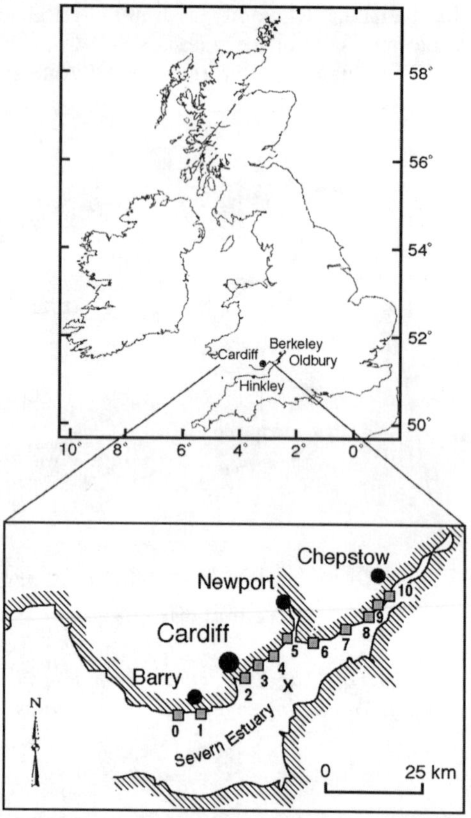

Figure 2 *Location of study area. Sediment sampling locations numbered (see Table 4 for site names). 'X' marks the approximate location of the discharge point.*

2.2 Analysis

2.2.1 Note on the definition of OBT

Organically-bound tritium (or OBT) is often quoted in the literature as distinct from tritiated water. The widely accepted definition of organically bound tritium (OBT) is that of tritium bound non-exchangeably to a carbon atom. However, analytically it is not possible to directly determine this component of tritium and it is more usual to measure the quantity of tritium that is removed via drying or distillation (taken as tritiated water but also including any volatile organic compounds) and the quantity of tritium remaining after drying or distillation. This is an arbitrary distinction bearing little relation to the accepted definition of OBT. In this study, a measure of the total ^{3}H content of the sample is made through total combustion of the sample. Where appropriate, a measure of the water-extractable ^{3}H content of the sample is also performed. The terms 'total' and 'water-extractable' tritium will be used in the following sections.

2.2.2 Water extractable ^{3}H ($^{3}H_{exch}$). Water extractable tritium was determined by mixing 5g of fresh sediment with 10ml of deionised water. The mixture was shaken for 1 hour, centrifuged for 20 minutes at 2500 RPM, and then filtered through a 0.45 μm membrane filter. The ^{3}H content of the filtrate was then determined.

2.2.3 Total tritium ($^{3}H_{total}$). All total tritium measurements were performed in a commercially-available 4 tube furnace (designed and fabricated by the Geosciences Advisory Unit, Southampton, UK). Total tritium was determined by combusting approximately 10g of fresh sediment. The sample was placed in a silica boat and transferred to the low temperature furnace zone. The furnace end cap was replaced and oxygen-enriched air passed over the sample. The temperature of the low temperature furnace was then ramped to 500°C at a rate of 5°C/min. The combustion products were passed over a Pt-alumina catalyst, heated to 800°C, to ensure the complete conversion of tritiated species to tritiated water. The tritiated water was then trapped in two 20ml Milli-Q™ water bubblers and the tritium content of these bubblers determined by liquid scintillation counting. Typically > 95% of the ^{3}H activity was trapped in the first bubbler.

Blank and tritiated sucrose standards were analysed after every five samples to monitor for increases in furnace background and catalyst failure

2.2.4 Liquid scintillation measurement. All tritium measurements were performed on a Wallac 1220 Quantulus low-level liquid scintillation counter. 8ml of aqueous sample were mixed with 12ml Gold Star scintillation cocktail in a 22ml polythene vial. The counter was calibrated for ^{3}H using a traceable tritiated water standard (Amersham QSA). Instrument backgrounds were determined by repeated counting of dead water samples collected from deep boreholes. Milli-Q™ water backgrounds were also measured and used routinely for background correction.

2.2.5 Recoveries of other ^{3}H-labelled compounds. Although ^{3}H-sucrose was satisfactory for the evaluation of routine furnace performance, the tracer may not accurately reflect the recovery of ^{3}H from other, more thermally stable, organic compounds. This could be seriously misleading considering the potentially large range of tritiated compounds that may be present in the sediments. Furnace performance was therefore also tested using a range of organic compounds labelled with tritium. These compounds were chosen to reflect to some degree the range of compounds that Amersham plc may discharge during routine operations. Furnace recoveries were determined using both a Pt-alumina catalyst and also a CuO-alumina catalyst (Table 1). Recoveries for all compounds were >90% when using the Pt-alumina catalyst.

Table 1 *Recovery of ^3H from various labelled compounds*

compound class	composition	Recovery Pt-alumina	Recovery CuO-alumina
Nucleotides	thymidine 5'triphosphate deoxyadenosine 5'-triphosphate (1:1)	95 %	82 %
Hydrocarbons	DMBA, benzopyrene (1:1)	93 %	79 %
Thymidine	Pure compound	98 %	99 %
Amino Acids	Leucine, lysine, phenylalanine, proline, tyrosine (1:1:1:1:1)	97 %	87 %
Methanol Waste	tritiated water : methanol 85:15	98 %	87%
Sugars	Sucrose, dextran, D-galactose (1:1:2)	98 %	90 %
HTO	Water	100 %	n/a

n/a – not analysed

Analytical reproducibility was also assessed through the replicate analysis of homogenised fresh sediment collected from Peterstone (Site 5). All measurements agreed within the estimated measurement uncertainty (Table 2).

Table 2 *Assessment of analytical reproducibility*

Replicate number	Activity Bq/g fresh sediment	Uncertainty
1	0.73	0.02
2	0.76	0.02
3	0.72	0.03
4	0.75	0.03
Mean value	0.74 ± 0.02	

3 RESULTS AND DISCUSSION

3.1 Spatial distribution of tritium in the Severn Estuary

Measurable levels of ^{3}H were routinely detected in sediments collected across the entire sampling region. The highest levels (up to 0.83 Bq/g fresh sediment) were detected in sites close to the discharge point (namely Sites 3, 4 and 5). The results agreed well with measurements of 0.50 – 0.98 Bq/g dry weight reported by McCubbin *et al* (2001) for Orchard Ledges (Sites 2 and 3). Levels of ^{3}H declined slowly away from the discharge point but were still detectable at Sites 0 (Leys Beach) and 10 (Beachley). The widespread distribution of sediment-bound ^{3}H reflects the highly dynamic and well-mixed nature of the Severn Estuary system.

Small-scale variability of ^{3}H activity concentrations was investigated at Orchard Ledges East and Maerdy Farm. At Orchard Ledges East (Site 3 in Figure 2), four samples were collected around an initial sampling point as shown in Figure 3. At Maerdy Farm (Site 4) a land-to-sea transect comprising of five samples was collected at 10m intervals (Table 3). In both instances, variability was found to be small, confirming that the local sediment system was well mixed.

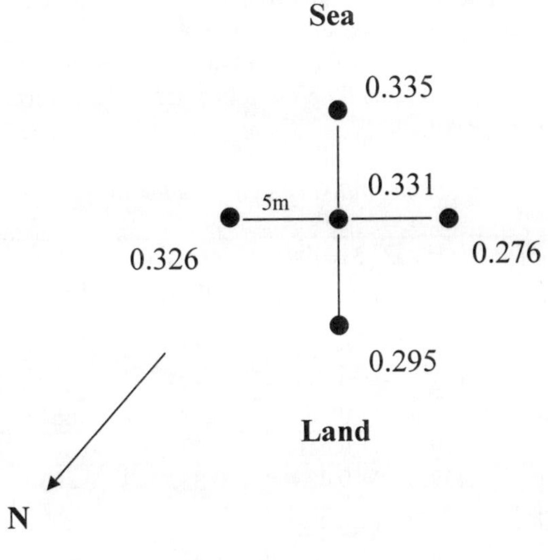

Figure 3 *Small scale variability in $^3H_{total}$ activities at Orchard Ledges East (Site 3). All results are in Bq/g fresh. Samples collected 30/8/01. Mean value for all results is 0.31 ± 0.03 Bq/g*

Table 3 *^{3}H activity concentrations in sediments collected along a sea/land transect at Maerdy Farm*

Site	Water-extractable ^{3}H ($^{3}H_{exch}$)	Total ^{3}H ($^{3}H_{tot}$)	Uncertainty (2 s.d.)
Maerdy Farm (Site 4) 31/7/00	<0.031	0.828	0.032
+10m seaward	<0.032	0.667	0.030
+20m seaward	<0.033	0.931	0.032
+30m seaward	<0.037	0.649	0.032
+40m seaward	<0.028	0.722	0.029
mean		**0.76 ± 0.12**	

Results are in Bq/g fresh weight

In all instances, water-extractable ^{3}H was below the detection limit (< 0.03 Bq/g fresh equivalent) indicating that the majority of ^{3}H was irreversibly bound within a species that was strongly adsorbed to the sediment. Particle size fractionation confirmed that the ^{3}H-species was mainly associated with the < 63μm clay fraction of the sediment. Analysis of freeze-dried sediment for $^{3}H_{total}$ confirmed that no ^{3}H had been lost during the freeze-drying process and that the ^{3}H was predominantly present in a non-volatile form.

Table 4 *Mean $^{3}H_{total}$ and maximum/minimum activities for sampling sites*

Site	Location	Mean $^{3}H_{total}$	Median	Geometric mean	Highest $^{3}H_{total}$	Lowest $^{3}H_{total}$
0	Leys Beach*	0.256				
1	Barry Island	0.176	0.137	0.145	0.533	0.083
2	Orchard Ledges west	0.192	0.171	0.172	0.338	0.077
3	Orchard Ledges east	0.399	0.364	0.380	0.666	0.27
4	Maerdy Farm	0.419	0.410	0.375	0.828	0.091
5	Peterstone Great wharf	0.437	0.488	0.322	0.803	0.015
6	St Brides Wentlooge	0.305	0.270	0.262	0.829	0.058
7	Goldcliff	0.198	0.209	0.113	0.312	0.004
8	Collister Pill*	0.152				
9	Sudbrook*	0.060				
10	Beachley*	0.163				

* = Only one data point

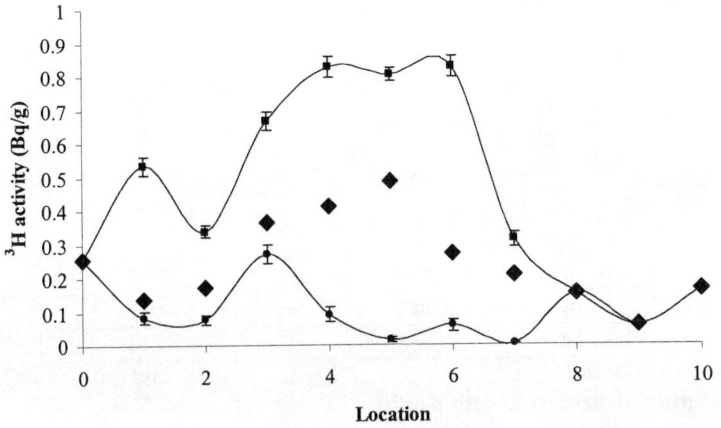

Figure 4: *Median $^3H_{total}$ (Bq/g) across all sampling sites. The data points represent the median $^3H_{total}$ (Bq/g) for each location (Figure 2 and Table 1). The bars show the range of values for $^3H_{total}$ measured at that site over the period Feb 2000 – Dec 2001.*

In addition to the studies performed in the Severn Estuary, samples of sediment were also collected in the vicinity of BNFL Sellafield. Marine discharges of ^{3}H from Sellafield (2260 TBq during the year 2000) are significantly higher than those from Amersham plc, Cardiff, (87TBq during the year 2000) and have occurred over a longer time period. However, it is believed that the ^{3}H discharged from Sellafield is in the form of tritiated water and hence it would be expected that direct uptake of ^{3}H onto sediment would be negligible. Conversion of tritiated water to non-aqueous tritium may, however, occur and studying this area permits the significance of tritiated water conversion to the long-term persistence of ^{3}H in the marine environment to be assessed.

Surface sediment samples were collected along the Cumbrian coastline at Maryport, Workington and Whitehaven. Salt marsh sediments were also collected from the Esk Estuary near Ravenglass. Analysis of these samples for $^3H_{tot}$ showed that levels of ^{3}H in sediment were extremely low and, with the exception of one sample, were below the limit of detection (Table 5). One sample collected from Ravenglass contained 0.017 Bq/g fresh weight of ^{3}H but this is almost an order of magnitude below the lowest ^{3}H activities observed in the Severn Estuary. It may therefore be assumed that the conversion of tritiated water to sediment-bound ^{3}H is not a significant route for the uptake of ^{3}H on sediment in the UK coastal environment.

Table 5 *Total ^{3}H activity along the Cumbrian coastline*

Site	^{3}H total Bq/g fresh
Maryport Harbour	<0.010
Workington Harbour	<0.009
Ravenglass salt marsh 1	<0.015
Ravenglass salt marsh 2	0.017 ± 0.012
Ravenglass salt marsh 3	<0.013
Ravenglass salt marsh 4	<0.011
Ravenglass salt marsh 5	<0.013
Whitehaven	<0.011

3.2 Tritium distribution with depth

A short core (to 5cm depth) was collected from Maerdy Farm (Site 4). The top fraction of this core consisted of non-consolidated, readily mobile sediment overlying a well-consolidated old marsh. The non-consolidated sediment had an apparently high turnover and the amount of this sediment present at a given sampling trip was highly dependent on prevailing weather conditions. It is this top fraction that is routinely sampled. Activity concentrations were a factor of ten greater in the unconsolidated sediment compared with the underlying marsh sediment (Table 6).

Table 6 *Maerdy Farm short core*

Fraction	Description	^{3}H total activity Bq/g fresh
1 - 0-1cm	Unconsolidated sediment	0.436 ± 0.037
2 – 1-2cm	Consolidated marsh sediment	0.045 ± 0.013
3 – 2-3cm	Consolidated marsh sediment	0.031 ± 0.013
4 – 3-4cm	Consolidated marsh sediment	0.043 ± 0.014
5 – 4-5cm	Consolidated marsh sediment	0.028 ± 0.014

A 40cm depth core was collected from the higher terrace of an established salt marsh at Peterstone (Site 5). The core was subdivided into 1cm fractions, which were analysed for $^{3}H_{total}$ and $^{3}H_{exch}$ activity although all $^{3}H_{exch}$ activity measurements were below the limit of detection.

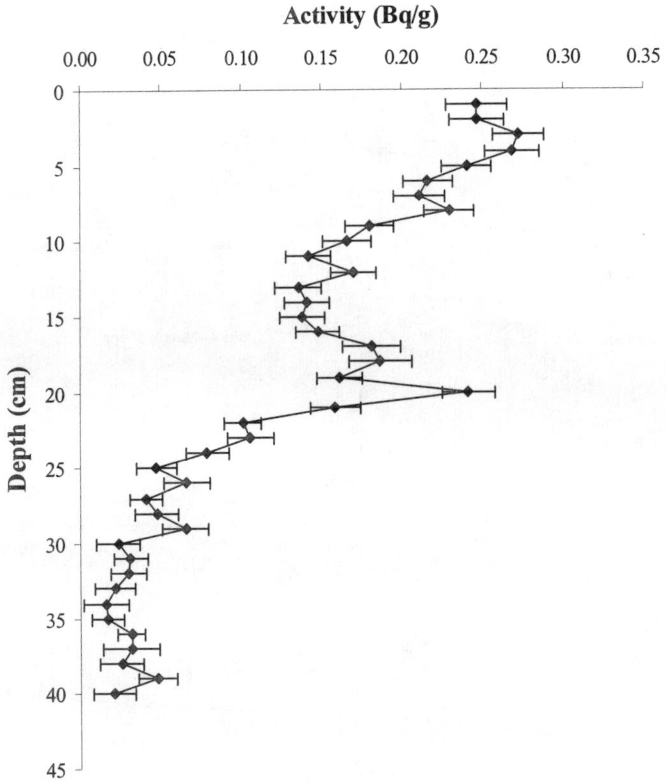

Figure 5: *Vertical distribution of total ^{3}H activity (Bq/g fresh weight) in the Peterstone core.*

The ^{3}H profile for the Peterstone core (Figure 5) shows a general decline with depth, with a superimposed broad peak between 15 and 25cm. The structure at depth down the core suggests that the profile does not simply reflect diffusion of mobile ^{3}H-species down the core and that either the core has preserved a chronology of ^{3}H discharges or that variable core geochemistry is controlling the mobility of ^{3}H down the core.

3.3 Temporal variation of tritium in the Severn Estuary sediments

No seasonal trends are observed in the data. The highest and lowest tritium activities at Orchard Ledges West (site 2) were found in February 2000 and February 2001 respectively. It is probable that the variability seen is produced by the weather or tides in the region for the week before the sample was collected. A general decline in sediment ^{3}H activities over the sampling period most probably reflects the decline in discharged ^{3}H activity since 1998.

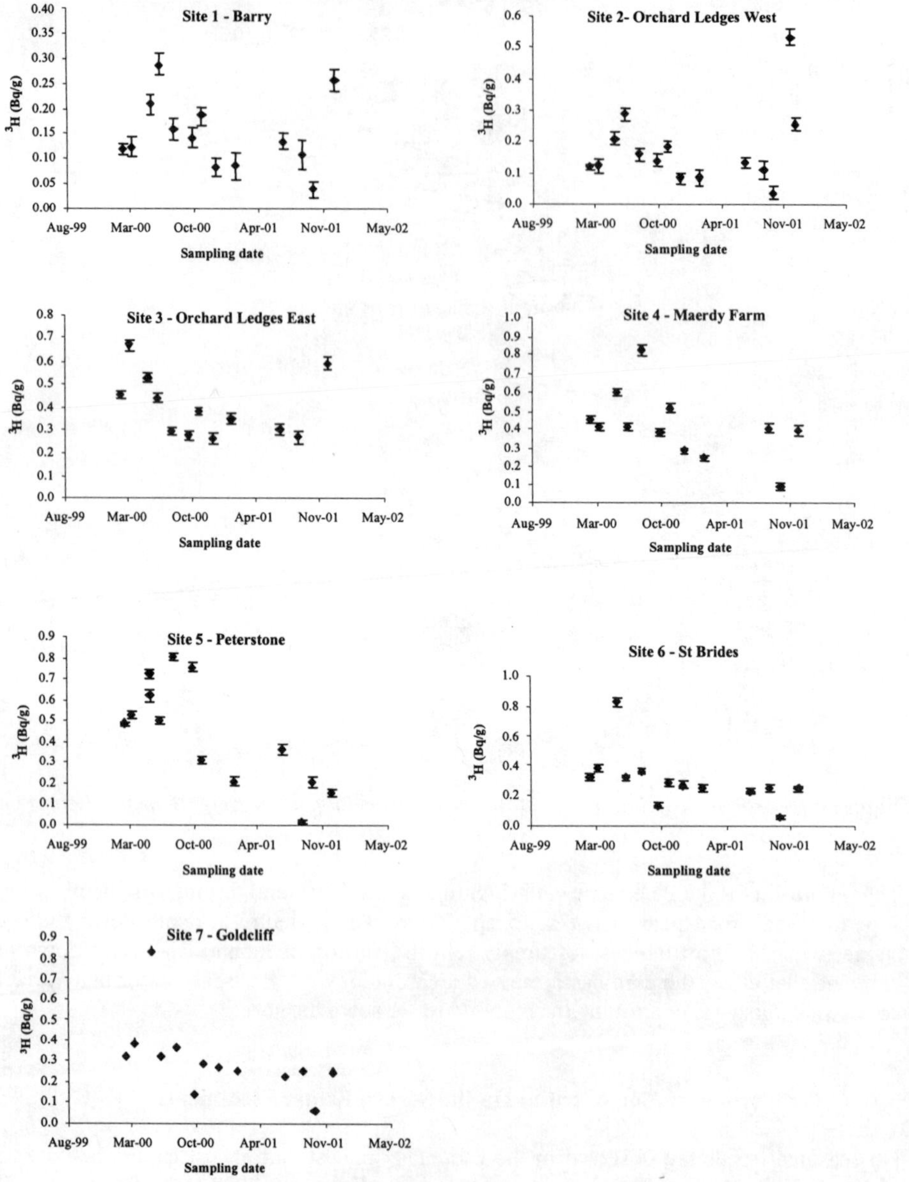

Figure 6: *Temporal variation in total ^{3}H activities at the 7 sites that were routinely sampled. ^{3}H activities are total ^{3}H in Bq/g fresh weight.*

4 CONCLUSIONS

Tritium has been detected in intertidal sediments collected along the northern edge of the Severn Estuary. The highest activities were found in the vicinity of the sewage discharge pipeline. However, elevated levels of ^{3}H were still detectable at significant distances away from this point indicative of the highly dynamic nature of the Severn Estuary. Small-scale variability in the ^{3}H activities of surface sediments was negligible. The tritium was present in a non-water extractable form (consistent with its long-term retention on the sediment). The ^{3}H was predominantly present associated with surface sediments that are rapidly deposited on, and resuspended from intertidal areas. Levels of ^{3}H in underlying consolidated marsh surfaces were significantly lower suggesting limited penetration of the ^{3}H. However, in accreting areas, a discrete profile of ^{3}H activity was observed. The levels of ^{3}H in intertidal sediments reflect only a small proportion of the total ^{3}H discharged from the Amersham site. However, the levels of ^{3}H are considerably higher than those found in intertidal sediments from the Cumbrian area even though the marine discharges from BNFL Sellafield are higher than those from Amersham plc. This suggests that the sediment-bound ^{3}H has originated from organically-bound ^{3}H discharged into the marine environment and not from the conversion of tritiated water into other particle-reactive species.

Further studies will be aimed at characterising the Peterstone core in more detail with the aim of understanding the underlying factors controlling the observed core profile. In addition, further temporal sampling will be performed to identify variations in sediment-bound ^{3}H activities following the introduction in 2002 of a new sewage treatment plant at Cardiff.

Acknowledgements

The authors would like to thank B Walters, S Conney, N Woods and B Gallani (FSA) for their support and useful suggestions. We would also like to thank J Williams (Amersham plc) for many interesting discussions and for providing much of the background information relating to Amersham's historical operations.

This study has been funded by the Food Standards Agency under contract R01034 .

References

1 Williams J.L., Russ R.M., McCubbin D. and Knowles J.F. *J. Radiol. Prot.*,2001 **21**, 337.

2 Radioactivity in food and the environment, 2000. RIFE 6. Food Standards Agency, London, UK (2001).

3 McCubbin D., Leonard K.S., Bailey T.A., Williams J. and Tossell P. *Mar. Pollut. Bulletin*, 2001, **42**, 852 .

HIGH PRECISION PU ISOTOPE RATIO MEASUREMENTS USING MULTI-COLLECTOR ICP-MS

I. W. Croudace, T. Warneke, P. E. Warwick, R. N. Taylor and J. A. Milton

Southampton Oceanography Centre, Southampton SO14 3ZH, UK

1. INTRODUCTION

Recent developments in multi-collector plasma mass spectrometry (MC-ICPMS) enable high precision isotope ratio data to be readily obtained for ^{239}Pu and ^{240}Pu at ultra-trace levels.[1] The presence of plutonium isotopes in the environment is primarily due to fallout from the atmospheric testing of nuclear weapons and from authorised discharges from military and civil nuclear establishments during the second half of the twentieth century. Plutonium is present in varying concentrations in soils, sediments and biota and measurement of the more abundant plutonium isotopes (^{239}Pu and ^{240}Pu) can provide information on the source of contamination, such as nuclear weapons production, weapon detonation, or reactor discharges.[2]

Table 1 *$^{240}Pu/^{239}Pu$ atom ratios for different sources*[3,4]

Source	$^{240}Pu/^{239}Pu$ Atom ratio
Sellafield Discharges (1950s-now)	0.05-0.23
Integrated weapons test fallout	0.18
Weapons grade plutonium	0.01-0.07
Chernobyl accident	0.40
MAGNOX Reactor (GCR)	0.23*
Pressurised Heavy Water Reactor (PHWR)	0.41*
Advanced Gas-cooled Reactor (AGR)	0.57*
Pressure Tube Boiling Water Reactor (RBMK)	0.67*
Boiling Water Reactor (BWR)	0.40*
Pressurised Water Reactor (PWR)	0.43*

* after fuel burn up

Fissile ^{239}Pu is present in weapons-grade plutonium at high relative abundance ($^{240}Pu/^{239}Pu$ typically 0.05 atom ratio) and at much lower abundances in mixed oxide fuels

($^{240}Pu/^{239}Pu$ approximately 0.4). The isotope is also produced during the detonation of weapons and in nuclear reactors from ^{238}U via neutron capture and subsequent beta decay of the resulting ^{239}U (t½~23.47 min) to ^{239}Pu. As well as being fissile, ^{239}Pu undergoes neutron capture, either briefly during weapons' detonation or for prolonged periods within a nuclear reactor, to produce ^{240}Pu and, through successive neutron capture, to generate the heavier isotopes of Pu. The ratio of $^{240}Pu/^{239}Pu$ will therefore depend on the composition of the source material and the subsequent irradiation history of the material. Nuclear weapon construction generally requires a low $^{240}Pu/^{239}Pu$ (<0.07). After detonation this ratio increases due to neutron capture, the exact value depending on the test parameters. For this reason the $^{240}Pu/^{239}Pu$ in specific weapon test fallout can vary between 0.10 and 0.35; the time integrated test ratio is about 0.18.[2] Weapons grade Pu, weapons fallout Pu and Pu produced in the nuclear fuel cycle, therefore, have $^{240}Pu/^{239}Pu$ ratios that are sufficiently different to permit discrimination of the various sources.

Table 2 *Summary of techniques used for $^{240}Pu/^{239}Pu$ determination* [1]

METHOD	ADVANTAGES	DISADVANTAGES	NOMINAL PRECISION @ ~50 fg (2 sigma)
Alpha Spectrometry	Not suitable because alpha energies interfere		
AMS	Can measure ~50fg	High potential cost, facilities rare	~ 18%
ICPMS (Quadropole)	High ionisation efficiency	Ion beam instability	> 30%
ICPMS (Sector) Single collector	High ionisation efficiency Good stability, good peak shape	Only a single collector	~ 3%
TIMS (Multicollector)	Stable ion beam Very good peak shape	Low ionisation efficiency No internal inter-element fractionation correction	~ 10%
MC-ICPMS	High ionisation efficiency Multicollection negates the unstable ion beam effect Interelement mass fractionation correction capability Can measure down to 5 fg		~ 1%

A variety of techniques[1] have been used to measure plutonium isotopic compositions and abundances (Table 2). Although the measurement of plutonium activities is commonly undertaken using alpha spectrometry it is not possible to distinguish ^{240}Pu and ^{239}Pu by this technique due to the similarity of their alpha energies. The remaining effective techniques involve atom counting and use a variety of mass spectrometric approaches. The effectiveness of the different approaches and their reproducibility at low concentrations depends on ionisation efficiency, multicollection and the energy spread of the ions being measured. A more detailed treatment of these issues are given in Taylor et al[1]. Thermal ionisation mass spectrometry (TIMS) has been an important method for the determination of plutonium isotope ratios and is capable of a reproducibility of better than 0.1% (2s) with relatively large sample loads (>1 ng Pu). At 200–500 fg the reproducibility is about 1.5% (2s) and about 10% at <50 fg. Quadrupole-based plasma-source mass spectrometers (ICP-MS) with single ion-counting detectors have been used to measure plutonium isotope ratios in environmental materials.[4] However, these instruments do not produce sufficiently flat-topped peaks to enable precise ratio measurement, with precision around 5% on solutions of about 3 pg ml^{-1}. The relatively recent ICP-MS that uses a sector-field mass analyser and a single-detector plasma instruments can generate more suitable peak shapes [5] and results in more reliable ratio determinations. A problem with plasma ionisation is the instability in signal intensity caused either by plasma ''flicker'' or by changes in the supply rate of analyte from the nebuliser to the ion source. Single collector instruments require that the ion beam at each mass be sequentially jumped into the detector. Thus, peak jumping is a limiting error on the precision of isotope ratio measurements. Multicollector instruments circumvent the problem of ion beam instability by acquiring data from all the required isotopes simultaneously. Thermal ionisation mass spectrometers have used multiple Faraday collector arrays for over 20 years, and have been the mainstay of high-precision isotope ratio analysis. In the mid-1990s multicollector, plasma source, sector field instruments (MC-ICP-MS) were developed. These spectrometers combine the advantages of multiple collectors with flat-topped peaks, sample introduction by solution and high ionisation efficiency. In the case of plutonium isotope measurement the amount of the element that is available and safe for analysis is usually quite low. Consequently, signal levels of ^{240}Pu and ^{242}Pu are generally too small to be reliably measured on Faraday collectors. Unless multiple ion-counting detectors (e.g. Daly detectors, Channeltrons) are available, it is necessary to revert to the determination of Pu isotope ratios by peak jumping each mass into a single ion counting detector, which negates the advantages of multicollection.

In this study a MC-ICPMS method has been developed in which an equal-atom uranium spike is added to separated plutonium samples before analysis to improve precision and which also uses a combination of Faraday and an ion-counting detectors. This method has been used to determine $^{240}Pu/^{239}Pu$ atom ratios in samples from a unique annual herbage archive and an Alpine ice core [3, 6] resulting in a new record of northern temperate latitude atmospheric deposition. Such a time-series for plutonium isotope ratios, which includes the background resulting from atmospheric nuclear testing, can be important for source-characterisation studies. The results presented here demonstrate that the MC-ICPMS can precisely measure plutonium isotope ratios at femtogram (microbecquerel) levels in environmental and other samples with a precision of 1% or better. This high sensitivity has implications for such areas as atmospheric transport models, recent geochronology, environmental studies and plutonium bioassay.

2 EXPERIMENTAL

The development of the methodology for measuring plutonium isotope ratio at high precision using MC-ICP-MS has been previously described [1] and is only briefly covered here. Following the chemical isolation of plutonium (and simultaneously of uranium if desired) the mass spectrometric analysis is generally made on about 1 ml of solution which, for example, in the case of a 100 fg/ml sample with a fallout $^{240}Pu/^{239}Pu$ of about 0.18, means a ^{239}Pu content of about 80 fg in the analyte. $^{240}Pu/^{239}Pu$ can be reproducibly measured to within 1.4% (2s) at 100 fg/ml and better than 0.3% at >3 pg/ml.

The MC-ICPMS technique employs a combination of Faraday and ion-counting detectors to eliminate imprecision caused by ion beam fluctuations. This method utilises a larger reference ion beam, suitable for Faraday collector acquisition, which is measured at the same time as each of the smaller objective ion beams. The measurement effectively becomes equivalent to a static multicollector analysis in which each isotope is counted simultaneously. Multicollector ICP-MS is not restricted to corrections using the same element, as is the case for TIMS analysis, for two reasons. Firstly, elements of similar mass and ionisation efficiency generally respond to ion beam fluctuations in the same way in a plasma source, and secondly, elements with these similar characteristics exhibit comparable levels of mass bias during a run due to the constant sample flux. The multicollector method used here detects both uranium and plutonium isotopes. Uranium is used as both a reference signal and as an in-run measure of mass bias. To achieve a precision of better than 1% at these concentrations an equal atom ^{236}U–^{233}U double spike was added to purified sample solutions to correct for drift in signal intensity between peak jump sequences. This double spike is also used to correct for instrument mass bias during each ratio determination.

2.1 Chemical separations

Samples were ashed and then a ^{242}Pu spike was added. This was followed by a double *aqua regia* acid digest. The separated liquid from the two leaches was evaporated and redissolved in 5-10 ml 8M HNO_3 with 1drop of concentrated HCl. Two ion-exchange columns were prepared for each sample. The first was a 6x0.7 cm i.d. column of Eichrom 1-X8 anion exchange resin (100-200 mesh) and the second a 2x0.7 cm i.d. column of Eichrom UTEVA™ resin. The UTEVA™ column was placed immediately below the anion exchange column and both columns were pre-conditioned with 10ml 8M HNO_3. The sample was transferred to the anion exchange column and the eluent passed directly onto the UTEVA™ column. The two columns were washed with 20ml 8M HNO_3 followed by 30ml of 3M HNO_3, and then separated. The anion column was washed with 30ml 3M HNO_3 followed by 25ml 9M HCl to remove thorium. Plutonium was then eluted with 50 ml of freshly prepared 1.2M HCl/H_2O_2 (50:1). The UTEVA™ column was washed with 10ml 3M HNO_3 followed by 10ml 9M HCl. The uranium was eluted with 10ml 0.02M HCl. Since the ^{238}U forms a hydride that interferes with the mass spectrometric measurement of ^{239}Pu the following additional separation of the plutonium is required. The eluent from the first Pu column was concentrated to 2 ml 9M HCl, 1 drop H_2O_2 and 7ml concentrated HCl were added to the concentrate before loading it onto an 4x0.7cm ID anion exchange pre-conditioned with 9M HCl. The column was washed with 10 ml 9M HCl, 40ml 7.2M HNO_3 and 15 ml 9M HCl. Plutonium was then eluted with 25 ml of freshly prepared ammonium iodide reagent (0.1M NH_4I/9M HCl). The eluent was evaporated to dryness and the residue was treated with *aqua regia* and concentrated HNO_3, before dissolving the residue in 2% HNO_3 (sub-boiled acid and MilliQ water) for mass spectrometry.

2.2 Instrumentation

Data was acquired using an IsoProbe ICP mass spectrometer (Micromass Ltd., Withenshaw, UK) at the Southampton Oceanography Centre, UK. The instrument comprises an argon plasma ion source, a Hexapole collision cell, a sector magnetic field and a multicollector array of nine Faraday detectors and an ion-counting Daly detector. The Daly detector is positioned after a retarding potential filter (WARP). Sample introduction to the argon plasma is via a desolvating nebuliser (MCN 6000, CETAC, Omaha, NE, USA). All samples were run using argon as collision gas (99.9999% purity; Air Products plc, Crewe, Cheshire, UK) admitted into the Hexapole at a rate of about 1.2 ml/min.

2.3 Measurement systematics

Pu isotope measurements were made using a combination of Daly and Faraday detectors in a peak-jump sequence. Pu isotope masses were measured with the Daly detector in the sequence ^{240}Pu ^{239}Pu ^{242}Pu. Faraday collectors were positioned at 3, 4 and 6 mass units lighter than Pu to receive the uranium ion beams. Each sequence was counted for 5 s, with a delay of 2 s before data acquisition after each mass jump. 50 cycles of the peak jump sequences were taken, each analysis lasting for about 18 min, during which about 1 ml of sample solution was aspirated. In this detector array ^{236}U is measured in a Faraday detector at the same time as ^{239}Pu, ^{240}Pu and ^{242}Pu. The fluctuations in signal intensity caused by plasma instability or ion beam drift will affect uranium and plutonium to the same degree, but by normalising each plutonium isotope to a simultaneously measured ^{236}U ion beam this instability is effectively eliminated. Efficiencies of the Faraday detectors were found to differ by <0.02%, and are thus a negligible source of error. Further details of the measurement array are given in Taylor et al.[1]

2.4 Blank measurement

Blank measurements were made on 0.3 M HNO_3 on each of the peaks and for each of the collectors in an identical array to the dynamic sequence used for sample measurement. The analysis time for the blank was set to be 20% of the sample analysis time. Blank peak intensities were subtracted from the matching sample peak intensity array. Errors in the blank measurement were <3% 2 SE for ^{239}Pu at count levels similar to those in Table 2. The errors on the blank measurements indicate that the limit of detection (3 s error on the blank) for ^{239}Pu is <4 fgml^{-1}, ^{240}Pu is <3 fg ml^{-1} and ^{242}Pu is <3 fg ml^{-1}.

2.5 Measurement test solutions

During the development of this technique, a plutonium test solution having a similar matrix to environmental samples was purified from 30 g of a marine sediment from the Irish Sea (Site 112) by acid leaching followed by 5 stages of anion-exchange chromatography. This sample has a ^{240}Pu/ ^{239}Pu ratio of about 0.22, which is close to environmental ratios expected from weapons fallout (about 0.18) and spent reactor fuel (about 0.23). A 120 ml 'mother' solution containing 70 ng of Pu was diluted with 0.3 M HNO_3 into fractions containing a range of concentrations between 100 fg ml^{-1} and the mother dilution of 580 pg ml^{-1}. The uranium double spike was added to each solution to provide a 1 ng ml^{-1} concentration of the ^{233}U and ^{236}U isotopes.

3 RESULTS AND DISCUSSION

3.1 Precision and accuracy

The high precision and reproducibility of the measurements are demonstrated by the data obtained from multiple measurements of the test sample solution (contaminated Irish Sea sediment; Site 112) (Table 2).

TABLE 2 *Measurement reproducibilities for $^{240}Pu/^{239}Pu$ atom ratios*[1]

Pu concentration	Pu Activity Equivalent	$^{240}Pu/^{239}Pu$ Atom ratio	+/- 2se	2 sd %	Number of measurements
100 fg ml^{-1}	0.3 mBq ml^{-1}	0.2265	0.0031	1.36	12
500 fg ml^{-1}	1.5 mBq ml^{-1}	0.2263	0.0021	0.94	13
1 – 5 pg ml^{-1}	3 - 15 mBq ml^{-1}	0.2262	0.0010	0.45	22
5-10 pg ml^{-1}	15 - 30 mBq	0.2262	0.0007	0.32	7
100 pg ml^{-1}	0.3 Bq ml^{-1}	0.2262	0.0003	0.11	10

The test sample used was purified from Irish Sea Sediment

The $^{240}Pu/^{239}Pu$ atom ratio is consistent across the range of concentrations with no systematic bias in the mean ratios for each concentration range. To assess the accuracy of the method, an equal atom Pu standard (UK-Pu-5, AEA Technology, UK) and three plutonium reference samples from the US New Brunswick National Laboratory were analysed several times. They were diluted to 10 pg ml^{-1} and spiked with $^{236}U/^{233}U$ at the same level as was used with the Site 112 solutions (Table 3). The UK-Pu-5 standard contains ^{236}U ($^{236}U/^{239}Pu$~0.0018704) so a correction was made to the measured $^{236}U/^{233}U$ accordingly. All measurements were made using the same dynamic measurement sequence as was used for the Site 112 solutions. The good agreements between certified and measured data shown in Table 3 clearly show that the accuracy of the developed method is high.

TABLE 3 *Accuracy for $^{240}Pu/^{239}Pu$ atom ratios* [3,6]

	Certified ^{240}Pu/^{239}Pu Atom ratio	Measured ^{240}Pu/^{239}Pu	
		0.5 ng/ml	5 ng/ml
NBL122	0.1320	0.1318 ± 0.001 (n=4)	0.1321 ± 0.0001 (n=3)
NBL 126	0.0209	0.0211 (n=1)	0.0204 (n=1)
NBL 128	0.0007	-	0.0007 (n=1)
UK-Pu-5	0.9662 ± 0.0011	-	0.9645 ± 0.0013 (n=7)

NBL –US New Brunswick National Laboratory; UK-Pu-5 - AEA Technology

3.2 ^{240}Pu/^{239}Pu in Rothamsted archive samples and an Alpine ice core

The methodology established for measuring plutonium isotope ratios was applied to determining the variation in plutonium content and the ^{240}Pu/^{239}Pu in two sets of environmental samples. The ultimate purpose was to establish a record of ground-level plutonium fallout for north-western Europe. The Rothamsted herbage archive and an Alpine glacier ice core (Dome de Gouter, France) were used as they are chronologically well-constrained [3, 5, 6]. The variations in ^{239,240}Pu specific activity found in both the Rothamsted grass and the Alpine ice (Figure 1) are virtually identical to previously reported fallout records [7, 8]. They reflect the testing of nuclear devices in the atmosphere from 1952 onward (assuming a time-lag of 1 year for deposition from high yield stratospheric tests). The large peak seen for 1963 reflects the final large tests carried out by the USA and the USSR before they ratified the International Test Ban Treaty. The ^{240}Pu/^{239}Pu atom ratio data for the grass and ice core vary in a very similar manner and their good agreement indicates that the record obtained is representative of ground level fallout for northern temperate latitudes. The grass data allow some individual nuclear tests to be identified because the ^{240}Pu/^{239}Pu ratio depends on the parameters of each individual test and is therefore expected to vary with time [9, 10]. The first measurable ratio observed is in grass cut in June 1952 but since the world's first thermonuclear test had not occurred until Oct 1952 (the US 10.4 MT 'Mike' Test [9]), it was clear that another test(s) was responsible. The most probable source of these tests would have been in the Nevada Desert where many low yield tropospheric tests were carried out from 1951 until 1962. Earlier studies[11] have shown that the deposition of fallout in the UK from a Nevada test could occur within 5 days if weather conditions were suitable. It is notable that only an isotope ratio method such as the one used here would be able to permit such an inference to be made. A more detailed discussion of the Rothamsted and Alpine ice core data is given in Warneke et al. [3,6].

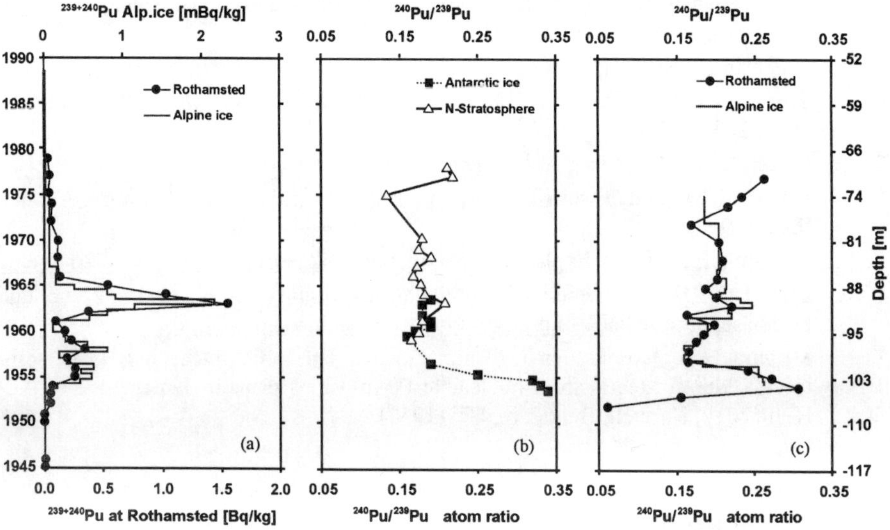

FIGURE 1 $^{240}Pu/^{239}Pu$ (atom ratio) and $^{239,240}Pu$ (Bq/kg) data for Rothamsted grass and an Alpine ice core [6]. The data are compared with records from Antarctic and Northern Stratosphere data [7, 8].

4 CONCLUSIONS

The multicollector ICPMS provides a powerful and rapid means of accurate and precise plutonium analysis. The procedure described involves adding a ^{236}U– ^{233}U double spike after the chemical separation of plutonium to improve reproducibility of the mass spectrometry measurement. The MS measurements involve using a dynamic counting array that includes using a Daly detector and seven Faraday detectors. Plutonium concentrations are measurable as low as 100 fg/ml with a precision of approximately 1% (2σ). This very high sensitivity capability has been used to investigate two chronologically well-constrained series of environmental samples (Rothamsted grass archive and an Alpine ice core). The results have provided a new record of ground-level fallout for northern temperate latitudes and have shown that tropospheric fallout from the Nevada Desert test site reached the UK from 1951 testing.

5 REFERENCES

1 R.N. Taylor,T. Warneke, J.A. Milton, I.W. Croudace, P.E. Warwick and R.W. Nesbitt, , *J. Anal. At. Spectrom.*, 2001, **16,** 279.

2 M. Koide, K.K. Bertine, T.J. Chow, E.D. Goldberg, *Earth and Planet. Sci.Lett.*, 1985, **72,** 1.

3 T. Warneke, High precision isotope ratio measurements of uranium and plutonium in the environment. PhD thesis Unpubl., University of Southampton, UK, 2002.

4 Y Muramatsu, S. Uchida, K. Tagami, S. Yoshida and T. Fujikawa, *J. Anal. At. Spectrom.*, 1999, **14**, 859.

5 C-S. Kim, C-K. Kim, J-I. Lee and K-J Lee, *J. Anal. At. Spectrom.,* 2000, **15**, 247.

6 T. Warneke, I.W. Croudace, P.E.Warwick and R.N. Taylor, *Earth Planet. Sci. Lett.*, 2002, 203,1047-1057.

7 Carter, M.W., Moghissi, A.A., *Health Phys.,*1977, **33**, 55.

8 Eisenbud, M. and Gesell, T. Environmental Radioactivity 4th edition, Academic Press (1997).

9 Lawson J.E., Nuclear Explosion Catalog, www.okgeosurvey1.gov/level2/nuke.cat.

10 US - DOE Environmental Measurements Laboratory, Stratospheric Radionuclide Database, http://cdiac.esd.ornl.gov/by_new/bysubjec.html#atmospheric.

11 Stewart, N.G., Crooks, R.N., Fisher, E.M.R., The radiological dose to persons in the UK due to debris from nuclear test explosions prior to January 1956. AERE HP/R 2017, Harwell, Berkshire, UK (1957).

MEASUREMENT OF GROSS ALPHA AND GROSS BETA ACTIVITY IN SEDIMENT: AN ANALYSTS' INTERCOMPARISON

J. Toole

Harwell Scientifics Ltd, 551 Harwell, Didcot, Oxfordshire OX11 0TD

1 INTRODUCTION

Over the last 21 years, the UK Analysts Informal Working Group has been meeting twice per year in members' laboratories to discuss various aspects relating to the measurement of radioactivity in environmental samples. Occasionally, the members have organised intercomparison exercises covering radiochemical analysis for radionuclides that are of topical interest or are particularly difficult or involve complex procedures. Some of these have been published previously.[1,2,3,4]

Towards the end of 1998, the Group decided that an intercomparison exercise on the measurement of gross radioactivity in a solid environmental sample would be informative for two reasons: firstly, such an exercise had not been carried out before and, secondly, it was a measurement that was frequently requested by clients.

The preparation of samples and measurement of gross activity in sediment or soil is a relatively straightforward and rapid process. It is therefore substantially cheaper than radionuclide-specific methods that are designed to separate, concentrate and purify the target radioelement(s). This speed and the low cost make it an attractive measurement for clients who may be generating large numbers of samples from e.g. contaminated land or site surveillance programmes.

The laboratories themselves, and indeed many of the clients, recognise that such gross activity measurements are only ***screening*** methods that can highlight areas which are contaminated above the expected background activity level, assuming this is known. However, it is also the experience of the laboratories that some clients can use the gross activity results in a way which is not justified by the analysis report. This can be as a result of not taking due account of reported uncertainties, by not taking account of the relevant background signal (not providing an appropriate background sample) or by not appreciating that the result is dependent upon the calibration radionuclides used by the contracted laboratory (the laboratory should state which these are).

Unlike the situation for water samples,[5] there is no standard UK procedure for measuring gross alpha and beta activities in sediment. Thus, since the laboratories participating were aware that a variety of sample preparation and counting procedures would be used, they recognised that a wide range of results might be observed. If true, this would be a useful demonstration of the *relatively* crude nature of this measurement as compared to, for example, the measurement of individual alpha-emitting radionuclides such as uranium or plutonium.

As usual, and consistent with other intercomparison exercises and proficiency testing schemes,[6] the participating laboratories have agreed at the outset that the measurement results reported here should be in an anonymous form.

2 CHOICE OF MATERIAL FOR THE INTERCOMPARISON

A sedimentary material can be prepared specifically for gross alpha and beta measurement by spiking an uncontaminated environmental sample with single or multiple alpha and beta-emitting radionuclides, followed by extensive work to homogenise and validate the homogeneity of the material. Alternatively, an already-prepared natural matrix reference material, whose radionuclide composition was reasonably well known, could be used.

We chose the latter option since this would be more efficient in terms of time and effort (the Group has no source of funds).

The raw material was an intertidal sediment collected in 1989 from the north-west coast of England by the Laboratory of the Government Chemist (LGC) as part of a contract from the Department of the Environment. It was prepared originally as one of four Natural Matrix Reference Materials selected to meet the requirements of the user community in the early 1990's.[7] At LGC, the collected sediment was freeze-dried, ground and finally sieved through a 150 μm aperture sieve. This material was then mixed, quartered and remixed before being subdivided into 50 gram lots sealed in plastic containers. LGC distributed samples to a number of international laboratories initially for measurement of gamma-emitting nuclides.

The sediment contains a wide range of fission and activation products originating from Sellafield's authorised discharges and the available radionuclide data have been compiled and published by the National Physical Laboratory (NPL).[8] NPL concluded that the international exercise to characterise the sediment had been successful and the activity concentrations of 21 radionuclides were certified.

3 DISTRIBUTION OF SAMPLES

The organising laboratory (AEA Technology, now Harwell Scientifics) distributed sub-aliquots of one of the LGC sample pots to six other laboratories in August 1998. As far as possible it was the aim to have all ten participating laboratories measure the gross activities on material from the same pot. However, there was not enough material in one pot so two laboratories had to use material from one of their own pots while another had to be provided with material from a second pot held by AEA Technology.

4 LABORATORY RESULTS

Ten laboratories agreed to measure the LGC intercomparison sediment. All ten labs returned gross α/β data, summarised in Table 1 below. The laboratories were asked to provide replicate analyses following their own sample preparation and sample counting

Table 1 *Gross alpha and gross beta activities from participating laboratories.*

Lab number	Reference nuclides	No. of analyses	UKAS Accredited method	mean gross alpha (Bq/g) [%RSD]	mean gross beta (Bq/g) [%RSD]	Counting system
1	^{241}Am, ^{137}Cs	5 α 5 β	No	3.5 [4.3]	1.8 [17.4]	LB 5500
2	^{242}Pu, ^{137}Cs	10 α 10 β	Yes	5.3 [11]	1.7 [19]	LB 770
3	^{241}Am, ^{137}Cs	2 α 2 β	No	7.8 [5.6]	1.7 [4.9]	LB 770
3	^{241}Am, ^{137}Cs	6 α 6 β	No	4.0 [18]	3.3 [4.2]	LB 770
4	^{239}Pu, ^{40}K	10 α 10 β	Yes	4.2 [3.6]	1.5 [16.8]	αscint 1588-4 GM MX123
5	^{241}Am, ^{40}K	10 α 10 β	No	5.8 [6.4]	1.9 [4.5]	LB 770
6	Unat, ^{137}Cs	1 α 1 β	Yes	1.8	3.1	α scint GM
7	^{241}Am, ^{137}Cs	7 α 7 β	No	1.5 [15.3]	3.2 [41.9]	LB 770
7	^{241}Am ^{3}H/^{14}C	4 α 4 β	No	1.5 [29]	3.5 [19]	1220 LSC
8	^{241}Am, ^{40}K	10 α, 2 β	Yes	3.0 [8.8]	1.7 [14.1]	LB 770
9	^{241}Am, ^{137}Cs	8 α 8 β	Yes	1.5 [17]	1.7 [5.0]	LB 770
10	^{241}Am, ^{36}Cl	1 α 1 β	No	2.82	1.11	1220 LSC

procedures. The mean of the repeat analyses carried out are given in the table as well as the type of counting system used. Participants also agreed to declare if their methodology was UKAS Accredited. The individual laboratory results are also shown in Figures 1 and 2, which show respectively the individual alpha and beta results reported by each laboratory. The numbers linked to each line signify the code number of the laboratory as listed in Table 1. Labs 3 and 7 provided two sets of alpha and beta data.

5 SAMPLE PREPARATION

There were a number of different approaches to the preparation of the counting sources. The quantity of material used to prepare a source ranged from 38 milligrams to 5 grams. Most labs slurried the received sample onto a counting tray or glass fibre filter paper using an alcohol or acetone and then dried the source by radiative heating or left to dry in air. One lab (4) dried the sample at 110°C before sieving, another lab (5) ground the sample to a fine powder; two labs (7 and 10) digested the sample in acid in a microwave oven, mixed

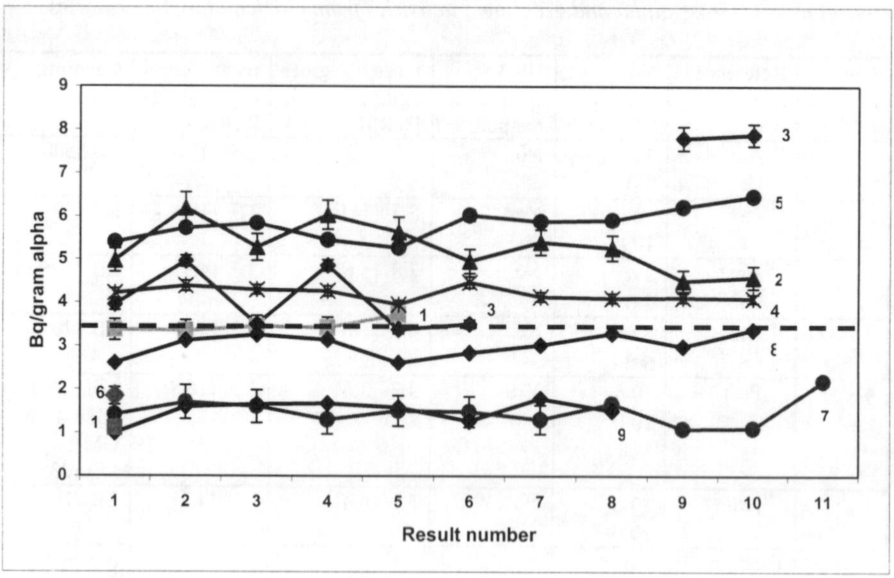

Figure 1. *Gross alpha activity reported by 10 laboratories for LGC sediment. Dotted line is calculated specific activity.*

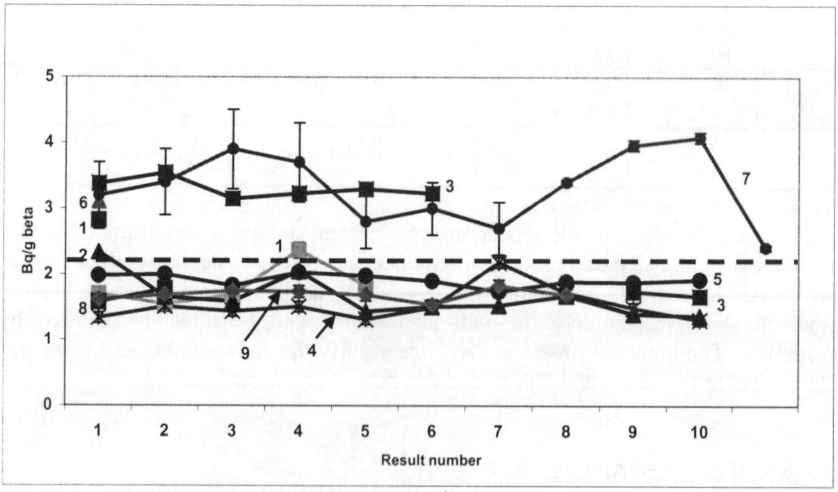

Figure 2. *Gross beta activity reported by 10 laboratories for LGC sediment. Dotted line is calculated specific activity.*

a portion of the solution with scintillation cocktail for LSC counting, use being made of alpha/beta discrimination and windowing out the ^{241}Pu counts. Lab 9 reacted the sample with sulphuric acid prior to drying, ashing and slurrying onto a tray. Based on information provided, it is thought that all labs preparing solid sources have used a source of infinite depth for alpha counting.

6 SAMPLE COUNTING

The most common system used was the Berthold low background LB 770 gas flow proportional counter, used by seven of the ten laboratories. One lab used a Tennelec LB 5500 gas-flow counter, two labs used an alpha scintillator and a Geiger-muller counter (for beta) and two labs used a Quantulus 1220 to count alphas and betas using pulse shape analysis to separate the decay types (and windowed out the Pu241 present at ca. 13 Bq g^{-1}). Counting times ranged from 60 minutes to 1000 minutes. One lab set a minimum count integral of 400.

In the calculation of results, four labs declared the use of in-house software.

7 DISCUSSION OF RESULTS

Using their own calibrated techniques, ten laboratories experienced in radiometry have produced gross alpha activities in sediment which ranged from 1 Bq g^{-1} to 7.9 Bq g^{-1}, i.e. a factor of almost 8. The gross beta results reported ranged from 1.3 Bq g^{-1} to 4.1 Bq g^{-1}, a factor of more than 3.

A degree of result variability is to be expected due to the use of different reference/calibration nuclides which will be detected at slightly different efficiencies. However, even comparing laboratories which have used not only the same reference nuclides but also the same detection system (e.g. labs 3, 7 and 9) reveals gross alpha results which are different by a factor 5 and beta results which differ by a factor of 2. The calibration of the source-detector system is clearly an area where there are difficulties, particularly for gross alpha measurements.

Another potential area of uncertainty is the degree to which radon gas is lost from a sample during manipulations (grinding, slurrying, drying) or in the degree of re-attainment of secular equilibrium thereafter. This is not a large contributor to the differences observed in the present case , as the gross alpha/beta activities are dominated by Pu and Am nuclides in the case of alpha activity, and by ^{137}Cs, ^{90}Sr/^{90}Y and ^{40}K in the case of beta activity.

The sedimentary material which was distributed for the exercise has been reasonably well prepared[7] and characterised for its radioactivity composition[8] with at least 16 of the 21 certified radionuclides showing no evidence of inhomogeneity. It is true that the certification values for individual radionuclides would have been based on gram quantities (10s or 100s of grams for gamma emitters) while sub-gram quantities have typically been used for the gross measurements presented here. However, two points need to be borne in mind.

Firstly, the replicate gross activity results for individual labs which prepared and counted at least 5 sources were internally consistent (repeatability better than 20% in all cases for alpha and beta).

Secondly, in a real situation, radioanalytical laboratories are seldom presented with nice dry, powdered homogenised environmental samples. More often the lab would receive a very inhomogeneous sample e.g. a bag of wet soil or sediment or sludge complete with stones/shells/biota/metal! Thus, the apparently poor between-lab agreement observed here may well be at the optimistic end of the scale. In addition, few clients wish to pay more for sample homogenisation and so a small portion of the inhomogeneous sample is often removed for source preparation, which must intuitively be sub-optimal. Fold in the uncertainties in the sampling procedure (normally not our responsibility) and one can see that large overall uncertainties can easily be generated.

Perhaps the analysts should insist on being provided with only enough sample for a duplicate analysis and for a repeat? Or perhaps the customer(s) should insist on the radioanalytical community using a standard method (there isn't one at the moment, although a draft ISO document has recently been circulated for comment).

Validation of gross activity in solid samples is difficult due to the absence of materials that are demonstrably homogeneous at very small aliquot size and whose gross alpha and beta activities are known. One can come close to the 'true' values by adding together the individual alpha and beta component radionuclides if the predominant ones have been measured with sufficient concensus. When one attempts to do this with the present material – LGC sediment – one still has to make assumptions about equilibrium status in the natural series and assume that no important contributory radionuclides have not been measured. The gross activities thus arrived at are:

Gross alpha ca. 3.4 Bq g^{-1}

Gross beta ca. 2.2 Bq g^{-1}

A method for the measurement of gross alpha and gross beta activities in soil or sediment using liquid scintillation counting (LSC) has been published recently[9]. (Laboratories 7 and 10 in the present exercise have used a similar procedure). While this method was backed up by extensive calibrations of the counting system, potential difficulties remained such as the inability of the microwave acid digestion to bring all of the alpha/beta-emitting nuclides into solution and the need to match closely the degree of quenching produced by the digest solution in the preparation of the calibration quench curves. The requirement for a microwave digestion step increases the time required to obtain a result; thus, where the client has a large sample throughput, rapidly accumulating material in a site remediation programme and needs to know if it will be sentenced as active waste, he may not be able to wait for such procedural steps. On the other hand, he would also want the results produced to be reliable and the results observed above demonstrate that the calibrations are not straightforward.

8 CONCLUSIONS

It can be seen that the provision of gross activity data for soil or sediment is not yet a fully-satisfactory measurement, for a number of reasons. The clients requesting such measurements need to discuss their requirements with the analytical service providers and agree what is the best approach for their particular programme. In addition, the regulatory agencies and waste acceptance organisations need to be satisfied about the interpretation of the results and that they fit the purpose of the programme.

REFERENCES

1 B.R. Harvey and A.K. Young,, *Sci. Tot. Environ.*, 1986, **69**, 13.

2 A.E. Lally, *Sci. Tot. Environ.*, 1986, **69**, 1.
3 D.S. Popplewell and G.J. Ham, *J. Radioanal. Nucl. Chem. Art.*, 1986, **115**, 191.
4 T.H. Bates, *Environ. Int.*, 1988, **14**, 283.
5 DoE Standing Committee of Analysts. Measurements for the Examination of Waters and Associated Materials. Measurement of alpha and beta activity of water and sludge samples, HMSO, London,1986.
6 M.J. Woods, S.M. Jerome, J.C.J. Dean and E.M.E. Perkin, *Appl. Radiat. Isot.,* 1996, **47**, 971.
7 A.S. Holmes, P.R. Houlgate, S. Pang and B. Brookman, Development of natural matrix reference materials for monitoring environmental radioactivity, DoE Report PECD 7/9/447 (1992)
8 I. Adsley *et al*, *Appl. Radiat. Isotop*, 1988 **49**, 1295.
9 R. Wong, W.C. Burnett, S.B. Clark and B.S. Crandall, in: *Environmental Radiochemical Analysis*, ed. G.W.A. Newton, Royal Society of Chemistry, 1998, Special Publication No. 234, 242.

THE RADIOLOGICAL SIGNIFICANCE OF FOODS COLLECTED FROM THE WILD

Norman Green

National Radiological Protection Board, Chilton, Didcot, Oxon OX11 0RQ, UK.

1 INTRODUCTION

The Food Standards Agency operates a monitoring programme to verify that levels of radionuclides present within foodstuffs are acceptable, and to ensure that the resulting public radiation exposure is within internationally accepted limits. The monitoring is independent of similar programmes carried out by nuclear site operators as a condition of their authorisation to discharge radioactive wastes. The results are published annually, together with those from the environmental monitoring carried out on behalf of the Scottish Environment Protection Agency (SEPA), in a report, Radioactivity in Foods and the Environment (RIFE). The latest report gives the results for the year 2001.[1] Within the monitoring programme there is some provision for the collection and evaluation of a limited number of foods growing wild, termed "free foods"; these are presently sampled on an opportunistic basis. Estimates of dose are then based on cautious, generic values for consumption rates since information on the type, availability and utilisation of free foods was scarce.

Two studies have recently been carried out to evaluate the radiological impact of free foods. The first was concerned with artificial radionuclides around four licensed nuclear sites in the south of the UK, and has been completed. The second was concerned with naturally occurring radionuclides in two areas remote from nuclear sites, one area where levels in soil were expected to be typical of the UK and one where levels should be considerably higher. Difficulties were encountered in the sampling of some samples due to the restrictions following the outbreak of Foot and Mouth disease in 2001. Consequently, this second study is still ongoing. Full results will not be available until the end of 2002.

The main objectives of both studies were:

- To identify the types of free food collected, and to determine typical and higher-than-average consumption rates.
- To identify foods that were important from the point of view of both numbers of collectors and quantities collected.
- To collect sufficient quantities of selected foods to enable meaningful radiochemical analyses to be carried out.
- To determine the activity concentrations of the relevant radionuclides in each foodstuff.

- To combine the activity concentrations with consumption rates and published dose coefficients to estimate the resultant doses, and to compare the doses with those estimated from consumption of cultivated foodstuffs.

The full results from the first study have already been published.[2,3] This paper presents the results of both studies in terms of habit data, and the radiological implications from the study around licensed nuclear sites.

2 HABIT SURVEY

2.1 Methodology

The areas chosen for study were based on nuclear sites, in the case of the first survey, and centres of population for the second. The bounds of each area were set about 10 km from the centre, using roads, tracks etc. where possible for ease of delineation. Two of the original sites were nuclear power stations, Hinkley Point in Somerset and Sizewell in Suffolk; the other two were the Atomic Weapons Establishment at Aldermaston in Berkshire and the radiopharmaceutical plant operated by Nycomed Amersham at Cardiff. The sites of interest for naturally-occurring radionuclides chosen for study were Chipping Norton in Oxfordshire, where activity concentrations of naturally-occurring radionuclides are typical of the UK, and Okehampton in Devon, where they are higher than average.

The habit survey was carried out by BMRB International Ltd. Individuals who make use of free foods on a regular basis were identified using local contacts, and interviewed by BMRB researchers. Ordnance Survey maps of scale 1:25000 were used to identify the exact location of collection points and other known areas of availability. Collectors were asked to estimate how much of each foodstuffs would normally be collected; standard sized containers were used to aid estimation, and the volume to mass ratio was measured on receipt of the samples where possible. Data were collected on all foods collected from the wild.

2.2 Results

About 200 regular collectors of free foods were readily located within the approximate 10 km radius around each site, the total being 1,202 people from the six sites. The individual values ranged from 173 (Cardiff) to 232 (Sizewell). Similar numbers of collectors were identified in an earlier study around the Sellafield nuclear reprocessing plant in Cumbria.[4] Taking the two present studies together, different types of free foods were collected. The top ten foods in terms of percentage of collectors are given in Table 1, the remainder were collected by 3% or less. The vast majority (85%) collected blackberries, which are also sampled in the ongoing monitoring programme.[1] For any given food, the percentage of collectors varied widely from site to site. For example cobnuts or hazelnuts were collected by about 6% from around Sizewell but by 36% from around Cardiff.

On average, each collector from around Aldermaston collected 2.2 free foods; the corresponding values from around Cardiff, Chipping Norton, Hinkley Point, Okehampton and Sizewell were 2.8, 1.8, 2.6, 1.7 and 2.6 respectively, with an overall average of 2.4. When considering overall doses from ingestion of cultivated foodstuffs, it is commonly assumed that an individual would consume no more than two foods at higher than average rates,[1,5] which in reference 1 are based on the 97.5th percentile value. In the first study, it was found that, taking all four sites together, 50 people collected more than one free food at higher than average rates. Of these, 33 collected two foods, 13 collected three foods,

Table 1 *Top ten free foods collected from around all six sites*

Foodstuff	*% of collectors*
Blackberry	85.4
Mushroom	34.1
Sloe	17.6
Cobnut/hazelnut	13.6
Chestnut	10.2
Elderberry	9.3
Elderflower	6.7
Crab apple	6.0
Rabbit	4.4
Bilberry	3.9

three collected four foods, and one collected five. Estimates of dose in this study were therefore based on the assumption that three most important contributors were consumed at higher than average rates. In practice, however, one food generally dominated the estimated doses, so any contribution from other foodstuffs would be less significant. In the second study, however, only eight people collected two foodstuffs at higher than average rates, and there was just one who collected three. It is anticipated, therefore, that estimated doses from naturally-occurring radionuclides will be based on the two most important contributors being consumed at higher than average rates. In addition, doses reported in the ongoing FSA programme are calculated using the cautious assumption that cultivated foodstuffs are consumed locally. In the case of free foods, this assumption is reasonable.

Consumption rates were estimated for all individual sites on the basis of the questionnaire in terms of fresh mass. The average and higher than average consumption rates (97.5th percentile) were used in the calculation of doses. The average rate was simply the total mass collected divided by the number of collectors, omitting those who did not specify how much they collected. The 97.5th percentile was the quantity collected by the 97.5th per cent collector. In reality, in these studies, the 97.5th percentile was only appropriate for collectors of blackberries; for foods collected by 20 or more people, the 97.5th percentile was taken as the next to highest consumer, and for most of the samples measured, the maximum rate was used.

It is important to emphasise that the application of the habit data elicited here need not be restricted to radiological assessments. Such data could also be used in studies for other potential contaminants where the impact of the use of free foods is of interest.

3 SAMPLE COLLECTION AND PREPARATION

Foodstuffs to be collected for measurement were chosen using two main criteria: those collected by most people and food collected in large quantities, regardless of how many people collect. In other cases, especially for the second study, some foods were chosen because of specific interest in that food. For instance, it is known that uptake of ^{137}Cs by mushrooms depends on the type of mushroom, with mycorrhizal mushrooms having uptakes between 1 and 2 orders of magnitude higher than saprophytic or parasitic species.[6] A wide selection of mushrooms was therefore collected during the second study in case this effect was also of importance for naturally occurring radionuclides. The samples collected from all six sites are shown in Table 2.

Table 2 *Foodstuffs sampled at each site and their preparation*

Foodstuff	*Aldermaston*	*Cardiff*	*Chipping Norton*	*Hinkley*	*Okehampton*	*Sizewell*
Apple			raw	raw		
Blackberry	raw	raw		raw	raw	raw
Bolete					raw	
Chanterelle		raw				
Chestnut	raw		raw	raw		raw
Cobnut		raw	raw		raw	
Crab apple	raw	raw	raw	jelly	raw	raw
Dandelion		wine				
Elderberry	raw	raw	raw	jam	raw	wine
Elderflower	wine	wine	raw	wine		wine
Field mushroom	raw		raw		raw	
Honey					raw	
Horse mushroom	raw				raw	
Nettle	raw	raw	raw	raw	raw	raw
Parasol		raw			raw	
Puffball	raw	raw			raw	
Rabbit	stew		stew	raw	stew	
Sloe	raw	raw	raw	raw	raw	raw
Watercress				raw		

People willing to collect samples of relevant foodstuffs had been identified in the questionnaire. They were requested to provide samples of about 2 kg of the chosen foodstuffs, the exceptions being mushrooms (0.5 kg) and rabbits (2 animals). In general, the results from the questionnaire indicated that most foods were reported as being eaten raw by at least one collector. For foods that were not eaten raw, the prepared state was requested. Hence, rabbits were stewed, elderflowers were made into wine and some elderberries were made into jam. In the second study, since it was expected that naturally-occurring radionuclides could be present in tap water, a sample of the water used for preparing the rabbit stew was also requested. In addition, a previous study on naturally-occurring radionuclides demonstrated the importance of direct deposition on the activity concentrations of the ^{222}Rn decay products ^{210}Pb and ^{210}Po in above ground crops.[7] For this reason, rainwater collectors were installed in Chipping Norton and Okehampton, and monthly samples of rain were collected during the months in which foodstuffs were sampled.

4 MEASUREMENT AND ANALYSIS

The radionuclides to be determined were agreed with the Food Standards Agency and depended to a large extent on the location being studied. In the initial study around nuclear sites, gamma-ray emitting radionuclides were determined in all samples. Tritium was also measured in all samples, in terms of total tritium in the cases of Aldermaston and Sizewell, and separated into tritiated water (HTO) and organically bound tritium (OBT) for Cardiff and Hinkley Point. Carbon-14 and ^{35}S were measured for all sites except Aldermaston. Because of the nature of the work at Aldermaston and Cardiff, some analyses specific to those sites were also carried out; ^{45}Ca was measured in samples from Cardiff only, and $^{239,240}Pu$, ^{241}Am and U were measured in samples from Aldermaston only. In the case of the study around Chipping Norton and Okehampton, the naturally-occurring radionuclides U, Th, ^{226}Ra, ^{210}Pb and ^{210}Po were measured in all samples with the exception of rainwater, when only ^{210}Pb and ^{210}Po were measured.

On receipt at the laboratory, the samples of foodstuffs were washed as necessary, and aliquots were taken for ^{3}H, ^{14}C, ^{35}S, ^{45}Ca and ^{210}Po analyses as soon as possible. These aliquots were frozen immediately and kept until required. The remainder was separated into edible and non-edible portions where necessary, for example, chestnuts were peeled and apples were cored and sliced. The edible portions were generally packed in a fresh state into suitable containers for measurement by gamma-ray spectrometry. For extra sensitivity, where necessary, samples were dried at 105° Celsius and ground before measurement. Samples used for the determination of actinides, U and Th were ashed at 450° Celsius.

Gamma-ray spectrometry was carried out using hyper-pure germanium detectors housed in a purpose-built, low background facility and suitably calibrated.

For analysis of HTO, an aliquot of fresh sample was heated to 120° Celsius and the water vapour condensed in a cooled test tube. For total ^{3}H the sample was dried and then combusted, and all the water vapour from both operations trapped as before. For samples where OBT content was required, separate aliquots of fresh sample were treated to both these processes, the difference in the results being ascribed to OBT. All tritium measurements were made using a liquid scintillation counter. Carbon-14 analysis was carried out by combustion of an aliquot of fresh sample and absorption of the carbon dioxide produced, before measurement using a liquid scintillation counter.

Measurement of ^{35}S was carried out on a separate aliquot of fresh sample using the procedure described by Evett,[8] modified by Wickendon.[9] Briefly, sulphur is oxidised to sulphate using Benedict's reagent, precipitated as barium sulphate and measured in a liquid scintillation counter. Calcium-45 was analysed using the method described by Humphreys.[10] Briefly, an ashed sample is dissolved in dilute hydrochloric acid, and calcium precipitated as calcium oxalate; this is converted to the perchlorate and dissolved in tri n-butyl phosphate. A solution of diphenyl oxazole in toluene was added, and the activity was measured using a liquid scintillation counter. Activities of both ^{35}S and ^{45}Ca were corrected for radioactive decay to the date of sampling, if known, or the date of receipt, if not.

Actinides, ^{226}Ra and isotopes of uranium and thorium were measured using methods in regular operation at the Board's laboratories. These analyses, together with those for tritium and gamma-ray emitting radionuclides, were carried out within a formal agreement with the United Kingdom Accreditation Service (accreditation numbers 1269 and 1502). Analyses for ^{14}C, ^{35}S and ^{45}Ca, whilst not covered by that agreement, were carried out under comparable schemes of quality assurance.

5 CALCULATION OF DOSES TO CONSUMERS

Calculation of doses to consumers has so far only been completed for the first study.[2]

5.1 Methodology

Consumption rates for most foods collected were calculated from the collection data using conversion factors from volume to mass originally developed by Fulker *et al*[4] and

Table 3 Conversion factors used in consumption data

Start product	*Approximate fresh mass, kg*	*Final product*
1 litre collected food	0.56	
50 g elderflower heads		1 litre wine
250 g elderberries		1 litre wine
1 rabbit	0.25	
1 kg fruit		2 kg jam/jelly

confirmed where possible by measurements on samples received. The data required to convert from the mass or volume collected to litres of wine were taken from recipes from a book on home wine making.[11] In the case of jams or jellies, typical recipes require equal quantities of raw material and sugar by mass,[12] so each kg of jam or jelly represented about 0.5 kg of raw material. The conversion rates used are given in Table 3.

Effective doses were calculated using the dosimetry recommended by the International Commission on Radiological Protection (ICRP) in its Publication 60;[13] for convenience "effective dose" is referred to as "dose" in this paper. The dose coefficients used were those published by ICRP in its Publication 72;[14] only adults were considered. The doses were calculated per kilogramme of each particular raw food to facilitate the identification of the relative contribution from each radionuclide measured. For those foods that were measured in a prepared state, the doses per kg collected took into account the conversion factors from raw to prepared state (Table 2). Where the activity concentration of a particular radionuclide was below the limit of detection, doses were calculated as though the activity concentration was equal to that limit. Although actual doses would therefore be below the reported values, this is the approach taken by the ongoing programme for cultivated foodstuffs operated by the Food Standards Agency[1] to avoid under-reporting of the doses. For each foodstuff sampled, the doses from each individual radionuclide were summed to give an overall dose per kilogramme of food. These values were then multiplied by the consumption rates appropriate to each group of consumers. Typical doses were estimated by assuming that an overall average consumer consumed all the free foods measured at the average rate; a higher than average consumer was assumed to consume the three most significant foods at the 97.5th percentile rate and all the others at average rates. (The second study will only use the two most significant foods at the higher than average rate.) Where foods were collected in different years, it was assumed that activity concentrations would have been the same in each year.

5.2 Radiological implications

The doses to average and higher than average consumers of free food around Aldermaston, Cardiff, Hinkley Point and Sizewell are summarised in Table 4; the foodstuff and radionuclide that contributed most to the dose are also given. For comparison, the corresponding estimates of doses from consumption of cultivated terrestrial foods for the relevant years of sampling [15,16] are also given. It should be stressed, however, that the higher than average consumers of free foods are not necessarily higher than average consumers of cultivated foods, so the doses cannot simply be added together.

Blackberries are the only free food collected on an opportunistic manner in many parts of the country as part of the ongoing Food Standards Agency monitoring programme. The Food Standards Agency programme indicated that blackberries were an important

Table 4 Summary of doses to consumers of free foods, μSv per year

	Aldermaston	*Cardiff*	*Hinkley*	*Sizewell*
Average consumer	2.9	4.0	3.5	1.2
Higher than average consumer	4.9	5.7	5.4	1.6
Terrestrial foods	<10	<15	~12	<10
Dominant foodstuff	puffball	elderflower nettles	elderflower	elderflower
Dominant radionuclide	^{137}Cs	^{14}C	^{14}C	^{14}C

contributor to the overall dose from the consumption of terrestrial foodstuffs around Cardiff and Hinkley Point. However, they were not the dominant contributor to dose in any of the areas in the first study (Table 4). Elderflowers appeared to be an important contributor to dose in most areas of the first study, although collected by only about 7% of people utilising free foods (Table 1). Elderberries were also important contributors to dose around Cardiff and Hinkley Point, with about 10% collecting them (Table 1). Elder trees are reasonably widespread and collection of elderflowers and/or elderberries could therefore be used in most areas as additional or alternative indicators to blackberries.

The estimated doses were low compared with the principal annual dose limit for members of the public of 1000 μSv,[13] and not of radiological importance. However, for at least three of the sites in the first study, the estimated doses were comparable with those implied from consumption of cultivated foodstuffs. In addition, the habit surveys indicated that a wide range of free foods are regularly collected by substantial numbers of people in all areas. The implications are, therefore, that when rigorous assessments of doses from the foodchain are required, then it would be prudent to take the collection of free foods into account. This could also be the case for other forms of contamination in the environment. However, in the case of nuclear sites, if cultivated foods are produced around the installation and if the doses from such foodstuffs are low, then comprehensive surveys of the type conducted in the present studies are not warranted.

6 CONCLUSIONS

Habit surveys have been conducted around four nuclear sites and in two areas remote from nuclear sites. A substantial number of people could be readily identified in each of the locations surveyed who make use of foodstuffs collected from the wild (free foods). Collectors were asked details of quantities collected, so that estimates of average and higher than average consumption rates could be derived. Radionuclides of interest were identified based on discharge data and radiological importance, and activity concentrations of these radionuclides were determined in samples of important foodstuffs. For the four licensed nuclear sites, the consumption rate data, the activity concentrations and the relevant dose coefficients were combined to estimate doses to average and higher than average consumers. The doses were comparable with those reported in the ongoing monitoring programme for cultivated foods. Continual comprehensive surveys of such foods are not warranted, although improvements to the current opportunistic sampling of free foods could be made. Results for the second study involving naturally occurring radionuclides will be published in due course.

The collection of foods from the wild seems to be widespread, and so when more rigorous assessments of doses from the foodchain are required, then contributions from free foods should be taken into account.

7 ACKNOWLEDGEMENTS

The first study was funded by MAFF Radiological Safety and Nutrition Division, the second by the Food Standards Agency. Thanks are extended to the enthusiastic interviewers, who endeavoured to identify so many collectors of free foods. We are also indebted to those who collected the samples, without whose diligence the work could not have been completed.

References

1 Food Standards Agency, Scottish Environment Protection Agency. Radioactivity in Food and the Environment, 2001. RIFE-7. Food Standards Agency, London, 2002.
2 N. Green, D.J. Hammond, M.F. Davidson, B.T. Wilkins, S. Richmond and S. Brooker. Evaluation of the Radiological Impact of Free Foods found in the Vicinity of Nuclear Sites. NRPB-M1018, NRPB, Chilton, 1999.
3 N. Green, D.J. Hammond, M.F. Davidson, B.T. Wilkins, S. Richmond and S. Brooker. *Rad. Prot. Dosim.*, 2001, **93**, 67.
4 M.J. Fulker, D. Jackson, D.R.P. Leonard, K. McKay and C. John. *J. Radiol. Prot.*, 1998, **18**, 1.
5 C.A. Robinson. Generalised Habit Data for Radiological Assessments. NRPB-M636, NRPB, Chilton, 1996.
6 C.L. Barnett, N.A Beresford, P.L. Self, B.J. Howard, J.C. Frankland, M.J. Fulker, B.A. Dodd, and J.V.R. Marriott. *Sci. Tot. Environ.*, 1999, **231**, 67.
7 G.J. Ham, L.W Ewers and B.T. Wilkins. Variations in Concentrations of Naturally Occurring Radionuclides in Foodstuffs. NRPB-M892, NRPB, Chilton, 1998.
8 T.W. Evett. Analytical Procedure for the Determination of Sulphur-35 in Milk , Herbage and Tacky-Shade Samples. Central Electricity Generating Board, SSD/SE/RN/20/77, 1977.
9 D.A. Wickendon. Development of a Sensitive Method for the Determination of sulphur-35 for a Food Surveillance Programme. Contract Report 1B023, MAFF, 1995.
10 E.R. Humphreys. *Int. J. Appl. Rad. Isotopes*, 1965, **16**, 345.
11 C.J.J. Berry. *First Steps in Winemaking*, 9th Edn., Nexus Special Interests, Hemel Hempstead, 1996.
12 *Good Housekeeping Book of Preserving*, ed. H. Southall, Ebury Press, 1991.
13 International Commission on Radiological Protection (ICRP). ICRP Publication 60, *Ann. ICRP*, 1991, **21**.
14 International Commission on Radiological Protection (ICRP). ICRP Publication 72, *Ann. ICRP*, 1996, **26**.
15 Ministry of Agriculture, Fisheries and Food, Scottish Environment Protection Agency. Radioactivity in Food and the Environment, 1996. RIFE-2. Food Standards Agency, London, 1997.
16 Ministry of Agriculture, Fisheries and Food, Scottish Environment Protection Agency. Radioactivity in Food and the Environment, 1997. RIFE-3. Food Standards Agency, London, 1998.

NATURAL RADIONUCLIDES MEASUREMENTS IN DRINKING WATER BY LIQUID SCINTILLATION COUNTING. METHODS AND RESULTS

Maurizio Forte,[1] Rosella Rusconi,[1] Elisabetta Di Caprio,[1] Silvia Bellinzona[2] and Giuseppe Sgorbati[3]

[1]ARPA Lombardia, Dipartimento sub-provinciale Città di Milano, Sezione Radioprotezione - Via Juvara 22, 20129 Milano (Italy)
[2] Università degli Studi di Milano, Facoltà di Fisica – Via Celoria 16, 20133 Milano (Italy)
[3]ARPA Lombardia, Settore Agenti Fisici – Via Restelli 1, 20124 Milano (Italy)

1 INTRODUCTION

Concern about total radionuclides content in water intended for human consumption has been brought to public attention by the recent Council Directive 98/83/EC,[1] subsequently enforced through an Italian law.[2] Parameter values have been fixed for Tritium content (100 mBq/l) and total indicative dose (0,1 mSv/year): the Directive points out that the total indicative dose must be evaluated excluding Tritium, ^{40}K, ^{14}C, Radon and its decay products, but including all other natural series radionuclides. Maximum concentration values for Radon are separately proposed in Commission Recommendation 2001/928/Euratom.[3]
Tritium determination follows a well established procedure, standardized by International Standard Organization.[4] On the contrary, total indicative dose evaluation requires more specific and cumbersome procedures for the measurement of radioactivity content, with special regard to natural series radionuclides. The large number of possibly involved radionuclides and the good sensitivities required make the application of traditional analytical techniques unsuitable in view of a large scale monitoring program.
World Health Organization guidelines for drinking water suggest performing an indirect evaluation of committed dose by measuring alpha and beta gross radioactivity and checking compliance to derived limit values;[5] the proposed limit values are 0,1 Bq/l for gross alpha and 1 Bq/l for gross beta radioactivity. Nevertheless, it is desirable to identify single radionuclides contribution to alpha and beta activity in order to perform more accurate measurements of committed dose.
Ultra-low level liquid scintillation counting coupled to extractive techniques and alpha-beta discrimination allows rapid and simple determination of all radiometric parameters relevant to dose evaluation, namely gross alpha and beta activity, uranium and radium isotopes content. For tritium and radon determination well established procedures, based on LSC, can also be used.
These techniques were applied to a preliminary monitoring program of tap waters in Lombardia; up to now, total alpha and beta activity and uranium isotope concentration have been measured. A Quantulus-Wallac scintillation counter has been used in this work. Some brands of bottled water were examined too, both for testing methods and because of the widespread use of mineral water by the Italian population. Mineral water brands are not reported here.

2 METHODS AND RESULTS

2.1 Gross Alpha and Beta

Gross alpha and beta activity is usually measured by counting the dry residue of a water sample. In US-EPA and ASTM methods an acidified amount of water is reduced in volume and evaporated to dryness on a steel planchet.[6, 7, 8] In ISO method 9696 and 9697 the residue is first sulphated by addition of sulphuric acid; a fixed amount of dry salts is then evenly dispersed on a steel planchet and counted by a proportional counter or other suitable counter (zinc sulfide scintillation counter for α emissions, plastic scintillation counter for β emissions). The availability of low-background liquid scintillation counters equipped with alpha-beta discrimination device provides an alternative for gross alpha and beta determination.[9, 10, 11, 12] The LSC method offers several advantages over the traditional procedure: 1) simultaneous alpha and beta measurement through alpha-beta discrimination technique - reduced counting times; 2) high (close to 100%) and rather constant detection efficiency for alpha emitters and for high energy beta emitters; 3) faster and more reproducible sample preparation; 4) spectral energy response through inspection of emission spectra.

It is possible, in principle, to verify compliance with WHO recommended values for alpha and beta activity content in water by ultra low level liquid scintillation counting without any previous treatment. Water is added to the scintillation cocktail in a proper amount, generally in a 8:12 ratio and counted for the time necessary to achieve desired sensitivity (1000 minutes): LLD of 80 mBq/l and 250 mBq/l for α and β activity respectively can be attained. A key point is the proper setting of the α/β discrimination parameter based on pulse shape analysis (PSA).[11]

Better sensitivities and reduced counting time can be achieved by sample preconcentration; both freeze drying technique and evaporation by heating have been used.[10, 12]

In the present work water samples were acidified (to avoid losses due to precipitations, polymerizations, colloid formations) and preconcentrated by slow evaporation on hot plate. 15 M bidistilled nitric acid was added to a 200 g sample up to pH 2,5 and the volume was reduced ten fold by heating; pH drops to 1,5 and in the same time all the dissolved radon is desorbed. Finally 8 g of the concentrated sample is transferred in the scintillation vial and 12 ml of Optiphase Hisafe 3 (Wallac) cocktail is added. No quenching effect of nitric acid was observed.

Detection efficiency was evaluated by measuring degassed pH 1,5 nitric solutions traced with ^{241}Am and $^{90}Sr/^{90}Y$ with activity concentrations similar to those of real samples. The alpha beta discrimination parameter (PSA) was set using the same standards: measurements were repeated increasing PSA value by 5 each time; optimum PSA value was found, corresponding to minimum α and β interference.

Alpha interference is the fraction of counts observed in the beta window with respect to the counts observed in alpha and beta windows when a pure alpha emitter is measured; the beta interference is the fraction of counts observed in the alpha window with respect to the counts observed in alpha and beta windows when a pure beta emitter is measured.

Since alpha and beta interference depends on sample quenching, interference curves were evaluated at different quenching values obtained by adding increasing amounts of CCl_4 to traced samples; quenching variation in real samples, however, was limited and had no influence on alpha-beta discrimination parameter setting.

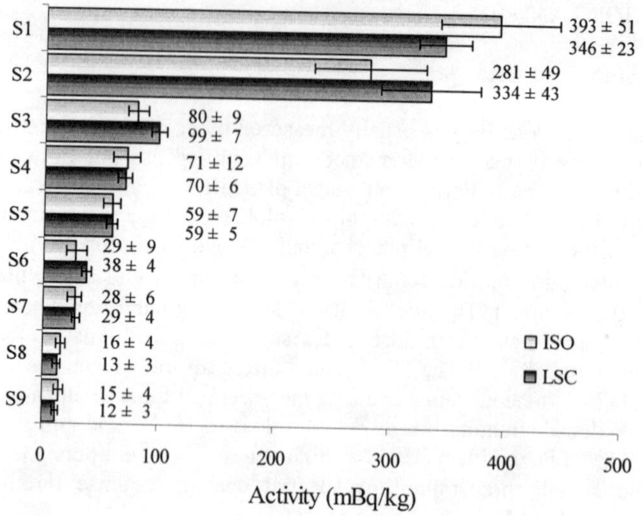

Figure 1 Gross alpha activities comparison

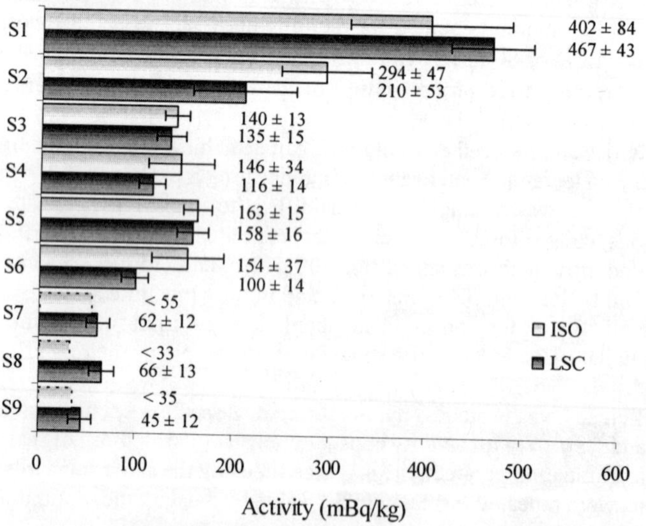

Figure 2 *Gross beta activities comparison*

LSC method was tested by comparing its outcomes with ISO procedure results for 9 water samples, 8 from bottled mineral waters and 1 from Milano tap water (S4) (Figures 1 and 2): a good agreement can be observed.

Repeatability was tested in a ten fold replication experiment; alpha measurement repeatability, as expressed by values distribution width, resulted to be 9 %; beta measurement repeatability resulted to be 16 %.

2.2 Uranium

A number of methods have been devised for total or isotopic determinations of uranium in water. Most widely used non-radiometric methods are fluorimetry, X rays fluorescence, ICP atomic emission or ICP mass spectrometry; the last one is growing in importance due to its rapidity, sensitivity and the possibility to perform isotopic composition evaluations. The main limitation is the cost of instruments, especially if extra sensitivity is needed for the more difficult determination of ^{234}U and ^{235}U isotopes besides the more abundant (in mass) ^{238}U.

In water analysis this is a quite crucial problem since isotopic equilibrium between ^{234}U and ^{238}U is generally not attained; uranium isotope disequilibrium can be due to transfer mechanisms from rocks to water and to the less stable position of ^{234}U in the lattice after recoil following alpha decay. The activity ratio $^{234}U/^{238}U$ generally varies between 1 and 1,5 but can reach much higher values, up to 7-8. The evaluation of total uranium activity from ^{238}U concentration can thus lead to underestimate total uranium content.

Radiometric methods, like semiconductor alpha spectrometry on electrodeposited samples, allow accurate determinations of all isotope concentrations thanks to good spectral resolution. They are nevertheless too cumbersome for a wide scale monitoring application.

LSC also offers an attractive option, especially when coupled to direct uranium extraction. Since '80s several researchers exploited the complexing power of phosphor or nitrogen compounds like TOPO (trioctyl-phosphin oxide), HDEHP (bis-2-etilhexyl-ortophosphoric acid) or TNOA (tris-N-octyl amine). These compounds can be added to a non water soluble scintillation cocktail giving an "extractive cocktail". By simply shaking the extractive cocktails with the water sample, the uranium moves into the organic phase. After phase separation, the extractive cocktail is ready to be counted.

Other actinides like thorium, plutonium and americium can be co-extracted; if necessary, complexing agents can be added to the water sample to suppress such interferences.[13]

Liquid scintillation counting coupled to selective uranium extraction was used in this work to assess uranium content in waters. Experimental conditions were optimized with regard to the cocktail selection, pretreatment and counting procedure. Extractive performances of four different scintillation cocktail were compared, two of them (C1 and C2) respectively used by others researchers[14, 15], the third realized in our laboratory (C3), the fourth (C4) prepared by adding HDEHP to the commercial Wallac cocktail (Optiphase Hisafe 3). HDEHP has always been used as the uranium complexing agent because of its low effect on quenching, especially when added to the scintillation cocktail in small amounts[14]. Except C4, extractive scintillation cocktails were prepared by adding a fluorescent substance (or a mixture of them), naphthalene (to enhance alpha-beta separation) and 5% HDEHP to an aromatic solvent (toluene or xilene) (Table 1).

Extraction yields were evaluated by measuring water samples acidified with nitric acid (0,7 M) and spiked with a known amount of natural uranium; the extraction yield was calculated as percentage of extracted uranium.

Extraction procedures were further investigated; different amounts of sample and cocktail were mixed and extraction conditions were slightly modified. Yields not far from 100 % were obtained by extracting in a separatory funnel (2 minutes shaking) 20 ml of test solution with 20 ml of cocktail. Extraction efficiency drops when using greater test solution volumes. Better results were obtained with a two-step extraction of 100 ml of test solution in 10+10 ml cocktail volume; similar extraction yields were obtained with the four cocktails (Table 2).

Table 1 *Composition of 1 liter extractive cocktail*

Cocktail	C1	C2	C3	C4* (Optiscint)
Solvent	Toluene	Xilene	Toluene	Diisopropil naphtalene
Fluo	PPO 4 g Bis-MSB 0,5 g POPOP 0,05 g	PBBO 4 g	PBBO 4 g	PPO Bis-MSB
Naphtalene	35 g	180 g	35 g	-
HDEHP	50 g	50 g	50 g	50 g

PBBO: *2-(4-biphenylyl)-6-phenyl-benzoxazole;* ***PPO***: *2,5-diphenyloxazole;*
Bis MSB: *1,4-bis(2-methystiryl)benzene;* ***POPOP:*** *1,4-bis(5-phenyloxazol-2-yl)-benzene.*
* *composition registered by Perkin Elmer-Wallac*

Since it is known that dissolved oxygen seriously affects both resolution and α/β discrimination, all samples were degassed after the extraction by sparging them with argon. This procedure could not be applied to the extractive cocktail C4 since gas bubbling caused foaming and subsequent cocktail spillover.

Counting characteristics of the four cocktails are summarized in Table 2. Besides background, optimal discrimination parameter setting (PSA) and alpha resolution, PSA plateau (PSA values range in which α and β interferences are lower than 1%) is also listed.

Resolution was calculated by the Horrocks formula.[16]

The C2 scintillation cocktail, which combines low background, good resolution and discrimination and a wide PSA plateau, was selected as the optimum cocktail.

Performances of glass, polyethylene, teflon and teflon coated polyethylene vials (20 ml) were compared. Glass vials gave poor results both for background and spectral alpha resolution; best results were obtained when using teflon and polyethylene vials. Teflon vials were discarded because of their high cost, while polyethylene vials are permeable to cocktail solvent. The best results were obtained with teflon coated polyethylene vials, which exhibit good resolution, low background and no solvent permeability.

In order to raise the analytical sensitivity, a sample preconcentration method was adopted too. One liter samples were first acidified with 5 ml of HNO_3 14 M in order to avoid uranium losses, then slowly evaporated on an hot plate to 100 ml; the final HNO_3 concentration is 0,7 M. Uranium was finally extracted by the selected procedure.

Table 2 *Extractive cocktails features*

Cocktail	Argon fluxed	Extraction yields	Background α window cpm	PSA	PSA plateau	Resolution (%) ^{234}U	Resolution (%) ^{238}U
C1	no	98,6 +/- 0,5	0,012	90	25	4,7	4,9
	yes	98,6 +/- 0,5	0,010	110	40	3,5	4,8
C2	no	98,2 +/- 0,8	0,026	130	55	3,3	4,9
	yes	98,2 +/- 0,8	0,009	130	70	2,7	3,7
C3	no	98,0 +/- 0,5	0,040	120	50	4,1	4,9
	yes	98,0 +/- 0,5	0,036	130	55	3,2	4,4
C4	no	98,4 +/- 0,8	0,053	130	55	3,6	3,9

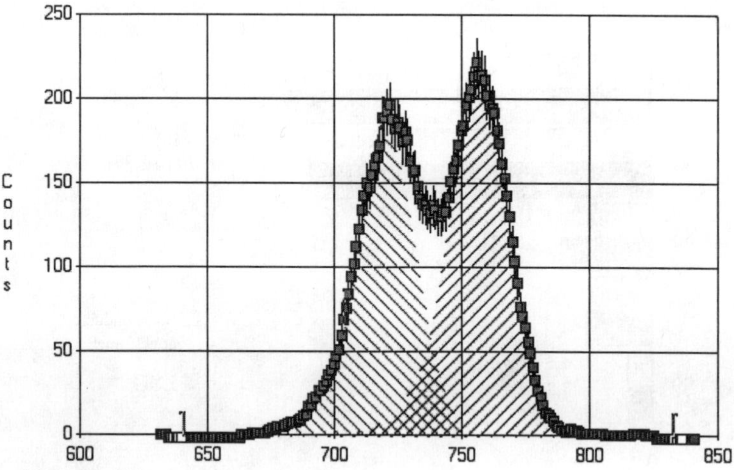

Figure 3: *^{238}U and ^{234}U α–peaks deconvolution*

Uranium measurements were made considering the alpha discriminated spectrum component (channels range 600-800).

^{238}U and ^{234}U content was evaluated applying spectral deconvolution of uranium alpha peaks; Canberra Genie 2k Interactive Peak Fit software was used to this purpose (Figure 3). ^{235}U contribution to total uranium was estimated to be lower than 2.5%, and was neglected when performing alpha spectra deconvolution.

The method was tested, with good results, by comparison with values obtained by two independent methods on some bottled water samples, namely: 1) semiconductor alpha spectrometry on electrodeposited samples; 2) ICP mass spectrometry (^{238}U alone). Results are shown in Figures 4 and 5.

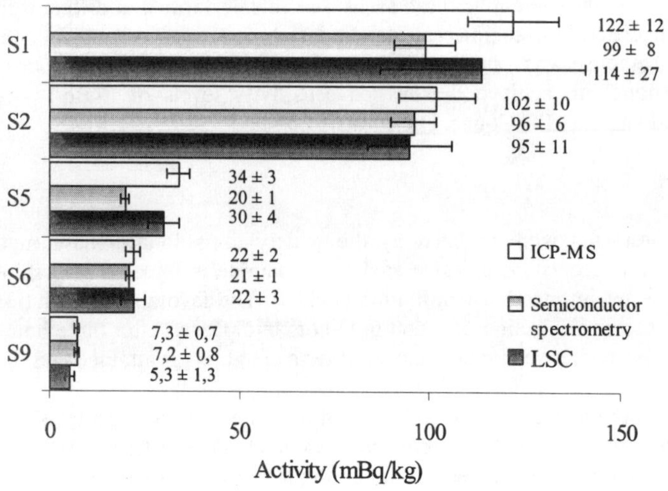

Figure 4: *^{238}U activities comparison*

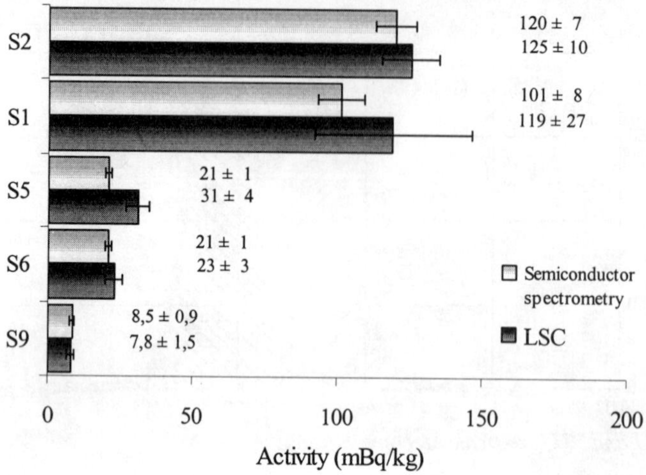

Figure 5: *^{234}U activities comparison*

2.3 Tritium

ISO 9698 method was used for tritium measurement. Water samples were distilled with a Vigreux apparatus in the presence of sodium carbonate and sodium tiosulphate. 8 ml of distilled water were then transferred in a teflon coated polyethylene vial, mixed to 12 ml of scintillation cocktail (Optiphase Hisafe 3 – Wallac) and stored for one day in the liquid scintillation counter sample holder to allow full decay of chemi-luminescence and photo-luminescence.
Detection efficiency was determined by measuring tritium standards (tritiated fructose pellets – Wallac) of different activity. Measurements performed at different times showed no change in efficiency, so internal standard technique was not used. The stability of instrumental response was checked in general periodic controls. One hour counting was considered enough to achieve the desired sensitivity levels of 5 Bq/l vs. a 100 Bq/l recommended value. (EEC Counc. Dir. 98/93/EC).

2.4 Radon

Radon measurements were realized by the widely used double-phase method.[17, 18, 19] According to this procedure an unaerated water sample is injected in a scintillation vial containing a water-immiscible scintillation cocktail. The favorable distribution coefficient causes the selective absorption of radon in the organic phase. After three hours the secular isotopic equilibrium between radon and short term daughters is attained and the sample can be counted.
Since in our experimental conditions it was not possible to prepare samples for scintillation counting straight on the field, we collected water in glass bottles.[20] A plastic tube was attached to the faucet and inserted in the bottle; a slow flux of water was maintained for about 10 minutes till no bubbles were present in the tube, leaving the water to spill over the bottle top. The bottle was tightly sealed, carried to the laboratory and analyzed within 24 hours.

Table 3 *Radon cocktails features*

Scintillation cocktail	Resolution %			Efficiency (%) average	Background cpm
	^{222}Rn	^{218}Po	^{214}Po		
Optifluor O	not resolved		1,58	415 +/- 20	2,62 +/- 0,05
NEF	3,01	2,92	1,71	475 +/- 25	2,25 +/- 0,05
Optiscint	0,98	0,96	0,44	430 +/- 20	2,26 +/- 0,05

Table 4 *Radon activities in directly measured and transported samples*

Container	Sampling	1° meas. Bq/kg	2° meas. Bq/kg	3° meas. Bq/kg	Mean value Bq/kg	Std. Dev. of mean value
None	immediate	5,82 ± 0,91	5,83 ± 0,92	5,57 ± 0,88	**5,74**	0,17
Glass	24 h	5,13 ± 0,88	4,93 ± 0,86	5,58 ± 0,94	**5,21**	0,38
Polyethylene	24 h	4,97 ± 0,85	4,39 ± 0,78	4,45 ± 0,78	**4,60**	0,37

Once in the laboratory, 10 ml of sample were drawn by a gas-tight syringe and delivered in a vial preloaded with 10 ml of scintillation cocktail and tared with an analytical balance. The exact sample amount was then determined by weighing.

Previous studies showed that teflon coated polyethylene vials were tight enough for radon especially when an organic solvent is present.[21, 22, 23]

Some different lipophilic scintillation cocktails are available; we compared performances of three of them, namely Optifluor O (Packard), Optiscint (Wallac) and NEF 957A (Packard). Results are reported in Table 3.Resolution was calculated by the the Horrocks formula.[16]

NEF cocktail has excellent counting properties (efficiency and background) and gives a quick and sharp phase separation. Optiscint adds some other useful characteristics to good detection features: like all di-isopropyl naphtalene (DIN) based cocktails, it exhibits low permeation through plastic vials, it is virtually odorless and gives little disposal problems because of its biodegradability; for these reasons it has been chosen for the present work.

Overall efficiency was calculated by measuring a vial filled with 10 ml of a ^{226}Ra standard solution and lipophilic cocktail after the radon buildup was complete (20 days) (ASTM 5072-98). Efficiency was evaluated as the average value of four different standards with activity ranging from 0,02 to 1 Bq.

It is well known that a main source of inaccuracy comes from radon leakage that may occur when sample transportation is involved[20, 24]. Analytical results of direct sampling and vial filling were compared to those obtained by collecting water in glass or plastic bottles, transporting them by car and analyzing them after 24 hours.

Results in Table 4 show an appreciable radon leakage in polyethylene bottle while the decrease of radon concentration in the glass bottle is smaller than experimental error.

2.5 Radium-226 and Radium-228

Radium isotopes determination is important because of their radiotoxicity and subsequently high contribution to committed dose.

Usually the most diffused isotope is alpha emitter ^{226}Ra, but relevant amounts of beta emitter ^{228}Ra and, sometimes, of alpha emitter ^{224}Ra can also be present.

LSC based techniques can be successfully applied in this field. Two strategies are possible: 1) radium purification and homogeneous phase counting; 2) sample preconcentration and heterogeneous phase counting.

The first approach allows in principle the measurement of all radium isotopes.
Radium purification can be accomplished with different methods: barium sulphate coprecipitation, selective scintillation cocktail extraction, filtration on selective membranes.[25, 26, 27, 28, 29, 30, 31]
The second method consists in an indirect measurement of ^{226}Ra alone through its short term decay products. As in radon measurement (see 2.4), the water sample is transferred to a scintillation vial preloaded with a water immiscible cocktail. ^{222}Rn produced by ^{226}Ra is absorbed in the organic phase and counted after the isotopic equilibrium is attained. The water sample can be preconcentrated by various techniques or analyzed without any previous treatment.[32, 33, 34, 35]

For ^{226}Ra measurement, in order to minimize pretreatment time the indirect method based on radon measurement was preferred for our routine controls.
An aliquot of the 10 fold preconcentrated water sample, previously prepared for gross alpha and beta measurement, was used for this purpose: 10 ml of preconcentrated sample were transferred in a teflon coated polyethylene vial and 10 ml of water immiscible cocktail Optiscint (Wallac) was added. The sample was measured after 21 days ingrowth without shaking the vial: radon diffusion was demonstrated to be quick enough to cause no differences between shaked and non shaked samples (< 3%). All vials were kept at constant temperature in the counter sample holder for the whole ingrowth period.
The method was tested on two water samples only. All other bottled water samples, previously measured by emanometry,[36] showed a radium content lower than minimum detectable activity (Table 5).

A preliminary work has been done to identify a method suitable for the contemporary measurement of ^{226}Ra and ^{228}Ra.
Previous studies showed that most Lombardia waters exhibit ^{226}Ra concentrations between 0,2 and 10 mBq/l.[36] Routine controls in Milano laboratory by γ spectrometry on resin concentrated 200 l tap water samples display ^{228}Ra medium concentrations of 1,2 mBq/l.
To increase the sensitivity of LSC analysis to a sufficient extent, preconcentration of at least 2 liters samples is necessary. To this purpose three different pretreatment methods have been compared:

a. Radium absorption by lead rhodizonate supported on charcoal;[38]
b. Filtration on Radium Rad Empore disks;[30, 31]
c. Selective absorption on cationic resins from a pH 5,5 EDTA (ethylen diammino tetra acetic acid) solution.

First tests gave high yields (near 100%) for a 2 liters spiked water preconcentration by using the three methods. The first two are not selective towards lead, so ^{210}Pb is observed in concentrated samples. The third one allowed isolation of radium isotopes only after pH 10 EDTA elution.
Concentrated samples were then purified by chromatography on a small cationic resin like in method c. and counted by LSC. α/β discrimination was also applied to achieve simultaneous ^{226}Ra and ^{228}Ra determination.

Table 5 *^{226}Ra activities comparison*

Water sample	LSC mBq/l	Emanometry mBq/l
Mineral water S1	188 ± 24	200 ± 40
Mineral water S2	104 ± 16	140 ± 28

Table 6 *Test methods performances*; * *combined efficiency (extraction + counting)*

	Sample volume (g)	**Meas. time** (min)	**Measure window** (channels)	**Spectrum**	**Background** (cpm)	**Efficiency** (%)	**LLD** (mBq/kg)
Gross α	80	1000	500-1000	α	0,099 ± 0,007	111 ± 2	**8**
Gross β	80	1000	500-1000	β	0,960 ± 0,020	73 ± 1	**24**
Total U	1000	1000	600-800	α	0,087 ± 0,004	98 ± 5 *	**0,4**
Tritium	8	60	1-250	α+β	1,8 ± 0,1	25 ± 1	**5000**
^{222}Rn	10	60	100-1000	α+β	2,5 ± 0,1	428 ± 20	**250**
^{226}Ra	100	1000	100-1000	α+β	2,5 ± 0,1	428 ± 20	**14**

2.6 Test methods summary

Table 6 resumes the main features of the above described test methods. Teflon coated polyethylene vials were always used.

Background samples were prepared in the following way:

- Gross α/β: 8 ml of radon free HNO_3 solution (pH 1.5) + 12 ml of Optiphase Hisafe 3
- Uranium: 20 ml of argon fluxed C2 cocktail
- Tritium: 8 ml of distilled dead water + 12 ml of Optiphase Hisafe 3
- ^{222}Rn and ^{226}Ra: 10 ml of boiled ultrapure (MilliQ) water + 10 ml Optiscint

Measurement uncertainty was evaluated according to ISO Guide to the Expression of Uncertainty in Measurement.[37] Uncertainty on sample amounts and on calibration standard activity as well as counting uncertainty were considered in the evaluation of the combined standard uncertainty; uncertainty on measurement results is always expressed in terms of expanded uncertainty (obtained by multiplying the combined standard uncertainty by a coverage factor k=2).

2.7 Applications to Environmental Samples

We used the above described methods to carry out radioactivity measurements of different water samples: bottled, surface and tap waters.

Only some representative examples are reported in this work (Figures 6-7 and Tables 7-8). Tap water samples were drawn in 13 of the largest Lombardia centers; gross alpha and beta activity and uranium isotope concentration were measured. Potassium chemical analysis was performed by ionic chromatography, ^{40}K activity was calculated taking into account its natural abundance (30,3 mBq per mg of K).

Southern Lombardia cities (Pavia, Cremona, Mantova) exhibited the lowest radioactivity concentrations, while northern (Sondrio, Lecco, Varese, Como) and north-eastern ones (Brescia, Bergamo) displayed medium-low levels. Higher values were found in Milano and surrounding areas (Parabiago, Lodi, Monza). Northern Lombardia is an alpine district; previous works showed a relevant dishomogeneity for both tap and bottled water produced in that area (52 samples were analyzed).[36] Thus samples collected in the main city should not be considered representative of the whole district.

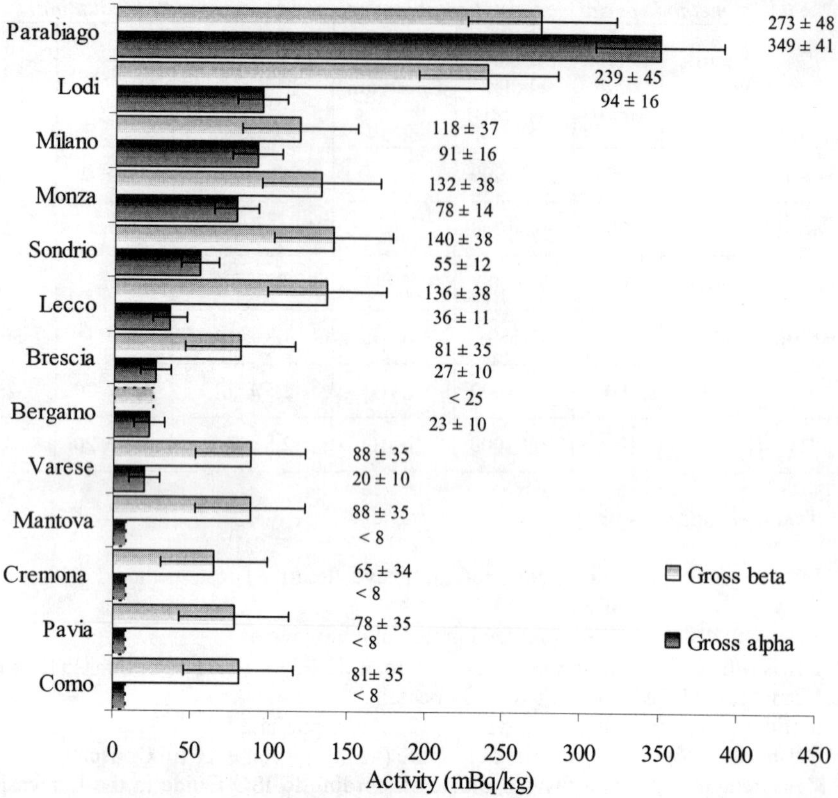

Figure 6: *Tap waters results – Alpha and beta activities*

A more detailed monitoring of waters from Milano and surrounding area is currently underway; a complete chemical and radiometric analysis (gross α and β, uranium, radium, tritium and radon) is being performed on samples drawn directly at wells; preliminary results show remarkable chemical and radiological differences in waters from same area wells; in 7 main wells from a small area, for instance, α activity ranges from 23 to 410 mBq/kg. Differences could be due to the wells depths; further analysis are still in progress.
Results in Figures 6 and 7 show that α activity in tap water is mainly due to uranium isotopes: $^{234}U/^{238}U$ ratio is generally close to 1.
Gross beta activity shows a more limited range of values; a major contribution to beta activity is due to ^{40}K (Table 7), especially in low activity waters. ^{40}K is not to be considered in committed dose evaluation, therefore, alpha activity values are more useful to identify critical situations.
Bottled mineral waters are mainly produced in northern and north-eastern part of Lombardia. Two of them, Mineral Water S1 and S2 exhibited high gross α activities which can not be entirely attributed to uranium isotopes (Table 8). In such cases ^{226}Ra measurements give concentration values consistent with the encountered difference (Table 5). This situation is rather common in water samples drawn in small towns of alpine districts.[36]

Table 7 *Gross beta and Potassium content; * after subtracting ^{40}K activity*

Tap water sample (departement)	**Gross β** mBq/kg	**^{40}K** mBq/kg	**Residual Gross β*** mBq/kg
Parabiago	273 ± 48	46 ± 2	227 ± 48
Lodi	239 ± 45	100 ± 5	139 ± 45
Milano	118 ± 37	46 ± 2	72 ± 37
Monza	132 ± 38	46 ± 2	86 ± 38
Sondrio	140 ± 38	76 ± 4	64 ± 38
Lecco	136 ± 38	100 ± 5	36 ± 38
Brescia	81 ± 35	30 ± 2	51 ± 35
Bergamo	< 25	7,0 ± 0,3	-
Varese	88 ± 35	61 ± 3	27 ± 35
Como	81 ± 35	42 ± 2	39 ± 35
Pavia	78 ± 35	48 ± 2	30 ± 35
Cremona	65 ± 34	30 ± 2	35 ± 34
Mantova	88 ± 35	64 ± 3	24 ± 35

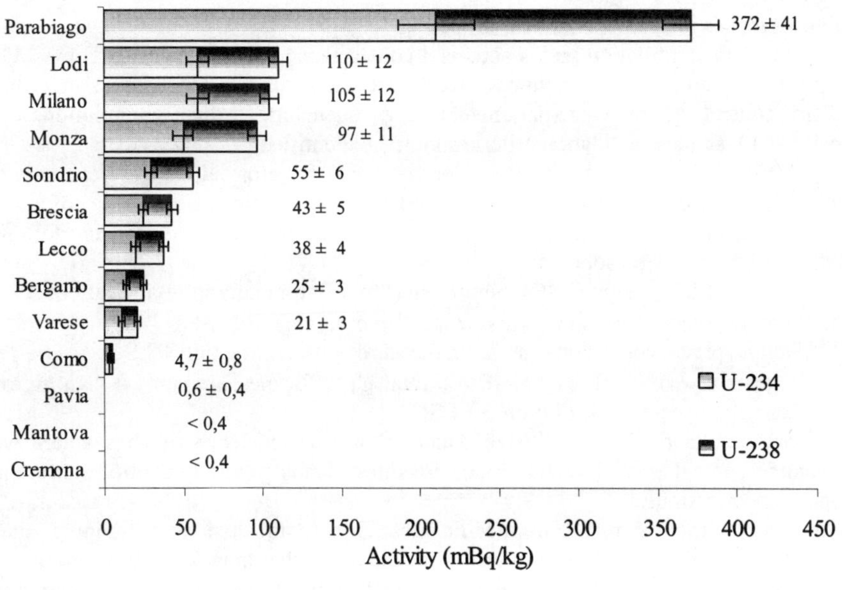

Figure 7: *Tap waters results-Total Uranium activities*

Table 8 *Bottled waters results*

Bottled water sample	**Gross α** mBq/kg	**U total** mBq/kg	**^{234}U** mBq/kg	**^{238}U** mBq/kg	**Gross β** mBq/kg	**^{40}K** mBq/kg
Min. water S1	346 ± 23	233 ± 29	119 ± 27	114 ± 27	467 ± 43	82 ± 4
Min. water S2	334 ± 43	220 ± 24	125 ± 10	95 ± 11	210 ± 53	48 ± 2
Min. water S5	59 ± 5	61 ± 7	31 ± 4	30 ± 4	158 ± 16	88 ± 4
Min. water S6	38 ± 4	44 ± 5	23 ± 3	22 ± 3	100 ± 14	64 ± 3
Min. water S9	12 ± 3	13 ± 2	7,8 ± 1,5	5,3 ± 1,3	45 ± 12	24 ± 1

Tritium and radon concentration values are not reported here in detail. Up to now tritium measurement always gave results lower than LLD (5 Bq/kg). Radon determination displayed high variability; values from 5 to 15 Bq/kg have been measured in Milano and surrounding area. Particular attention has to be paid to the presence of treatment water plants (by active charcoal for instance) which can strongly influence radon concentration.

3 CONCLUSIONS AND FURTHER DEVELOPMENTS

Liquid scintillation has proven to be a quick, versatile and accurate tool for radiometric investigation both on surface and on drinking waters. Thanks to high detection efficiency and low instrument background, alpha and beta emitting isotope activities can be measured with good sensitivities.

Small differences in sample chemical properties (e.g. pH value, amount of oxygen present, etc.) can modify scintillation yields and, as a consequence, measurement outcomes. Great care must be paid in defining sample treatment and counting procedures: in order to maintain control of relevant parameters (e.g. quenching value, chemiluminescence intensity, etc.), suitable validation criteria should be identified.

World Health Organization proposes derived limits for gross alpha and beta activities; compliance to these limits should ensure compliance to committed dose value enforced by Italian law. In order to check full compliance to Italian law requirements, the following analytical scheme may be adopted:

a. Tritium (LSC), radon (LSC), potassium (Ionic Chromatography) measurements
b. Sample preconcentration – gross α and β measurements (LSC)
c. Sample preconcentration – uranium measurements (Extraction + LSC)
d. If gross α > 100 mBq/kg or if α > total U, ^{226}Ra measurements (^{222}Rn ingrowth from preconcentrated solution b + LSC)

Preliminary results on Lombardia tap waters show the existence of sites where water radionuclide content exceed WHO proposed values; this is generally due to high uranium isotopes concentrations.

Previous works showed that,[36] in specific areas, a relevant dose contribution is due to radium isotopes for which dose conversion factors are higher than uranium ones. In some of these critical situations WHO proposed derived limits are not exceeded but committed dose is higher than 0.1 mSv/y with special regard to group age < 1 year.

Pre-existent data on water radionuclide content are incomplete, and often do not consider ^{228}Ra contribution to committed dose. It is therefore desirable to perform a full preliminary screening on Lombardia tap waters, in order to check suitability of WHO proposed derived limits to our local situation.

Therefore in order to:

- check adequacy of WHO gross alpha and beta derived limits to our situation
- identify correlations between waters radioactivity content and aquifers characteristics to identify critical areas

we are planning to apply the above described methods to a monitoring program on the whole Lombardia area.

References

1 EEC Council Directive 98/93/EC of 3 November 1998 on the quality of water intended for human consumption. Off. J. L330, 05/12/98.

2 D.Lgs. 2 febbraio 2001 n. 31: Attuazione della direttiva 98/93/CE relativa alla qualità delle acque destinate al consumo umano. G.Uff. 52, 3/3/2001.

3 EEC, Commission Recommendation of 20 December 2001 on the protection of the public against exposures to radon in drinking water supplies. 2001/928/Euratom

4 ISO 9698, Water quality: Determination of tritium activity concentration – Liquid scintillation counting method -1989

5 WHO, Guidelines for drinking water quality 2nd Edition Vol. 1 (1993) Vol.2 (1996).

6 ISO 9696, Measurement of gross alpha activity in non saline water. Thick source method - 1992.

7 ISO 9697, Measurement of gross beta activity in non saline water. Thick source method - 1992.

8 EPA 00-0, Radiochemical determination of gross alpha and gross beta particle activity in water. *Radiochemistry Procedure Manual*. USEPA 520/5-84-006.

9 L. Salonen, *2° International Conference on Analytical Chemistry in Nuclear Technology*; Karlsruhe 5-9/6/1989

10 L. Salonen, *First-Joint Finnish-Soviet Symposium on Radiochemistry*, Helsinki 16/5/1990.

11 J.A. Sanchez-Cabeza et al., *Radiocarbon*, 1993, 43.

12 J.A.Sanchez-Cabeza and L. Pujol, counter. *Health Physics*, 1995, **68 (5),** 674.

13 M. Abuzeida, B.H. Arebi, Y.A. Zolatarev, N.A. Komarov, *J. of Radioanal. and Nucl. Chem., Articles,* 1987, **116**, 285.

14 Yu-fu Yu, B. Salbu, H.E. Bjornstad, H. Lien, *J. of Radioanal. Nucl. Chem. Letters*, 1990, **145**, 345.

15 J.R.Cadieux, S. Clark, R.A. Fjeld, S. Reboul, A. Sowder, *Nucl. Instr. and Meth. in Phis. Res.*, 1994, **A353**, 534.

16 D.L. Horrocks,. *the Review of Scientific Instr.*, 1964, **35**, 334.

17 H.M. Prichard, T.F. Gesell, *Health Physics*, 1977, **33**, 577.

18 ASTM D 5072-98, Standard test method for radon in drinking water.

19 E. Vitz, *Health Physics*, 1991, **60**, 817.

20 N. Kinner et al., *Env. Science and Techn.*, 1991, **25**, 1165.

21 P. Belloni al., *7th International Symposium on Environmental Radiochemical Analysis*, Bournemouth, Sept. 1994.

22 P. Belloni, G. Ingrao, G.P. Santaroni, R. Vasselli, Urbino, giugno 1995.
23 F. Schönofer, technical sheet, Wallac Oy - Sept 1989.
24 J.H. Hightower, J.E. Watson, *Health Physics,* 1995, **69**, 219.
25 S. Chalupnik, J.M. Lebecka, *Radiocarbon,* 1993, 397.
26 K. Sato, T. Hashimoto, *J. of Environm. Radioact.,* 2000, **48**, 247.
27 Yong-Jae Kim, Chang-Kyu Kim, Jong-In Lee, *Appl. Rad. Isotopes,* 2001, **54**, 275.
28 S. Möbius et al., *Radiocarbon,* 1993, 413.
29 M.P. Blanco Rodriguez, F.Vera Tomè, J.C. Lozano, V. Gomez Escobar, *Appl. Rad. Isotopes,* 2000, **52**, 705.
30 S.Möbius, K.Kamolchote, T.Rakotoarisoa, *Advances in Liquid Scintillation Spectrometry,* Karlsruhe, May 2001.
31 F. Schönofer, *Radioact. Radiochem.,* 2001, **12**, 33.
32 Kuo et al., *Appl. Rad. Isotopes*, 1997, **48**, 1245.
33 H. Higuchi et al., *Anal. Chem.*, 1984, **56**, 761.
34 M. Roveri, Enel CRTN, Servizio Ambiente, E0/87/01/MI –1987.
35 F. Schönofer, *Analyst,* 1989, **114**, 1345.
36 G. Sgorbati, M. Forte, G. Gianforma, R. Rusconi, C. Margini, A.S.L. Milano Città – 1998.
37 ISO Guide to the expression of uncertainty in measurement -1992
38 M.T. Valentini-Ganzerli, L. Maggi, V. Crespi Caramella, G. Premoli, *J. Radioanal. Nucl. Chem.*, 1997, **221**, 109.

^{210}Po CONCENTRATIONS IN UK SEAFOOD*

A.K. Young[1], D. McCubbin[1], K. Thomas[2], W.C. Camplin[1], K. S. Leonard[1] and N. Wood[2]

[1]CEFAS Lowestoft Laboratory, Pakefield Road, Lowestoft, Suffolk NR33 OHT, UK
[2]Radiological Protection and Research Management Division, Food Standards Agency, 125 Kingsway, London WC2B 6NH.

1 INTRODUCTION

There has been a growing awareness since the mid 1980s of the radiological significance of natural radionuclides, particularly ^{210}Po ($t½$ = 138 days), in marine and terrestrial foodstuffs.[1,2,3,4,5] A detailed analysis of ^{210}Po in the duplicate diet study of the Portuguese population indicates that the average ingestion rate is ~1.2 Bq day^{-1} per person.[3] The effective dose from ^{210}Po in the diet was shown to vary from 25 μSv $annum^{-1}$ in a person consuming no seafood to 1,000 μSv $annum^{-1}$ in a heavy consumer of molluscs.[3]

Enhanced concentrations of natural series radionuclides can occur in discharges from non-nuclear industries, particularly wastes released by phosphate fertiliser plants.[6] In the UK, large volumes of waste were discharged into the eastern Irish Sea at Saltom Bay (near Whitehaven) from the Rhodia Consumer Specialities Ltd. (formerly Albright and Wilson) phosphoric acid production plant between 1954 and 1992. This resulted in significantly enhanced levels of natural radionuclides in the local shellfish and hence dose to the local seafood consumers.[7,8,9] Since 1992, concentrations of natural radionuclides in the local environment have decreased concomitant with changes in discharge practices.[10]

Estimates of the impact of anthropogenic natural radionuclide inputs require an accurate knowledge of ambient baseline concentrations and the extent of natural fluctuations. At present, information for UK seafood is sparse. However, signatories to the OSPAR convention, such as the UK, must ensure that discharges of radioactive substances are near background values for naturally occurring radioactive substances and close to zero for artificial radioactive substances.[11] In this investigation, attention was focussed upon the edible fraction of the most commonly consumed marine species to link in with the surveillance programme run by the Food Standards Agency (FSA). The analytical data from this wider programme, and associated dose assessments, are reported in the annual Radioactivity in Food and the Environment (RIFE) series of reports that the FSA jointly publishes with the Scottish Environment Protection Agency (SEPA).[12]

2 SAMPLING AND ANALYSIS

2.1 Sample Collection

About one hundred samples of twelve different marine species were collected from more than thirty locations over three years, from July 1999 to August 2001 (Figure 1). Sampling sites were selected to be distant from known anthropogenic inputs of natural radionuclides. For a balanced perspective, the sampling locations were distributed around the coastline of England and Wales with a small number of additional sites in Northern Ireland, southern Scotland and the North Sea.

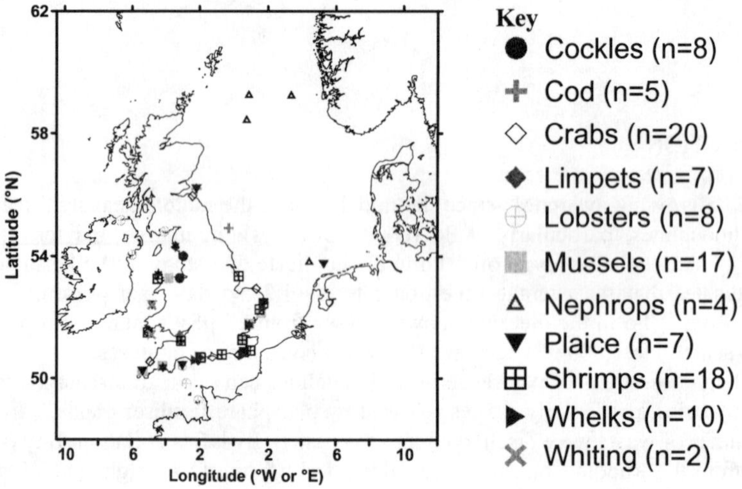

Figure 1 *Location of individual sampling sites. n indicates total number of each species analysed.*

To ensure that samples were representative of the area from which they were collected, approximately 5 kg of fresh live weight, commercially sized individuals of each species were taken from each location. Collection was carried out by local fishermen, fisheries officers and members of the public. For practical reasons, samples were collected throughout the year, although the majority were obtained in the summer months. Ideally, all sampling would have been carried out within a narrow time frame (one month) to remove the potential for temporal variations to complicate data interpretation.[13]

2.2 Preparation and analysis

Analyses were restricted to edible material for all species, to link in with the FSA surveillance programme. Food preparation practices can have a significant impact upon the radionuclide content, particularly of particulate-associated radionuclides in shellfish.[14,15] In the present study, it was assumed that preparation methods used by critical group consumers do not involve depuration. With the exception of *Nephrops*, all shellfish species were cooked in artificial seawater. Samples were subsequently dissected to remove all edible material (e.g. in the case of crabs both white and brown meat) for radiochemical analysis.

Dried samples were spiked with a ^{209}Po yield tracer and wet-oxidised using a combination of nitric acid and hydrogen peroxide. Following evaporation to dryness and dissolution in weak hydrochloric acid (~0.5 M), Po was allowed to autodeposit onto polished silver discs at ~90°C for 4 hours. The silver discs were removed and α-counted on Si surface barrier detectors. The minimum detectable ^{210}Po activity was about 0.1 mBq. Counting errors were generally maintained below 5 % (1σ). Recoveries of the ^{209}Po yield tracer were typically in the range of 75-95 %

3 RESULTS

The range and median values for the ^{210}Po concentration in all the sampled species is shown in Figure 2.

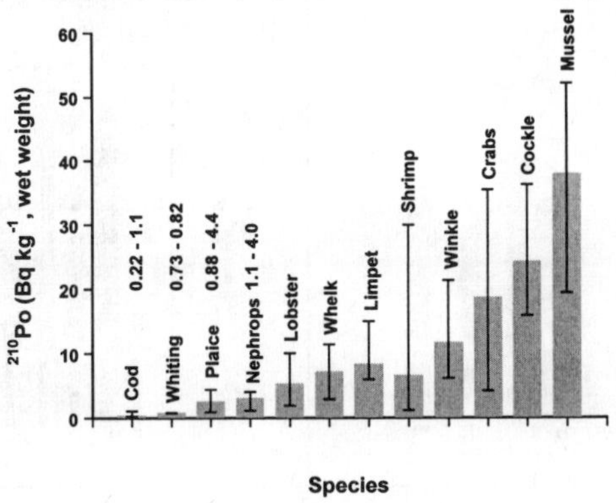

Figure 2 *Median ^{210}Po concentrations in edible fractions for marine species collected from various UK sites (July 1999-August 2001). Error bars indicate range of values observed in individual samples. For clarity, ranges are listed for selected organisms*

The data indicate median ^{210}Po concentrations in individual species ranged from 0.42 Bq kg^{-1} in cod up to 38 Bq kg^{-1} in mussels (i.e. a variation of ~90 fold). Clearly, the dose from ^{210}Po, to man resulting from seafood consumption is extremely sensitive to variations in diet. The error bars show that there was marked variation in the ^{210}Po content between individual samples of the same organisms (typically by ~ 4 fold and up to 26 fold for shrimps). ^{210}Po concentrations increased in the order cod ~ whiting < plaice ~ *Nephrops* < lobster ~ shrimp ~ whelk ~ limpet < winkle ~ crab < cockle < mussel.

The data in Figure 2 were further assessed to examine whether the variations in concentrations between individual samples of the same organisms could be accounted for by differences in levels between separate sampling sites. Selected results are provided in Figures 3c and 3d for mussels (species containing the largest radionuclide concentrations) and shrimps (species exhibiting the largest variation in radionuclide concentrations), respectively.

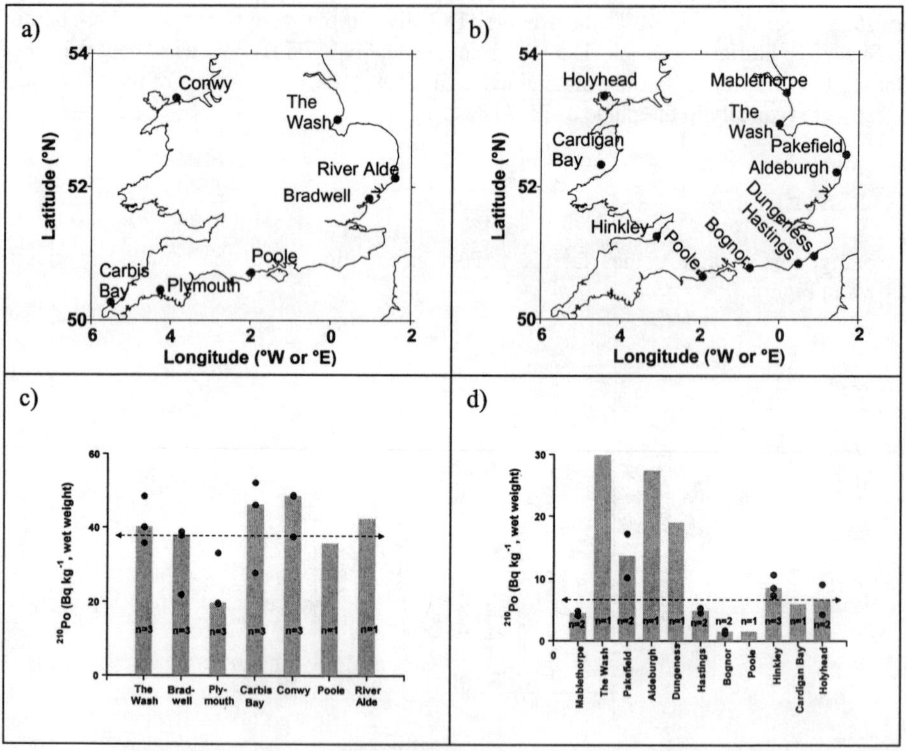

Figure 3 *Variation in ^{210}Po concentrations between individual sampling sites (July 1999-August 2001). Dashed arrows indicate overall UK median value from all sites (as shown in Figure 2). Dots indicate values for individual samples and n indicates number of samples collected at each site: a) Location of mussel sampling sites, b) Location of shrimp sampling sites, c) ^{210}Po in mussels, d) ^{210}Po in shrimps.*

The data in Figure 3c indicate that, at each sampling site, ^{210}Po mussel concentrations varied measurably (typically by 1.6 fold). The scatter in the mussel data at an individual site was, therefore, only slightly reduced compared with that observed in the pooled data for all sampling sites (variation between maximum and minimum mussel values was ~2.7 fold, Figure 2). Given the variability in the data, and the very limited numbers of samples analysed, no firm conclusions regarding variations in ^{210}Po mussel concentrations between sampling sites can be drawn. Nevertheless, levels at Plymouth were possibly marginally lower than those observed at other locations. A similar pattern was noted for cockles, limpets and whelks (data not shown). The variability in ^{210}Po concentrations at the individual sites where more than one sample was collected was typically two fold, compared with a scatter of 3-4 fold observed in the pooled data for all sampling sites.

^{210}Po shrimp concentrations (Figure 3d) at the individual sites where more than one sample was collected exhibited a typical variability of 1.5 fold which was similar to that noted for mussels (1.6 fold). The scatter in the shrimp data for the individual sites was, however, markedly reduced compared with that observed in the pooled dataset (variation between maximum and minimum shrimp values was ~26 fold, Figure 2). The assessment of differences in ^{210}Po shrimp bioaccumulation between individual sampling sites is, however, complicated by the paucity of data. The two highest values (observed at the Wash and Aldeburgh) are unfortunately based on the analysis of single samples. Excluding these and other sites from which single samples were collected, ^{210}Po shrimp concentrations decreased in the order Pakefield (Lowestoft) >Hinkley~Holyhead>Hastings~Mablethorpe>Bognor. Insofar as it is possible to interpret this small dataset, the site differences do not appear to be related to sampling times. The variation between the median value at Pakefield and Bognor was almost 10 fold. These results provide a preliminary indication that ^{210}Po shrimp concentrations may vary significantly between separate sampling sites. Possibly, differences in ^{210}Po content between individual sampling sites may also occur for crabs and winkles (data not shown).

4 DISCUSSION

The results presented here indicate large differences in ^{210}Po concentrations between edible fractions of different seafood species as well as significant variations in levels between individual samples. For some species (e.g. crabs and shrimps), there appear to be significant differences in ^{210}Po concentrations between individual sites indicating that the application of generic UK wide values for 'baseline' concentrations of naturally occurring radionuclides in seafood, for dose assessment purposes, may not be appropriate.

To account for the variations in ^{210}Po between individual species/sampling locations requires an understanding of, amongst other factors, ^{210}Po distribution between different organs and its bioaccumulation pathways. ^{210}Po is known to be non-uniformly distributed within fish and benthic organisms.[16,17,18,19] The highest ^{210}Po concentrations occur in organs involved in digestion and metabolism such as the digestive gland of marine invertebrates[18] and the intestines, stomach, spleen and pyloric caecal of fish[17]. The lowest levels are found in muscle and bone. Consequently, part of the reason for the low ^{210}Po concentrations in fish is because the edible fraction consists of muscle tissue only whereas that for the other species includes the digestive gland, a known biological hotspot for ^{210}Po accumulation.

Uptake of ^{210}Po by marine organisms is thought to occur primarily from food.[20] Since ^{210}Po is very insoluble in seawater the majority of ^{210}Po in the water column is bound to suspended particulate material.[21] Consequently, differences in ^{210}Po bioaccumulation between individual species occur as a result of variations in absorption efficiency and feeding behaviour. The ^{210}Po absorption efficiency in fish is low (~0.05), whereas that in prawns is significantly greater (~0.35).[20] This partly accounts for the gross differences (Figure 2) between ^{210}Po levels in fish and shellfish. The more subtle variations between ^{210}Po concentrations in the individual fish species (median concentration in plaice ~4 fold greater than that for cod and whiting) was presumably because of their varying diet. The predominant prey of adult cod and whiting is fish, whereas plaice feed on a range of benthic species.

The greatest ^{210}Po concentrations (Figure 2) were observed in mussels and cockles. These species are filter and suspension feeders, respectively, and graze on phytoplankton and other suspended matter in the near bottom water.[22,23] The ventilation rate of the mussel is high (~ 1.5 litre hr^{-1} for a typical 5 cm organism). This and the fact that levels of ^{210}Po in the digestive gland of (Colwyn Bay) mussels have been shown to be strongly correlated with changes in levels of suspended particulate material[19] suggests that the majority of ^{210}Po in these organisms is derived by leaching from sedimentary particles.

^{210}Po concentrations in shrimp may vary between sites (Figure 3d). Shrimps (and crabs) are opportunistic feeders and indiscriminately take food from the benthic zone.[24] Consequently, differences in available food sources between different sampling sites could perhaps account for some of the variability. Indeed, it has been reported that there appears to be a clear relationship between diet and ^{210}Po concentrations in penaeid and carid shrimps from the north-eastern Atlantic.[25] The variability in ^{210}Po concentrations is large, even among individuals of the same species. For example, considerably higher levels have been found in *Sergestes* spp. compared with *Pasiphaea* spp. although these two shrimps live at similar depths and occupy the same environmental niche.[16] It was subsequently suggested that, given the relatively short ^{210}Po half-life, the ageing of food might account for the discrepancy since the feeding of these benthopelagic amphipods is irregular.[26] It is acknowledged that the suggestions for ^{210}Po variability arising from the present data are necessarily speculative, in the absence of any ancillary measurements.

5 CONCLUSIONS

Median ^{210}Po concentrations in the edible fractions of commonly consumed marine species ranged from 0.42 Bq kg^{-1} in cod up to 38 Bq kg^{-1} in mussels (i.e. a variation of ~90 fold). Marked variations occurred in levels of ^{210}Po between individual samples of the same organisms (typically by ~ 4 fold). The greatest variability was observed in shrimps (up to 26 fold) and may be due to differences in environmental conditions (e.g. food availability) between sampling locations. Given the extent of natural fluctuations, caution needs to be exercised in the application of generic UK wide values for 'baseline' concentrations of naturally occurring radionuclides in some species.

Acknowledgements
We are grateful for the funding provided by the Ministry of Agriculture, Fisheries and Food (MAFF) and the Food Standards Agency (FSA) during the period of contract RO3010. We wish to thank Dr. D. J. Swift for commenting on the draft manuscript and the contribution provided by an FSA co-project officer, Dr. Will Munro.

References

1 P. McDonald, S.W. Fowler, M. Heyraud and M.S. Baxter, *J. Environ. Radioact.*, 1986, **3**, 293.
2 R.J. Pentreath, and D.J. Allington, Dose to man from the consumption of marine seafoods: a comparison of naturally-occurring ^{210}Pb with artificially produced radionuclides, in *Proceedings of VII international IPRA congress*, Sydney, 10-17 April, 1988, **3**, 1582.
3 F.P. Carvalho, *Health Physics*, 1995, **69**(4), 469.

4 A. Aarkrog, M.S. Baxter, A.O. Bettencourt, R. Bojanowski, A. Bologa, S. Charmasson, I. Cunha, R. Delfanti, E. Duran, E. Holm, P. Jeffree, H.D. Livingston, S. Mahapanyawong, H. Nies, I. Osvath, Li. Pingyu, P.P. Povinec, A. Sanchez, J.N. Smith, and D.J. Swift, *J. Environ. Radioact.*, 1997, **34**(1), 69.

5 D. Pollard, T.P. Ryan, and A. Dowdall, *Rad. Prot. Dosim.*, 1998, **75**(1-4), 139.

6 A.J. Poole, D.J. Allington, A.J. Baxter and A.K. Young, *Sci. Total. Environ.*, 1995, **173/174,** 137.

7 M. McCartney, P.J. Kershaw, D.J. Allington, A.K. Young and D. Turner, *Rad. Prot. Dosim.*, 1992, **45**(1-4), 711.

8 S.F.N. Rollo, W.C. Camplin, D.J. Allington and A.K. Young, *Rad. Prot. Dosim.*, 1992, **45**(1-4), 203.

9 W.C. Camplin, A.J. Baxter and G.D. Round, *Environ. Int.*, 1996, **22**, Suppl. 1, 5259.

10 M.. McCartney, C.M. Davidson, S.E., Howe and G.E. Keating, *J. Environ. Radioact.*, 2000, **49**(3), 279.

11 Sintra Statement of the Ministerial Meeting of the OSPAR Commission, dated 23 July 1998.

12 FSA and SEPA, Radioactivity in food and the environment. RIFE-6 report, London, UK. See also previous reports in this series.

13 P. McDonald, G.T. Cook and M.S. Baxter, 'Natural and artificial radioactivity in coastal regions of the UK' in *Radionuclides in the Study of Marine Processes*, eds P.J. Kershaw and D.S. Woodhead. Elsevier Applied Science, London, 1991, pp. 329-339.

14 W.A. McKay, C.M. Halliwell and C. Rose, *J. Radiol. Prot.*, 1997, **17**(2), 115.

15 D. Jackson and A. Rickard, *Radiat. Prot. Dosim.*, 1998, **75**(1-4), 155.

16 M. Heyraud, and R.D. Cherry, *Mar. Biol.*, 1979, **52**, 227.

17 B. Skwarzec, *J. Environ. Radioact.*, 1988, **8,** 111.

18 D.J. Swift, D.L. Smith, D.J. Allington and M.J. Ives, *J. Environ. Radioact.,* 1994, **23**, 213.

19 M.A. Wildgust, P. McDonald and K.N. White, *Sci. Tot. Env.*, 1998, **214,** 1.

20 F.P. Carvalho and S.W. Fowler, *Mar. Ecol. Prog. Ser.*, 1994, **103**, 251.

21 N. Tanaka, Y. Takeda and S. Tsunogai, *Geochim. Cosmochim. Acta.,* 1983, **47**, 1783.

22 P.R. Boyle, 'Molluscs and Man', *The Institute of Bilogy's studies in biology*, Edward Arnold, London, 1981, **134**, 60 pp.

23 D.S. McLusky, The estuarine ecosystem-(Tertiary level biology). 1. Estuarine ecology, Blackie and Son Ltd., Glasgow, UK, 1981.

24 G.F. Warner, The Biology of crabs, Paul Elek (Scientific Books), London, 1977.

25 M. Heyraud, P. Domanski, R.D. Cherry and M.J.R. Fasham, *Mar. Biol.*, 1988, **97**, 507.

26 S. Charmasson, P. Germain and G. Leclerc, *Rad. Prot. Dosim.*, 1998, **75**(1-4), 131.

A STUDY OF COLLOIDS IN THE NEAR-FIELD WATERS AT THE LOW-LEVEL WASTE REPOSITORY SITE AT DRIGG.

P. Warwick[(1)], S. J. Allinson [(1)], A. Eilbeck [(2)]

(1) Centre for Environmental Studies, Department of Chemistry, Loughborough University, Loughborough, Leicestershire UK LE11 3TU.

(2) Environmental Risk Assessments, Research and Technology, BNFL, Sellafield, Seascale, Cumbria UK CA20 1PG.

1 INTRODUCTION

Extensive investigations have been carried out on the interaction of colloids with pollutants, such as pesticides, radionuclides, heavy metals etc and, their transport through the terrestrial environment [1,2]. Work at the Savannah River Site [3] and the Nevada Test Site [4] has indicated increased mobility of radionuclides when associated with colloids. Colloids may sorb large amounts of radionuclides due to their large surface area to mass ratios. The fate of colloids within the environment is a function of their size, morphology and composition as well as the chemistries of the surrounding water body and the geological material [5]. Colloids may therefore play an important role in the transport of radionuclides from a nuclear waste repository.

To assess the possible effect of colloids on radionuclide transport in and around the BNFL owned low-level radioactive waste site at Drigg near Sellafield in Cumbria samples of groundwaters were extracted from the near-field, i.e. from the trenches, of the site to determine (i) the types and populations of colloids and (ii) whether radionuclides are associated with colloids. Water samples were taken anaerobically, at low flow-rates, by use of a submersible pump at depths between 8 and 11 metres below ground level. The samples were ultrafiltered (12 μm, 1 μm, 30kD and 500 kD) under nitrogen and the membranes and the filtrates were analysed for radioactive and colloid content. This paper presents an overview of the properties of the trench groundwaters with respect to the physical characterisation of the waters, colloid population, colloid characterisation and radionuclide loading of colloids. Groundwater samples were also allowed to go aerobic to determine the distribution between the resulting precipitates and the residual waters.

2 EXPERIMENTAL

2.1 Groundwater sampling

Samples of groundwaters were extracted from the trenches by using a micro-purge low-flow procedure [6]. The equipment was purged fully with nitrogen before taking samples of water in order to maintain anaerobic conditions. Conductivity, Eh, pH, temperature, dissolved oxygen concentration (DOC) and iron content measurements were monitored throughout the procedure. A sample of trench water was collected when these measurements were constant.

2.2 Sample treatment

The groundwater samples were ultrafiltered sequentially under nitrogen. 12 μm, 1 μm, 30,000 Dalton (D) and 500 D membranes were used to separate the size fractions of the particles present into "particulate" (> 1 μm), "colloidal" (1 μm – 30,000 D), "small colloidal" (30,000 D – 500 D) and "ionic" (< 500 D). Small volumes of groundwater were filtered (~ 20 cm^3) and the 30,000 D membranes were dried and then analysed by scanning electron microscopy (SEM) to estimate colloid population. SEM-EDS (energy dispersive X-ray analysis) was employed to analyse the elemental composition of individual colloids on the membranes. Aliquots (~ 500 cm^3) of the extracted water and of each filtrate were taken for radioactivity measurements. Tritium measurements were made by distilling a sample of the water and mixing it with Ultima Gold LLT liquid scintillant before counting in a Packard 2750 TR/LL liquid scintillation counter. The counting efficiency of each sample was determined by spiking the sample with standard tritiated water and recounting. Gross alpha and gross non-tritium beta activities were determined by evaporating 500 cm^3 of sample to dryness and weighing a sample of the residue into a 6 cm-diameter planchette. Each sample was then counted in a FAG FHT650K1 gas flow proportional counter. Am and K standards were used to determine the counting efficiencies for beta and alpha particles. Gamma activities were determined using an EG & G Ortec GEM series high purity Germanium coaxial detector and an ORTEC Maestro II program. The counting efficiencies were determined using a mixed radionuclide standard (National Physical laboratory, Teddeington, UK). Microwave acid digestion was used to digest the membranes and precipitates and the resulting liquor used for radioactivity measurements [6].

A portion of each anaerobic filtrate was allowed to go aerobic and the resulting precipitate separated by filtering through a 1.0 μm membrane. The aerobic supernatant fraction of each filtrate was taken for radioactivity measurement. Microwave acid digestion was used to digest the membranes and precipitates and the resulting liquor used for radioactivity measurement.

Aliquots of the unfiltered groundwater samples were passed through strong anion and strong cation exchange resins (Amberlite IR-120 and IRA-400 respectively) and radioactivity measurements were performed on the eluents.

3 RESULTS AND DISCUSSION

The trench groundwater sampling details are shown in Table 1. Three samples were taken from standpipe 3/3 at different times to determine possible seasonal variation. Samples P4/5p1/LA51 and P4/4p1/L74 were taken from trench four so that comparisons between samples from the same trench could be drawn.

Table 1 *Water, sediment and sampling depths for groundwater samples*

Sample	Date	Water level (m)	Sediment depth (m)	Pump height (m)
P3/3p1/K516	November 1999	7.55	9.28	8.70
P3/3p1/K538	March 2000	7.08	9.23	8.20
P3/3p1/LA52	August 2000	7.50	9.24	8.50
P4/5p1/LA51	August 2000	9.32	10.72	10.00
P4/4p1/L74	November 2001	8.77	10.17	9.30
P6/1p1/L78	January 2001	6.46	9.61	7.90
P5/4p1/L77	January 2001	10.33	11.33	10.50

Sample P3/3p1/LA52 became aerobic during analysis and therefore could not be fully analysed. Standpipe P5/4 was found to be non-transmissive and therefore the sample was taken aerobically and consequently could not be filtered.

3.1 groundwater characterisation

The ranges (from start to finish of extraction) of conductivity, Eh, pH, temperature, dissolved organic concentration (DOC) and iron content during the sampling of the groundwaters are shown in Table 2.

Table 2 *Variation in conductivity, Eh, pH, temperature, DOC and iron content during sampling.*

Sample	Conductivity ($\mu S.cm^{-1}$)	Eh (mV vs. SHE)	pH	Temperature (°C)	DOC ($mg.dm^{-3}$)	Fe content ($mg.dm^{-3}$)
P3/3p1/K516	1106 - 855	32.7 – (- 7.3)	5.10 – 6.83	11.1 – 12.1	Not measured	9.73 – 13.35
P3/3p1/K538	818 - 734	62.9 – 59.4	7.01 – 7.03	12.0 – 12.0	Not measured	1.31 – 1.88
P3/3p1/LA52	150 - 259	80.4 – 70.5	6.99 – 6.95	14.5 – 17.3	- 0.4 – (- 0.2)	24.1 – 49.4
P4/5p1/LA51	270 - 424	91.5 – 72.9	6.97 – 7.00	14.5 – 17.9	1.4 – 0.9	3.61 – 5.01
P4/4p1/L74	380 - 203	99.5 – 97.8	6.98 – 7.02	12.0 – 10.7	1.2 – 1.0	36.2 – 30.4
P6/1p1/L78	786 - 717	136.1 – 56.4	7.13 – 7.14	8.4 – 9.7	2.2 – 1.9	12.3 – 23.0

The conductivity, DOC, Eh, pH and iron content measurements of the Drigg groundwater samples indicate that the waters contain medium to high total dissolved solids (TDS) (220 – 1000 mg dm^{-3}) [7] and that they are pH neutral and reducing. The groundwaters are relatively low in DOC. The iron content is < 17 % of the TDS.

The average size measurements and zetapotentials of the groundwaters (Table 3) were measured using a Zetamaster S (Malvern Instruments). The results show that the average size of particles in the groundwaters are < 1 μm in diameter, i.e. colloids, and that the particles are negatively charged.

Table 3 *Particle size analysis and zetapotential of the groundwaters.*

Sample	Particle size (nm)	Zetapotential (mV)
P3/3p1/K516	788.8 ± 164.2	-
P3/3p1/K538	909.8 ± 246.7	- 13.1 ± 1.5
P3/3p1/LA52	981.0 ± 198.8	- 15.4 ± 0.5
P4/5p1/LA51	804.1 ± 258.8	- 10.2 ± 1.3
P4/4p1/L74	853.2 ± 220.0	- 12.0 ± 1.0

3.2 Colloid population.

Colloidal population was determined by counting the number of colloids on sections of the 30,000 D membranes observed by SEM. Table 4 shows the number of sections of membrane analysed and the calculated colloid population of five of the groundwater samples. Colloid population is in the order of 10^9 - 10^{11} colloids dm^{-3}, which is in agreement with values reported in other work [8].

Table 4 *Colloid population of Drigg groundwater samples.*

Sample	No of sections	Population (colloids dm^{-3})
P3/3p1/K538	9	$1.07 \pm 0.33 \times 10^{11}$
P3/3p1/LA52	11	$7.91 \pm 4.74 \times 10^{9}$
P4/5p1/LA51	9	$5.78 \pm 5.38 \times 10^{11}$
P4/4p1/L74	32	$5.11 \pm 2.72 \times 10^{10}$
P6/1p1/L78	25	$1.01 \pm 0.95 \times 10^{10}$

3.3 Colloid characterisation.

Elemental composition of individual colloids was determined by SEM-EDS. The most abundant elements found were iron and silicon, which were present in almost all of the colloids analysed.

Qualitative analysis was performed using a Cambridge Stereoscan 360 with Link System Energy Dispersive Analyser. Analysis of 12 μm, 1 μm and 30,000 D membranes for samples P4/5p1/LA51 and P3/3p1/LA52 identified iron and silicon as the major components of the colloids. Other (minor) elements identified were Na, Mg, Al, S, Cl, K, Ca, Mn, Ti, Ni, Cr, Cu, and Pb. The results also showed that the elemental composition was the same regardless of particle size.

Quantitative analysis was performed using a Leo 1530VP SEM with EDAX Phoenix EDS analyser which showed good correlation with ICP-AES results for kaolinite and montmorillonite [9]. Oxygen was omitted from the analysis calculations so as to prevent interference from oxygen present in the membranes themselves. The quantitative results confirmed that iron and silicon were the major components of the colloids. Other elements identified were Na, Mg, Al, P, S, Cl, Ca, Cu, and Zn. Table 5 shows the Si:Fe ratios of the colloids separated on 30 kD membranes. The majority of the particles analysed had a Si:Fe ratio of > 0.5 and up to 15.5% of the colloids analysed were 100% silicon. Colloidal aggregates and single colloids were also compared for elemental composition and were found to show no significant difference thus their tendency to aggregate or remain separate is not a function of their elemental composition. The presence of 100 % silicon colloids implies that the majority of colloids are silicon based with an iron coating. The anaerobic to aerobic transition of the groundwaters caused the precipitation of an amorphous red-brown solid produced by the oxidation of Fe(II) to Fe(III).

Table 5 *Details of Si:Fe ratios for groundwater samples*

Sample	Si:Fe => 0.5	Si = 100%
P3/3p1/K538	77.8 %	0 %
P3/3p1/LA52	93.3 %	0 %
P4/5p1/LA51	30.8 %	7.7 %
P4/4p1/L74	97.3 %	1.4 %
P6/1p1/L78	47.9 %	15.5 %

3.4 Activity analyses.

Table 6 shows the tritium, gross alpha, gross non-tritium beta and gamma activities for the groundwater samples. All activities are Bq dm^{-3}. The measurements are performed in triplicate for the tritium activity and in duplicate for all other measurements.

The majority of the radioactivity is due to tritium and is in the order of $10^3 - 10^5$ Bq dm^{-3}. Gross alpha activity levels are generally less than 1 Bq dm^{-3}. Gross non-tritium beta and ^{137}Cs activity levels are of the order of $10^2 - 10^3$ and $10^0 - 10^2$ Bq dm^{-3} respectively. ^{40}K and ^{60}Co activity levels are generally below the minimum detectable amount. Seasonal variation accounts for the difference in the activity levels for the three samples taken from standpipe P3/3p1. The activity levels for samples P4/5p1/LA51 and P4/4p1/L74 are of the same order of magnitude showing that samples taken from the same trench have similar levels of activity.

Table 6 *Radioactive measurements (Bq dm^{-3}) of the groundwater samples*

Sample	Tritium	Gross alpha	Gross non-tritium beta	^{137}Cs	^{40}K	^{60}Co
P3/3p1/K516	1202.0 ± 42.0	0.165 ± 0.068	128.08 ± 20.16	-	-	< 0.3
P3/3p1/K538	1932.5 ± 102.5	0.659 ± 0.018	125.84 ± 0.24	-	-	< 0.3
P3/3p1/LA52	6904 ± 70	0.162 ± 0.018	197.43 ± 0.85	4.82 ± 0.14	< 3.8	< 0.3
P4/5p1/LA51	105406 ± 5220	0.212 ± 0.016	543.45 ± 1.05	63.69 ± 0.39	< 3.8	< 0.3
P4/4p1/L74	39308.8 ± 747.6	0.309 ± 0.016	470.25 ± 0.11	42.68 ± 0.35	< 3.8	< 0.3
P6/1p1/L78	871740.8 ± 22542.5	0.301 ± 0.012	729.48 ± 0.71	159.90 ± 0.65	< 3.8	1.44 ± 0.16
P5/4p1/L77	706079.60 ± 81673.07	9.953 ± 0.215	1083.59 ± 1.95	289.58 ± 0.86	25.62 ± 1.79	3.09 ± 0.18

3.4.1 Measurement of gross alpha activities. Filtration of the groundwaters allows the size fractions of the particles in the waters to be separated and analysed. In the following, "filtrate" refers to the groundwater that has passed through the indicated membrane, apart from the unfiltered water where "filtrate" indicates the total groundwater. "Membrane" indicates the activity retained by the indicated membrane.

"Aerobic supernatant" and "precipitate" refers to the activity present in the solid and liquid fractions formed in each filtrate once allowed to go aerobic. Values for supernatant and precipitate are not given for the 500 D filtrate, as there was no observable precipitate formed. Activity measurements for the filtrates are performed in duplicate. All other measurements are single values.

Figure 1 shows an example of the gross alpha activities for a groundwater sample (P4/4p1/L74). Errors are large in these analyses due to the low activity levels being measured.

Similar results were obtained from measurements of other samples. The filtrate results show some decrease in alpha activity due to filtration and some retention on the 1 μm and 30,000 D membranes. Up to 16.0 and 9.8 % respectively. In all samples half of the alpha activity was retained on the 500 D membrane and half was present in the 500 D filtrate. The 500 D membrane is the only membrane showing quantifiable levels of alpha activity. This indicates that half of the alpha activity is associated with the colloidal (1 μm – 30,000 D) size fraction and half is associated with the ionic (< 500 D) size fraction.

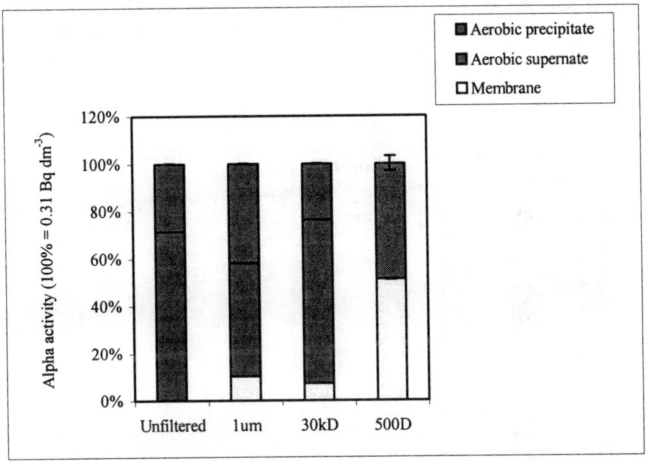

Figure 1 *Measurements of the gross alpha activities for groundwater sample (P4/4p1/L74)*

Once the filtrates and total solution have gone aerobic the alpha activity distributes between the supernatant and precipitate. Samples P3/3p1/K538 and P4/4p1/L74 showed generally more activity in the supernatant than in the precipitate. With the exception of P3/3p1/LA52, where the distribution was equal, the other samples showed more alpha activity within the precipitate than in the supernatant. Once aerobic, the redox-sensitive particles/colloids will form a precipitate and the stable particles/colloids will remain in solution. The alpha activity distributes between the supernatant and precipitate in varying degrees although as no precipitate formed within the 500 D filtrate all the alpha activity from the < 500 D size fraction remains in solution. Alpha spectrometric analysis of sample P5/4p1/L77, after passing the solution through a TRU-resin column, showed that ^{238}U and progeny were present in the waters.

3.4.2 Measurement of Gross Non-tritium beta activities. Figure 2 shows an example ((P4/4p1/L74)) of gross non-tritium beta analysis. Similar results were obtained from other samples.

The results show that, within errors, there is little decrease in the non-tritium beta activity with filtration. Less than 1 % of the activity is retained on the 1 μm and 30,000 D membranes. There is evidence (up to 6.4 %) for the retention of slightly more activity on the 500 D membrane. This suggests that the majority o f the non-tritium beta activity is present within the ionic size range (< 500 D). Once the filtrates and total sample were allowed to go aerobic the activity analyses showed that the majority of the activity remained in the supernatant. With the exception of the unfiltered P4/5p1/LA51 sample, in which approximately half of the activity was associated with the precipitate, up to 24.0 % of non-tritium beta activity was associated with the precipitate. This again suggests that the majority of the non-tritium beta activity is associated with the ionic size range. The activity levels measured in the supernatant are frequently higher than those measured for the total sample. The increase may be due to the cleaning of the matrix afforded by the removal of the precipitate from the solution leading to a higher counting efficiency.

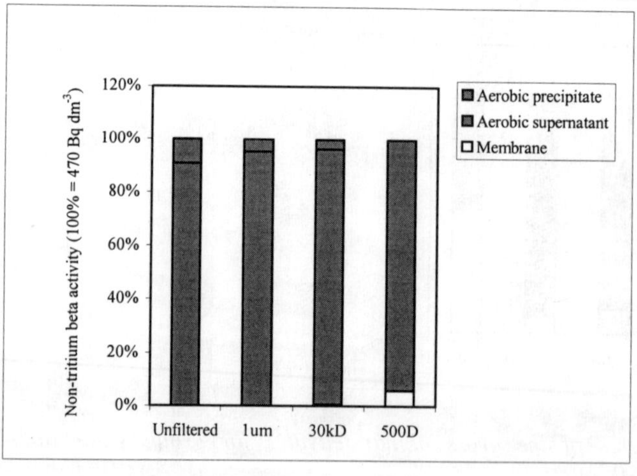

Figure 2 *Measurements of the non-tritium beta activities for groundwater sample (P4/4p1/L74)*

3.4.3 Gamma spectrometry analysis. Examples (from P4/4p1/L74) of the results from measurements of ^{137}Cs activities are shown in Figure 3.

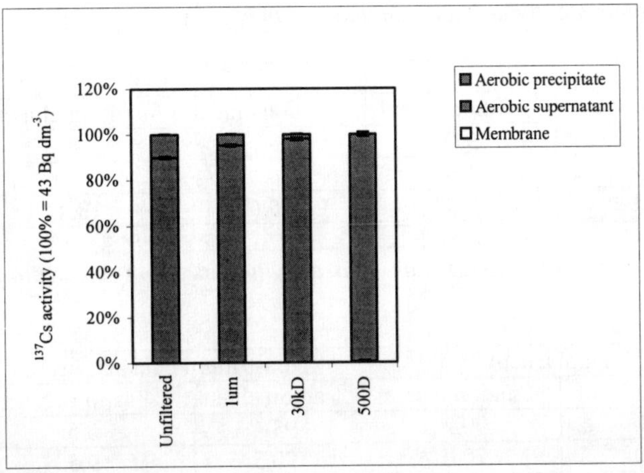

Figure 3 *Measurements of ^{137}Cs activities for groundwater sample (P4/4p1/L74).*

Gamma activity measurements showed that ^{137}Cs and ^{40}K were present in all samples. ^{60}Co was quantifiable in samples P6/1p1/L78 and P5/4p1/L77.

Filtration results show that there is little decrease in the ^{137}Cs activity after filtration. Less than 1 % of the activity is retained on any of the membranes. This suggests that the majority of the ^{137}Cs activity is associated with the ionic size fraction (< 500D). Once the filtrates and total samples were allowed to go aerobic the activity analyses showed that the majority of the ^{137}Cs activity remained in the supernatant and therefore was not associated with redox-sensitive colloids.

The majority of the ^{40}K activity was below the minimum detectable amount, 3.8 Bq dm^{-3}, with the exception of sample P5/4p1/L77. Other radionuclides identified, but not at quantifiable levels within samples P4/4p1/L84, P6/1p1/L78 and P5/4p1/L77, were ^{210}Pb (^{238}U decay chain) and ^{212}Pb (^{232}Th decay chain). Additional radionuclides present in sample P5/4p1/L77 were ^{140}Ba, ^{155}Eu, ^{214}Pb and ^{214}Bi (^{238}U decay chain), ^{228}Ac (^{232}Th decay chain) and possibly ^{223}Ra and ^{211}Bi (^{235}U decay chain), ^{210}Tl (^{238}U decay chain), ^{234}Th and ^{239}Pu.

3.4.4 Ion-exchange experiment. The percentagse of activities passing through either a cation or an anion exchange resin are shown in Tables 7 and 8. Only P5/4p1/L77 contained quantifiable levels of ^{40}K. The results showed that ^{137}Cs, ^{40}K and non-tritium beta activity were retained by the cation exchange resin but not by the anion exchange resin. The results indicate that the activity is associated with positively charged species. The alpha activity is associated with both positively and negatively charged species. The exact distribution is difficult to determine due to the low levels of activity present.

Table 7 *Percentage of gamma activities in anion and cation eluents*

	^{137}Cs activity		^{40}K activity	
	% anion eluent	% cation eluent	% anion eluent	% cation eluent
P4/4p1/L74	91.0	9.0	-	-
P6/1p1/L78	98.2	1.8	-	-
P5/4p1/L77	99.1	0.9	86.6	13.4

Table 8 *Percentage of alpha and non-tritium beta activities in anion and cation eluents.*

	Gross alpha activity		Non-tritium beta activity	
	% anion eluent	% cation eluent	% anion eluent	% cation eluent
P4/4p1/L74	9.8	90.2	93.4	6.6
P6/1p1/L78	30.0	70.0	90.5	9.5
P5/4p1/L77	55.5	44.5	96.2	3.8

4. CONCLUSIONS

The groundwaters sampled from the trenches at Drigg in Cumbria were low in DOC, pH neutral and reducing. They contained 220 – 1000 mg dm^{-3} of TDS of which less than 17 % was due to iron. Particles present in the samples were negatively charged and were less than 1 μm in diameter on average. Colloids were present at populations of 10^9 – 10^{11} colloids dm^{-3}. Elemental analysis showed that the colloids were predominantly iron coated silica at a ratio of < 0.5. Other elements present were Na, Mg, Al, P, S, Cl, K, Ca, Mn, Ti, Ni, Cr, Cu, Zn and Pb. The activities of the samples were dominated by tritium (10^3 – 10^5 Bq dm^{-3}). The alpha activities in the samples are only significantly (approximately 50% of the gross alpha activities) retained on the 500 D membrane. The alpha activities distribute between the supernatant and the precipitate once the samples have been allowed to go aerobic. The alpha activities are associated with both positively and negatively charged species. The ^{137}Cs and non-tritium beta activities are not retained significantly by any of the membranes and are thus classed as ionic. Up to 24 % of the non-tritium beta and less than 1 % of the ^{137}Cs activity is associated with redox-sensitive colloids. The non-tritium beta and ^{137}Cs activities are associated with positively charged species which should not be mobile within the environment. Radionuclides from the ^{238}U, ^{235}U, ^{233}U and ^{232}Th decay chains were identified.

Acknowledgements

The authors are grateful to BNFL for sponsoring this work.

References

1 J. N. Ryan and M. Elimelech, *Colloid Mobilisation and Transport in Groundwater,* 1996, Colloids and Surfaces A: Physicochemical and Engineering Aspects **107**, 1-56.

2 M. B. McGechan and D. R. Lewis, *Transport of Particulate and Colloid-Sorbed Contaminants Through Soil, Part 1: General Principles,* 2002, Biosystems Engineering **83(3)**, 255-273.

3 D. I. Kaplan, P. M. Bertsch, D. C. Adriano, K. A. Orlandini, *Actinide Association with Groundwater Colloids in a Coastal Plain Aquifer*, 1994, Radiochimica Acta, **66-7**, 181-187.

4 A. B. Kersting, D. W. Efurd, D. L. Finnegan, D. J. Rokop, D. K. Smith and J. L. Thompson, *Migration of Plutonium in Ground Water at the Nevada Test Site*, 1999, Nature, **397 (6714)**, 56-59.

5 S. W. Swanton, *Advances in Colloid and Interface Science,* 1995, **54**, 129-208.

6 P. Warwick, S. Allinson, K. Beckett, A. Eilbeck, A. Fairhurst, K. Russell-Flint and K Verrall, *Sampling and Analysis of Colloids at a Low-Level Radioactive Waste Repository Site*, 2002, Journal of Environmental Monitoring, **4**, 1-7.

7 S. E Kegley and J. Andrews, *The Chemistry of Water*, University Science Books, Sausalito, California, 1998.

8 Karen Verrall, *Extraction and Characterisation of Colloids in a Waste Repository Leachate,* Doctoral Thesis, Loughborough University, 1998.

9 Andrew Fairhurst, *The Role of Inorganic Colloids in the Transport of Toxic Metals Through the Environment*, Doctoral Thesis, Loughborough University, 1996.

LOW LEVEL LIQUID SCINTILLATION COUNTING IN A DEEP UNDERGROUND LABORATORY: BACKGROUND REDUCTION ASPECTS

F. Verrezen and C. Hurtgen

Department of Low Level Radioactivity Measurements, Radiation Protection Division, Belgian Nuclear Research Centre (SCK•CEN), B-2400 Mol, Belgium.

1 INTRODUCTION

In low level liquid scintillation measurements the inherent system background (B) of the spectrometer, together with the counting efficiency (ε), is the limiting factor for the achievable sensitivity (LLD). As counting efficiency is mostly determined by sample preparation and counting vial composition and therefore fixed for any given routine analytical procedure, background reduction is often the only way to gain sensitivity, even though the LLD is proportional to the counting efficiency but proportional only to the square root of the background rate as formulated in equation (1).

$$LLD = \frac{C_1 + C_2\sqrt{B}}{C_3 \cdot \varepsilon} \qquad (1)$$

In this equation C_1 is a constant to account for zero blank case corresponding to a 5% probability of false negatives (frequently used value for $C_1 = 2.71$). C_2 is a constant to account for a 5% probability of making Type I or Type II errors (most frequently $C_2 = 4.65$ according to Currie) and C_3 combines the factors allowing the expression of LLD in terms of units (containing factors such as counting time, chemical yield, aliquot size, ...).[1] B stands for the total number of counts acquired for an appropriate blank sample and ε is the counting efficiency for the given isotope in the given geometry.

The system background specific for a liquid scintillation counter consists of several components, such as natural radioactivity in material outside the counters main shielding, natural radioactivity incorporated in the shielding material and incorporated in the detector material (PMT's, scintillation cocktail, vials, ...); radon and its progeny diffusing inside the counter or the detectors and protons, neutrons and secondary photons induced by muons.[2,3] Some of these components are intrinsic to the counter and are therefore difficult to alter (e.g. radioactive material inside the shielding or detector material).

The cosmic radiation component (muon flux and muon induced radiation) is probably the most interesting factor for optimization of liquid scintillation counter performance. One way of reducing the importance of the cosmic radiation is by increasing the counters shielding by installing the counter in a (deep) underground measurement facility. At SCK•CEN we have both an above ground and underground laboratory to our disposal. We installed a liquid scintillation counter (Packard TriCarb 2250CA) in the underground facility and compared measurement results with the above ground laboratory.

2 METHOD AND RESULTS

2.1 Experimental Setup

The entire experiment consists of two parts. In the first part a set of twelve counting vials with different but well known sample composition is measured in the Packard TriCarb 2250CA liquid scintillation counter, installed in the above ground laboratory for liquid scintillation measurements (situated on the first and topmost floor of one of the buildings at the SCK•CEN site). The type of vial used and its composition is presented in Table 1. The set consists of various background vials (ranging from self prepared blanks in polyethylene and Teflon vials to commercially available, unquenched background samples in sealed glass vials) and various reference vials containing known amounts of ^{3}H or ^{14}C. The polyethylene vials are prepared in our laboratories, the glass sealed vials were purchased from Packard Life Sciences long before the start of the experiment.

Table 1 *Composition of liquid scintillation vials and vial type used*

Vial Code	Contents	Vial Type
HS	OptiPhase HiSafe 3 cocktail (20 ml)	PE, capped, 20 ml
UG	Ultima Gold™ XR cocktail (20 ml)	PE, capped, 20 ml
HS/H_2O	OptiPhase HiSafe 3 and Reversed Osmosis water mixture (10 ml each)	PE, capped, 20 ml
UG/H_2O	Ultima Gold™ XR and Reversed Osmosis water mixture (10 ml each)	PE, capped, 20 ml
BGD89	Unquenched, sealed BGD-vial (1989/05/12)	Glass, sealed, 20 ml
BGDR382	Unquenched, sealed BGD-vial (1972/08/14)	Glass, sealed, 20 ml
Teflon/Cu	Self made Teflon vial with Teflon stopper and copper cap, 5 ml 'dead' benzene	Teflon, Cu capped, 5 ml
HS/^{3}H	OptiPhase HiSafe 3 (19 ml) and ^{3}H reference (1 ml, 1630 ± 1.5% Bq/ml on 1998/08/12) mixture	PE, capped, 20 ml
UG/^{3}H	Ultima Gold™ XR (19 ml) and ^{3}H reference (1 ml, 1630 ± 1.5% Bq/ml on 1998/08/12) mixture	PE, capped, 20 ml
STD^{3}H	Unquenched, sealed ^{3}H standard, (255600 ± 1.3% dpm on 1989/06/07)	Glass, sealed, 20 ml
STD^{14}C	Unquenched, sealed ^{14}C standard, (127700 ± 1.7% dpm on 1986/10/14)	Glass, sealed, 20 ml
STD^{3}HR382	Unquenched, sealed ^{3}H Standard, (152100 ± 1.5% dpm on 1972/08/14)	Glass, sealed, 20 ml

Each vial is counted for 100 minutes and the raw spectral data are recorded onto the counters hard disc. The set of vials remains in the temperature controlled sample changer compartment of the Packard 2250CA counter (at about 15 ± 2 °C) for the entire duration of the experiment. To verify the stability of the signal and the composition of the counting vials and to exclude environmental fluctuations of the background radiation field the measurement is repeated approximately every ten days over a period of almost a year (33 measurement cycles from January 2001 until November 2001). For the purpose of comparison the set of vials was initially also measured once in three different liquid scintillation counters, installed in the same room of the above ground laboratory (Packard TriCarb 1900CA, Wallac 1414 Guardian and Wallac 1220 Quantulus).

In the second phase of the experiment the Packard 2250CA counter and the set of counting vials is transferred to the underground laboratory located at the beginning of a 39 m long gallery, in the Underground Research Facility (URF) of HADES. The gallery has an inside diameter of 3.5 m and is dug out at a depth of 224 m in the upper third of a 100 m thick clay layer, covered with about 180 m of glauconitic soil containing sand. At an average density of 2.22 g cm^{-3}, this depth corresponds to about 500 m w.e.. Inside the gallery, the air ventilation maintains the radon concentration at about 50 Bq m^{-3}. At the depth of 500 m w.e. the nucleonic component of the cosmic radiation is reduced by at least five to six orders of magnitude and the muon flux is reduced by a factor of about 800 (to less then 0.2 m^{-2} s^{-1}), compared to ground level.[4] Once the counter is installed, the measurement of the twelve vials is resumed under identical counting conditions (except for counter surrounding) with a frequency of approximately once every three weeks again over a period of almost a year (12 measurement cycles from November 2001 until August 2002).

2.2 Results and Discussion

2.2.1 Sample Stability. Measuring the set of counting vials over a period of one year guarantees a representative result for true instrumentation background and counting characteristics, little influenced by long term environmental fluctuations. However, due to the extended storage time of the counting vials, sample (and thus signal) stability becomes a very important aspect in the possibility to compare measurement results. The slope of the signal evolution in time is the most appropriate means of investigating signal degradation or signal drift. Table 2 summarizes the observed slopes (expressed in cpm per year) and calculated uncertainty for all counting vials over the one year measurement time both in the above ground and the underground laboratory. The relative signal drift (Δ_{signal}) is also stated. Figure 1 gives an illustration of the signal evolution of two polyethylene and one sealed glass sample vial with the results obtained in the above ground laboratory (upper part) compared with the results from the underground laboratory (lower part).

The data in Table 2 confirm the stability of the measured signal for all background vials, as the uncertainty on the slope is almost always greater or equal to the value of the slope itself and the signal drift (Δ_{signal}) has the same absolute value as the uncertainty on each individual measurement. In addition two other spectral parameters (SIS and tSIE values) are monitored for their time dependency and found to be stable within experimental uncertainty.

A similar analysis is made for the measured count rates of the vials containing a known amount of ^{3}H or ^{14}C. In the case of ^{3}H the observed count rate was corrected for radioactive decay ($T_{½} \sim 4500$ d) prior to comparison. As can be seen clearly from the data in Table 2, the results for the sealed glass vials are very stable in time, where the results for the polyethylene vials show a clear signal decrease. This is most likely caused by diffusion of ^{3}H through or in the vial wall. However, even with sample storage times exceeding one year, signal degradation is about 5 %.

Table 2 *Signal variation over the one year storage time for the set of counting vials as obtained with the Packard TriCarb 2250CA LSC counter both above ground and in the underground laboratory.*

Vial Code	Above ground measurement Slope (cpm/y)	Above ground measurement Δ_{signal} (%/y)	Underground measurement Slope (cpm/y)	Underground measurement Δ_{signal} (%/y)
HS	1.6 ±1.4	4.1	0.3 ± 0.6	2.6
UG	1.6 ± 1.3	3.9	0.1 ± 0.6	0.7
HS/H_2O	1.6 ± 1.6	3.9	-0.8 ± 0.7	-7.4
UG/H_2O	2.4 ± 1.3	5.6	0.2 ± 1.1	1.5
Teflon/Cu	-0.2 ± 0.7	-1.2	0.03 ± 0.22	1.6
BGD89	-1.6 ± 1.5	-4.2	0.1 ± 0.6	0.8
BGDR382	-0.8 ± 1.0	-2.4	-0.3 ± 0.7	-2.4
HS/^{3}H	-980 ± 140	-4.7	-1470 ± 500	-7.0
UG/^{3}H	-880 ± 140	-3.6	-1590 ± 860	-6.3
STD^{3}H	70 ± 580	0	50 ± 320	0
STD^{14}C	-190 ± 510	-0.2	-630 ± 740	-0.5
STD^{3}HR382	18 ± 390	0	-1450 ± 1580	-1.4

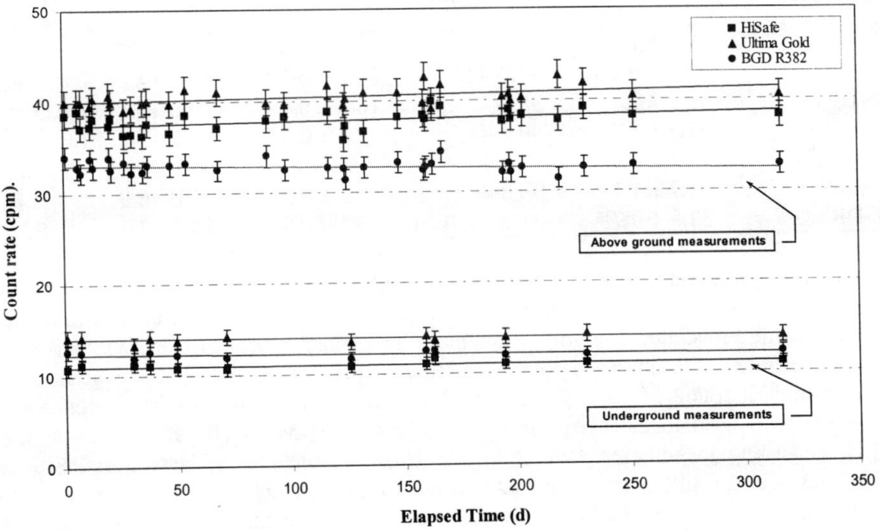

Figure 1 *Stability of measurement signal for HiSafe 3 and Ultima Gold (polyethylene vials) and BGDR382 (sealed glass vial) for above ground and underground measurements.*

Therefore, it can be concluded that the results obtained in the first year of the measurement campaign (above ground laboratory) can be compared with the results obtained in the second year (underground laboratory) with respect to background reduction and instrument performance enhancement.

2.2.2 Background Reduction. For every individual measurement performed on each of the counting vials, both above ground and in the underground laboratory, the total accumulated signal is obtained by summing the observed counts over all 4000 counting channels (full spectrum integration). Next, for each counting vial the average and standard deviation of the individual full spectrum integrals for each measurement campaign (above ground and underground) is calculated. Table 3 gives a summary of these results. As can be expected, a clear reduction of the background is observed for the measurements in the underground laboratory. For the polyethylene vials a background reduction factor of about 3.4 can be achieved, while the reduction factor for the Teflon vial with copper cap is almost twice as high. For the (sealed) glass vials the reduction factor is slightly less (about 2.5). Probably the lower value for the background reduction is caused by the intrinsic radioactivity of the vial material (e.g. ^{40}K in the glass), which causes a partial background contribution that is not affected by the counters location. Therefore only a part of the total background signal is subject to attenuation by the sand and clay shield of the underground laboratory.

Table 3 *Comparison of the background signal (full spectrum integral) in the above ground and underground laboratory for the Packard TriCarb 2250CA LSC counter.*

Vial Code	Above ground measurement		Underground measurement		BGD reduction
	Average and StDev (cpm)	Range (cpm)	Average and StDev (cpm)	Range (cpm)	
HS	38.9 ± 1.0	35.7-39.9	11.1 ± 0.3	10.6-11.7	3.4
UG	40.5 ± 0.9	39.0-42.9	14.0 ± 0.3	13.4-14.2	2.9
HS/H_2O	40.2 ± 1.1	38.3-42.4	10.5 ± 0.4	10.0-11.1	3.8
UG/H_2O	42.5 ± 1.0	40.8-44.5	12.4 ± 0.5	11.5-13.1	3.4
Teflon/Cu	13.1 ± 0.4	12.3-13.9	1.9 ± 0.1	1.8-2.2	6.8
BGD89	36.7 ± 1.1	34.6-40.2	15.6 ± 0.3	15.0-15.9	2.4
BGDR382	32.9 ± 0.7	31.2-34.7	12.2 ± 0.3	11.7-12.7	2.7

A similar analysis is made for the counting vials containing a known amount of ^3H or ^{14}C (reference vials). These results are used to calculate the integrated full spectrum cpm-values and the associated counting efficiencies. No enhancement nor degradation of the counting efficiency is observed for measurements performed in the above ground and underground laboratory. Finally the results of the background vials and the reference vials are paired and used to calculate the figure of merit ($FOM = E^2/B$) and detection limit (LLD, according to equation (1)). It is important to note that all detection limits are calculated over the full spectrum signal and not in an optimized window. The calculated values are given in Table 4. For the purpose of comparison this table also contains the values for FOM and LLD for three other types of commercially available liquid scintillation counters: Packard TriCarb 1900CA, Wallac 1414 Guardian and Wallac 1220 Quantulus.

From the data in the table it can be concluded that all counters perform more or less equally well above ground, with the exception of the Wallac 1220 Quantulus, which offers a better FOM and LLD. The counter performance of the Packard TriCarb 2250CA, when placed in the underground laboratory, increases to a level comparable with the Wallac 1220 Quantulus.

Table 4 *Comparison of counter performance for the Packard TriCarb 2250CA LSC counter with other LSC counters (values between brackets are for the measurements in the underground laboratory).*

Vial Code	Packard TriCarb 2250CA		Packard TriCarb 1900CA		Wallac 1414 Guardian		Wallac 1220 Quantulus	
	FOM	LLD (Bq/l)	FOM	LLD (Bq/l)	FOM	LLD (Bq/l)	FOM	LLD (Bq/l)
HS/H_2O HS/^3H	11 (43)	23 (12)	10	25	14	21	93	8
UG/H_2O UG/^3H	15 (53)	20 (11)	14	21	19	18	110	8
STD^3H BGD89	110 (260)	8 (5)	110	7	120	7	240	5
STD^3HR382 BGDR382	130 (310)	7 (4)	140	7	170	6	350	4
STD^{14}C BGD89	260 (590)	5 (3)	280	5	290	5	550	3

3 CONCLUSION

As can be expected, putting a commercially available liquid scintillation counter in a deep underground measurement facility such as the HADES underground laboratory (shielded with about 500 m w.e.) increases the counters performance. This performance increase is caused by background reduction only (due to the shielding against cosmic radiation) and not by increased counting efficiency. Despite background reduction factors for gamma spectroscopy measurements reported in literature ranging from 50 to a few hundreds the present study indicates that the reduction factor for liquid scintillation counting is much smaller (about 2.5 to 3.5).[5,6,7] The same values for background reduction are reported in a similar study performed in the Gran Sasso underground facility (2.7 to 4.7 at 3800 m w.e.) indicating that additional shielding will most likely not contribute to any further reduction of the background.[8] Therefore about two thirds of the above ground background signal is induced by cosmic radiation.

Since the detection limit is only proportional to the square root of the background, installing the counter in a deep underground laboratory yields only an improvement for the LLD of about 30-50% (reduction factor 1.5 to 2), with no improvement on other performance parameters such as counting efficiency. Therefore in most cases the gain in counter performance will be out weighted by the inconveniences caused by reduced accessibility of the counter.

References

1 L. A. Currie, Limits for detection and quantitative determination, *Analytical Chemistry*, 1968, **40**(3), p. 586-593.

2 P. Theodorsson, Radiometric ^{14}C dating: new background analysis, basis of improved systems, *Radiocarbon*, 1998, **40**(1), p. 157-166.
3 P. Theodorsson, Analysis of background of low-level β/γ systems, 1997, Science Institute, University Iceland, Reykjavik, Iceland, 1997, p. 9-12.
4 D. Mouchel and R. Wordel, in *Proceedings of the Methods and Applications of Low-Level Radioactivity Measurements Workshop,* ed. J. Fietz, Forschungszentrum Rossendorf FZK, Rossendorf/Dresden, 1996, IRMM Low Level Underground Laboratory HADES, p. 25-29.
5 S. Niese, M. Kohler and B. Gleisberg, Low-level counting techniques in the underground laboratory "Felsenkeller" in Dresden, *J. Radioanal. Nucl. Chem.*, 1998, **233**, p. 34-38.
6 S. Niese and M. Köhler, in *Proceedings of the Methods and Applications of Low-Level Radioactivity Measurements Workshop,* ed. J. Fietz, Forschungszentrum Rossendorf FZK, Rossendorf/Dresden, 1996, Low-level counting techniques in the underground laboratory "Felsenkeller" in Dresden, p. 34-38.
7 G. Heusser, in *Proceedings of the Methods and Applications of Low-Level Radioactivity Measurements Workshop,* ed. J. Fietz, Forschungszentrum Rossendorf FZK, Rossendorf/Dresden, 1996, Analysis of background components in Ge-spectrometry and their influence on detection limits, p. 5-8.
8 W. Plastino, L. Kaihola, et al., Cosmic Background reduction in the radiocarbon measurement by liquid scintillation spectrometry at the underground laboratory of Gran Sasso, *Radiocarbon,* 2001, **43**(2A), p. 157-161.

COMPLETE DISSOLUTION OF SOIL SAMPLES BY FUSION*

A.V. Harms

National Physical Laboratory, Queens Road, Teddington TW11 0LW, UK

1 INTRODUCTION

The accurate determination of the concentration of a radionuclide that does not emit any significant gamma-ray requires the complete dissolution of the sample matrix.[1] Complete decomposition[2] of the matrix is essential to achieve equilibrium between the radionuclide and the yield tracer, which is generally added to the sample in order to correct for losses during the analytical isolation procedure.[3,4] However, complete dissolution is especially difficult for quadrivalent and pentavalent elements (e.g., actinides) which may form very refractory oxides and carbides. Leaching of the sample matrix with various mineral acids (e.g., HCl or HNO_3), often successful for mono-, di- and tervalent elements, is generally not quantitative for both quadrivalent and pentavalent elements. Moreover, when compounds such as refractory (aluminium) silicates or oxides are present in the sample matrix, like in soil and sediment samples, even the recovery of the lower valent elements may be incomplete. Hydrofluoric acid will dissolve silica and some silicates, forming or SiF_6^{2-} ions in solution or volatile SiF_4.[5–8] However, some soil and sediment samples contain refractory minerals, such as chromite ($FeCr_2O_4$), rutile (TiO_2), and zircon ($ZrSiO_4$), which are not dissolved completely by HF or HF/$HClO_4$ mixtures. More rigorous methods such as fusion with a flux at elevated temperatures (e.g., 1000 °C) may be more successful.[9] Fluxes are intermediate reagents, as the solidified sample/flux mixture (melt) generally needs to be dissolved in order to allow any subsequent treatment (e.g., liquid chromatography). Fluxes for rock, soil and sediment samples include carbonates, hydroxides, peroxides, borates, pyrosulfates, bisulfates and fluorides.

The traditional flux for dissolving silicates is sodium carbonate[10] (melting point 851 °C) which at elevated temperatures breaks down complex silicates to Na_4SiO_4. The resulting mixture is soluble in acid, but silica-monomers may transform into insoluble polymeric species. This polymerisation depends on pH, concentration and temperature. These silicate species are commonly in dynamic equilibria that may be quite rapidly established, within hours. Because Na_2CO_3 is a relative mild flux, refractory oxide compounds may not dissolve completely.

Of the lithium borates, $Li_2O{\cdot}nB_2O_3$, the metaborate $LiBO_2$ ($n = 1$) and the tetraborate $Li_2B_4O_7$ ($n = 2$) will react with refractory oxides and silicates at elevated temperatures.[5,10–14] $LiBO_2$ (melting point 845 °C) is preferred for dissolution of acidic oxides (e.g., silica) while $Li_2B_4O_7$ (melting point 920 °C) is preferred for basic oxides.[13] The glassy melt resulting from the fusion between the sample and the borate will generally dissolve in dilute acid. However, acid strength should be controlled within rather close limits. There must be enough to react stoichiometrically with the Li_2O in the flux; however, too much acid results in less complete solution of silica.[14] When H_3BO_3 is combined with Na_2O_2 (or when Na borate is combined with H_2O_2), a compound is formed which can be formulated as $NaBO_3$. This peroxoborate compound can be used together with $LiBO_2$ to dissolve several inert minerals such as chromite and rutile.[15] As with carbonate, peroxide and hydroxide fluxes, silicates dissolve in molten alkali borates; however, they are not eliminated from the reaction mixture and the formation of silica colloids or even precipitates may cause problems in subsequent steps (e.g., liquid chromatography).

Effective compounds to react with siliceous refractory compounds are lithium fluoride (melting point 845 °C) and potassium fluoride (melting point 846 °C) which will form SiF_6^{2-} or volatile SiF_4 with refractory silicates at high temperatures. The fluoride anion forms soluble complexes with most metal ions (however, not with e.g., Ca, Sr and the rare earths). A combination of a potassium fluoride and a pyrosulfate fusion has been used by Sill et al. [4,9,16–20] to dissolve soil and sediment samples containing refractory silicates and oxides. Sodium pyrosulfate ($Na_2S_2O_7$) is formed *in situ* by heating Na_2SO_4 with H_2SO_4 and can be used at temperatures up to nearly 800 °C.[21] Residual fluorides (HF and SiF_4) are volatilised simultaneously, simplifying subsequent procedures. Moreover, the sulfate anions forms soluble anionic complexes with ter-, and quadrivalent actinides. Finally, $Na_2S_2O_7$ at this temperature is also a relative powerful oxidising agent: the last traces of very resistant organic matter and carbides are oxidised more rapidly and completely than by treatment with perchloric acid.

In this work, attempts are described to dissolve the IAEA certified reference material Soil-6[22] and the mineral zircon[23] completely by means of fusion with sodium carbonate, alkali borates ($LiBO_2$, $NaBO_2$ and $Li_2B_4O_7$), potassium fluoride, and sodium pyrosulfate fluxes.

2 METHODS

2.1 General

The certified reference material Soil-6 was obtained from the IAEA (Vienna, Austria). The approximate chemical composition of this soil is 38.5% SiO_2, 22.9% CaO, 8.9% Al_2O_3, 3.7% Fe_2O_3, 2.9% K_2O, 1.9% MgO, 0.6% Na_2O, 0.5% TiO_2 and 0.3% SO_3 (with 19.8% not specified).[24] Zircon ($ZrSiO_4$) was obtained from Johnson-Matthey (Royston, UK).

All reagents were of analytical grade or better and were obtained from Aldrich-Sigma (Bornem, Belgium). Milli-Q Plus (Millipore, Milford, MA, USA) water was used for all solution preparation. A Thermolyne (Dubuque, Iowa, USA) Type 6000 Furnace or a Bunzen burner (propane/air) were used to heat the sample/flux mixture to 1050 and 980 °C, respectively. The heating was continued for 15 min. A 50-mL Pt-crucible (97/3 Pt/Ir) was used to contain the reagents at these temperatures.

Poly(tetrafluoroethylene) [PTFE] labware was used, when working with fluoride-containing solutions. HT-200 Tuffryn membrane filters (0.2 μm; 25 mm Ø) [Pall Gelman Sciences, Ann Arbor, Missouri, USA] were used for filtration, unless otherwise indicated. Weights of residues refer to the weight after drying for 2 h at 60 °C, unless otherwise indicated. In the calculation of the residual yield, it is assumed that the insoluble residue originates solely from the soil or sediment sample and not from the reagents added. This may lead to an overestimation as part of the insoluble residue may originate from the reagents added (e.g., precipitation of CaF_2 due to the addition of HF).

Both the Soil-6 and the zircon samples were dried at 60 °C for at least 24 h, and the Soil-6 samples were dry-ashed at 600 °C before use in order to remove any traces of organic material, unless otherwise indicated. The yield for Soil-6 after ashing was 97.2 ± 0.2 % (w/w). The samples were carefully mixed with the flux before fusion. In general, a small amount of $NaNO_3$ (approximately 0.1 g per g of ash) was added to the sample before fusion in order to facilitate the oxidation of the last organic traces and carbides. In some cases, a small amount of NaCl (less than 0.1 g per g of ash) was added to the sample before fusion to reduce sticking of the melt to the Pt-crucible.

2.2 Na_2CO_3 fusion

A sample of 1.1 g of Soil-6 ash was weighed into a Pt-crucible and mixed with 3.0 g of Na_2CO_3. A green homogeneous melt (3.2 g) resulted after heating at 1050 °C. To this melt was added 125 mL of 1 M HNO_3. After a few days, 0.6 g (58%) of insoluble residue was collected by filtration. This residue did not dissolve in 20 mL of 14 M HNO_3. Subsequently, a second sample of Soil-6 ash was fused with Na_2CO_3 in a similar way. The resulting melt was treated several times with concentrated HNO_3 and HF solutions. However, the resulting residue did not dissolve completely in 2 M HNO_3.

A sample of 2.6 g of zircon was weighed into a Pt-crucible and mixed with 8.2 g of Na_2CO_3. A white inhomogeneous melt (9.7 g) resulted after heating for 1.5 h at 1000 °C.

2.3 Alkali borate fusion

A sample of 0.5 g of Soil-6 was weighed into a Pt-crucible and mixed with 3.1 g of $LiBO_2$, 0.7 g of $Li_2B_4O_7$ and 0.3 g of $NaNO_3$. A glassy melt resulted after heating at 980 °C. 3.8 g of the melt did not dissolve completely in 200 mL of 2.3 M HCl. The solution was extremely difficult to filter. Two other samples of Soil-6 ash were fused with $LiBO_2$, $Li_2B_4O_7$ and $NaNO_3$ in a similar way. One of the melts did dissolve partly in 50 mL of 0.5 M HNO_3. However, within a few days, the solution turned into a gel. The other melt was fused with Na_2CO_3. The resulting melt was dissolved in 50 mL of 4 M HCl. After filtration, 30 mg (5%) of white residue remained. However, soon after the filtration a bulky white precipitate was formed in the filtrate. Within a month the solution turned into a gel.

Finally, a sample of 0.5 g of Soil-6 was weighed into a Pt-crucible and mixed with 2.2 g of $NaBO_2$, 0.3 g of Na_2O_2, 0.5 g of H_3BO_3 and 0.8 g of $LiBO_2$. A glassy melt resulted after heating at 980 °C. The melt (3.4 g) dissolved almost completely in 210 mL of 0.7 M HNO_3/2.5% tartaric acid. 16 mg (3%) of insoluble residue remained.

2.4 Combination of an alkali borate fusion with HF treatment

A sample of 2.1 g of Soil-6 ash was weighed into a Pt-crucible and mixed with 4.7 g of $NaBO_2$, 0.8 g of $Li_2B_4O_7$ and 0.2 g of $NaNO_3$. A glassy melt resulted after heating at 1050 °C. To 1.6 g of this melt was added 25 mL of 10 M HCl en 15 mL of 29 M HF. After evaporation to near dryness, it was attempted to dissolve 2.7 g of the light yellow precipitate in 50 mL of 4 M HCl. After filtration, 35 mg of insoluble residue remained. This residue was dissolved completely in 15 mL of 4 M HCl.

A second sample of 2.0 g of Soil-6 ash was weighed into a Pt-crucible and mixed with 4.5 g of $LiBO_2$, 0.6 g of $Li_2B_4O_7$, 0.1 g of NaCl and 0.2 g of $NaNO_3$. A glassy melt resulted after heating at 1050 °C. To 6.3 g of the melt was added 10 mL of 10 M HCl and

50 mL of 29 M HF. Subsequently, 5.4 g of H_3BO_3 and 90 mL of 4 M HCl were added. After filtration, 1.4 g of white residue was collected. 5 mL of 14 M HNO_3, 1.0 g of H_3BO_3 and 6 mL of 18 M H_2SO_4 were added to the residue. After evaporation to near dryness, the residue was dissolved in 450 mL of 4 M HCl. After filtration, 20 mg (1%) of white insoluble residue remained.

A sample of 1.0 g of zircon was weighed into a Pt-crucible and fused for 2 h at 1100 °C with 6.0 g of $LiBO_2$. The resulting melt dissolved partly in 200 mL of 1 M HNO_3/6 M HF/6% H_2O_2. After evaporation to near-dryness, a white residue was obtained. This residue was fused with 7.0 g of Na_2SO_4 and 4 mL of concentrated H_2SO_4. The resulting melt was dissolved completely in 200 mL of 0.8 M H_2SO_4.

2.5 Alkali fluoride fusion

A sample of 1.0 g of Soil-6 was weighed into a Pt-crucible and mixed with 0.07 g of $NaNO_3$ and 3.4 g of $KF \cdot 2H_2O$. A blue-green homogeneous melt resulted after heating at 980 °C. 1.0 g of H_3BO_3 was added. A brown homogeneous melt resulted after fusion at 980 °C. This melt did not dissolve completely in either 3 M HNO_3 or 4 M HCl.

A second sample of 3.4 g of Soil-6 was weighed into a Pt-crucible and mixed with 5 mL of 14 M HNO_3 and 5 mL of 28 M HF. After evaporation to near dryness 12.1 g of $KF \cdot 2H_2O$ was added. The melt was heated to 980 °C, after which 9 mL of 18 M H_2SO_4 was added. After heating for a short period, 4.9 g of Na_2SO_4 was added. The mixture was heated to 980 °C to form a red-hot solution. After cooling, 18.9 g of the melt did dissolve almost completely 300 mL of 1.7 M HCl. After some time, a very fine precipitation (<10 mg or 0.3%) appeared in the solution.

A third sample of 1.1 g of Soil-6 ash was weighed into a Pt-crucible and mixed with 2.4 g of $LiBO_2$, 1.1 g of $Li_2B_4O_7$, 0.1 g of NaCl and 0.4 g of $NaNO_3$. 4.4 g of a glassy melt resulted after heating at 980 °C. 1.7 g of this melt was fused at 980 °C with 5.3 g of $KF \cdot 2H_2O$. To the resulting 4.8 g of melt was added 140 mL of 4 M HCl. After filtration, 1.7 g of white residue remained. After heating to 980 °C for 15 min, 11.1 mg (3%) of insoluble residue remained.

A fourth sample of 3.1 g of Soil-6 was weighed into a Pt-crucible and mixed with 3 mL of 14 M HNO_3 and 1 mL of 28 M HF. After evaporation to near dryness, 9.5 g of $KF \cdot 2H_2O$ was added. The mixture was heated to 980 °C, after which 2.7 g of $LiBO_2$ was added. The mixture was heated to 980 °C resulting in 10.8 g of melt. It was attempted to dissolve the melt in 350 mL of 2.4 M HCl. The resulting solution was extremely difficult to filter.

A sample of 2.6 g of zircon was weighed into a Pt-crucible and mixed with 4.8 g of LiF, 15.0 g of 18 M H_2SO_4 and 6.0 g of Na_2SO_4 and was slowly heated to 980 °C. After cooling, the white melt did not dissolve completely in 200 mL of 0.7 M HNO_3.

3 RESULTS

3.1 Na_2CO_3 fusion

Samples of Soil-6 seemed to dissolve completely in molten Na_2CO_3 resulting in the formation of apparently homogeneous melts. However, the melt did not dissolve completely in dilute acid. Repeated treatment of the melt with HNO_3/HF did also not result in complete dissolution. Fusion of zircon with Na_2CO_3 did not result in a homogeneous melt. Apparently, Na_2CO_3 as a flux is too mild to dissolve zircon and refractory components in soils.

3.2 Alkali borate fusion

Samples of Soil-6 seemed to be soluble in molten alkali borate, resulting in the formation of glassy and apparently homogeneous melts. However, these melts did not dissolve completely in dilute acid. A major part of the melt initially dissolved only to form either a fine precipitate or a viscous gel after standing for some time. Presumably, initially soluble silicates polymerised into insoluble species.[14] Combination of an alkali borate fusion with a Na_2CO_3 fusion initially resulted in only a small amount of insoluble material, but after some standing the solution turned into a gel. It is again likely that soluble silicates, formed during the fusion, eventually polymerised into insoluble species.

The addition of a small amount of $NaBO_3$ (a compound formed *in situ* from H_3BO_3 and Na_2O_2) to the flux combined with the addition of tartaric acid to the final solution led to a significant decrease in the relative amount of precipitation, while no gel was formed.

3.3 Combination of an alkali borate fusion with HF treatment

Subsequently, it was tried to dissolve the glassy melt resulting from the borate fusion in a mixture of either HF/HCl or HF/HNO_3 in an attempt to volatilise silicate species and borate species as SiF_4 and BF_3, respectively. A sample was fused with alkali borates and subsequently treated with a mixture of concentrated HCl/HF. After evaporation to near dryness, the residue was completely soluble in 4 M HCl. However, upscaling of the procedure by either increasing the melt/solution or the sample/flux ratio, resulted in an incomplete dissolution of the melt. The relative insolubility of LiF compared to NaF is probably not the cause for the observed incomplete dissolution, as Ca^{2+}, a major component in soils, forms a far lesser soluble compound with the fluoride anion, i.e., CaF_2. The precipitates formed in these procedures consisted of white powders, which were easy to collect by filtration, and they did not have the tendency to form gels upon standing, unlike the precipitates formed after the alkali borate fusion without HF treatment (see above).

It has been known for a long time that CaF_2-precipitation may be prevented or even be undone by addition of boric acid (H_3BO_3) which forms a complex with the fluoride anion. The resulting BF_4^- anion is very soluble in aqueous solutions. Although borate species were already present in the reaction mixture, it was decided to study the effect of the addition of H_3BO_3. It resulted in the near complete dissolution (1% of insoluble residue). Zircon was dissolved completely in diluted H_2SO_4 after fusion with $LiBO_2$, treatment with a mixture of HNO_3, HF and H_2O_2 and a second fusion with $Na_2S_2O_7$.

3.4 Alkali fluoride fusion

A Soil-6 sample was fused with KF and H_3BO_3, and it was attempted to dissolve the resulting homogeneous melts in dilute acid. However, a large amount of insoluble residue was obtained. Subsequently, a combination of a KF and pyrosulfate fusion was tried. This was successful: after a KF/pyrosulfate fusion, only approximately 0.3% of insoluble residue remained. Finally, the combination of alkali borate fusion and KF fusion was studied. One order of fusion proved to be rather successful (alkali borate / KF fusion), as only 3 % of the sample did not dissolve. Reversing the order (KF fusion followed by a $LiBO_2$ fusion) was not successful as the solution obtained was extremely difficult to filter, due to a fine precipitation.

Zircon was fused with LiF/pyrosulfate resulting in a homogeneous melt. However this melt did dissolve completely in dilute acid.

4 CONCLUSION

Generally, the Soil-6 samples did dissolve completely during fusion with the molten fluxes Na_2CO_3, alkali borates, KF and $Na_2S_2O_7$. Generally, a homogeneous mixture was formed at 1000 °C, a prerequisite for the equilibrium between an analyte and its yield tracer. However, it proved to be difficult to dissolve the resulting melts completely. Two procedures, (i) alkali borate fusion combined with HF/H_3BO_3 treatment and (ii) $KF/Na_2S_2O_7$ fusion, resulted in near-complete dissolution (>99 %) of Soil-6 samples. Procedures based on fusions with Na_2CO_3 or alkali borates without any subsequent HF-treatment did result in either partly insoluble melts or precipitation and gel-formation due to the polymerisation of silicates.

Zircon was dissolved completely with a procedure consisting of a combination of $LiBO_2$ fusion, $HNO_3/HF/H_2O_2$ treatment and $Na_2S_2O_7$ fusion. Fusions of zircon with Na_2CO_3 or $LiF/Na_2S_2O_7$ did not result in a subsequent complete dissolution of zircon.

Acknowledgements

The author thanks Simon Jerome, Timos Altzitzoglou and Dietmar Reher for the valuable discussions during the preparation of the manuscript and Satwant Johal for technical assistence with part of the work. Part of the work described here was performed at the Institute for Reference Materials and Measurements (IRMM), Geel, Belgium.[22]

References

1 S. Bajo in *Preconcentration Techniques for Trace Elements*, eds., Z.B. Alfassi and C.M. Wai, CRC Press, Boca Raton, USA, 1992, Chapter 2.
2 R. Bock, *A Handbook of Decomposition Methods in Analytical Chemistry*, International Textbook Co., Glasgow, 1979.
3 C.W. Sill, *Health Phys.*, 1975, **29**, 619.
4 C.W. Sill, in *NBS Special Publication 422, Accuracy in Trace Analysis: Sampling, Sample Handling, and Analysis*, 1976, pp. 463-490.
5 Z. Kowalewska, E. Bulska and A. Hulanicki, *Fresenius J. Anal. Chem.*, 1998, **362**, 125.
6 M.H. Hiatt and P.B. Hahn, *Anal. Chem.*, 1979, **51**, 295.
7 B. Bernas, *Anal. Chem.*, 1968, **40**, 1682.
8 R.F. Anderson and A.P. Fleer, *Anal. Chem.*, 1982, **54**, 1142.
9 C.W. Sill and D.S. Sill, *Radioact. Radiochem.*, 1995, **6**, 8.
10 P.E. Jackson, J. Carnevale, H. Fuping and P.R. Haddad, *J. Chrom. A*, 1994, **671**, 181.
11 C.O. Ingamells, *Anal. Chim. Acta*, 1970, **52**, 323.
12 I. Friberg, *J. Radioanal. Nucl. Chem.*, 1999, **241**, 549.
13 H. Bennett and G.J. Oliver, *Analyst*, 1976, **101**, 803.
14 I. Croudace, P. Warwick, R. Taylor and S. Dee, *Anal. Chim. Acta*, 1998, **371**, 217.
15 C. Feldman, *Anal. Chem.*, 1983, **55**, 2451.
16 L.L. Smith and J.S. Yaeger, *Radioact. Radiochem.*, 1996, **7**, 35.
17 C.W. Sill, F.D. Hindman and J.I. Anderson, *Anal. Chem.*, 1979, **51**, 1307.
18 C.W. Sill, K.W. Puphal and F.D. Hindman, *Anal. Chem.*, 1974, **46**, 1725.
19 C.W. Sill, *Anal. Chem.*, 1961, **33**, 1684.

20 L.L. Smith, F. Markun and T. Ten Kate, Argonne National Laboratory, Argonne Report ANL/ACL-92/2, 1992.
21 N.Y. Chu, *Anal. Chem.*, 1971, **43**, 449.
22 A.V. Harms, Internal Report EC-JRC IRMM, GE/R/RN/02/00, 2000.
23 A.V. Harms, S.M. Jerome, D.K. Tyler and M.J. Woods, *Radiat. Prot. Dosim.*, 2001, **97**, 137.
24 IAEA Report on Soil-6, IAEA/RL/111, 1997.

LSC: FROM THE ROUTINE COUNTER TO THE REAL SPECTROMETER

I.A. Kashirin [1], A.I. Ermakov[1], S.V. Malinovskiy[1], K.M Efimov[2], V.A. Tikhomirov[1], A.I. Sobolev[1]

[1]Scientific-Industrial Association "Radon", 7-th Rostovsky lane 2/14,119121 Moscow, Russia.
[2]Institute of Ecology-Technological Problems, Krivorozhskaya st. 33, 113638 Moscow, Russia.

1 INTRODUCTION

The simultaneous analysis of multi-component radionuclide mixture taken from environmental or technological systems represents a very complicated problem in terms of both the uncertainty of the mixture composition and overlapping of the beta-spectra. Often in order to solve this problem the combined use of γ,-α- and β-spectrometry analysis is needed including laborious and expensive radiochemical isolation of alpha and beta components. Liquid scintillation counting (LSC) in a number of cases can do it alone as a consequence of the techniques unique capacity to detect with high efficiency practically all kinds of radiation.

However, being provided with standard software, LSC allows examining mainly the single-component system, or in some cases the two- or three-component mixture with known composition and activity ratios of the same order of magnitude. This may be enough for medical or biochemical tasks but is not very useful for radiation surveillance and control purposes.

In order to increase the capabilities of the LS counters we propose a new approach to LS spectrometry. It is based on special software named "RadSpectraDec" and a spectra library, which is capable of the simultaneous identification and quantification of the components of complex mixtures with unknown composition and consisting of the most abundant radionuclides without regard to their activity ratio[1-3]. The developed approach can be applied to all models of modern LS counter with linear or logarithmic amplification that have file output in ASCII format and allows analysis of the complex mixtures of β-, α-, and X-ray emitters in wide range of quenching parameters.

2 METHOD AND RESULTS

2.1 Reagents

All chemicals were of reagent grade. LS cocktails Ultima Gold AB, Ultima Gold XR, Ultima Gold F, Hionic-Fluor, Pico-Fluor and Insta-Fluor were from Packard Instrument Company, USA.; OptiPhase HiSafe 3 and OptiPhase TriSafe were from PerkinElmer Life Sciences, Wallac Oy, Finland.

Bis(2-ethylhexyl)phosphoric acid (HDEHP) from Fluka had a purity about 60%. Standard radioactive solutions ^{3}H, ^{14}C, ^{22}Na, ^{32}P, ^{33}P, ^{35}S, ^{45}Ca, ^{54}Mn, ^{55}Fe, ^{57}Co, ^{60}Co, ^{63}Ni, ^{65}Zn, ^{85}Kr, ^{88}Y, ^{85}Sr, ^{89}Sr, ^{90}Sr/^{90}Y, ^{99}Tc, ^{103}Pd, ^{106}Ru, ^{133}Ba, ^{134}Cs, ^{137}Cs, ^{125}Sb, ^{125}I, ^{129}I, ^{131}I, ^{144}Ce, ^{147}Pm, ^{148}Gd, ^{152}Eu, ^{154}Eu, ^{226}Ra, ^{236}Pu, ^{237}Pu, ^{238}Pu, ^{239}Pu, ^{241}Pu, ^{242}Pu, ^{232}U, ^{233}U, ^{234}U, ^{235}U, ^{236}U, ^{238}U, ^{232}Th, ^{237}Np, ^{241}Am, ^{243}Am and ^{244}Cm were obtained from All-Russian Scientific Research Institute of Metrology (Russia), Packard Instrument Company (USA) and Amersham International (UK). Their purities were of about 95-97%.

2.2 Apparatus

"Tri-Carb 2550 TR/AB" and "Tri-Carb 2700 TR" liquid scintillation analyzers ("Packard", USA), "1220Quantulus" and "Guardian" liquid scintillation spectrometers (PerkinElmer Life Sciences, Wallac Oy, Finland), "Triathler" liquid scintillation counter (Hidex Oy, Finland) and "SKS-07P" –spectrometer (Green Star, Russia) were used in this study. All measurements were carried out with low diffusion polyethylene vials.

2.3 Creation of Analytical Database and Sample Preparation

The analytical database (the library of quenched spectra) for each radionuclide were obtained for at least 10 quench levels with CCl_4 and $FeCl_3$ as chemical and colour quench agents, respectively, using LS cocktails Ultima Gold AB or OptiPhase HiSafe 3. All spectra in the radionuclides library are normalized and convoluted into the groups of channels.

Separation and purification of ^{226}Ra, ^{210}Bi, ^{210}Pb, ^{210}Po, ^{90}Sr and ^{90}Y from corresponding equilibrium standard radioactive solutions was performed using chromatography on VS-15X resin on the base of crown ether[4]

In order to reduce the quenching and to improve the alpha-peak resolution the majority of α-emitters were extracted by 0.2 M HDEHP solutions in Ultima Gold F or Insta-Fluor LS cocktails from the appropriate acidic mediums. Before the measurement each counting sample was deoxygenated by sparging with dry oxygen-free argon.

Liquid samples for counting were prepared by direct mixing of aliquots with LS-cocktails or after preliminary concentration using evaporation or extraction. The counting samples from soil or sediment were prepared using transformation of the solution after leaching into colorless pyrophosphate form[2].

2.4 Development of the Program

"RadSpectraDec" is written in C++ using Microsoft Visual Studio 6.0 and operates under Windows 95/98/ME/NT/2000. It fits the counter spectrum of the sample to the elemental spectra of individual radionuclides obtained from a previously created nuclide library. The main idea of the program "RadSpectraDec" is that a mathematical algorithm decodes the measured and convoluted spectrum by recursive minimization of a special functional, representing the difference between a model spectrum and the experimental spectrum[1].

Also the algorithm of the group convolution was developed. It is specific for each kind of LS-counter and takes into account the mode of amplification[1,5].

2.5 Testing of the Developed Approach with β-Radionuclides

2.5.1 The LS analyzers with linear *amplification.* The main part of this work has been performed with a "Tri-Carb 2550 TR/AB" analyzer (linear amplification) and then supplemented with investigations using the "Tri-Carb 2700 TR"(linear), "1220 Quantulus"(logarithmic), "Guardian"(logarithmic), "Triathler" (linear and logarithmic) and also the SKS-07P-spectrometer (linear). It was shown that the method is suitable for using with different LS analyzers and LS cocktails. For analyzers with linear amplification in order to convolute the real spectrum into the groups of channels we used the quasi-arithmetic progression[1,3]. The pilot investigations were carried out with the model mixtures, prepared from the standard radioactive solutions and measured on the "Tri-Carb 2550 TR/AB" analyzer. The results obtained are presented in Table 1 and show that the method allows analysis of the complex radionuclide mixtures with different activity levels.

Table 1 *The results of the analysis of the model mixtures, measured on "Tri-Carb 2550 TR/AB*

Mix	Bq/sample, obtained **(actual contents)**									
№	^{55}Fe	^{3}H	^{63}Ni	^{14}C	^{35}S	^{137}Cs	^{90}Sr/^{90}Y	^{60}Co	^{125}I	^{241}Am
1				145±40 **(120)**	1225±120 **(275)**					
2	3.0±0.6 **(3.1)**		62±6 **(62)**							
3	432±110 **(540)**			720±140 **(830)**			4.8±0.8 **(5)**			
4		874±175 **(1000)**	1.4±0.4 **(1)**							
5						103±18 **(100)**	0.8±0.2 **(1)**			
6	1.03±0.21 **(1)**					0.19±0.04 **(0.2)**	0.09±0.02 **(0.1)**			0.58±0.12 **(0.5)**
7		460±78 **(500)**	1.1±0.2 **(1)**			0.67±0.21 **(0.5)**	0.17±0.03 **(0.2)**			0.47±0.09 **(0.5)**
8		0.3±0.1 **(0.2)**	0.21±0.05 **(0.2)**	0.28±0.08 **(0.2)**		0.25±0.06 **(0.2)**	0.19±0.05 **(0.2)**	0.10±0.04 **(0.2)**	0.19±0.05 **(0.2)**	0.22±0.05 **(0.2)**

It seems that many difficult radioanalytical tasks may be solved without chemical separation by the joint use of analyzer and this program e.g. simultaneous determinations of ^{14}C and ^{35}S in mixtures as well as the soft emitting corrosion radionuclides ^{55}Fe and ^{63}Ni with different activity ratios. Using the developed software it is possible to identify and quantify the components of complex mixtures of low- and high-energy emitting radionuclides. One example of spectrum decoding for the more complicated eight-components radionuclides mixture 8 (Table 1) is presented in Figure 1.

As we obtained good results in the model tests, the developed approach has been applied to the analysis of real technological and environmental samples without radiochemical separation of the individual components. There are two examples of spectral decoding for a real waste technological mixture and liquid sample, taken from the borehole in the region of subterranean nuclear tests shown in Figures 2 and 3.

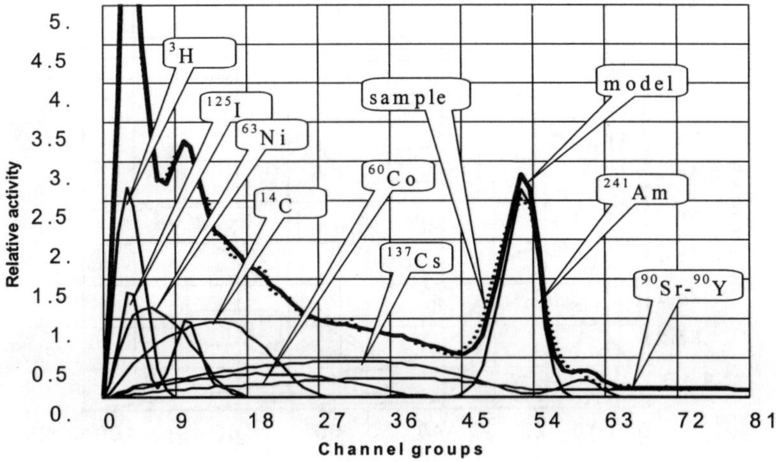

Figure 1 *The spectra decoding for the model 8-components mixture (0.2 Bq-each.)*

Tests with the "SKS-07P spectrometer" (Green Star, Russia), which is equipped with two photomultiplier tubes (R331, Japan) and has a linear mode of amplification, suggested the possibility of using this LS counter and the software "RadSpectraDec" as well as the "Tri-Carb" models. Also the developed approach is suitable for application with the "Triathler" counter in the linear mode of amplification. However, it is less able than above mentioned spectrometers as a consequence of using only one photomultiplier tube and the impossibility of determining isotopes with low energy such as ^{55}Fe.

2.5.2 *The LS analyzers with logarithmic amplification.* For the logarithmic amplification of the "Quantulus", "Guardian" and "Triathler" we also proposed dividing a counter spectra into groups of channels. Each group consists of ten consistent channels and thus the whole scale presents about 100 groups[5]. This allows us to get more statistically reliable information for calculations and increases the accuracy of the analysis. In order to take into account the logarithmic mode of amplification it was necessary to rebuild the existing analytical database and to slightly modify the program. The testing was performed with the use of model mixtures prepared from standard radioactive solutions with OptiPhase HiSafe 3 LS cocktail and next with real complex technological mixtures. It was shown, that both the "Quantulus" and "Guardian" with the assistance of software "RadSpectraDec" are capable of solving the most complicated radioanalytical problem as well as the spectrometers with the linear mode of amplification ("Tri-Carb"). Also the "Triathler" counter in the logarithmic amplification count mode may be successfully used for analysis of a number of radionuclides mixtures with their components energies greater or equal to tritium (18.6 keV). As an example, the obtained results of the comparative testing of the "Quantulus" and "Tri-Carb" spectrometers using the "RadSpectraDec" program are presented in Table 2. The process of spectra decoding for this analyzed 6-components mixture measured on the "Quantulus" is shown in Figure 4.

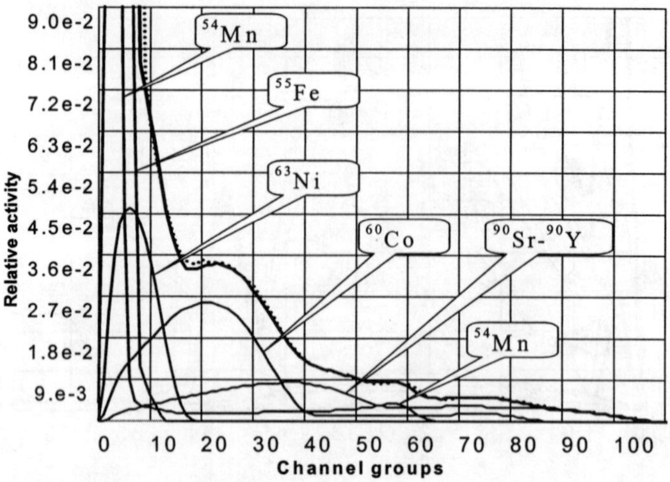

Figure 2 *The spectra decoding for the waste technological mixture at a stage of the* ^{63}Ni *preparation:* ^{54}Mn *- 650 ± 13 Bq;* ^{55}Fe *- 60000 ± 600 Bq;* ^{60}Co*-260 ± 8 Bq;* ^{63}Ni *- 200 ± 8 Bq; (*^{90}Sr*+*^{90}Y*)- 140 ± 7 Bq.*

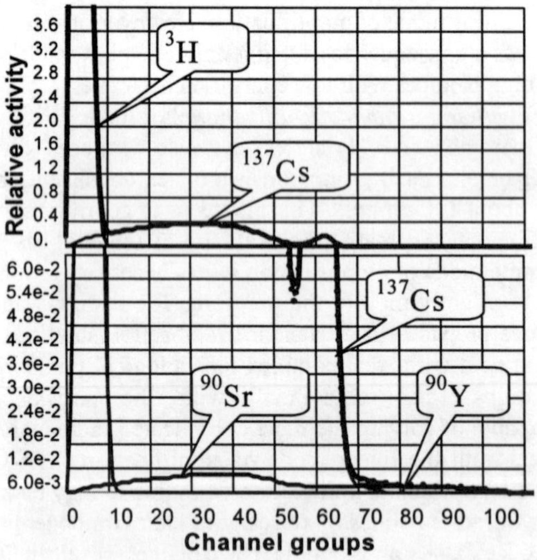

Figure 3 *The spectra decoding for the liquid sample taken from the borehole in the region of subterranean nuclear tests:* ^{3}H *- 5700 ± 60 Bq;* ^{137}Cs *- 450 ± 5 Bq; (*^{90}Sr*+*^{90}Y*) - 9± 2 Bq.*

Table 2 *The results of testing the 6-components mixture on the "Quantulus" and "Tri-Carb" using "RadSpectraDec".*

Mixture composition	Activity, Bq/sample		
	Obtained		Actual
	1220 Quantulus	Tri-Carb 2550 TR/AB	
^{3}H	11±2	9±2	11
^{35}S	30±6	26±6	32
^{60}Co	15±3	16±3	12
^{63}Ni	9±2	11±2	10
$^{90}Sr+^{90}Y$	25±5	25±2	22
^{137}Cs	11±2	11±2	10

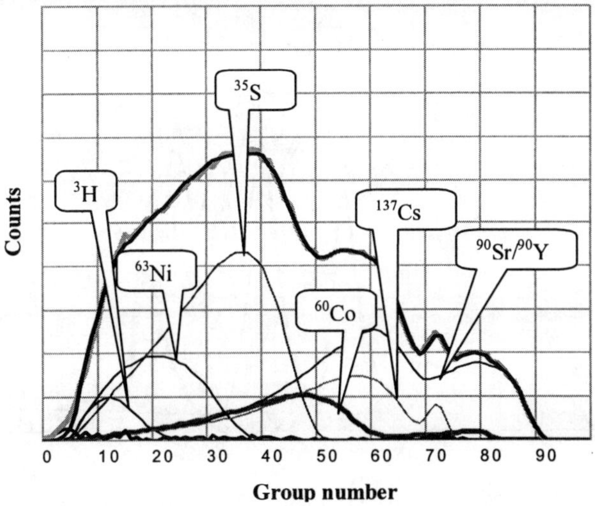

Figure 4 *The spectra decoding for the model 6-components mixture (^{3}H, ^{63}Ni, ^{35}S, ^{60}Co, $^{90}Sr/^{90}Y$, ^{137}Cs) measured on the "Quantulus".*

2.6 Testing of Developed Approach with α-Radionuclides

It is known, that determination of the alpha-emitting radionuclides by liquid scintillation spectrometry presents a very difficult problem because of poor resolution of α-peaks. In the most cases it is possible to correctly determine only a single α-radionuclide in the mixture.

The developed approach can be used for simultaneous identification and quantification of alpha-emitting radionuclides in the complex mixtures as well as beta-emitters. In order to minimize the quenching effects and to produce the best energy resolution during LS counting we used the liquid-liquid extraction of the α-radionuclides from HNO_3 solutions by a scintillating-extractive cocktail that contains bis(2-ethylhexyl)phosphoric acid (HDEHP) as extractive molecule and Ultima Gold F or Insta-Fluor as LS cocktails.

It was noted that a commercially available HDEHP from Fluka has a purity about 60% and may contain up to 40% mono-2- ethylhexylphosphoric acid. This crude mixture had a low specificity for extraction of the most α-radionuclides. Data obtained has shown that isotopes of U, Th, Pu, Am, Cm, Np and Gd can be extracted from 0.5 M HNO_3 solution with yields of about 100%. The joint application of the liquid-liquid extraction with crude HDEHP, LS counting and "RadSpectraDec" software may be used for "screening" of samples. This provides for each sample a list of α-radionuclides for performing individual radiochemical separations with monitoring of chemical yield, to evaluate the required activities of tracers, and, in a number of cases, immediately obtain the values of radionuclide activities in the sample. For example, the process of spectrum decoding of the α-emitting radionuclides mixture after HDEHP (crude) extraction from 0.5M HNO_3 solution with the use of "RadSpectraDec" in Figure 5.

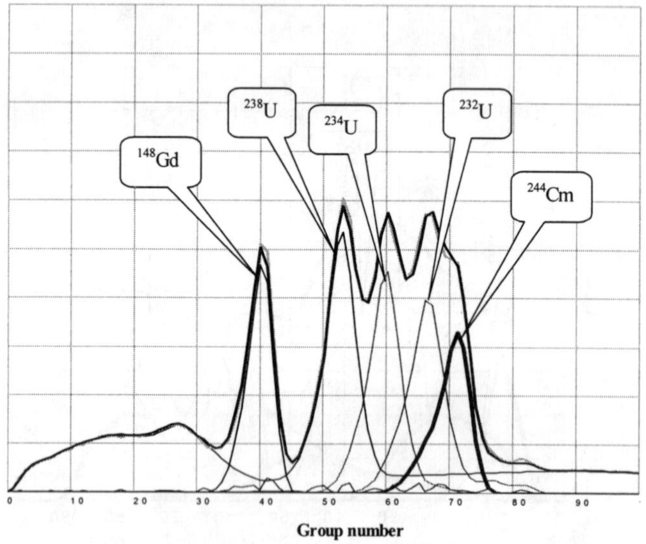

Figure 5 *The spectrum decoding of the α-emitting radionuclides mixture after HDEHP (crude) extraction.*

It must be noted that, as a rule, ionization quenching is present in the mixtures of α-emitting radionuclides in addition to traditional chemical and optical quenching. This may make it difficult to correctly decode the spectrum because of unexpected shifts of the spectra owing to ionization. In order to avoid this problem we suggest using ^{148}Gd (Eα 3.18 MeV) as the standard for quenching value correction because of its small energy of emission which considerably differs from the energies of most commonly used alpha-radionuclides. Its localization allows us to determine the exact quenching value and to take into account the existing spectrum shift for correct identification and quantification of the α-emitters.

Purified HDEHP as an extractive molecule becomes more specific and can be used for the selective extraction of isotopes of U, Th, Pu, Am etc. from environmental or technological mixtures under the certain conditions[6-8]. This approach is widely used in PERALS technology where relative resolution (e.g., fwhm/E_α) of α-peaks is about 4-7%[9]. Using extraction with HDEHP and counting on a conventional Tri-Carb analyzer it is possible to increase the relative resolution to 9-10% and it is sufficient for the correct spectrum decoding of the mixtures α-emitters with the help of "RadSpectraDec" program. An example of spectra decoding for a mixture of U-isotopes and the data obtained are presented in Table 3 and Figure 6.

Table 3 *The results of the analysis of a U- isotopes mixture after HDEHP-extraction.*

Radionuclide composition of the mixture	Activity, Bq/sample	
	Actual contents	Obtained contents («RadSpectraDec»)
^{232}U	5.0	5.1±1.3
^{234}U	5.0	4.1±1.1
^{235}U	6.0	4.9±1.3
^{238}U	6.0	5.9±1.2

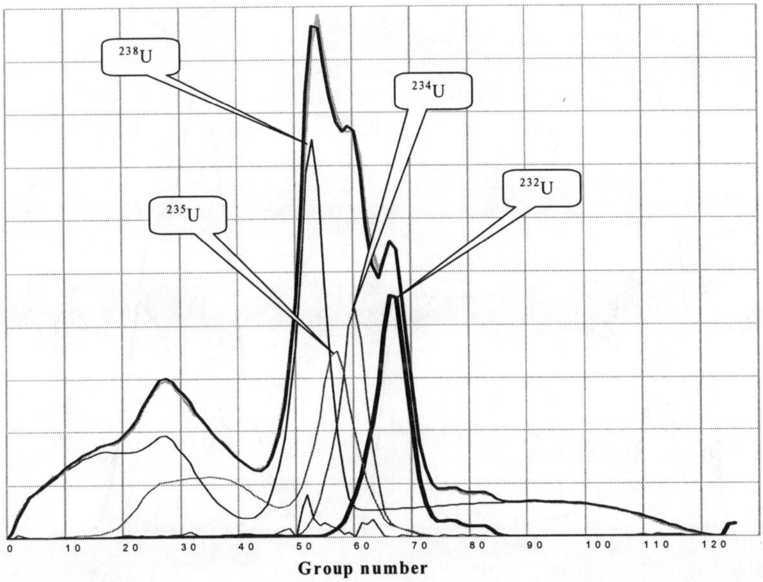

Figure 6 *The spectrum decoding of a U- isotopes mixture after HDEHP extraction from 0.5 HNO_3 solution.*

Using the HDEHP-extraction, LS counting and "RadSpectraDec"-processing it was possible to simultaneously determine 6 isotopes of Pu both beta and alpha-emitting. The results obtained are shown in Figure 7 and Table 4.

Table 4 *The results of the analysis of a Pu- isotopes mixture with ^{148}Gd as quenching value standard.*

Radionuclide composition of the mixture	Activity, Bq/sample	
	Actual contents	Obtained contents («RadSpectraDec»)
^{236}Pu	2.7	2.7±0.5
^{237}Pu	3.6	3.6±0.6
^{238}Pu	2.7	3.2±0.6
^{239}Pu	4.0	4.3±0.7
^{241}Pu	2.8	2.4±0.5
^{242}Pu	3.0	3.5±0.7
^{148}Gd	3.0	2.9±0.3

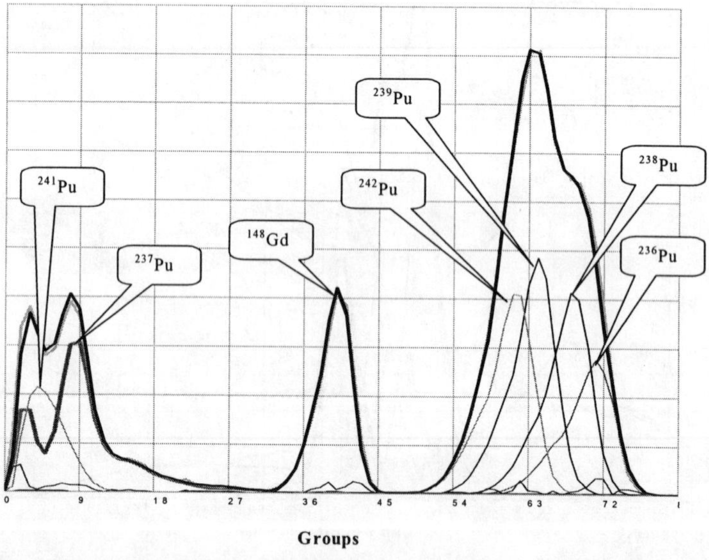

Figure 7 *The spectrum decoding of a Pu- isotopes mixture after HDEHP extraction. ^{148}Gd was used as the standard for the quenching value correction.*

This developed approach was successfully applied to the analysis of real environmental and technological samples and good results have been obtained. An example of the analysis of a potable water sample taken from the borehole using extraction with HDEHP followed by LS counting and spectra decoding with ^{232}U as the tracer of radiochemical yield is presented in Figure 8 and Table 5.

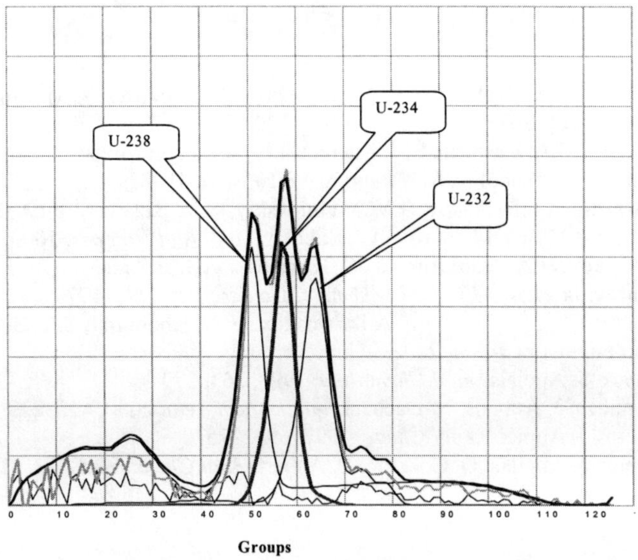

Figure 8 *The spectrum decoding of the real environmental sample (potable water, VZU-8, borehole) after HDEHP- extraction from 0.5M HNO_3 solution.*

Table 5 *The results of the analysis of the U- isotopes mixture from the potable water, VZU-8, borehole)*

Radionuclide composition of the mixture	Activity, Bq/L	
	Obtained contents («RadSpectraDec»)	α-spectrometry data
^{234}U	0.53±0.05	0.55±0.06
^{238}U	0.51±0.05	0.47±0.06

3 CONCLUSION

A new method to simultaneously measure the activity of complex radionuclide mixtures using liquid scintillation counting (which is applicable to the analysis of environmental and technological samples) is presented. From our point of view, it has been useful to create a real liquid scintillation spectrometer, which allows a radionuclide assessment for each sample, and in a number of cases, immediately obtain the values of radionuclide activities in sample. The elimination of time-consuming stages such as the radiochemical separation of the individual radionuclides considerably simplifies the analysis. The precision of the described method depends on the quality of reference solutions, chemical and radionuclide mixture composition and counting statistics.

References

1 S.V. Belanov, I.A. Kashirin, S.V. Malinovskiy, M.E. Egorova, K.M. Efimov, V.A. Tikhomirov, A.I. Sobolev, *RF Patent* № 2 120 646, 1997.
2 S.V. Belanov, I.A. Kashirin, S.V. Malinovskiy, A.I. Ermakov, K.M. Efimov, V.A. Tikhomirov, A.I. Sobolev, *RF Patent* № 98 106 407, 1998.
3 I.A. Kashirin, A.I. Ermakov, S.V. Malinovskiy, S.V. Belanov, Yu.A. Sapozhnikov, K.M. Efimov, V.A. Tikhomirov, A.I. Sobolev, *Appl.Rad.Isotopes*, 2000, **53**, 303.
4 A.I. Ermakov, I.A. Kashirin, A.V. Kovalev, V.B. Ribalka, A.I. Sobolev, V.A. Tikhomirov, *In Theses 13th Radiochemical Conference*, 1998, 177.
5 S.V. Malinovsky, I.A. Kashirin, A.I. Ermakov, V.A. Tihomirov, S.V. Belanov, A.I.Sobolev, *Radiocarbon*, 2002, in press.
6 N.Dacheux, J. Aupiais, *Anal.Chem.Acta*, 1998, **363**, 279.
7 C. Veronneau, J. Aupiais, N. Dacheux, *Anal.Chem.Acta*, 2000, **415**, 229.
8 N.Dacheux, J. Aupiais, *Anal.Chem.*, 1997, **69**, 2275.
9 N.Dacheux, J. Aupiais, O. Courson, C. Aubert, *Anal.Chem.*, 2000, **72**, 3150.

SIMULTANEOUS DETERMINATION OF LONG-LIVED RADIONUCLIDES IN ENVIRONMENTAL SAMPLES

N. Vajda, Zs. Molnár, É. Kabai and Sz. Osváth

Institute of Nuclear Techniques, Budapest University of Technology and Economics
Budapest, 1521-Hungary

1 INTRODUCTION

The majority of long-lived radionuclides produced in the nuclear fuel cycle can be regarded as "difficult-to-determine nuclides" (DDN) due to the low specific activities and the absence of γ radiations of medium or high energies in the decay schemes. Most of the transuranium nuclides are α emitters, fission products e.g. ^{90}Sr, ^{129}I and ^{99}Tc emit almost exclusively β particles, and some of the activation products e.g. ^{59}Ni, ^{63}Ni also decay by pure β emission processes. The radionuclides to be discussed in this paper as well as their major nuclear properties are listed in Table 1.

Table 1 *The nuclear properties of the radionuclides under investigation*

Nuclides	Half-life	Decay mode	Radiation to be detected
Radionuclides/tracers used for recovery determination			
^{85}Sr	60 days	E.C.	E_X = 13.4 keV
^{242}Pu	3.7x10^{5} years	α	$E_\alpha \approx$ 4.9 MeV
^{243}Am	7380 years	α	$E_\alpha \approx$ 5.3 MeV
daughter:			
^{239}Np	2.35 days	β	E_γ = 103, 106 keV
^{232}U	70 years	α	$E_\alpha \approx$5.3 MeV
daughter:			
^{228}Th	1.9 years	α	$E_\alpha \approx$ 5.4 MeV
^{99m}Tc	6 hours	β	E_γ = 140 keV
^{131}I	8 days	β	E_γ = 364 keV
activation product of I:			
^{126}I	13 days	β	E_γ = 388 keV

Nuclides	Half-life	Decay mode	Radiation to be detected
Nuclides to be determined			
^{90}Sr	28 years	β	$E_{\beta Max}$ = 550 keV
daughter:			
^{90}Y	64 hours	β	$E_{\beta\ Max}$ =2.27MeV
^{89}Sr	50 days	β	$E_{\beta\ Max}$ = 1490 keV
^{239,240}Pu:			
^{239}Pu	2.44×10^{4} years	α	$E_{\alpha} \approx$ 5.1 MeV
^{240}Pu	6600 years	α	$E_{\alpha} \approx$ 5.2 MeV
^{238}Pu	86.4 years	α	$E_{\alpha} \approx$ 5.5 MeV
^{237}Np	2.2×10^{6} years	α	$E_{\alpha} \approx$ 4.7 MeV
^{241}Am	432 years	α	$E_{\alpha} \approx$ 5.5 MeV
^{244}Cm	7.9 years	α	$E_{\alpha} \approx$ 5.8 MeV
^{242}Cm	163 days	α	$E_{\alpha} \approx$ 6.1 MeV
^{63}Ni	100 years	β	$E_{\beta\ Max} \approx$ 67 keV
^{59}Ni	7.5×10^{4} years	E.C.	Kx= 6.93 keV
^{238}U	$4.5.10^{9}$ years	α	$E_{\alpha} \approx$ 4.2 MeV
^{235}U	7.0×10^{8} years	α	$E_{\alpha} \approx$ 4.5 MeV
^{234}U	2.4×10^{5} years	α	$E_{\alpha} \approx$ 4.8 MeV
^{232}Th	1.4×10^{10} years	α	$E_{\alpha} \approx$ 4.0 MeV
^{230}Th	7.5×10^{4} years	α	$E_{\alpha} \approx$ 4.7 MeV
^{99}Tc	2.13×10^{5} years	β	$E_{\beta\ Max}$ = 293 keV
^{129}I	1.57×10^{7} years	β	
activation product:			
^{130}I	12 hours	β	E_{γ} = 536 keV

The major objective of the present work was to obtain a method that is applicable for the simultaneous determination of many long-lived nuclides of the nuclear fuel cycle. These isotopes represent long-term hazards for the environment via releases from nuclear facilities, contamination due to nuclear explosions or accidents.

A combined radiochemical separation procedure has been developed that enables the simultaneous determination of several DDNs from a single sample aliquot. The technique is faster, simpler and less labour consuming than parallel analyses of sub-samples – by reducing troubles of sample destruction –and it is appropriate for the determination of several nuclides in tiny samples that cannot be further divided as e.g. hot particles.

Several radiochemical procedures have been developed and published that are appropriate for the determination of single radionuclides and groups of isotopes e.g. those of Pu and Am; Pu, Am and Sr;[1] Pu, Am, Np and Sr;[2,3] the actinides,[4] etc. Standard procedures have been prepared for the analysis of Sr, Pu, Am, I, Tc nuclides etc. in different matrices.[5,6] The present work focuses on the optimal combination of selected procedures that have been partially developed and discussed by excellent radiochemists in different laboratories, partially studied in our laboratory co-operating with the IAEA's Laboratories at Seibersdorf. The selective extraction chromatographic materials were developed and tested by the group of P. Horwitz.[7] They developed the extraction chromatographic procedure for separation of Am from the matrix components.[8] The Am-lanthanide separation procedure was adopted from the Eichrom Guide.[9,10] Our studies comprise ^{90}Sr analysis,[11] simultaneous determination of ^{129}I and ^{99}Tc isotopes,[12] as well as that of ^{237}Np together with other actinides.[13,14] The recent developments about ^{237}Np separation are concisely discussed in this paper.

The present procedure is based on a series of co-precipitation and extraction/extraction chromatographic separations followed by nuclear measuring methods i.e. α, β, γ, X-ray spectrometries and neutron activation analysis. The basic structure of the method as well as "tailoring" problems arising while assembling the single procedures are discussed. The method was validated by analysing standard reference materials, an example is given for the analysis of all listed nuclides in a single sample aliquot. Some application examples for the analysis of groups of radionuclides in environmental samples are shown.

2 MATERIALS AND METHODS

The following extraction chromatographic materials were used:
UTEVA: supported di-pentylpentyl phosphonate
TRU: supported N,N-octylphenyl-di-i-butylcarbamoylmethyl phosphine oxide
TEVA: supported methyl-octyl-di-decyl ammonium nitrate
SR.SPEC: supported bis-(t-butylcyclohexano)-crown(18,6)ether
DMG: supported di-methylglyoxime
The first four materials were purchased from Eichrom Inc., the last one was prepared in our laboratory. Column sizes for UTEVA, TRU and TEVA were: length 34 or 48 mm, diameter 7 mm, for Sr.Spec and DMG were: length 100 mm, diameter 9 mm.
Analytical grade chemicals, certified radioactive solutions and standard reference materials were used.

α, β, γ, X-ray spectrometric measurements were performed with calibrated PC-based multichannel analysers attached to the relevant detectors, i.e. Si, liquid scintillator, well type HP Ge and Si(Li). All equipment is operated and maintained according to ISO 17025 guidelines.

For neutron activation measurements samples were irradiated at the Budapest Research Reactor facility of the KFKI AEKI, at a thermal neutron flux of about 10^{14} /s/cm^2.

The radiochemical procedure is based on co-precipitations used for pre-concentration and source preparation and on extraction or extraction chromatography for selective separation/purification purposes. The flow-chart of the procedure is shown in Fig. 1.

Exact experimental conditions for the separation of all the radionuclides from 1 g of exhausted ion exchange resin are given below. Results of analysis are shown in the next chapter under "Discussion".

1. The following tracers and carriers were added to 1 g of resin in a closed destruction vessel: about 0.1 Bq ^{232}U-^{228}Th, 0.1 Bq ^{242}Pu, 0.2 Bq ^{243}Am-^{239}Np, 10 mg Sr, 6 mg Ni and 2 mg I. The sample was boiled/destroyed with 50 ml 98% sulfuric acid, adding dropwise 15 ml 65% nitric acid. Volatile components including iodine were passed through 2 traps filled with 100 mL 3M NaOH and 100 mL 0.6 M NaOH.
2. Elemental I_2 was extracted from the acidified trap solution after the addition of 0.5 g of $NaNO_2$ with 3x10 mL CCl_4. Iodide was reduced and back-extracted with 3x5 mL distilled water containing 10 µL hydrazine each. The aqueous phase was evaporated and transferred to high purity quartz vials in the presence of 20 µL 0.1M LiOH and 1 mL 25% NH_3. Closed ampoules were irradiated together with standards of ^{129}I and I^- carrier for 10 hours followed by cooling of about 12-15 hours. Ampoules were broken with a special tool, residues were washed and dissolved with about 60 mL 0.2 M NaOH containing 50 µL hydrazine. I_2 was

extracted from the acidified solution after the addition of 0.5 g of $NaNO_2$ with 2x5 mL CCl_4. The γ spectrum of the combined organic phase was determined.

3. Tc was extracted from the sulfuric acid solution after dilution to about 3M with 5x10 mL tri-butyl phosphate followed by its back-extraction with 3x10 mL 2M NaOH. Tc was purified by extraction chromatography using TEVA column. It was loaded from 1 M HCl, the column was washed with 0.1 M HNO3 and Tc was stripped with 15 mL 8M HNO3. Finally Tc was co-precipitated with about 6 mg Mn as MnS, dissolved with 1 mL 2M HCl and mixed with scintillation cocktail. The β radiation of ^{99}Tc was detected by LSC. The source purity was controlled by γ spectrometry.
4. Actinides were pre-concentrated from the diluted sulphuric acid solution with ferrous hydroxide. About 1g Mohr salt and 2 mL hydrazine were added to the solution and the pH was adjusted to 7 using 25% NH_3.
5. The precipitate was dissolved and oxidized by evaporating with 3x10 mL 65% HNO_3. The residue was taken up in 10 mL 1M HNO_3. Oxidation state was adjusted by addition of 0.5 g $NaNO_2$. After standing for overnight the acidity was also adjusted to 8M and the solution was loaded on a UTEVA column. The time of separation was reported. Pu was selectively stripped with 15 mL 9M HCl-0.1M NH_4I, Th and Np were stripped together with 10 mL 4M HCl and finally U was eluted with 20 mL 0.1M HCl. Thin α sources of each separated element were prepared by micro co-precipitation with lanthanide fluoride according to Sill:[15] To 20 ml of 1M HCl or HNO_3 50 µg of Nd and 5 mL of 40% HF were added.
Small amounts of reducing agents (e.g. 100 mg Mohr salt) were used to transfer actinides to ter- or tetra-valent oxidation states. Co-precipitated nuclides were filtered on 0.2 µm pore size Teflon membranes.
Th and Np were separated from each by the following procedure: The strip solution was evaporated, chlorides were converted to nitrates by evaporation with 3x5 mL 65% HNO_3, residues were taken up in 20 mL 1M HNO_3 and warmed for 2 hours with 250 µL 0.01 M $KMnO_4$. A co-precipitated Th/NdF_3 source was prepared. In the filtrate Np was reduced with 1 mL 5M NH_2OH followed by its co-precipitation resulting in Np/NdF_3 source that was measured by γ spectrometry immediately. All the thin sources were counted by α spectrometry.
6. The effluent of UTEVA was combined with the filtrate of $Fe(OH)_2$. 1 g of $CaCl_2.2H_2O$ and 5 g oxalic acid were added and the pH was adjusted to 5 with 25% NH_3. Sample was filtered.
7. The oxalate precipitate was destroyed by evaporation with 3x10 mL 65% HNO_3. The residue was taken up in 10 mL 1M HNO_3. About 300 mg ascorbic acid was added and the sample was loaded on a conditioned TRU column. Column was eluted with 10 mL 2M HNO_3 and 1 mL 9M HCl. Am and Cm were eluted with 15 mL 4M HCl. Separation from lanthanides was not performed. α source was prepared by micro co-precipitation with neodymium fluoride and counted by α spectrometry.

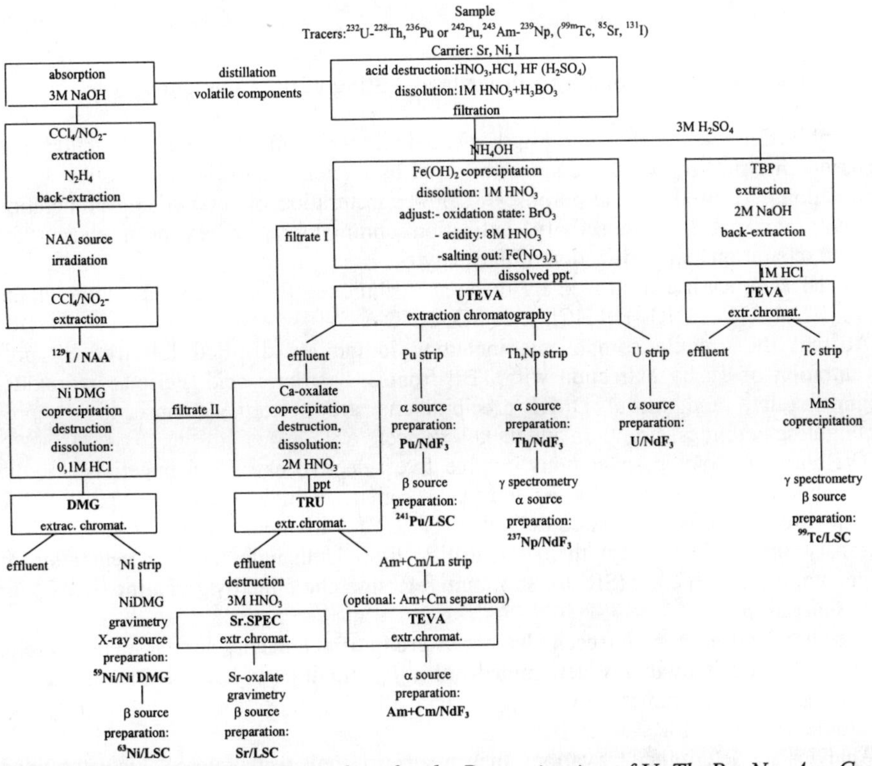

Figure 1 *Flow chart of the Procedure for the Determination of U, Th, Pu, Np, Am, Cm, Sr, Ni, I and Tc isotopes*

8. The effluent and the eluate of the TRU column were evaporated to dryness, destroyed with 3x5 mL 65% HNO_3 and the residue was taken up in 30 mL 3M HNO_3 and loaded on a conditioned Sr.SPEC column. The column was eluted with 80 mL 3M HNO_3, Sr was stripped with 30 mL distilled water. A Sr oxalate source was prepared after the addition of 300 mg oxalic acid and adjusting the pH to 9 with NH_3. Yield was determined by gravimetry. Sample was dissolved in a LS vial with 1 mL 1M HNO_3 and mixed with LS cocktail followed by repeated liquid scintillation countings.
9. Ni was precipitated from the calcium oxalate filtrate with about 0.1 g di-methyl glyoxime. The precipitate was filtered and dissolved with 1M HCl and purified on a DMG column. Ni was finally precipitated with DMG, filtered and the recovery was determined by gravimetry. X-ray spectrometry was applied to analyse ^{59}Ni. Then, the DMG complex was destroyed with 3x2 mL 65% HNO_3 and the residue was dissolved with 1 mL 0.5 M HCl, mixed with LS cocktail and counted by LSC to determine ^{63}Ni.

3 DISCUSSION

3.1 The Structure of the Combined Procedure, Fitting of the Components

The combined procedure shown in Fig. 1 is based on precipitations, solvent extractions and extraction chromatographic separations followed by nuclear measuring techniques. While precipitations are used for the purposes of pre-concentration or final source preparation thus not requiring high selectivity, extraction chromatography has been chosen for selective separation and purification of the analytes.

Spiked samples are destroyed by different techniques fitted to the character of the matrix (HNO_3-HCl, HNO_3-HF-HCl, or H_2SO_4-HNO_3).

At first the volatile components including iodine are distilled followed by pre-concentration of Tc by extraction with TBP from 3M sulfuric acid. Ferrous hydroxide, calcium oxalate and nickel DMG precipitations are performed consecutively pre-concentrating actinides, strontium and nickel, respectively.

The chromatographic separations involve five selective extraction materials, i.e. the organic phosphonate (UTEVA) for tetra-hexa valent actinides (Th, U, Np, Pu), the carbamoylmethyl phosphine oxide derivative (TRU) for trivalent actinides (Am,Cm), the quaternary amine (TEVA) for their purification from lanthanides and the separation of technetium, the crown ether (SR) for strontium retention and dimethylglyoxime (DMG) for nickel separation.

The chemical recovery for each element or group of chemically similar elements (e.g. Am and Cm) are individually determined either by gravimetry (e.g. Sr, Ni) or nuclear spectroscopies (γ spectrometry for I, Np, occasionally Tc, α spectrometry for the majority of α emitters).

Analytes are determined by various nuclear spectroscopic techniques. ^{129}I is determined by neutron activation analysis. The procedure includes the post-irradiation separation of I. The β radiation of ^{99}Tc is detected by LSC. Thin α sources are prepared by micro co-precipitation with lanthanide fluorides and detected by α spectrometry. The β spectrum of ^{90}Sr together with ^{90}Y daughter (and occasionally with ^{89}Sr) is determined by LSC. X-ray spectrometer is used to analyse ^{59}Ni and LSC to determine ^{63}Ni.

Components, sub-procedures have to be carefully selected and adjusted to each other to avoid interferences, cross-talks during measurement, to meet the sometimes contradictory demands and to achieve the highest sensitivity and accuracy.

- The sequence of the steps was chosen according to the nature of the DDNs and the matrices: volatile components were removed before any destruction in open systems. After precipitation of iron hydroxide at pH 7 oxalates were precipitated followed by NiDMG if NH_3 had been used previously to adjust the pH forming soluble Ni complexes. If samples contain lot of alkaline earth metals (e.g. corals) they begin to precipitate with iron hydroxide carrying some Sr. In this case UTEVA effluents are combined with the iron hydroxide filtrate before oxalate precipitation. For samples of low Ca and Sr content (e.g. nuclear wastes) the UTEVA effluent is used for Am-Cm analysis and Sr is determined from the iron hydroxide filtrate.
- Tracers were also carefully selected. The ^{243}Am tracer served as yield monitor for Am and Cm and via its ^{239}Np daughter for ^{237}Np. Its amount was selected so that both the parent nuclide could be determined by α, and the daughter by γ spectrometry. ^{232}U tracer unless it has been recently purified contains ^{228}Th that

offers possibilities for the determination of Th recoveries. Using carriers for yield calculations the natural content of the carrier in the sample has to be taken into account/correction.

- Chemical procedures were adjusted to the demands of nuclear measuring techniques. α sources were prepared by micro co-precipitation with NdF_3. This technique offered a further purification step that was used to separate Np from Th (see Paragraph 5 in Materials and Methods) or similarly Pu from traces of Th. LSC measurements require small quenches that can be assured by colorless samples, low acid and salt contents. That's why e.g. bright purple NiDMG complexes were destroyed and Tc was co-precipitated with MnS from 8M nitric acid before mixing them with cocktail.

The combined procedure is composed of sub-procedures for the separation of iodine, technetium, nickel, strontium and actinides that can be applied not only simultaneously according to the total flow chart but basically independent of the other sub-procedures. In the latter case some pre-concentration steps can be occasionally omitted.

- Separation of Sr starts with Ca oxalate co-precipitation and the Fe(OH)x scavenge is not necessary to obtain high source purity mainly due to the high selectivity of the crown ether.
- On the contrary, to obtain properly high decontamination factor for Ni the Fe(OH)x scavenge is desirable, and can help improve the separation from Co isotopes that have high activities in certain waste samples.
- The highest decontamination factors are needed for the analysis of low-levels of ^{99}Tc where traces of impurities would inhibit the accurate evaluation of the beta spectra. ^{60}Co impurities were experienced in some Tc sources despite of performing all the separation steps. (The decontamination factor for ^{60}Co was about 10^4.)
- To determine Am the preliminary removal of actinides of higher oxidation states by the UTEVA is necessary, because tetravalent actinides would be preferentially retained by the TRU column causing possible interferences. The Am-lanthanide separation can be omitted for those samples which do not contain interfering amounts (less than 0.1mg in each sample) of lanthanides.
- The determination of the actinides should start with ferrous hydroxide co-precipitation. Ferric hydroxide is not efficient to concentrate actinides of higher oxidation states, especially U. Thus, the basic procedure for U or Pu analysis consists of ferrous hydroxide co-precipitation, extraction chromatographic separation on UTEVA and preparation of co-precipitated α sources. Both U and Pu are retained on UTEVA from 8M HNO_3-0.1M $Fe(NO_3)_3$. Oxidation state adjustment is required to extend the procedure for the retention of Np.

3.2 Extension of the Procedure to Separate Np

The central part of the procedure is the separation of actinides using UTEVA. According to the product catalogue and previous studies, the di-pentylpentyl phosphonate has similarly high distribution coefficients (k'~300-1000) from 8M HNO_3 for Th(IV), Np(IV), Pu(IV) and U(VI), but nothing was known about the behaviour of other species. It is almost impossible to keep all actinides in the best retained tetravalent form. It is likely that Np is pentavalent when Pu is tetravalent, and while Np is tetravalent, Pu is reduced to trivalent. Neither trivalent nor pentavalent forms are strongly retained. With a strong oxidizing agent – on the contrary- it is possible to convert Np and Pu together with U to hexavalent state while Th remains tetravalent, studies were performed on the retention properties of these species. BrO_3^- and NO_2^- were selected as oxidizing agents.

In a series of experiments with test solutions distribution coefficients, sorption kinetics and elution chromatographic properties of Np species on UTEVA were studied and published elsewhere.[14] It was found that Np can be retained on UTEVA, k' values between 600-1000 were attained from 8M HNO_3-0.1M $Fe(NO_3)_3$. In the absence of the salting out agents the k' was an order of magnitude lower.

Elution chromatograms determined with test solutions are shown in Fig. 2. The oxidation state of Np was adjusted with $KBrO_3$ as well as with $NaNO_2$. Test samples were loaded on the columns in 5 ml solution followed by their elution.

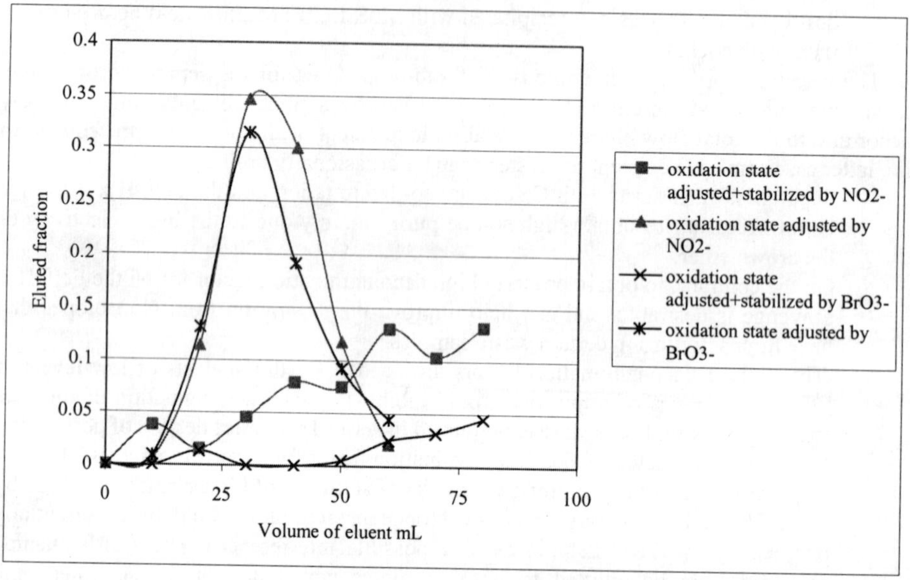

Figure 2 *Elution chromatograms of Np on UTEVA form 8M HNO_3-0,1M $Fe(NO_3)_3$ in the presence and absence of valence state stabilizers i.e. potassium bromate or sodium nitrite (column height 34 mm, inner diameter 7 mm)*

Elution chromatographic studies showed that the adjustment and stabilization of the higher oxidation states of Np are necessary conditions for Np retention. When oxidizing agents were not added to the eluent Np was completely eluted with about 50 mL solution. Np probably as hexavalent species is well retained on UTEVA when the oxidation state is stabilized with the addition of the oxidizing agent. About 80 mL of load can be passed through the small columns with less than 1-2 % losses. Bromate is preferred to nitrite.

Under the selected conditions for Np separation the behavior of other actinides on UTEVA was also tested. To 50 mL 1M HNO_3 solution containing $Fe(NO_3)_3$ U, Th, Pu and Np tracers were added, the desired oxidation state was adjusted with $KBrO_3$ and $NaNO_2$, respectively. The nitric acid concentration was adjusted to 8M and the sample was loaded on the column that was eluted with further 10 mL of 8M HNO_3-0.1M $Fe(NO_3)_3$ containing 0.1 g of oxidizing agent, followed by 5 ml 8M HNO_3 and 3 mL 9M HCl. Pu was stripped as Pu(III) with 15 mL 9M HCl-0.1M NH_4I, Th and Np were stripped together with 10mL 4M HCL and U with 20 mL 0.1 M HCl. From the strip solutions α sources were prepared by co-precipitation that were analysed by α and γ spectrometry. Results of the test experiments are shown in Fig. 3.

Studies showed that U and Pu are well separated on the column, losses are due to source preparation and some releases of the nuclides from the column. Np and Th are quantitatively recovered in the same fraction. They can be easily separated from each other by oxidizing Np selectively, while Th is co-precipitated with NdF_3. A co-precipitated Np source is prepared after its consecutive reduction.

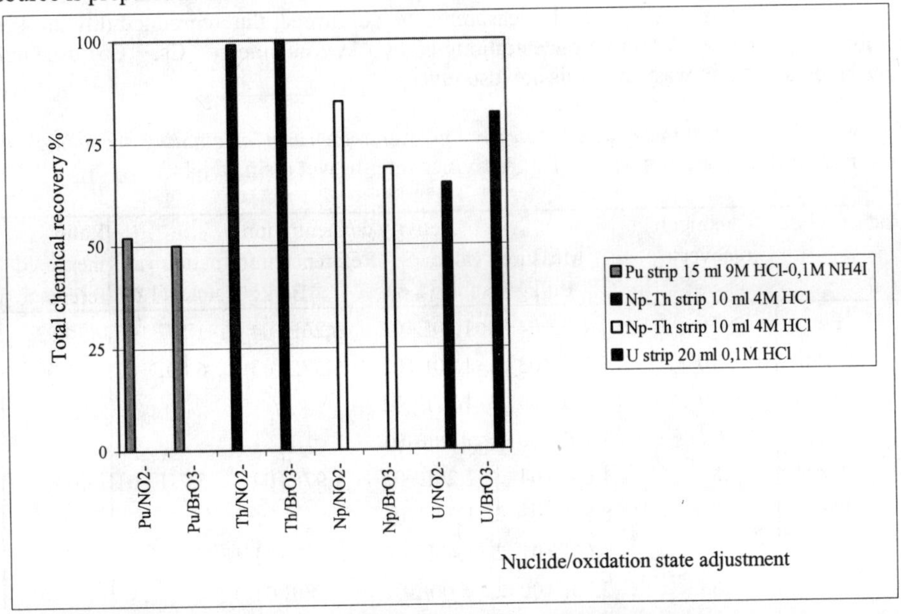

Figure 3 *Actinide separation on UTEVA and preparation of α sources, column height: 34 mm, inner diameter: 7 mm, load solution: 100 ml 8M HNO_3-0,1M $Fe(NO_3)_3$. Valence state was adjusted by a) $NaNO_2$, b) $KBrO_3$*

Chromatographic conditions that favour Np separation by adjusting Np(VI) and Np(V) oxidation states result in some losses of U and Pu, it is likely that nitrite reduces U to some extent and is responsible for reduced recoveries of U while bromate keeps U in hexavalent state that is well extracted. Some losses of Pu are recognized probably due to the moderate oxidation both by nitrite and bromate. Nevertheless, both methods make possible the simultaneous determination of all four actinides. Recoveries of real samples were often higher than those experienced under test conditions because the volume of the load solutions was typically smaller than 100 mL.

3.3 Validation of the Combined Procedure

There is a lack of certified reference materials that provide reference values for the long-lived DDNs listed in this paper. An exhausted ion exchange resin originating from a nuclear power plant was selected for validation purposes because it contained all the radionuclides to be analysed. The homogenized sample was distributed among 5 laboratories that participated in the intercalibration of analytical methods within the European Commission project (F12W-0034) on "Inventory and Characterization of Important Radionuclides for Safety of Storage and Disposal" in 1995. Results obtained by the analysis of a single sub-sample are given in Table 2. The experimental conditions were described in the previous chapter.

Reference values were established only for the α emitter Pu and Am nuclides and the β emitter Sr and Ni nuclides. Measured values acceptably agree with those reference ones. Information values were obtained from the measurements of single laboratories. Our results were obtained by using radioactive tracers and carriers, almost all recoveries were higher than 50%. Rarely studied radionuclides e.g. ^{241}Pu, ^{237}Np, ^{59}Ni, ^{129}I and ^{99}Tc could be detected. Results of the repeated measurements confirmed the reproducibility of the combined procedure. Activity concentrations of two key nuclides (^{137}Cs, ^{60}Co) used for correlation studies in waste analysis are also given.

Table 2 *Long-lived radionuclides in an ion exchange resin originating from nuclear power plant (*Sample mass was 1,01 g. Activities refer to wet resin weight.)

Radionuclide	Chemical recovery %	Activity concentration: Measured value Bq/kg		unc. (1 σ)	Reference/information value Bq/kg		unc. (1 σ)	Ratio measured/ reference
^{238}Pu	70	1,17E+04	±	1,60E+02	1,26E+04	±	1,57E+03	0,93
^{239,240}Pu	70	8,54E+03	±	1,20E+02	8,57E+03	±	6,80E+02	1,00
^{241}Pu	70	2,86E+05	±	1,57E+02				
^{237}Np	72	1,79E+00	±	1,90E-01				
^{241}Am	82	1,68E+04	±	2,29E+02	1,97E+04	±	5,71E+01	0,85
^{244}Cm	82	8,85E+03	±	1,24E+02				
^{234}U	63	2,60E+01	±	1,40E+00	2,98E+01	*		
^{235}U	63	2,70E+00	±	5,00E-01	9,30E-01	*		
^{238}U	63	7,10E+00	±	6,00E-01	9,84E+01	*		
^{63}Ni	54	2,32E+07	±	1,00E+06	3,30E+07	±	9,90E+06	0,70
^{59}Ni	54	3,20E+05	*		6,50E+04	*		
^{90}Sr	46	1,04E+06	±	2,08E+04	1,00E+06			1,04
^{129}I	54	1,44E+01	±	1,64E+00	210, ?500	*		
^{99}Tc	51	1,29E+04	*					
^{137}Cs		2,88E+08	±	2,88E+06	3,11E+08	±	8,71E+06	0,93
^{60}Co		2,81E+08	±	2,81E+06	3,16E+08	±	1,61E+07	0,89

Reference date: 1/1/1995 except for ^{241}Pu and ^{241}Am that is 1/9/2001
* information value

3.4 Fields of Application

The combined procedure as a whole has been rarely used because of lack of interest for all listed nuclides in given samples and problems in selecting the desirable sample size and the suitable destruction method. The method was used almost excludingly for the analysis of radioactive wastes, i.e. ion exchange resins and evaporation concentrates where most of the nuclides are above detection level. Mixed solid wastes were analysed for Ni and Sr radionuclides after an acidic leaching procedure.

Table 3 *Analysis of uranium, neptunium and plutonium in environmental samples (valence states were adjusted with *bromate, **nitrite)*

Radio-nuclide	Measured value Yield %	Act.conc. Bq/kg	σ Bq/kg	"Reference value" Confidence interval Bq/kg
IAEA-135: sediment from Sellafield *				
^{239,240}**Pu**	69	189,8	3,3	**205-225.8**
^{238}**Pu**		44,2	1,6	**41.6-45**
^{238}**U**	66	18,95	0,67	27-36.5
^{234}**U**		17,11	0,62	20.9-32
^{237}**Np**	80	0,32	0,07	0,88
IAEA-368: sediment from Mururoa *				
^{239,240}**Pu**	92	28,59	0,44	**29-34**
^{238}**Pu**		8,35	0,33	**7.6-8.9**
^{238}**U**	58	27,7	0,57	**25-33**
^{234}**U**		32,7	0,63	21.5-44.8
^{237}**Np**	98	0,083	0,018	0,013
IAEA-384: sediment from Fangataufa**				
^{239,240}**Pu**	78	94,4	6,2	**105,4-108,9**
^{238}**Pu**		33,8	3	**38,1-39,6**
^{238}**U**	92	31,52	1,4	**31,41-33,33**
^{234}**U**		30,25	1,3	34,2-43,5
^{237}**Np**	69	<LD		

Bold values refer to reference ones.

Methods derived from the combined procedure have been used for the analysis of a great variety of samples. Sr and Pu nuclides have been determined in ground water after pre-concentration 40-50 L samples by ion exchangers in order to monitor tentative releases of the Nuclear Power Plant Paks. Detection limits of 10^{-4} and 10^{-6} Bq/L have been reached for ^{90}Sr and ^{239,240}Pu, respectively. The concentration of ^{129}I in well and underground water was determined by treating 50 L samples with the intention to establish baseline levels for a future waste disposal site. Iodine was pre-concentrated by co-precipitation with AgCl. The detection limit of the method was about 10^{-6} Bq/L. Solid environmental samples have been analysed for U, Np, and Pu nuclides, sometimes including Am as well. To illustrate the capabilities of the technique, results obtained for a couple of standard reference materials are shown in Table 3. Typically 5 g ash were taken for the analysis and wet destruction procedure using HNO_3, HF and HCl was applied.

All results have been obtained at chemical recoveries above 60%. Some of the measured concentrations are slightly lower than the confidence interval probably due to the use of old tracers, while most of the results are within the given confidence interval. For Np there are no established reference values. Measured data are close to the detection limit.

4 CONCLUSIONS

A combined radiochemical procedure has been developed for the simultaneous determination of the following long-lived "difficult-to-determine" nuclides: ^{90}Sr, ^{89}Sr, $^{239,240}Pu$, ^{238}Pu, ^{237}Np, ^{241}Am, ^{244}Cm, ^{242}Cm, ^{63}Ni, ^{59}Ni, ^{238}U, ^{235}U, ^{234}U, ^{232}Th, ^{230}Th, ^{99}Tc, ^{129}I. The procedure is based on precipitations and selective extraction chromatographic separations followed by nuclear measuring techniques. Chemical recoveries are typically above 50%. Sub-procedures for the analysis of sub-groups of radionuclides can be easily derived. Major achievement of the method is its extension to the analysis of ^{237}Np.

References

1 J.J. LaRosa, E.L. Cooper, A. Ghods-Esphahani, V. Jansta, M. Makarewitz, S. Shawky and N. Vajda, *J.Environ. Radioactivity,* 1992, **17**, 183.
2 J.J. Moreno, J.J. LaRosa, P.R. Danesi, K. Burns, N. Vajda and M. Sinojmeri, *Radiactivity and Radiochemistry*, 1998, **9** (2), 35.
3. J. Moreno, N. Vajda, P.R.Danesi, J.J.LaRosa, E.Zeiller and M.Sinojmeri, *J. Radioanal. Nucl. Chem.*, 1997, **226**/1-2, 279.
4 R. Pilviö on Separation of U, Th, Pu, Am and Cm with TRU and UTEVA Resins, *Eichrom Users' Group Meeting*, Karlsruhe, 10 March 1997.
5 EML Procedures Manual, HAS-300, U.S. Atomic Energy Commission, 28th edition, 1997.
6 Measurement of Radionuclides in Food and the Environment, Guidebook, IAEA Technical Report Series No. 295, Vienna, 1989.
7 E.P. Horwitz, *New Chromatographic Materials for Analysis of Actinides, Sr, and Tc in Environmental, Bioassay, and Nuclear Waste Samples*, Research report under the auspices of U.S. Department of Energy, contract No. W-31-109-ENG-38.
8 *Resin Product Guide*, Eichrom Technologies Inc., in www.eichrom.com
9 *Eichrom Analytical Products Description*, Eichrom Industries Inc., 1995.
10 E.P. Horwitz and D.G. Kalina, *Solvent Extraction Ion Exchange*, 1984, **2**, 179.
11 N. Vajda N., A. Ghods-Esphahani, E. Cooper and P. Danesi, *J. Radioanal. Nucl. Chem.*, 1992, 162, 307.
12 E. Kabai, N. Vajda, in *Proc. 14th Radiochemical Conference*, Marianske Lazne, Czech Republic, 14-19 April 2002.
13 N. Vajda, in *Eichrom Europe's Vienna Users' Group Meeting*, Vienna, 7th June 2002, available from website www.eichrom.com.
14 N. Vajda, Zs. Molnár, É. Kabai, P. Zagyvai, in *Proc. II Eurasian Conference on "Nuclear Science and Its Application"*, Almaty, Kazakhstan, 16-19 Sept 2002.
15 C.W. Sill, *Nucl. Chem. Waste Management*, 1987, **7**, 201.

ANALYSIS OF Am, Pu AND Th IN LARGE VOLUME WATER SAMPLES IN THE PRESENCE OF HIGH CONCENTRATIONS OF IRON

M Schultz[1], W Burnett[2], T Hinton[3], J Alberts[4], M Takacs[4].

[1] ORTEC (Ametek, Inc.), 801 S. Illinois Avenue, Oak Ridge, TN 37831 USA
[2] The Florida State University, Department of Oceanography, Tallahassee, FL USA
[3] The University of Georgia, Savannah River Ecology Laboratory, Aiken, SC USA
[4] The University of Georgia, School of Marine Sciences, Sapelo Island, GA USA

1 INTRODUCTION

In far-field studies of actinide behavior, concentrations of analytes approach the detection limits of alpha-spectrometry (within reasonable count times). As actinide concentrations approach zero, larger sample sizes can improve the sensitivity of measurements. In many cases, large water samples do not create a great analytical challenge, as pre-concentration may be achieved by a number of methods[1-2]. However, in cases where the concentration of iron (Fe) is great, such as is the case for sampling of bottom waters in seasonally anoxic basins, it may be necessary to remove Fe from the matrix prior to radiochemical separations. In this paper, a method is presented for the analysis of americium (Am), plutonium (Pu), and thorium (Th) in large water samples (20+kg) in the presence of high concentrations of Fe and organic matter (OM). The study site is a seasonally anoxic lake (Pond B), located at the Savannah River Site (a U. S. Department of Energy Facility) in the southeastern United States. Seasonal anoxic conditions in the bottom water of Pond B result in high concentrations of Fe and Mn from early spring (April) through mid-autumn (September) each year. The method presented here was developed as part of a broader study of the association of actinides with the monomictic cycle in Pond B[1]. The field procedure is based on a technique for analysis of large seawater samples[2]. The laboratory technique draws from several sources[3-7]. The concentrations of actinides in Pond B are quite low (20 μBq to 200 μBq per 50 L sample range) — representative of far-field actinide concentrations and typical concentrations of naturally-occurring thorium isotopes. The method produced excellent chemical separations — samples were counted for one-week counting periods to quantify the activity of the actinides. In no case was the presence of cross contamination observed for n = 264 counting sources. Chemical recoveries are quite good — average recoveries were 91 +/- 8 % (Am), 84 +/- 6 % (Pu), 67 +/- 14 % (Th) respectively.

2 EXPERIMENTAL

2.1 Field Sampling

Samples were collected via a high-speed electrical pump, using an industrial grade garden hose, with depth demarcations in one-meter increments. The samples were pumped directly to (pre-weighed) new or acid-cleaned 20-L carboys (equipped with spigots for removal of supernatant solution) or 50-L drums. The volumes of solutions presented here presume a 20-L sample size. Samples were acidified immediately (100 mL concentrated HCl, pH 1-2) and weighed. A gas generator was used to supply electrical power to a digital balance to allow weighing of samples in the field. Once acidified, tracer solutions and 15 mL 0.5 M manganous chloride ($MnCl_2$) were added to the carboys. The carboys were sealed and shaken vigorously and the solutions were allowed to stand for at least one hour. A 15-mL aliquot of freshly-prepared-saturated potassium permanganate ($KMnO_4$) was then added and the samples were shaken vigorously again to ensure adequate mixing. The bright purple color o f the $KMnO_4$ solution served as a visual clue that mixing was complete (in some cases, the color disappeared quickly as divalent Fe rapidly reduced the Mn). The $MnCl_2/KMnO_4$ procedure is designed to complete a redox cycle for Pu[2]. Although this may be true for low-Fe content samples, the rapid disappearance of the purple color indicates that large amounts of divalent Fe control the redox state of Pu. Since the presence of Fe^{2+} suggests reducing conditions in the water samples, Pu species present are likely to be the reduced forms (i.e., Pu^{3+}/Pu^{4+}) that co-precipitate readily with MnO_2,. Thus, Pu co-precipitation via the $MnO_2/Fe(OH)_3$ is effective under these conditions, despite the difficulties in cycling redox conditions for samples.

Solutions were allowed to stand one hour. A fter being allowed to stand, a 150 mL aliquot of concentrated ammonium hydroxide (NH_4OH) was added to adjust the pH of the solutions to about 9. The pH of solutions was verified using pH strips. The carboys were sealed and shaken vigorously and finally let stand (24-36 hours) to allow for settling of the resultant manganese dioxide/iron hydroxide ($MnO_2/Fe(OH)_3$) precipitate.

After the settling period, the supernatants were decanted and discarded via a spigot. The precipitates were transferred to acid-cleaned 1-L polyethylene reagent bottles and taken to the laboratory for processing.

2.2 Laboratory Procedure

2.2.1 Tracers and Reagents. Tracer solutions were diluted from Standard Reference Materials (SRM's) obtained from the United States Department of Commerce National Institute of Standards and Technology (NIST). Dilutions were prepared gravimetrically using a Mettler-Toledo Model 240 digital balance. The gravimetric dilutions were prepared by diluting to volume in volumetric flasks (and measuring by mass), with a stated uncertainty of ± 2% (two sigma). Tracer solutions were prepared at a 100 $\mu Bq\ mL^{-1}$ nominal concentration (1 mL added per sample by volumetric pipet) to approximate the activity level of 20-50 kg water samples from Pond B. Appropriate measurements (n = 5) of diluted SRM's were made to verify the dilution factor of each standard. Tracers were ^{242}Pu, ^{243}Am, and ^{229}Th. Corrections were made for known quantities of ^{241}Am (0.16% correction of ^{243}Am tracer, based on the NIST certificate) and

^{230}Th and ^{232}Th, based on blank measurements (n = 4, with one outlier) that indicated a reproducible reagent blank of 0.41% ± 0.05% for ^{232}Th and 1.1% ± 0.2% for ^{230}Th (uncertainty based on standard deviation of replicates). Reagent blanks did not indicate a significant contribution of $^{239/240}Pu$ that required a correction for Pu activity calculations.

New labware was purchased for this study. Extensive efforts were made to ensure that no contamination was introduced during sampling and processing. Reagents chosen were American Chemical Society (ACS) Grade, diluted with distilled and de-ionized water.

2.2.2 Laboratory Precipitation Techniques. Precipitates were transferred to acid cleaned 2-L glass beakers (that were reserved for this low-level project) and acidified with 6 M HCl (to pH <2). Samples were then diluted to 750 mL and heated gently. As heating began, a 25-mL aliquot of 0.1 g 10 mL^{-1} ascorbic acid ($C_6H_8O_6$) was added to reduce Mn and Fe. This produces a clear to yellowish green solution (depending upon the concentration of Fe). Samples were allowed to heat gently for between 30 minutes and one hour. Next, a 25-mL aliquot of 0.1 g 10 mL^{-1} sodium nitrite ($NaNO_2$, freshly prepared) was added to oxidize Fe to the +3 state (while Mn remains divalent). The solutions turn brownish to yellow in color (depending on the concentration of Fe).

Concentrated NH_4OH was added slowly to pH 8 to precipitate amorphous $Fe(OH)_3$, while Mn^{2+} remains with the supernatant solution. The $Fe(OH)_3$ precipitation worked best when solutions were boiled vigorously for several minutes as this produced a much denser amorphous solid that settled rapidly with cooling.

The supernatant solutions were decanted and discarded. Samples were then wet-ashed three times using a mixture of concentrated HNO_3 and 30 % H_2O_2. The precipitates were first acidified with concentrated HNO_3. Next, a 10-mL aliquot of 30% H_2O_2 was added and the samples were allowed to react at low-heat and taken to dryness (this reaction should be carried out at low heat). This procedure was repeated three times and the samples were then dissolved (and transferred to 50-mL polypropylene centrifuge tubes) with 25 mL 4 M HCl. In some cases, the Fe concentration was sufficiently great to require the use of 250-mL centrifuge bottles. In these cases, the volumes of reagents described here were scaled accordingly.

In the next step, the samples were to be dissolved in HCl and transferred to centrifuge tubes for precipitation with LaF_3. However, upon dissolution, the presence of a whitish flocculent indicated the presence of amorphous silica gel (Si) in most samples. These were transferred with a minimum amount of 3 M HNO_3 to disposable plastic beakers set-up in double boiler configurations for gentle heating. To volatilize the Si, 2 mL concentrated hydrofluoric acid (HF) was added and the samples were taken slowly to complete dryness. This procedure was repeated a second time to ensure complete removal of Si. A 5-mL aliquot of 3 M HNO_3 was then added and the samples were taken to complete dryness to remove excess fluoride. This was repeated a second and third time to ensure complete removal of excess fluoride.

The samples were then transferred to 50-mL polypropylene centrifuge tubes with 25 mL 4 M HCl. The solutions were diluted with 10 mL 1 g 10 mL^{-1} $NH_2OH{\bullet}HCl$ solution (which reduces Fe^{3+} to Fe^{2+} and ensures that Pu is in the +3/+4 states). The actinides were next precipitated (and Fe is removed from the matrix) by the addition of 2 mg La^{3+}, followed by 3 mL concentrated hydrofluoric acid (HF) to coprecipitate LaF_3. Samples were centrifuged for five minutes at approximately 2500 rpm and the solution phase was discarded by decanting. This step serves to remove from the matrix not only Fe, but also natural U, which does not co-precipitate with the tri-valent rare earth in the absence of a much stronger reducing agent (such as titanous chloride $TiCl_3$ or chromous chloride $CrCl_3$).

The translucent-grayish LaF_3 precipitate was then dissolved in 20 mL 3 M HNO_3-2.5 % boric acid (H_3BO_3) in preparation for column separations.

2.2.3 *Column Separations via TEVA/TRU.* The method for separation of Am, Pu and Th is similar to that presented by Burnett et al. (1996)[2] and Schultz (1996)[8]. Best results were obtained using a tandem column arrangement using TEVA and TRU resins (Eichrom Technologies, Darien, IL USA). The methods were refined to minimize the need for redox reagents that are applied frequently to adjust the redox state of Pu. A number of variations of the TEVA/TRU tandem column arrangement have been presented[5,8]. The arrangement is an attractive alternative to a single column arrangement not only because of time savings during the column separations (elutions are carried out simultaneously), but more importantly, because the method provides for redundant separation steps — this inherently improves elemental separations.

In general, the TEVA/TRU methods that have been presented can be divided into two groups:

- Methods that reduce Pu in the load solution. Pu (and Am) passes through the TEVA column and is retained on the TRU column.

- Methods that do not adjust the oxidation state of Pu in the load solution. Pu (and Th) is retained by the TEVA column, while Am passes through the TEVA and is retained on TRU.

Method 1 is attractive because it provides a redundant separation of Pu and Th, while Method 2 provides this same benefit for the separation of Am and Pu (since Pu is retained on the TEVA column, while Am passes through).

Although each of the above approaches is theoretically sound, experience has shown that trace cross-contamination may occur in each case. In the case of Method 1, Maxwell, et al. (1997)[5] report that for precise work (such as radiobioassay), trace amounts of Th were found pervasively in Pu spectra. The alternative suggested, for precise work, was to add a second TEVA column to remove trace Th from Pu spectra. On the other hand, when using Method 2, trace amounts of Am tend to be found in Pu spectra as the elution of Am tends to leave trace amounts on the TRU column. Since Pu is eluted in the tri-valent state, using Method 2, any Am that remains on the TRU column is likely to be eluted with Pu actinide fractions is provided as well, enhancing the purity of the final counting sources. In the alternative method, presented below, the elution volume for Am was increased to 30 mL from 15-20 mL suggested in previous methods. Since Pu is retained on the TEVA column, an increase in the elution volume for Am can be used without regard for Pu leakage into the Am fraction.

2.2.4 *The Alternative Approach to the TEVA/TRU Method.* The method presented here combines characteristics of Method 1 and 2 to produce an approach that proved effective in obtaining excellent chemical recoveries, while providing for outstanding elemental purity and spectral resolution. To summarize, the TEVA and TRU columns are arranged in tandem, so that the load solution (3 M HNO_3) passes through the

TEVA column and directly into the TRU reservoir (Fig. 1). Pu is expected to be predominantly in an oxidized form (likely +4) and is retained (with Th^{4+}) on the TEVA column. Americium (as Am^{3+}) passes through the TEVA column and is retained on the TRU column (as are trivalent rare-earth elements). A solution of 3 M HNO_3-0.05 M $NaNO_2$ is passed over the tandem arrangement to ensure that Pu is maintained in the +4 oxidation state. The addition of $NaNO_2$ ensures that Am can be removed cleanly from TRU for much the same reason. After rinsing the columns of excess NO_2^-, the columns are separated (and can be run simultaneously for elution of Am and Th (Fig. 2-3). A 30-mL (3 x 10 mL) elution volume (4 M HCl) removes quantitatively Am from the TRU column, while an equal volume (in 5-mL increments) of 9 M HCl removes virtually all Th from TEVA. However, since trace quantities of Th are known to be retained on TEVA (likely due to adsorption to the resin beads, rather than extraction into the organic phase[5]), a redundant step is needed to ensure that Pu is removed cleanly. This is accomplished by recombining the TEVA and TRU columns in the tandem arrangement (Fig. 4). Pu is eluted using a reductant (hydroquinone in 4 M HCl). The inclusion of the TRU column in the elution of Pu retains any residual Th that is retained on TEVA, thus providing excellent purity for the Pu fraction. Americium is then separated from La by using a second TEVA column (Fig. 5).

2.2.5 Lanthanum/Americium Separation on TEVA. Following elution from the TEVA/TRU tandem column arrangement, La^{3+} and Am^{3+} were separated via TEVA resin using a solution of ammonium thiocyanate (NH_4SCN) in dilute HCl (Fig. 5). Tri-valent rare earths and tri-valent actinides (trans-plutonium) are known to exhibit remarkably similar behavior and present a unique challenge to ion-exchange and extraction chromatography[9-11]. Cation exchange procedures have been used to separate lanthanides and tri-valent actinides. However, due to the similarity in chemistries of these elements, the cation-exchange techniques require large resin beds and generate substantial quantities of acid waste. On the other hand,

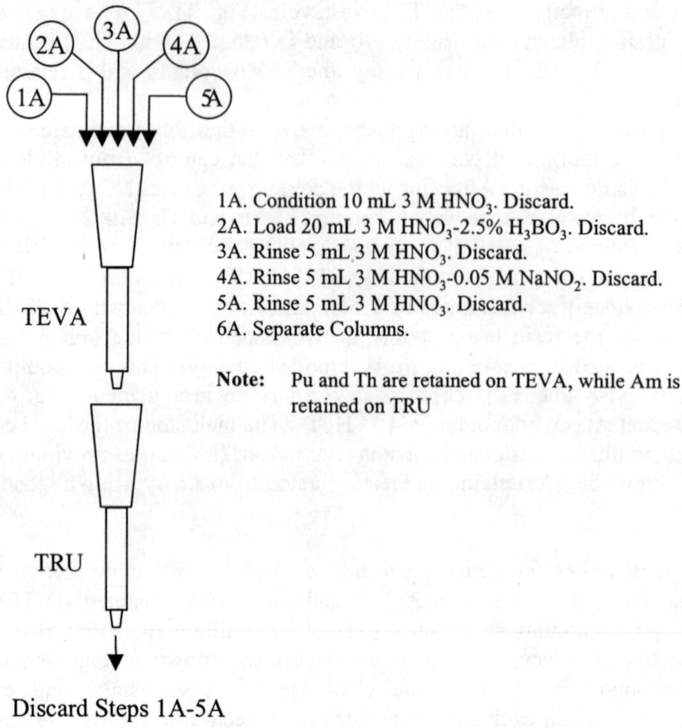

1A
2A
3A
4A
5A
TEVA
TRU
1A. Condition 10 mL 3 M HNO_3. Discard.
2A. Load 20 mL 3 M HNO_3-2.5% H_3BO_3. Discard.
3A. Rinse 5 mL 3 M HNO_3. Discard.
4A. Rinse 5 mL 3 M HNO_3-0.05 M $NaNO_2$. Discard.
5A. Rinse 5 mL 3 M HNO_3. Discard.
6A. Separate Columns.
Note: Pu and Th are retained on TEVA, while Am is retained on TRU
Discard Steps 1A-5A

Figure 1 *Tandem arrangement of the TEVA and TRU resin columns. Eluents from steps 1A through 5A are discarded.*

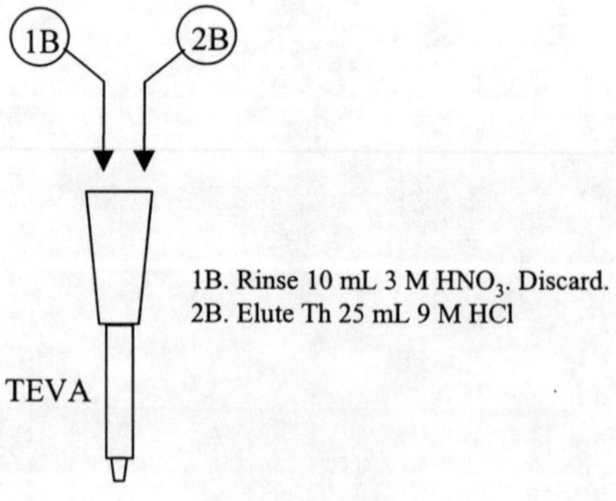

Figure 2 *Elution of Th from the TEVA column after the columns are split.*

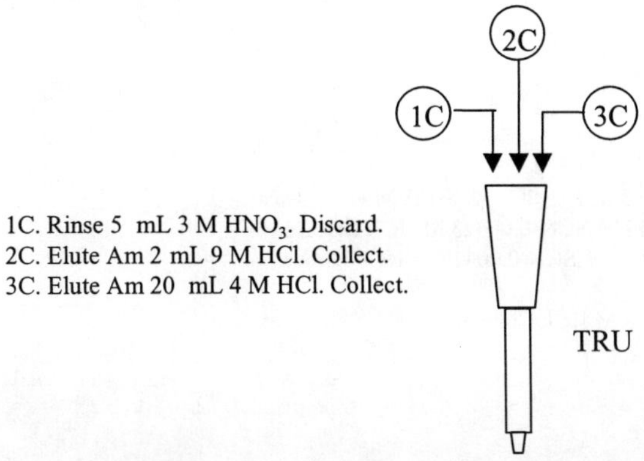
2C
1C
3C
1C. Rinse 5 mL 3 M HNO_3. Discard.
2C. Elute Am 2 mL 9 M HCl. Collect.
3C. Elute Am 20 mL 4 M HCl. Collect.
TRU

Figure 3 *Elution of Am (and La) from the TRU column. Pu is retained on the TRU column during this step.*

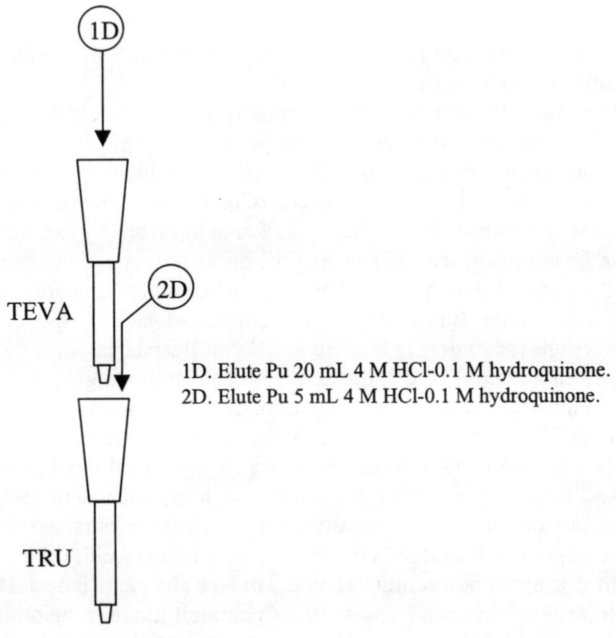
1D
2D
TEVA
1D. Elute Pu 20 mL 4 M HCl-0.1 M hydroquinone.
2D. Elute Pu 5 mL 4 M HCl-0.1 M hydroquinone.
TRU

Figure 4 *TEVA and TRU columns are restacked for elution of Pu with hydrochloric acid and hydroquinone.*

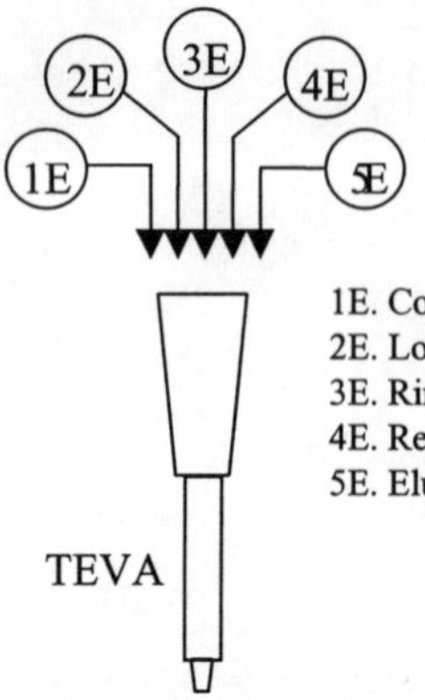

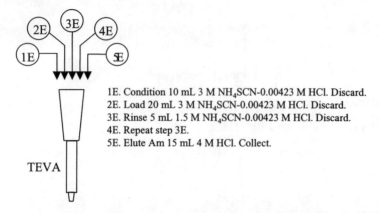

1E. Condition 10 mL 3 M NH_4SCN-0.00423 M HCl. Discard.
2E. Load 20 mL 3 M NH_4SCN-0.00423 M HCl. Discard.
3E. Rinse 5 mL 1.5 M NH_4SCN-0.00423 M HCl. Discard.
4E. Repeat step 3E.
5E. Elute Am 15 mL 4 M HCl. Collect.

Figure 5 *Separation of Am and La (rare earth carrier) using TEVA resin in a dilute hydrochloric acid/ammonium thiocyanate medium.*

extraction chromatographic techniques can be performed using standard 2-mL resin beds and relatively smaller quantities of acid waste are generated.

The method of choice has been the use of a thiocyanate/formic acid solution to effect the separation[9]. These authors studied the potential for separation of europium (Eu) and Am, using a number of quaternary ammonium salts, including Aliquot-336 (general formula R_3CH_3NCl, where R is a mixture of C_8 and C_{10} carbon chains, with C_8 predominating[10]. Aliquot-336 thiocyanate/formic acid was found to effect, by far, the best separation. The reason for the improved separation using the thiocyanate media is found to be the tendency of the thiocyanate ligand to form inner sphere complexes more readily with actinides than with lanthanides (in weakly acidic solutions) and in the greater tendency of ions with 5f electrons (actinides) to bond to soft donor ligands[11].

The original method utilized a liquid/liquid extraction approach employing Aliquot-336 as the organic phase and the thiocyanate/formic acid solution as the acid phase[9]. Since the extractant impregnated on the TEVA resin is Aliquot-336, the approach was easily adapted to the use of this extraction chromatography resin. However, for practical environmental radiochemistry applications, following a TRU-resin separation from other actinides, the formic acid medium does not dissolve quantitatively residues that are obtained as the elution volume is taken to dryness. It seemed reasonable that another acid, with better solvation properties, at the correct concentration, should produce the same effect as the formic acid. Hydrochloric acid was chosen because of its common usage in laboratories and its qualities as a solvent for many substances. The appropriate concentration (pH) was calculated by using the acid constant K_a for HCOOH (1.8 E-08), which translates to a pH of 2.37 and a concentration of HCl of 4.23 E-03 M. In initial tests, the approach proved to be effective in achieving high chemical recoveries and excellent resolution for Am spectra (Average Chemical Recovery 86%, standard deviation 12%, n=3). Variability was reduced by preparing precisely the HCl/NH_4SCN solutions to ensure that the pH of the load and rinse solutions was correct (Average Chemical Recovery 96%, standard deviation 2%, n=5). Results were excellent also for complete analyses for the work presented in this paper — for Am in large water samples in the presence of large amounts of Fe, where the use of the LaF_3 co-precipitation was effective in removing Fe from the matrix, but necessitated a quantitative separation of La and Am.

To separate Am and La, the eluted solutions from the TRU column were taken to near complete dryness. A single wet-ashing step was used to oxidize organic material, which is dissolved from the TRU resin bed. Five mL of concentrated HNO_3 was added to the cooled samples in 50-mL glass beakers. To this solution was added a 2-mL aliquot of 30% H_2O_2 and the solution was heated gently (covered) to complete dryness (but not baked). Once dry, a 5-mL aliquot of 9 M HCl was added to remove the remaining HNO_3 and the solution was heated to complete dryness (uncovered). Next, a 2-mL 1 M HCl was added to ensure that all remaining HNO_3 was removed and the samples were taken slowly to very near complete dryness (uncovered). Once the HNO_3 was removed, the samples were dissolved in 10 mL of 0.00846 M HCl (covered, with gentle heating to achieve complete dissolution). To the cooled samples was added 10 mL 6 M NH_4SCN and the samples were swirled and allowed to stand for several minutes to ensure complete mixing (the NH_4SCN solution is quite viscous) and complete dissolution. Trace amounts of Fe were evidenced (in some samples) by the presence of a reddish color — Fe forms a blood-red complex with thiocyanate in solution. The separation was then carried out by washing the column with two 5-mL washes with 3 M NH_4SCN-0.00423 M HCl to remove La. Finally, Am was eluted with a 15 mL rinse with 4 M HCl. Four drops of 10 mg mL^{-1} Fe^{3+} solution were added (to inhibit adsorption to the glass beaker) and the eluted Am solutions were taken to complete dryness by gentle heating (uncovered) in preparation for source preparation by NdF_3 micro-precipitation.

2.2.6 Source Preparation using NdF_3 microprecipitation. Alpha-counting sources for this work were prepared by neodymium fluoride microprecipitation. Solutions eluted from columns were taken to complete dryness and treated in the same manner as described above for Am. Following the addition and drying step, residues were dissolved in 5 mL 2 M HCl and heated gently. The solutions were transferred to 50-mL polypropylene centrifuge tubes with 5 mL DIW. To the solutions was added 50 μg Nd^{3+} solution and the solutions were swirled and allowed to stand several minutes. Plutonium samples were reduced by dropwise addition of ascorbic acid (noted by a color change of the solutions from yellow to clear, as Fe is reduced). The presence of ascorbic acid and reduced Fe ensures that Pu exists in solution as Pu^{3+}/Pu^{4+} — the oxidation states that are coprecipitated with the rare earth fluoride. In the presence of reduced Fe, it is expected that the predominant oxidation state of Pu will be Pu^{3+}. Next, a 2-mL aliquot of concentrated HF was added and the solutions were let stand 30 minutes to allow complete precipitation of NdF_3. Sources were then prepared by filtration, using 25-mm diameter (0.1 μm mesh) polypropylene Gelman™ filters, pre-rinsed with 80% ethanol (EtOH). The centrifuge tubes were rinsed with ~5 mL EtOH solution (transferred over the filters to rinse out excess F^-). The filters were dried, at 50°C for about 20 minutes and mounted to self-adhesive 2-inch metal planchets for counting.

3. METHOD SUMMARY AND EVALUATION

In this paper, a method is presented for the analysis of actinides Am, Pu and Th in large water samples (20+ kg) — in presence of high concentrations of Fe. Initial pre-concentration of actinides, from the bulk water samples, can be accomplished by an alkaline $MnO_2/Fe(OH)_3$ precipitation step. Manganese is then removed by selective precipitation of Fe as a hydroxide. Actinides are then separated from Fe using a LaF_3 co-precipitation technique. The rare-earth fluoride is dissolved in a mixture of nitric and boric acids and the actinides are separated by use of extraction chromatography resins TEVA and TRU. Finally, La and Am are separated using TEVA resins (by use of NH_4SCN in dilute HCl). Precise preparation of the NH4SCN/HCl solution at pH 2.37 produces best results. Counting sources are prepared by micro-precipitation with NdF_3. The method produced excellent chemical separations — samples were counted for one-week counting periods to quantify the activity of the actinides. In no case was the presence of cross contamination observed for n= 264 counting sources. Chemical recoveries are quite good — average recoveries were 91 +/- 8 % (Am), 84 +/- 6 % (Pu), 67 +/- 14 % (Th) respectively.

Acknowledgements

This research was partially supported by Financial Assistance Award Number DE-FC09-96SR18546 from the U.S. Department of Energy to the University of Georgia Research Foundation.

References

1 M. K. Schultz, W. C. Burnett, T. Hinton, J. J. Alberts, and M. Takacs, 2002, In preparation, *J. Env. Rad.*

2 J. J. La Rosa, W. C. Burnett, S. H. Lee, I. Levy, J. Gastaud, and P. P. Povinec, 2000, **248**, 3, *J. Rad. Nuc. Chem.*

3 D. Nelson, *38th Annual Conference on Bioassay, Analytical and Environmental Radiochemistry*, Santa Fe, NM., 1992.

4 E . P. Horwitz, R. Chiarizia, H. Diamond, R. C. Gastrone, S. D. Alexandratos, A. Q. Trochimczuk, and E. W. Crick, *Solvent Extr. Ion Exch.* 1993, **11**.

5 S. L. Maxwell, Radioact. & Radiochem. 1997, **8**, 4.

6 S. L Maxwell, S T. Nichols, Radioact. And Radiochem.2000, **11**, 4.

7 H. Dulaiova, G. Kim, W. C. Burnett, and E. P. Horwitz, Radioact. and Radiochem. 2001, **12**, 3

8 M.K Schultz, 1996, *MS Thesis,* The Florida State University, Tallahassee, Florida, USA.

9 R. Chiarizia, R. C. Gatrone, and E. P. Horwitz, Solvent Extraction and Ion Exchange. 1995, **13**, 4.

10 T. C. Lo, M. H. I. Baird, and C. Hanson, eds., Handbook of Solvent Extraction, J. Wiley and Sons, NY, NY, 1983.

11 G. R. Choppin and J. Kettels, J. Inorg. Nucl. Chem.1965, **27**.

IN-SITU RADIONUCLIDE RETARDATION IN GROUNDWATER CONDUCTING SYSTEMS – OVERVIEW OF THE RESEARCH CARRIED OUT AT NAGRA'S GRIMSEL TEST SITE, CENTRAL SWITZERLAND

Biggin, C.[1], Möri, A.[2], Alexander, W.R.[1], Ota, K.[3], Frieg, B.[1], Kickmaier, W.[1] and McKinley, I.[1]

[1] Nagra (National Co-operative for the Disposal of Radioactive Waste), Switzerland [2]
[2]Geotechnical Institute Ltd, Bern, Switzerland
[3]JNC (Japan Nuclear Fuel Cycle Development Institute), Japan

1 INTRODUCTION

Beginning two decades ago, Nagra (Swiss National Co-operative for the Disposal of Radioactive Waste) has developed an understanding of contaminant retardation in waste repository host rocks that integrates input from geological, hydrogeological and geochemical studies on the composition of the rock and groundwater [1].
The GTS consists of a series of underground tunnels located approximately 450 metres beneath the east flank of the Juchlistock mountain in the crystalline rocks of the Aare Massif in the central Swiss Alps (Figure 1). The tunnel system is at an altitude of 1749-1789 metres above sea level and branches off a pre-existing access tunnel to two hydro-electric power stations. It was mainly excavated using a full-face tunnel boring machine (TBM) with a diameter of 3.5 metres. The environs of the GTS have been characterised in terms of geology, hydrology, hydrogeology, mineralogy, petrography, geochemistry and natural decay series radiochemistry.

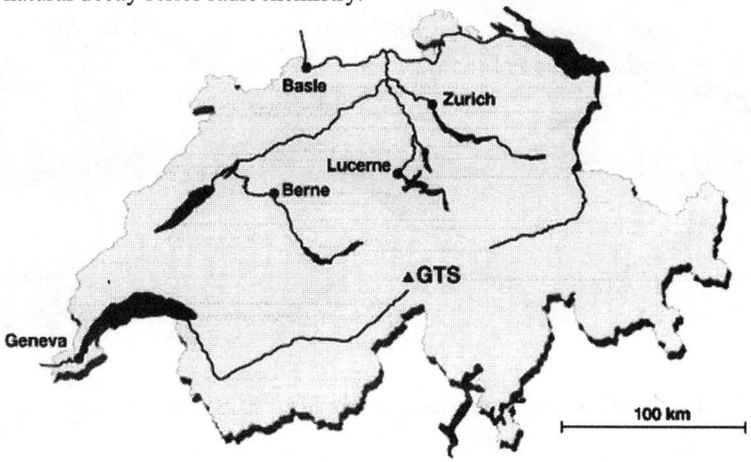

Figure 1: Map of Switzerland. Nagra's Grimsel Test Site (GTS) lies in the crystalline Aare Massif in the central Swiss Alps

Work on understanding the in-situ retardation of radionuclides in the rock of the GTS began in 1985 with the hydrogeological characterisation of a water conducting shear zone in granodiorite. This continued with a large series of *in situ* radionuclide retardation experiments, increasing in complexity from simple, non-sorbing radionuclide (fluorescein dye, ^{82}Br, ^{123}I, ^{3}He and ^{3}H) through various weakly sorbing radionuclides ($^{22\ 24}$Na, ^{85}Sr and ^{86}Rb), to a long-term experiment with more strongly sorbing $^{134,\ 137}$Cs [2, 3, 4, 5, 6, 7, 8]. The radiochemistry experiments are carried out in an IAEA level B radiation controlled zone within the GTS tunnels (Figure 2 [9]. This allowed the use of chemically complex tracers ($^{95m,\ 99}$Tc, ^{113}Sn, ^{75}Se, ^{234,235,238}U, ^{237}Np, ^{60}Co and ^{152}Eu) to be applied to *in situ* experiments. Subsequent overcoring of a part of the experimental shear zone allowed these nuclides to be recovered [9, 10, 11]. This paper will discuss the *in situ* radionuclide tracer tests carrier out at the GTS. This covers four projects, the first of which, the Migration Experiment (MI), began in 1986, up to the current experiment investigating the effect of cement leachates from a repository on the behaviour of radionuclide retardation in the surrounding rock.

1.1 The aims of the in-situ radionuclide tests at the GTS are thus

To derive parameters on the *in situ* behaviour of radionuclides in fractured rock which can be combined with laboratory and natural analogue[a] studies, to produce and test computer models of radionuclide transportation and retardation in a repository host rock which will be used to assess potential release scenarios from repositories. As noted elsewhere [12, 1], simply basing a model of radionuclide transport in fractured rocks on laboratory and natural analogue may not represent the conditions found in a repository. Thus it can be seen that the most appropriate data for these models will be produced from a structured combination of *in situ*, laboratory and analogue studies

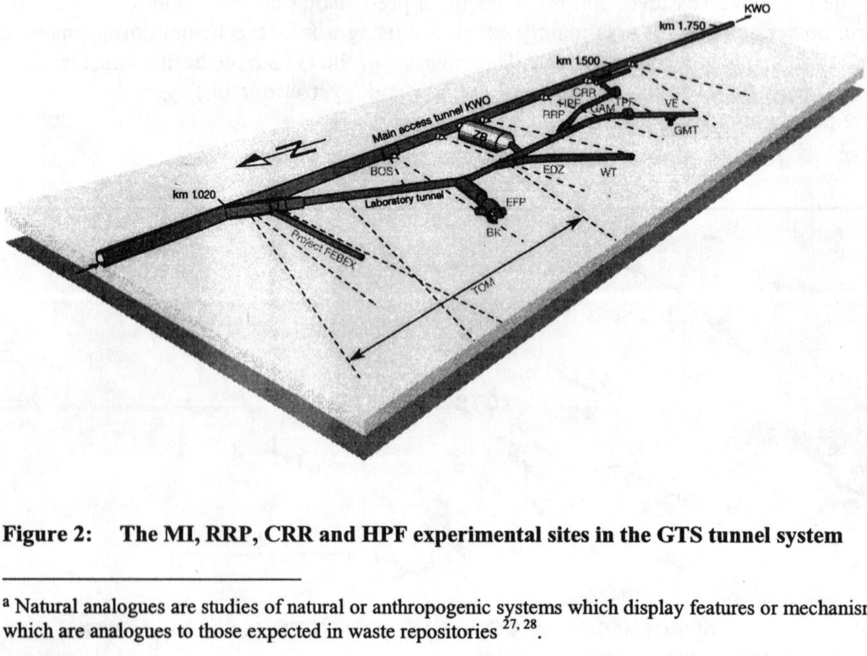

Figure 2: The MI, RRP, CRR and HPF experimental sites in the GTS tunnel system

---

[a] Natural analogues are studies of natural or anthropogenic systems which display features or mechanisms which are analogues to those expected in waste repositories [27, 28].

2 THE BASIS OF GEOSPHERE TRANSPORT MODELLING

In accordance with current understanding of contaminant transport in the geosphere [13], the dominant processes governing solute transport in fractured rocks are generally assumed to be (see also Figure 3):

- advection and dispersion within the water-conducting features and
- retardation due to matrix diffusion into the rock matrix and sorption onto mineral surfaces.

In the formal assessment of the likely long-term performance of a waste repository (PA), these processes are assumed to operate in extensive heterogeneous networks of water-conducting features, although a detailed, small-scale understanding of the structure of the features is also required in order to model matrix diffusion [14]. As illustrated in Figure 4, the processes (including the geometrical parameters and the parameters that are used to define the rates and spatial extent of processes) are derived from a broad base of information which includes the results of a range of characterisation techniques and general scientific understanding. The information is interpreted, in terms of transport-model parameters, by means of various supporting hypotheses and models. For example, measured transmissivities are converted to advection parameters via groundwater flow models [15] and sorption measurements are converted to transport model parameters via a K_d or sorption isotherm model [16].

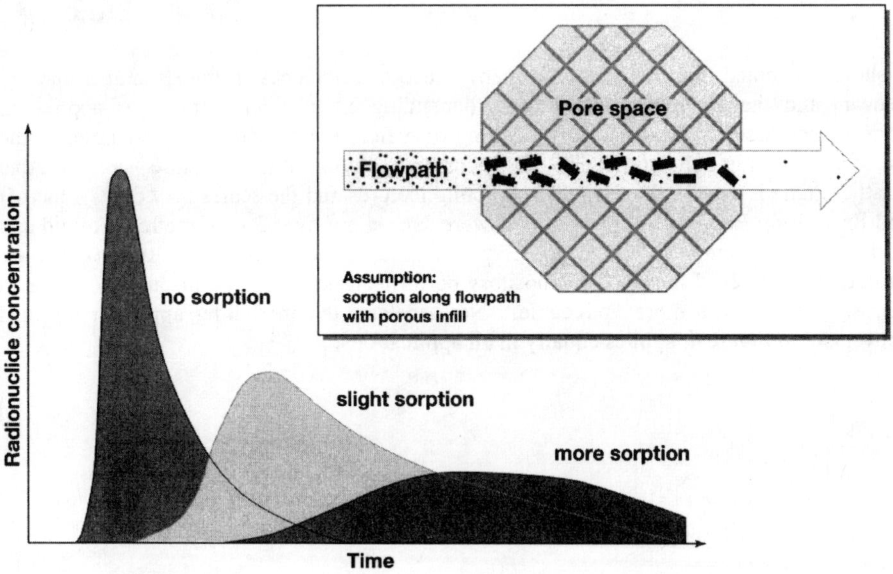

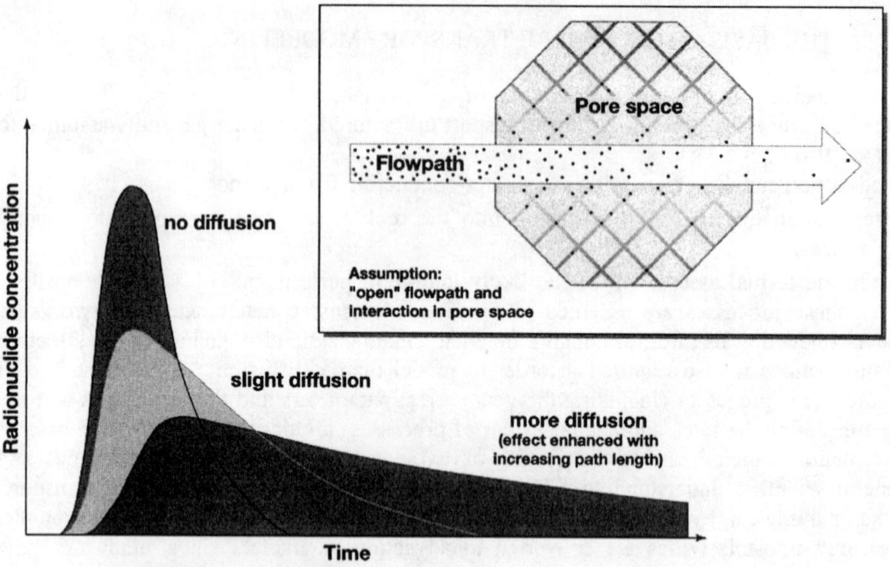

Figure 3: Illustration of the effects of different retardation mechanisms on radionuclide tracers in the geosphere [1]

It should be noted that there may be many specific differences in the structures that are relevant and the parameter values used, depending on whether a model is applied in performance assessment or to the modelling of a field experiment. For example, in the experimental shear zone at the GTS, the focus was on an individual feature – a shear zone – rather than on a network of water-conducting features and the scales may differ - the MI and Excavation Project (EP) experiments were carried out over 2 - 15 metres, Colloid and Radionuclide Retardation (CRR) and Hyperalkaline Plume in Fractured Rock (HPF) are carried out over 2 - 4 metres. In repository performance assessment, distances of 100s to 1000s of metres are more applicable. Nevertheless, the modelling approach outlined above, and in Figure 4, applies equally in all applications.

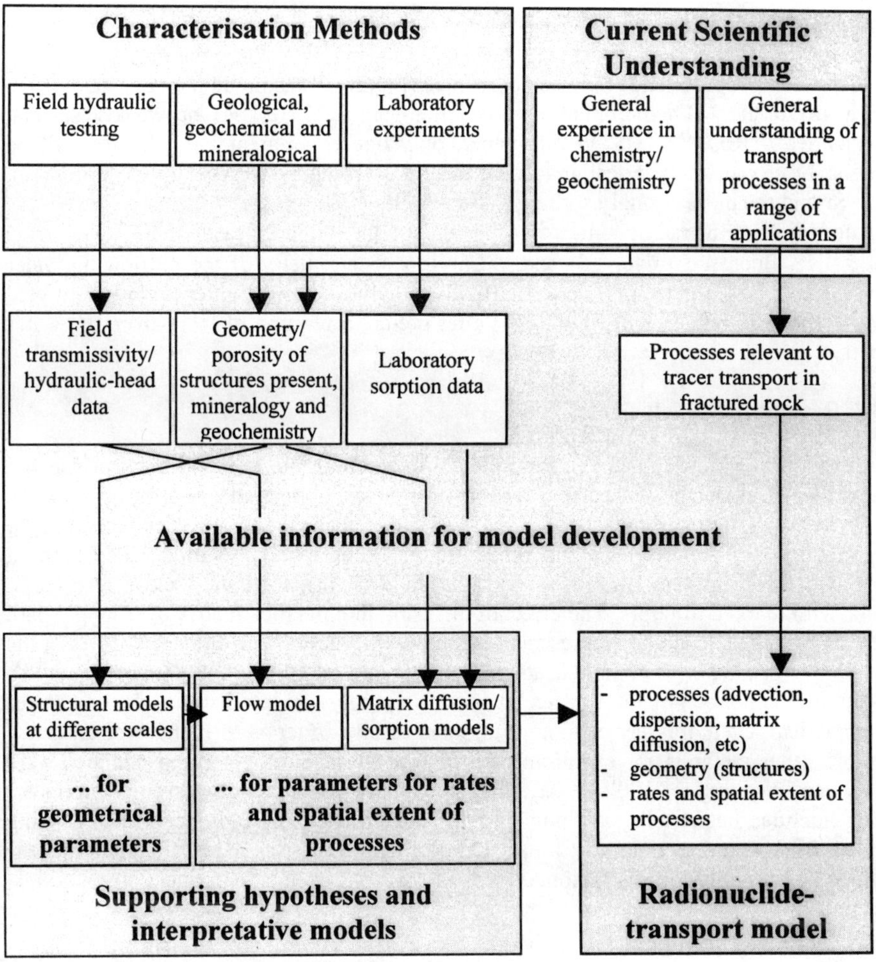

Figure 4: The use of supporting hypotheses and interpretative models to interpret field and laboratory data in terms of input parameters of a transport model[8]

3 THE MIGRATION EXPERIMENT (MI)

The central component of MI was a series of radionuclide-transport tests performed in well-defined dipole flow fields created via an injection and extraction boreholes (with a higher rate of extraction than injection) [17], supported by a range of field and laboratory investigations and modelling studies aimed at understanding and, ultimately, predicting the results of the radionuclide tests. The tests were performed within a single, approximately planar, shear zone that was selected on the basis of a wide range of experimental criteria [4]).

3.1 Experimental Procedure

MI was based on a well characterised shear zone in what was then an IAEA level C controlled zone at the GTS (Figure 2). The physical set up of the experiment is shown in

Figure 5a. The tunnels cut the shear zone and a series of boreholes were sunk from both sides of the AU tunnel to intersect the plane of the shear zone. These boreholes were then used to establish dipole fields between boreholes (Figure 5b) as noted above. Much of the early work in the MI experiment involved non-sorbing (^{3}He, ^{3}H) or weakly sorbing radionuclides (^{86}Rb, $^{22,24}Na$). In this paper, only one experimental run of MI will be discussed briefly, namely a test involving the non-sorbing fluorescent dye, fluorescein, ^{22}Na, ^{85}Sr and the more strongly sorbing ^{137}Cs.

After injection of radionuclides into the shear zone, online measurement was carried out for ^{22}Na, fluorescein dye, ^{85}Sr and ^{137}Cs via γ spectrometry of the outflowing water (both NaI and HPGe detectors were used). Eventually, ^{137}Cs activities were determined offline at the Paul Scherrer Institute (PSI) after pre-concentration of ^{137}Cs from the water samples.

3.2 Results and conclusions

The measured breakthrough curves from the 1.7 m and 5 m dipole experiments are shown in Figure 6. With high flow rates over short distances, compared to variously strongly sorbing cations, non-sorbing anions show similar peak times, less enhanced peak concentrations and rapidly decaying breakthrough. For weakly to moderately sorbing tracers with breakthrough over a longer flow field (5 m), there is a clear retardation coupled with marked dilution. The effect of diffusion into the rock matrix over longer path lengths is clear for ^{22}Na and fluorescein, as shown in Figure 3. In the case of ^{137}Cs, the retardation over the 5 m flow field is so great that even after 10,000 hours (over 400 days), breakthrough is not complete. For nuclides more strongly sorbing than Cs, the timescales are impractical, even 400 day of monitoring is excessive in terms of manpower and cost. Therefore other methods for examining the *in situ* retardation of sorbing nuclides were investigated. Waiting for 100's of days may eventually produced breakthrough curves for sorbing nuclides but it tells nothing of how and where these radionuclides are being removed from solution onto the rock. A new approach was need for studying the retardation of strongly sorbing radionuclides - this was the excavation project.

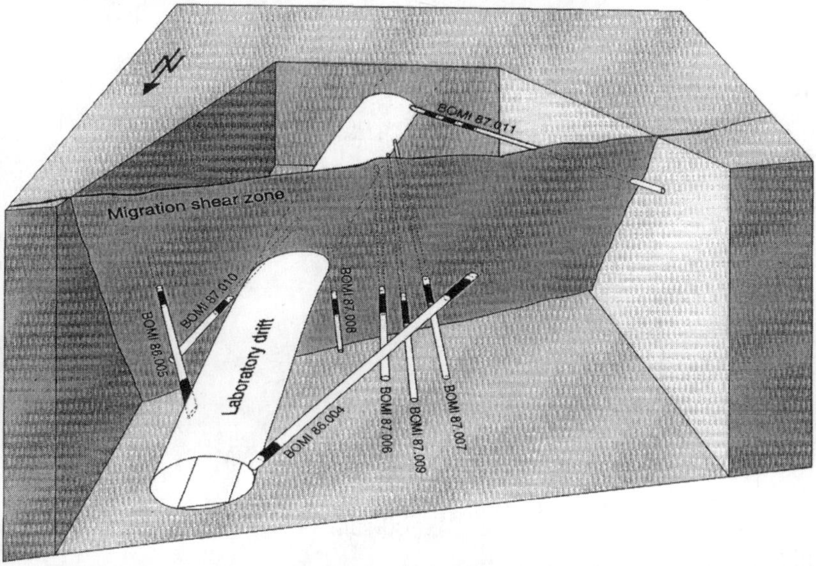

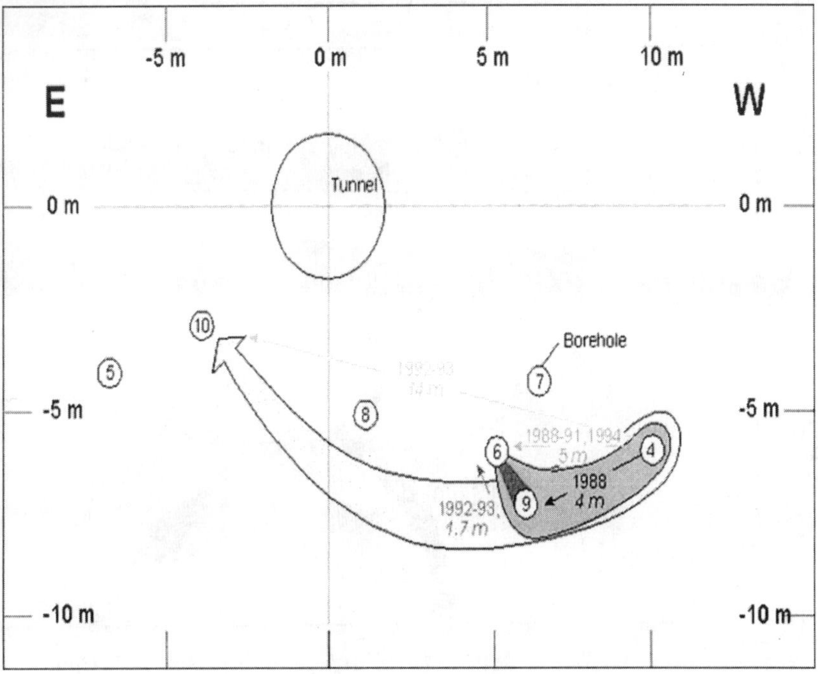

Figure 5a: Schematic layout of the experimental shear zone in the GTS
Figure 5b: Location of the boreholes in the plane of the shear zone [4]

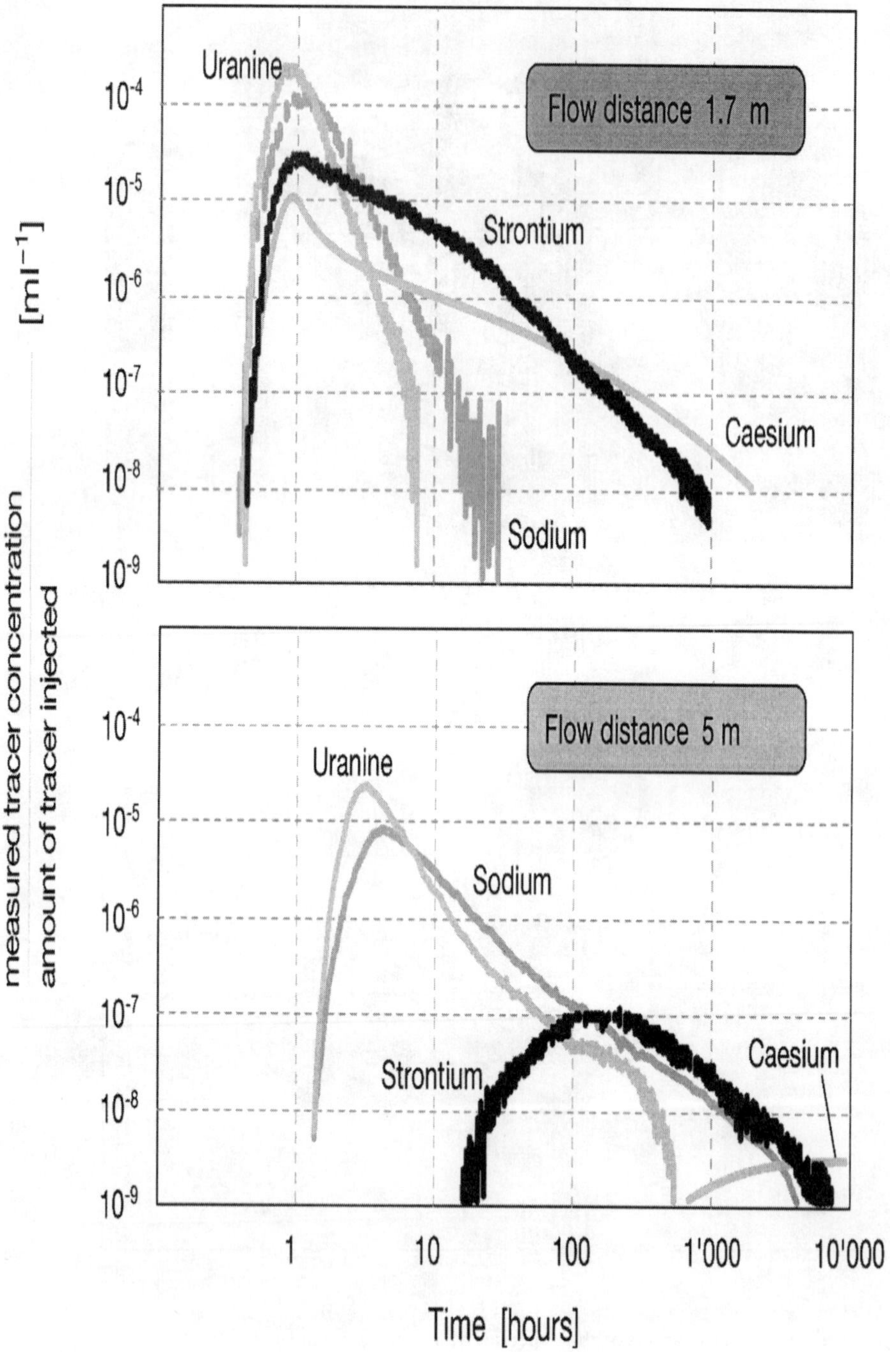

Figure 6: Examples of measured breakthrough curves from the Migration Experiment [8]

4 THE EXCAVATION PROJECT (EP)

4.1 Aims with respect to transport model testing

The main aim of EP was to extend the form of model testing developed in MI[18]. At present, studies such as MI produce data for the overall flow system 'seen' by the radionuclide breakthrough curves. These data are then modelled to produce a set of 'best fit' bulk or average values for the retardation parameters, such as rock/water distribution coefficient (or Kd) values, flow-wetted surface area and matrix diffusion depth, by comparing calculated model curves with the experimental break-through curves. However, the model values obtained may not be a unique solution and, from this process, no detailed information is obtained about the actual sites of retardation. Thus is it very difficult to extrapolate from one site or flow system to another where properties are different. This is a key issue as, in practice, the geosphere around a radioactive or chemo-toxic waste repository will not be exhaustively explored, due to the necessity to maintain favourable characteristics in as unperturbed a state as possible. Hence it is essential that extrapolations from nearby or similar sites, where such restrictions do not apply, can be justified by a thorough understanding of the factors influencing radionuclide transport [10].

4.2 Overview of experimental procedure in the Excavation Project (EP)

This focuses on the structure of the shear zone and on the behaviour of radionuclides that are relevant to repository post-closure safety (see Figure 7), but are so strongly retarded by interaction with the shear-zone rock that they are not expected to pass through the dipole flow field in experimentally reasonable times. Consequently, a "post-mortem" analysis of the shear zone has been performed, in which the entire dipole flow field has been immobilised using a newly developed resin injection technique, excavated (see Figure 8) and taken back to the laboratory for analysis of its three-dimensional structure and of the distribution of injected radionuclides within the flow field (Figure 9) [9, 10, 11, 19].

4.3 Radiochemical techniques at the GTS during injection of shear zone

Injection of the shear zone was carried by a direct injection technique that had been perfected during the injection of non-sorbing radionuclide cocktails. A glass vial containing the radionuclide cocktail (^{60}Co, ^{75}Se, ^{113}Sn, ^{152}Eu, ^{234}U, ^{235}U, ^{233}Pa and ^{237}Np) was placed into the injection interval of a triple packer system. Once the packer was introduced into the injection borehole, an internal device in the packer broke open the vial and released the cocktail to the shear zone dipole flow field.

To allow determination of radionuclides in the outflowing solution, the tubing containing the outflowing solution was wrapped around a P-type Ge detector. This allowed the real time behaviour of the injected isotopes to be observed and permitted a mass balance to be made. In conjunction to the on-line detection system, conventional γ and α spectrometers were used off-line, the γ spectrometers for counting of bulk samples when the outflowing activity was close to the detection limits of the on-line system. Determination of ^{234}U, ^{235}U ^{237}Np/^{233}Pa was also carried out off-line due to the lower activities of these nuclides and their low counting efficiencies by γ spectrometry.

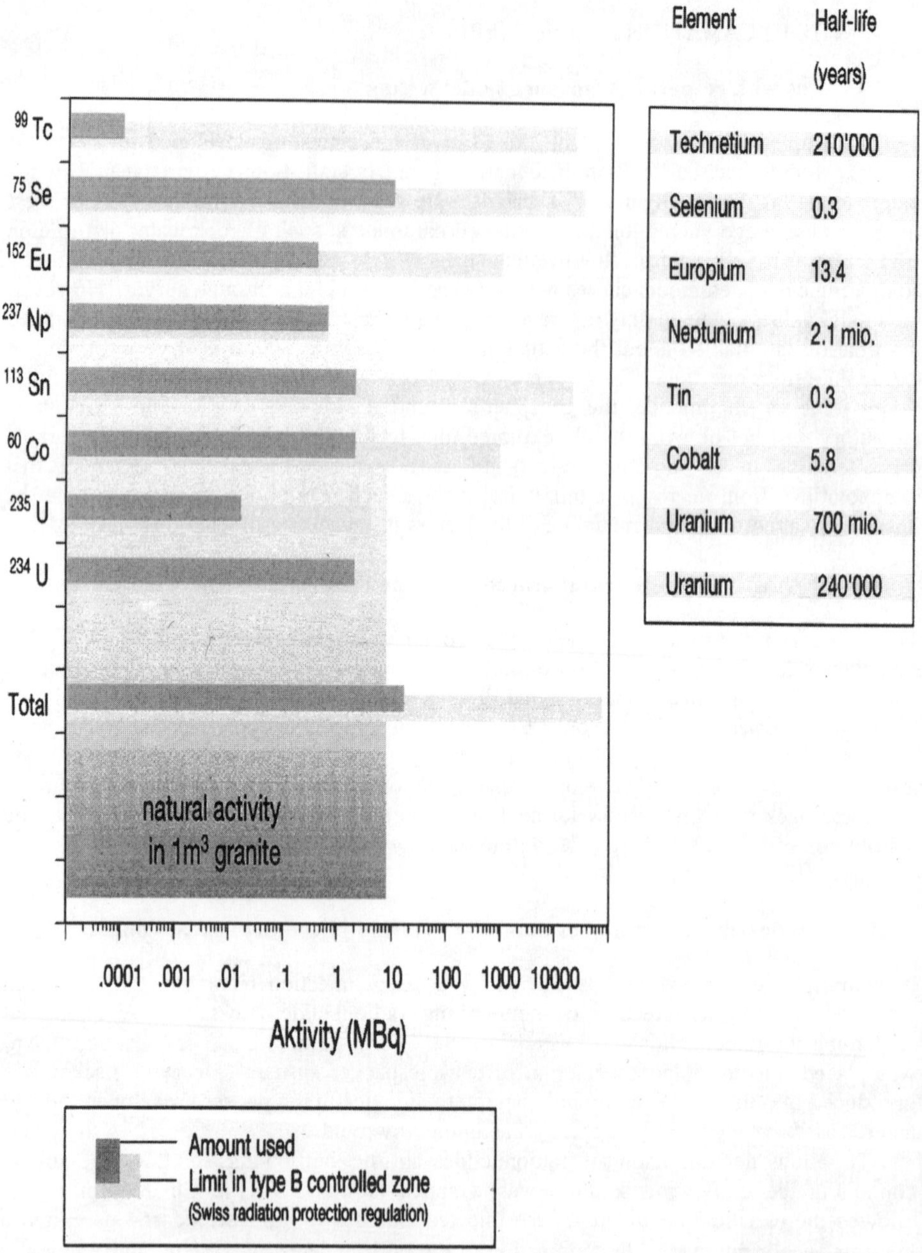

Figure 7: Activities of the safety relevant radionuclides injected into the experimental shear zone in the EP experiment [20]

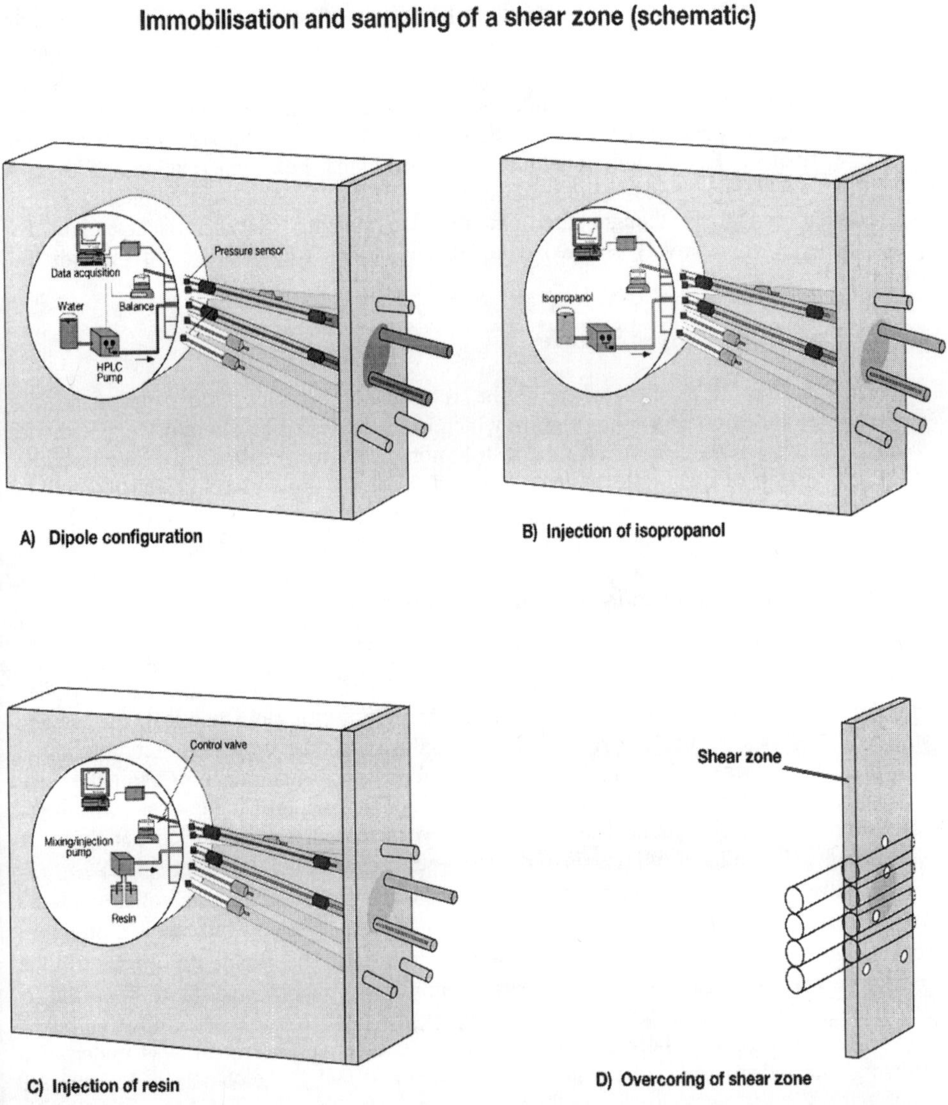

Figure 8: **Excavation of the experimental shear zone during the EP experiment. A: injection of the radionuclides into a dipole flow field. B: the water in the dipole is replaced by isopropanol to ensure adequate wetting of the rock by the resin which follows. C: specially developed resin is injected into the experimental shear zone and allowed to polymerise and harden. D: the immobilised (radioactive) part of the shear zone is recovered for analysis by overlapping triple barrel drill cores, drilled parallel to the shear zone [9]**

The detection limits for the on-line and off-line γ spectrometers were around 1 and <0.01 Bq ml^{-1} [21]. Analysis of α emitters was carried out by evaporation of 5 ml aliquots onto stainless steel planchets. To account for lowering of the α counting efficiency due to sample self adsorption, ^{244}Cm was added to several samples as an efficiency tracer. As expected, the recoveries of the radionuclide cocktail in the outflowing solution were low, with more than 75 % of the individual tracers trapped in the shear zone (excluding ^{75}Se and ^{234}U, ^{235}U). Although added in the reduced form as HSe, ^{75}Se was the only nuclide not to be retarded and it is thought that oxidation to selinate (SeO_4^{2-}) occurred in the groundwater, the negatively charged complex could then pass un-retarded through the rock.

4.4 Excavation of the shear zone

A technique was developed whereby the rock could be successfully stabilised for extraction by injection of an epoxy resin, which has been tested in laboratory experiments to ensure there was no disruption to the distribution of the previously sorbed radionuclides [22]. Investigation of drilling techniques [19] found that only triple-barrel coring techniques were suitable. This was necessary to guarantee full core recovery and preserve the fault gouge in the cores [10].

4.5 Radiochemical techniques in the EP project

After overcoring, the extracted cores (380 mm external diameter and weighing about two tonnes) were cut into smaller core sections of about 30 cm. These were then sub-sectioned into thinner slices of about 4 cm and shipped to laboratories for radiochemical, surface and structural analysis. Only the radiochemical analysis will be discussed here but more information on these cores can be found in reference 22. A range of radiochemical techniques were utilised to observe the distribution of the radionuclide cocktail within the rock slices. The radiochemical analysis of the extracted cores was carried out by AEA Technology, Harwell and full details of the procedures can be found in reference 22.

Initially, the samples were screened for α activity by α-autoradiography using CR-39 Track Plastic. Areas of high α activity (i.e. from ^{234}U and ^{237}Np) were observed on a mm scale (Figure 9). β-radiography was also carried out but the high β activities present in the rock resulted in smearing of the tracks which gave poorer resolution than the α autoradiography. The autoradiography techniques provided an initial screening and based on these results a more thorough investigation of the active core areas was undertaken utilising Secondary Ionisation Mass Spectrometry (SIMS) and Nuclear Microprobe techniques. These generally focussed on the fault gouge material within the sample, the higher porosities of which make it more likely to act as sites of radionuclide uptake [11]. Ion images of the fault gouge were produced with SIMS analysis. Use of Ag instead of Au as a sample coating allowed U and Np distributions to be distinguished in the samples. Rutherford Back-Scattering (RBS) techniques were useful for analysis for heavier atomic elements (i.e. actinides) on a lighter substrate such as silicates. The problems of poor mass resolution were overcome by combining RBS with SIMS. Finally, the total mass of activities was determined by total dissolution techniques followed by γ spectrometry, α spectrometry or High Resolution-Inductively Coupled Plasma-Mass Spectrometry (HI-ICP-MS).

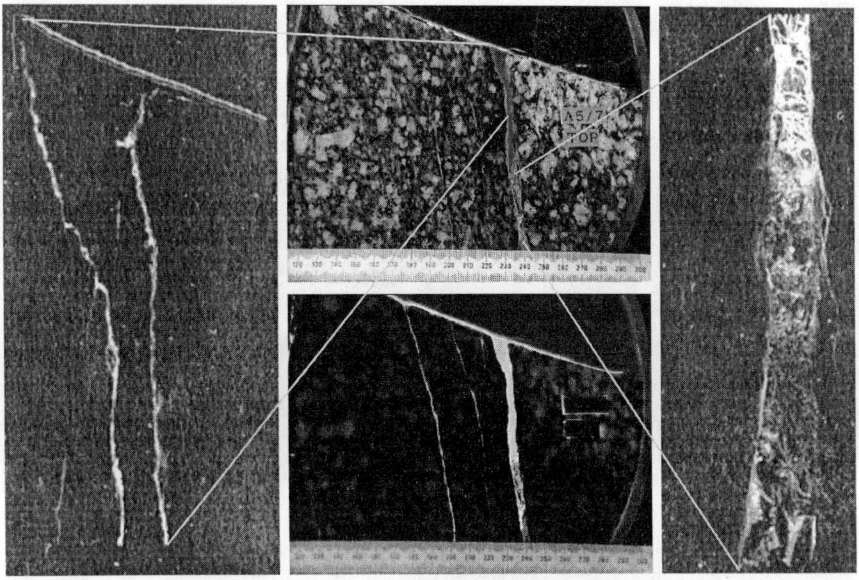

Figure 9: Top middle picture shows a resin impregnated fault in the shear zone materiel, directly beneath this is the same section in UV light to highlight the flow paths. To the left and right of these photos is the α -autoradiographs of the faults, the areas in while (along the side of the fault) showing the most alpha activity [11]

4.6 The main conclusions from the radiochemical analysis of the EP

The impregnation and excavation technique were successfully applied to the shear zone. Visualisation of the flow field revealed a three dimensional image of the flow paths within the dipole and the relevant retardation sites of the radionuclides within the shear zone could be detected.

- Highest activities were found on the surface as well as within the well preserved fault gouge which filled the fracture or just coated the fracture walls
- Evidence of radionuclide dispersion from main flow path via minor channels running parallel to main flow fracture.
- Some evidence of dispersion into mylonites in the core.
- ^{237}Np, ^{113}Sn and Mo (stable) strongly sorbed near injection borehole
- Some migration of U, Cs and Co??

The results of the MI and its successor, the EP project gave confidence that the current understanding of the basic processes involved in migration and retardation of radionuclides in the geosphere was appropriate and confirmed the conceptual models for radionuclide transport modelling. With the experience and confidence gained in these relatively simple projects, it was clearly time to begin tackling some of the more obdurate open questions associated with the deep geological disposal of wastes.

5 THE RECENT WORK: CRR AND HPF

5.1 Introduction

The Engineered Barrier System (or EBS) of a repository will slowly degrade and eventually fail, potentially releasing small quantities of radionuclides to the host rock. It is expected that most nuclides will decay within the EBS or be retarded in the host rock surrounding the repository. However, failure of the engineered barriers is also likely to produce changes to the existing conditions in the repository host rock and the following two projects examine such scenarios.

5.2 Colloid and Radionuclide Retardation (CRR)

5.2.1 Introduction. In the case of high-level radioactive wastes, the EBS consists of vitrified wastes in steel containers that are placed into another steel canister, which is then surrounded by bentonite clay and placed deep underground in a repository (Figure 10).

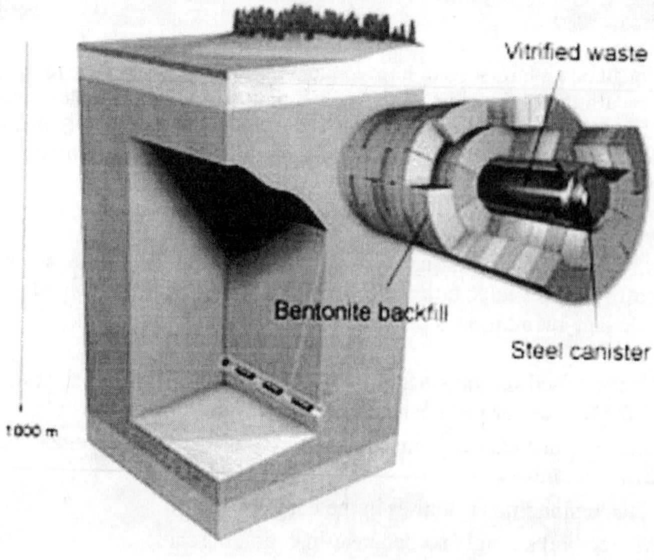

Figure 10: Diagram of the Engineered Barrier System (EBS) for a HLW repository. The vitrified waste is placed into steel containers which are then surrounded by bentonite clay

The CRR experiment is dedicated to study the *in situ* migration behaviour of safety-relevant actinides and fission products in absence and presence of bentonite colloids in a granitic shear zone. Following failure of the canisters, radionuclides are expected to slowly diffuse through the bentonite barrier and some may even reach the EBS/host-rock interface where erosion of the bentonite could produce colloids, which could take up some of the released radionuclides and transport them through the host rock to the biosphere (see Figure 11), mainly along water-conducting features such as shear zones. For many regulators, the effect of colloid-facilitated transport of strongly sorbing radionuclides generated at the EBS/host-rock interface remains very much an open question. For example, HSK, the Swiss Federal Nuclear Safety Inspectorate, noted in 1998 that "...the generation of colloids at the boundary of the EBS cannot be excluded...." and that "...colloid facilitated transport of strongly sorbing radionuclides at the EBS/host-rock boundary remains very much an open question.."

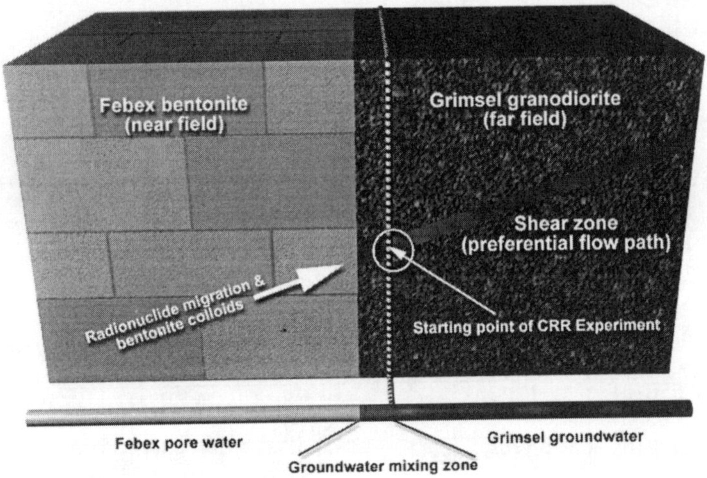

Figure 11: Concept behind the CRR project: radionuclides released from the bentonite could be transported through the far-field by colloids eroded from the bentonite backfill [23]

The broad objective of the CRR experiment is to provide an improved understanding of the stability and in-situ retardation of colloid associated, safety relevant radionuclides in the vicinity of the EBS/host-rock interface. The project tests the results from laboratory tests and modelling studies against *in situ* dipole radionuclide retardation tests at the GTS.

Before starting the field tests in the summer of 2000, a programme of laboratory and modelling work was performed to assess the feasibility of the project and to provide important supporting data on the *in situ* solubility of repository relevant radionuclides. The stability of bentonite colloids and the sorption behaviour of the various radionuclides on the bentonite and host rock was also examined. Classical batch experiments were carried out with Am, Pu, U, Tc, Cs and Se with Grimsel granodiorite, fracture infill (fault gouge) material, bentonite colloids and the equipment to be used in the experiment. Timescales were kept short to investigate the sorption kinetics on similar timescales to the *in situ* experiment and Grimsel groundwater was used as the solute. Full details of the laboratory studies can be found in Möri [23].

The experiment was conducted in a water-bearing shear zone in the IAEA level B radiation controlled zone at the GTS. The main component of this experiment were the two injections of sorbing radionuclides, the latter of which would take place in the presence of 20 mg l^{-1} bentonite colloids. Several runs of non-sorbing ^{131}I were injected before, between and after the two injections of the strongly sorbing nuclides. Details of the nuclide injections are shown in Table 1. Different isotopes for U, Pu and Am were used for the two cocktails to allow differentiation between the effects of the two different experimental runs. Colloid concentrations in the outflowing solutions were measured by Laser Induced Breakdown Detection (LIBD) and the radiochemical analysis of the outflowing solutions was similar to that of the excavation project. On line γ spectrometers were again used for the analysis of the outflowing solution in conjunction with off line γ and α spectrometers and ICP-MS analysis. Full details of the set up used can be found in [21]

Table 1: Details of the injections for CRR#31 and CRR#32 (in the presence of 20 mg l^{-1} bentonite colloids and the methods of analysis

CRR#31	Injected Activity	Method of detection	CRR#32	Injected Activity	Method of detection
No Colloids	Bq		20mgL-1 bentonite colloids	Bq	
^{85}Sr (II)	9.65E+04	γ spec	^{85}Sr (II)	8.24E+04	γ spec
^{131}I (-I)	7.57E+04	γ spec	^{99}Tc (IV)	6.56E+01	ICP-MS
^{232}Th (IV)	1.07E-03	ICP-MS	^{131}I (-I)	5.56E+04	γ spec
^{238}U (VI)	2.86E-01	ICP-MS/ α spec	^{137}Cs (I)	6.07E+05	γ spec
^{237}Np (V)	5.91+02	ICP-MS/α spec	^{232}Th (IV)	1.03E-03	ICP-MS
^{238}Pu (IV)	6.08+02	α spec	^{233}U (VI)	7.22E+03	ICP-MS/α spec
^{242}Pu (IV)	3.55E+01	ICP-MS	^{237}Np (V)	6.72E+02	ICP-MS/α spec
^{243}Am (III)	1.08E+03	ICP-MS/ α spec	^{238}Pu (IV)	7.20E+02	α spec
			^{244}Pu (IV)	1.11E-01	ICP/MS
			^{241}Am (III)	2.04E+03	ICP-MS/α spec

5.2.2 Results of in situ test in the CRR experiment. At the time of writing the full results from the CRR are not yet complete. Some preliminary observations are given below but full results should be published in the spring of 2003[a].

The bentonite colloids were found to be stable at the given chemical conditions of the Grimsel groundwater and to migrate through the test shear zone with only minimal retardation (recoveries of about 90 % were calculated from the outflowing solutions). The tri- and tetravalent actinides (AM and Pu) were found to be associated with the colloids and this enhanced their mobility through the shear zone. They arrived slightly earlier than ^{131}I and the penta- and hexavalent actinides (Np and U). The recovery of Am and Pu increased from about 20-30 % to 60-80 % in the presence of bentonite colloids. Cs showed two distinct peaks in the outflowing solution, the first peak is a result of a small amount of Cs associated with the bentonite colloids and the second peak is "true dissolved" and retarded Cs. The results so far are summarised in Figure 12.

[a] In the meantime, interested readers are recommended to visit the GTS web-site at www.grimsel.com

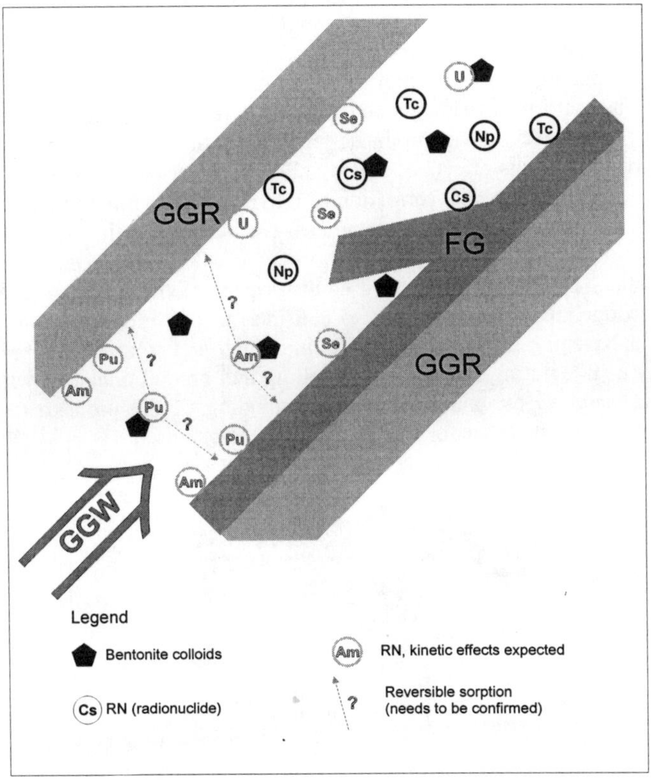

Figure 12: Schematic diagram of the preliminary findings of the CRR experiment

5.3 Hyperalkaline plume in fractured rock (HPF)

5.3.1 Introduction. In the case of low and intermediate level radioactive wastes (and some very long-lived intermediate level wastes) and many chemo-toxic wastes, most current repository designs envisage the use of large volumes of cementitious materials to immobilise the waste and to backfill the repository [24]. One reason for this is the expected very low solubilities of most contaminants under the hyperalkaline conditions prevalent in cementitious systems. However, there is also a negative side to such designs and it has long been recognised [25] that the use of such large quantities of cementitious materials in an underground (deep or near-surface) radioactive or toxic chemical waste repository could, in the long term, give rise to a 'plume' of hyperalkaline water displaced by groundwater flow from the repository into the host rock . Interaction of the rock with this hyperalkaline porewater could be detrimental to the performance of the host rock as a barrier to radionuclide migration [26]. While some potential processes, such as enhancement of fracture porosity by dissolution of host rock minerals and sealing of matrix porosity due to secondary mineral precipitation, are likely to be deleterious, other simultaneous processes may counteract these or even outweigh them to produce a positive effect on repository performance. Such positive processes include precipitation of secondary minerals with improved sorption characteristics (when compared to the primary mineralogy) and sealing of fractures by secondary minerals inhibiting water movement and radionuclide transport.

Modelling, laboratory experiments and studies of natural cement leachates carried out over the last few years have produced much useful, information and increased understanding of the various effects of the so-called high pH plume (see, for example, the conceptual model for the potential formation of a hyperalkaline plume in a repository host rock illustrated in Figure 13) but these studies have been inconclusive in predicting the overall result with respect to host rock performance. This has been, at least in part, because each study concentrated on a particular aspect of the problem more or less in isolation and no one study has attempted to consider all the information in an integrated manner [12]. Consequently, the HPF (Hyperalkaline Plume in Fractured rock) project should be seen as way of integrating laboratory, modelling and natural analogue with the results from the *in situ* studies, to gain a better understanding of the processes relevant to a cementitious nuclear waste repository.

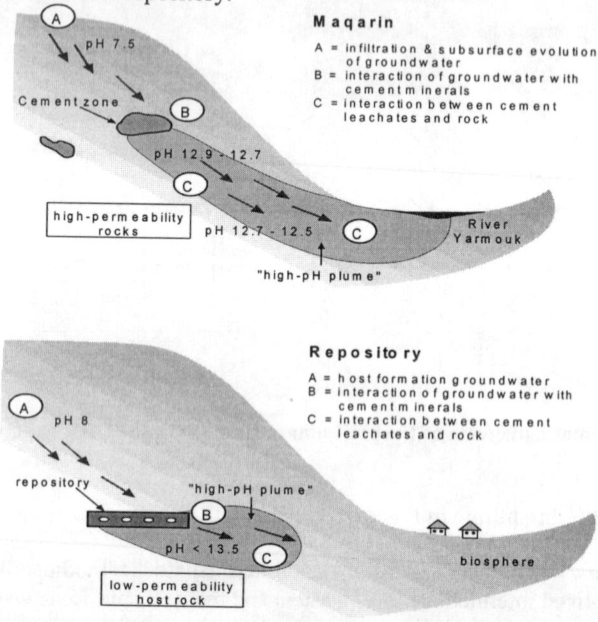

Figure 13: The potential formation of a hyperalkaline plume
Above: observations from the Maqarin Natural Analogue site, northern Jordan
Below: conceptual model for a repository host rock [24]

5.4 Experimental Work

The experiment is located at the GTS in the IAEA Level B controlled zone and the experiment is similar in general design to others carried out at the GTS, involving injection and extraction of tracers through a water bearing shear zone. Where this experiment differs from MI and CRR is that there is continual pumping of a hyperalkaline fluid (pH 13.3; representing the leachates from a cementitious repository) through the shear zone. Combined with the introduction of a cement fluid to the shear zone is the injections of radionuclide solutions. To date, the radionuclides have been limited to non-sorbing nuclides, namely ^{22}Na, ^{82}Br and ^{131}I as the flow characteristics of the shear zone are still changing due to the interaction of the hyperalkaline leachates and the shear zone. However, by January 2003, a series of long lived and strongly sorbing nuclides will be injected into the flow field. As with the EP experiment, the shear zone will be overcored and the specific sites of retardation of the strongly sorbing (repository relevant) nuclides will be examined. It is planned to introduce ^{134}Cs , ^{60}Co, ^{152}Eu as the sorbing tracers and ^{131}I as a conservative tracer to monitor the extracted cement waters via on-line γ spectrometry.

Currently, the tracer tests with ^{22}Na, ^{82}Br and ^{131}I are measured with an online γ spectrometer as described for the excavation project, no offline detectors are required as the activities added are sufficiently large to allow easy detection over the course of the experimental run. Changes to the detection system (from MI, EP and CRR) have been in the form of software upgrades and switching to electrically cooled γ spectrometers. This is more suitable for long term monitoring as the equipment can be left to run without the need for repeated filling of liquid nitrogen.

5.5 Future work at the GTS

Planning has already begun for the next phase (Phase VI) of work at the GTS. As more confidence has developed in our understanding of the processes controlling the behaviour of radionuclides in groundwater systems, so the complexity of the experiments has increased. By building on the successes (and failures) of the past, the radionuclide projects at the GTS have become more and more experimentally complex and have moved from basic research into the behaviour of tracers in the environment to answering very specific open questions regarding the likely long-term behaviour of a deep geological repository. Thus, the next phase of experiments at the GTS will continue this trend and investigate additional repository-relevant issues. The work carried out to date has been on a longer temporal and spatial scale than laboratory experiments, but the groundwater flow rates in these experiments has been around one million times faster than would be expected in a suitable repository host rock. Therefore the next phase of work will consider significantly longer time scales than has been the case so far and focus on semi-stagnant groundwater systems, to try to better match the conditions in and around a waste repository. Although at an early stage, current proposals consider experimental durations of up to 50 years, more than an order of magnitude longer than has been the case in any rock laboratory anywhere in the world to date. Further information and regular updates will be available on www.grimsel.com.

6 DISCUSSION AND CONCLUSIONS

6.1 The value of field experiments in model testing

Given the scales of time and space that must be considered when assessing the long-term behaviour of a deep geological repository for waste (either radioactive or chemo-toxic), direct testing of the realism, or conservatism, of a model in the system of interest is impossible. Field experiments, such as those performed at the GTS, are consequently of great value in that:

- the fundamental transport processes that operate in the system are expected to be the same or similar to those relevant to any repository host rock;
- the structures present, though differing in detail, are also similar to those of potential repository host rocks;
- the scales of time and space over which the experiments operate, though often considerably smaller than those of true repository relevance, are larger than those achievable in the laboratory;
- the degree of characterisation of the system that is possible is greater than that achievable at a repository site (due to the smaller spatial scales involved in many field experiments and the need, at a repository site, to avoid perturbing the favourable properties of the system);
- field experiments, being performed *in situ*, are less subject to experimental artefacts than laboratory experiments.

Field experiments can never show beyond doubt that a model applied in performance assessment is "correct", given, for example, the differences in scales of space and time involved. They can, however, add to the body of observations and experiments with which the model is consistent and thus build confidence in its application in performance assessment. This can be further strengthened by carrying out a series of complementary and inter-related experiments as has been the case at the GTS with the MI, EP, CRR and HPF projects. These have utilised a wide range of supporting field, laboratory and natural analogue studies [18].

7 ACKNOWLEDGEMENTS

A task as large as the joint Nagra-JNC Radionuclide Migration Programme has many players and the authors would like to extend their thanks to all of our colleagues world-wide who have contributed in any form to the lessons learnt over the last 15 years. While the CRR and HPF projects would not now be in as advanced a state as they currently are without the input from their forefathers (namely MI and EP), the teams involved here (CRR: ANDRA, BMWi/FZK, ENRESA, JNC, Nagra and USDoE/Sandia. HPF: ANDRA, JNC, Nagra and SKB) are warmly thanked for their efforts over the last two years.

References

1 W.R. Alexander, P.A. Smith and I. G. McKinley (2002a). Modelling radionuclide transport in the geological environment: a case study from the field of radioactive waste disposal. Ch.5 *in* E. M. Scott (ed), Modelling Radioactivity in the Environment, Elsevier, Amsterdam, The Netherlands (*in press)*.

2 I. G. McKinley, W.R. Alexander, C. Bajo, U. Frick, J. Hadermann, F.A. Herzog and E. Hoehn (1988) The radionuclide migration experiment at the Grimsel rock laboratory, Switzerland. Sci. Basis Nucl. Waste Manag. XI, pp179-187.

3 W. R. Alexander (1995) Natural cements: How can they help us safely dispose of radioactive wast? Radwaste Magazine 2, vol 5, Sept. 1995, pp 61-69.

4 U. Frick, W. R. Alexander et al., (1992) The radionuclide migration experiment-overview of investigations 1985-1990. Nagra Technical Report Series NTB 91-04, Nagra, Wettingen, Switzerland.

5 W. Heer and J. Hadermann (1996) Modelling radionuclide migration field experiments. Nagra Technical Report NTB 94-18, Nagra, Wettingen, Switzerland.

6 K. Ota, W.R. Alexander, P.A. Smith, A Möri, B. Frieg, U. Frick, H. Umeki, K. Amano, M. M. Cowper and J.A. Berry (2001) Building confidence in radionuclide retardation programme. Sci. Basis Nucl. Waste Manag. XXIV, 1033-1041.

7 P.A. Smith, W. R. Alexander, W. Heer, T. Fierz, P.M. Meier, B. Baeyens, M. H. Bradbury, M. Mazurek and I. G. McKinley (2001a) The Nagra-JNC *in situ* study of safety relevant radionuclide retardation in fractured crystalline rock I: The Technical Report Series NTB 00-09, Nagra, Wettingen, Switzerland.

8 P.A. Smith, W. R. Alexander, W. Kickmaier, K. Ota, B. Frieg and I.G. McKinley (2001b) Development and Testing of Radionuclide Transport Models for Fractured Rock: Examples from the Nagra/JNC Radionuclide Migration Programme in the Grimsel Test Site, Switzerland. J. Contam. Hydrol. 47, 335-348 *Also published in Japanese in JNC Technical Review 11, JNC TN1340 2001-006, JNC, Tokai, Japan.*

9 W. R. Alexander, B. Frieg, K. Ota and P. Bossart (1996) The RRP project: investigating radionuclide retardation in the host rock. Nagra Bulletin No. 27 (June 1996), pp43-55.

10 W. R. Alexander, K. Ota and B. Frieg (2002b) Eds. The Nagra-JNC *in situ* study of safety relevant radionuclide retardation in fractured crystalline rock II: the RRP project methodology development, field and laboratory tests. Nagra Technical Report Series NTB 00-06 *(in press)*, Nagra, Wettingen, Switzerland.

11 A Möri, B. Frieg, K. Ota and W. R. Alexander (2002a) Eds: The Nagra-JNC *in situ* study of safety relevant radionuclide retardation in fractured crystalline rock III: the RRP project final report. Nagra Technical Report NTB 00-07 *(in prep)*, Nagra, Wettingen, Switzerland.

12 W. R. Alexander, K. Ota, B. Frieg and I. G. McKinley (1998a) The assessment of radionuclide retardation in fractured crystalline rocks. Extended abstract *in* Sci. Basis Nucl. Waste Manag. XXI, 1065-1066.

13 NEA (1999). Characterisation of water-conducting features and their representation in models of radionuclide migration. Proceedings of the 3rd GEOTRAP Workshop, Barcelona, Spain, 10-12 June 1998 *(in prep)*. NEA/OECD, Paris, France.

14 Nagra (1994) Kristallin-1, Nagra Technical Report Series, NTB 93-22, Nagra, Wettingen, Switzerland.

15 M. Mazurek, A. Gautschi, S. Vomvoris (1992): Deriving input data for solute transport models from deep borehole investigations: ap approach for crystalline rocks; Proc. Int. Symp. On Geologic disposal of spent fuel, high-level and alpha-bearing wastes, IAEA-SM-326/31, pp55-67, Antwerp, October 1992.

16 I. G. McKinley and W. R. Alexander (1992) Constraints on the applicability of in situ distribution coefficient values. J. Environ. Rad. 15, 19-34.

17 E. Hoehn, J. Eikenberg, T. Fierz, W. Drost and E. Reichlmayr (1998) The Grimsel Migration Experiment: field injection-withdrawal experiments in fractured rock with sorbing tracers. J. Contam. Hydrol. 34, 85-106.

18 W. R. Alexander, A Gautschi and P. Zuidema (1998b) Thorough testing of performance assessment models: the necessary integration of *in situ* experiments, natural analogues and laboratory work. Extended abstract *in* Sci. Basis Nucl. Waste Manag. XXI, 1013-1014.

19 B. Frieg, W. R. Alexander, H. Dollinger, C. Bühler, P. Haag, A. Möri and K. Ota (1998) In situ impregnation for investigating radionuclide retardation in fractured repository host rocks. J. Contam. Hydrol. 35, 115-130.

20 W.R. Alexander and W Kickmaier (2000) Radionuclide retardation in water conducting systems – lessons learned in the research programme in the Grimsel Test Site/Radionuklidretardation in wasserführenden Systemen – Erfahrungen aus den Untersuchungsprogrammen im Felslabor Grimsel. Proceedings of the Conference on 'Sanierung der Hinterlassenschaften des Uranerzbergbaus (Chancen und Grenzen der geochemischen und Transport-Modellierung bei der Verwahrung von Uranbergwerken und bei der Endlagerung radioaktiver Abfälle)', 5th Workshop (18 May, 2000), Dresden, Germany.

21 J. Eikenberg, M. Rüthi, W. R. Alexander, B. Frieg and T. Fierz (1998) The excavation project in the Grimsel Test Site: *in situ* high resolution gamma and alpha spectrometry of ^{60}Co, ^{75}Se, ^{113}Sn, ^{152}Eu, ^{235}U and ^{237}Np/^{233}Pa. Sci. Basis Nucl. Wsate Manag. XXI, 665-662.

22 A Möri, M. Schild, S. Siegesmund, A. Vollbrecht, M. Adler, M. Mazurek, K. Ota, P. Haag, T. Ando and W. R. Alexander (2002b). The Nagra-JNC *in situ* study of safety relevant radionuclide retardation in fractured crystalline rock IV: The *in situ* study of matrix porosity in the vicinity of a water-conducting fracture. Nagra Technical Report NTB 00-08 *(in press)*, Nagra, Wettingen, Switzerland.

23 A. Möri, H. Geckeis, T. Missana, J. Guimera, C Degueldre, P. Meier, K. Ota, H. Appenguth, W. R. Alexander and P Hernan (2002c) The colloid and radionuclide retardation experiment (CRR) at the Grimsel Test Site – preliminary evidence for colloid mediated radionuclide transport in fractured rock. Sci. Basis Nucl. Waste Manag. J. *(in press)*.

24 W. R. Alexander (1995) Natural cements: How can they help us safely dispose of radioactive waste? Radwaste Magazine 2, vol 5, Sept. 1995, pp 61.69.

25 F. F. Ewart, S. M. Harland and P. W. Tasker (1985). The chemistry of the near-field environment. Sci. basis Nucl. Wsate Manag. IX, 539-546.

26 A. H. Balth, N Christofi, C. Neal, J. C. Philp, M. R. Cave, I. G. McKinley and U. Berner (1987) Trace element and microbiological studies of alkaline groundwaters in Oman: a natural analogue for cement pore-waters. Nagra Technical Report NTB 87-16, Nagra, Wettingen, Switzerland.

27 N. A. Chapman and I. G. McKinley (1987), The Geological Disposal of Nuclear Waste; John Wiley and Sons, London, UK.

28 W. M. Miller, W. R. Alexander, N. A. Chapman, I. G. McKinley and J. A. T. Smellie (2000) Geological disposal of radioactive wastes and natural analogues. Waste management series, vol. 2, Pergamon, Amsterdam, The Netherlands.

ELUTION BEHAVIOR OF TECHNETIUM AND RHENIUM THROUGH A Tc-SELECTIVE CHROMATOGRAPHIC RESIN COLUMN

S. Uchida and K. Tagami

Environmental and Toxicological Sciences Research Group, National Institute of Radiological Sciences, Anagawa 4-9-1, Inage-ku, Chiba 263-8555, JAPAN

1 INTRODUCTION

Technetium-99 is produced by fission of ^{235}U or ^{239}Pu with a relatively high yield of about 6%. The sources of environmental ^{99}Tc are releases from nuclear weapons tests and the nuclear fuel cycle which includes nuclear fuel reprocessing plants and enrichment facilities.[1, 2] Since ^{99}Tc has a long half-life of 2.11 x 10^5 y, the radionuclide has been accumulating in the environment. It has high mobility in the soil-water system and also high plant-availability through roots; ^{99}Tc is one of the most important radionuclides for dose assessment.[3, 4]

Determination of ^{99}Tc in environmental samples has been studied to clarify Tc behaviour in the environment.[5-10] To measure ^{99}Tc in environmental samples, chemical separation and purification of the nuclide from interfering elements are required, because ^{99}Tc is a soft beta-emitting radionuclide and its concentration in the environment is very low. The use of a yield tracer is necessary to monitor the chemical yield through the separation steps. There are several yield tracers for ^{99}Tc in use including ^{99m}Tc ($T_{1/2}$ = 6.01h), ^{97m}Tc ($T_{1/2}$ = 91d), ^{95m}Tc ($T_{1/2}$ = 61d) and Re. Among the Tc isotopes which are available, ^{95m}Tc is thought to be the most useful isotope to obtain chemical yields for ICP-MS. However, commercially available ^{95m}Tc solution may contain ^{99}Tc.[11] We successfully produced and separated carrier-free ^{95m}Tc from a niobium target,[12] however, in general, it is difficult to get a constant carrier-free ^{95m}Tc supply. Moreover, if the addition of a yield monitor is necessary at sampling sites, e.g., on a ship for seawater sampling, a radioactive tracer cannot be used there. Thus, from a practical viewpoint, Re is a good yield monitor for ^{99}Tc in environmental water samples. Some reports have shown that Re is a very close analogue of Tc, that is, the behaviours of the two elements are almost the same through the separation procedures.[13-15]

Recently, a Tc-selective chromatographic resin (TEVA resin) was developed by Eichrom Industries Inc., Darien, IL, U.S.A. This resin has been shown to retain Tc efficiently and selectively from aqueous solutions[16, 17] and also to sorb Re strongly.[18] However, there have been no reports comparing Re and Tc behaviors during the Tc separation procedure. In this study, we performed tracer experiments to compare the elution behaviours of Tc and Re through the TEVA resin column under our earlier proposed conditions.[19] Also, we paid attention to Mo and W concentration in the final solution which was introduced into the ICP-MS for measuring ^{99}Tc and Re, because the isotopic abundance of ^{98}Mo is 24.13% and those of ^{184}W and ^{186}W are 30.64% and 28.43%, respectively; these isotopes might interfere with the counting of ^{99}Tc, ^{185}Re, and ^{187}Re as hydrides. Ruthenium as well was noted for potential interference in the ^{99}Tc counting, because of its nuclide at mass 99 with an isotopic abundance of 12.7%. However, the Ru concentrations in environmental waters are usually lower than those of Mo. For example, Ru and Mo concentrations in seawater are 7 x 10^{-4} ng/ml and 10 ng/ml, respectively.[20]

We also studied the possibility of using an isotopic dilution method, that is, using an enriched ^{185}Re spike instead of a Re standard solution in which Re isotopic abundance was natural (^{185}Re: 37.4%, and ^{187}Re: 62.6%). Since Re is one of the rarest elements in the earth's crust, the data are limited for environmental samples. Moreover, as the geological behaviour of Re is similar to that of Tc,[21] knowledge of Re could also be used to improve dose assessment models for Tc in the natural environment.

2 EXPERIMENTAL

2.1 Reagents

Prepacked TEVA resin columns were used in the experiments. The resin is an aliphatic quaternary amine sorbed on Amberchrom CG71. The resin grain size was 100 – 150 μm and the volume of the resin in the column was about 2 ml. The length and diameter of the column were 4.2 cm and 0.8 cm, respectively. The water (>18 MΩ) used was prepared with a Milli-Q water system (Millipore Co.) and nitric acid was ultra pure grade (Tama Chemicals, AA-100).

2.2 Standard and Spike Solutions

Technetium-99 standard solution (Japan Isotope Association, NH_4TcO_4 form in 0.1% ammonia solution) was used for the tracer experiments and for standard solutions to make a calibration curve by ICP-MS. For Re, we used multi-element standard solution (SPEX : XSTC 8) which contained 10 μg/ml of Re (natural isotopic abundance). This standard solution also contained 10 μg/ml of Mo. The XSTC 8 standard solution was also used to make calibration curves for Re and Mo.

A Re spike, enriched ^{185}Re isotope, was acquired from Oak Ridge National Laboratory. The atomic percent for ^{185}Re was 96.74. The metal was dissolved in 1M HNO_3 solution then diluted to obtain a 5.3 μg/ml Re solution. The stock solution was stored in a PTFE bottle.

2.3 Sample Pretreatment

In order to evaluate the characteristics of the sorption and the elution behaviors of Tc and Re on the TEVA resin, pure water and tap water were used. The tap water was collected at our laboratory in Chiba Prefecture, Japan. The pH of the tap water was 7.4 and the electrical conductivity was 0.22 mS/cm. Two hundred and fifty ml of each solution were adjusted to 0.1M HNO_3 with concentrated HNO_3. Next, each solution was spiked with ^{99}Tc and the multi-element standard solution. The concentration of ^{99}Tc in the prepared solution was 0.4 mBq/ml and the Re and Mo concentrations were 40 pg/ml. Concentrations of these elements were high enough for their determinations by ICP-MS. These sample solutions served for the column experiment.

2.4 Procedures

Prior to using the TEVA resin column, the resin was rinsed with 10 ml of 0.1M HNO_3. A 250-ml portion of the sample solution was eluted through the column at a flow rate of ca. 1.5 ml/min. Next, the column was washed with 40 ml (8 ml x 5) of 1, 2, 4 or 8M HNO_3 (Figure 1). Finally, Tc and Re retained in the column were stripped with 10 ml (5 ml x 2) of 12M HNO_3. The loading solutions and the nitric acid solutions were allowed to drain completely and the eluate was collected into PTFE vials for the measurement of ^{99}Tc, Re and Mo by ICP-MS.

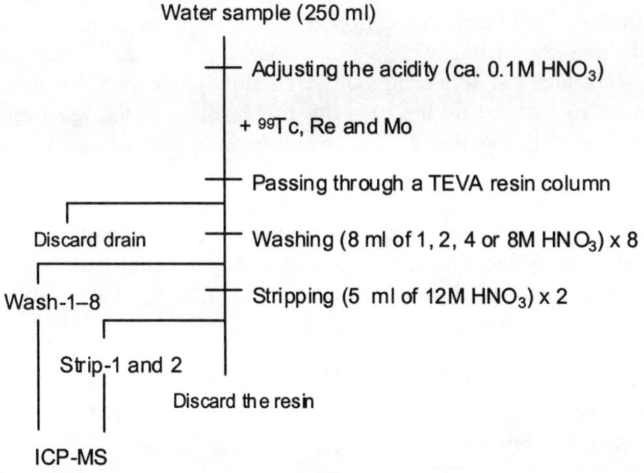

Figure 1 *Experimental scheme for determination of the elution behaviors of Tc, Re and Mo in pure water and tap water samples passed through TEVA resin columns.*

2.5 Measurements

A quadrupole type ICP-MS (Yokogawa: PMS-2000) was used for the determination of ^{99}Tc, Re and Mo in the solutions. For the study of the Re and Mo behaviours in a TEVA resin column, the stable isotopes ^{185}Re and ^{98}Mo were measured. Rhenium-185 has a natural abundance of 37.4% and ^{98}Mo has the highest natural abundance of 24.13%. Measurements were carried out five times per sample. Each isotope was scanned 500 times. Total counting times for ^{99}Tc, Re and Mo were 20 s, 2 s and 2 s, respectively. Under these conditions, the detection limits for ^{99}Tc, ^{185}Re and ^{98}Mo were 0.04 mBq/ml, 0.7 pg/ml and 0.8 pg/ml, respectively.

For the isotopic dilution study for Re measurement, masses of 185 and 187 were scanned 1000 times and total counting times were 15 s. Tungsten-184, which has the highest abundance in W isotopes as 30.64%, was also monitored for 3 s during the Re measurement, because W may bond with H which in turn would interfere with the Re counting. For instance, $^{184}WH^+$ and $^{186}WH^+$ cannot be differentiated from ^{185}Re and ^{187}Re, respectively, during measurements by the ICP-MS.

3 RESULTS AND DISCUSSION

3.1 Tc , Re and Mo Elution Behaviors through a TEVA Resin Column

For the determination of ^{99}Tc in environmental waters, the resin manufacturer recommended washing the TEVA resin column with 1M HNO_3 after introducing a solution into the column, and then, using 12M HNO_3 to strip Tc from the column.[22] However, this procedure was proposed to prepare samples for liquid scintillation counting.

In our previous paper, we proposed a new Tc separation procedure for ICP-MS measurement samples. This procedure combined use of 2M HNO_3 for washing the column and 8M HNO_3 for stripping Tc.[19] Under this proposed procedure, we found that Ru was effectively removed from the solution. As mentioned above, ^{98}Mo might interfere with the ^{99}Tc counting and in environmental waters Ru concentrations are lower than Mo concentrations, e. g., Ru content is five orders of magnitude lower than Mo in seawater. Thus, in this study, we also focused on the behavior of Mo through the resin column. If ICP-MS were to be applied, ^{99}Tc together with Re could be measured using short counting times.

The elution behaviours of Tc, Re and Mo are listed in Table 1. The counting errors were less than 5% for all of the samples. There were no differences in each nuclide's behaviour between pure water and tap water. When the sample solution which had a nitric acid concentration of 0.1M was introduced into the column, Re was extracted onto the resin

Table 1 *The differences of each recovery of ^{99}Tc, Re and Mo in wash and strip solutions when nitric acid concentrations for wash were 1, 2, 4 and 8M. The sample solutions were pure water and tap water. Counting errors were less than 5%.*

Pure water	Volume (ml)	Mo-98				Tc-99				Re-185			
		1M	2M	4M	8M	1M	2M	4M	8M	1M	2M	4M	8M
Load*	250	0.92	0.92	0.92	0.92	0.00	0.00	0.00	0.00	0.00	0.00	0.00	0.00
Wash-1	8	0.01	0.01	0.02	0.02	0.01	0.01	0.01	0.98	0.00	0.00	1.00	0.98
Wash-2	8	0.00	0.01	0.01	0.01	0.00	0.00	1.00	0.00	0.00	0.00	0.00	0.00
Wash-3	8	0.00	0.01	0.01	-	0.00	0.00	0.00	-	0.00	0.10	0.00	-
Wash-4	8	0.00	0.00	0.00	-	0.01	0.00	0.00	-	0.00	0.50	0.00	-
Wash-5	8	0.00	0.00	-	-	0.00	0.01	-	-	0.00	0.39	-	-
Strip-1**	5	0.02	0.02	-	0.02	1.00	1.04	0.01	0.01	0.98	0.04	0.00	0.00
Strip-2**	5	0.04	-	-	-	0.02	-	-	-	0.00	-	-	-
Tap water	**Volume (ml)**	**Mo-98**				**Tc-99**				**Re-185**			
		1M	2M	4M	8M	1M	2M	4M	8M	1M	2M	4M	8M
Load*	250	1.03	1.03	1.03	1.03	0.00	0.00	0.00	0.00	0.00	0.00	0.00	0.00
Wash-1	8	0.01	0.01	0.02	0.02	0.00	0.00	0.03	0.98	0.00	0.00	1.00	1.00
Wash-2	8	0.00	0.00	0.00	0.00	0.00	0.00	0.95	0.00	0.00	0.00	0.00	0.00
Wash-3	8	0.00	0.00	0.00	-	0.00	0.00	0.00	-	0.00	0.10	0.00	-
Wash-4	8	0.00	0.00	-	-	0.00	0.00	-	-	0.00	0.45	-	-
Wash-5	8	0.00	0.00	-	-	0.00	0.00	-	-	0.00	0.39	-	-
Strip-1**	5	0.04	0.03	0.02	0.02	0.95	1.03	0.01	0.01	1.05	0.06	0.00	0.00
Strip-2**	5	0.02	0.02	0.03	0.02	0.01	0.01	0.01	0.01	0.00	0.00	0.00	0.00

(Notes) *: The acidity of the loading solution was 0.1M HNO_3.
**: The 12M HNO_3 was used as strip solution.
-: Not measured.

completely the same as Tc was, while most of the Mo passed through the column with the sample solution; the column was then washed with 40 ml (8 ml x 5) of 1, 2, 4 or 8M HNO_3. When 1M HNO_3 was used, Re was well retained on the resin. With 2M HNO_3, Re was gradually removed and it was found in the third 8-ml wash fraction. When higher concentrations of nitric acid solutions than 4M HNO_3 were used for washing the column, Re was removed from the resin easily and we found it in the first 8-ml fraction. In our previous report,[19] Tc was strongly sorbed onto the resin column even when 40 ml of 2M HNO_3 were introduced into the column for washing. The same result for Tc was obtained in this study. The experimental results of Tc and Re could be explained by our previous study using batch experiments;[18] the sorption characteristics of Tc and Re onto the TEVA resin were different. Since the distribution coefficient of Re between the resin and nitric acid solution was almost half the value of that of Tc over a wide range of nitric acid concentrations, from 0.1M HNO_3 to 8M HNO_3, Re could be removed with less than 40 ml of 2M HNO_3. Thus, when the TEVA resin is applied for Re separation, less than 1M HNO_3 should be used for washing the column to avoid any Re loss. For stripping, we had previously found 12M HNO_3, recommended by the manufacturer,[22] was too strong for stripping Tc from the resin.[19] As seen in Table 1, 8 ml of the 8M HNO_3 could completely remove both Tc and Re from the resin.

For Mo, only a small part of the initial amount remained on the resin after passing the sample solution through the column. About 2% of the initial amount was removed with a wash solution of 8 ml 8M HNO_3, though ca. 2-4% were found in the strip solution of 12M HNO_3. Thus, Mo could be almost completely eliminated through the TEVA resin separation steps and the isobaric interference due to Mo could be decreased for ^{99}Tc measurement by ICP-MS.

To measure Re precisely, it is necessary to eliminate the effects of isobars that might be involved in forming oxides, chlorides and hydrides with rare earth elements (REEs) or W at masses 185 and 187, even in plasma. Therefore, mass spectrum data were necessary to check whether there were isobars or not. In this study, we especially focused on getting W hydride generation percentages. Figure 2 shows the mass spectra of 1 μg/ml W with and

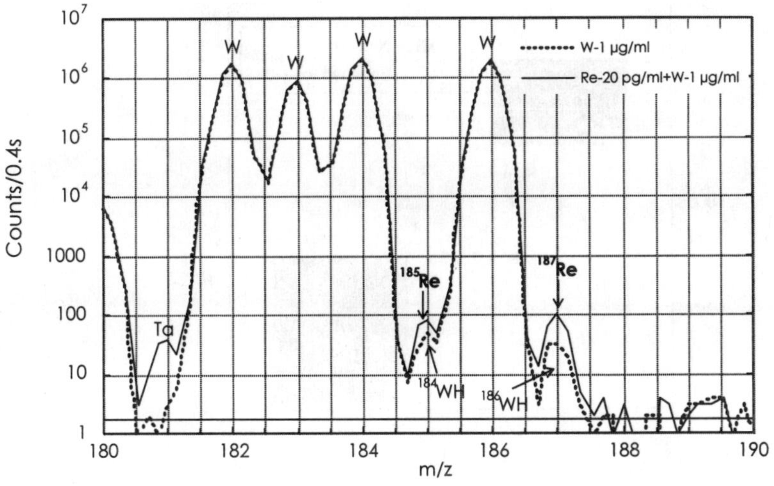

Figure 2 *Mass spectra of 1 μg/ml W solution with and without Re 20 pg/ml.*

without 20 pg/ml Re. Although the Re content was much lower than W, Re could be measured if WH counts were subtracted at masses of 185 and 187. The percentage of WH^+ generation compared with W^+ was less than 0.003% under the measurement conditions. In the natural environment, W concentration in stream water was estimated as 30 pg/ml[23] while Re in river water was 0.2-48 pg/ml in South America[24] and 0.9-6.5 pg/ml in Japan.[25] Thus, W hydrides give negligible interference for Re.

3.2 Application of Isotope Dilution Method to Measure Tc and Re in the Same Sample

Isotope dilution methods have been used for Re determination in environmental samples.[e.g. 24, 26] From the tracer experiment in this study, we thought an enriched ^{185}Re spike could be used instead of a Re standard solution having natural isotopic abundances, such that Re could be measured as well as Tc from the same sample. As a quadrupole type ICP-MS has one detector, it had been said in the past that the precision would be much worse than that of thermal ionization mass spectrometry. However, due to the development of ICP-MS machines, the 185 and 187 masses can now be scanned rapidly.

Figure 3 shows the counts per second of Re isotopes (at masses of 185 and 187) by ICP-MS; the sample contained ca. 50 pg/ml enriched ^{185}Re. The relative standard deviation for the sample determination was less than 2% for each mass, thus, the calculation for the natural Re was not affected by any error originating from the counting. The determined natural Re concentration was 50 pg/ml and the detection limit of this method was 1.5 pg/ml using ^{187}Re. The TEVA resin showed high selectivity for Re which was similar to that for Tc. When appropriate concentrations of nitric acid solutions, 1M and 8M, were used for washing and stripping, respectively, the elution behaviors of Tc and Re were the same. This meant that Re could be used as a yield tracer of Tc in environmental water samples and both elements could be concentrated in an 8-ml portion of 8M HNO_3 with more than 98% recovery. Also, because the resin could eliminate other matrix elements including Mo under the above mentioned combined use of nitric acid solutions, it should be applicable to determinations of Tc and Re concentrations by ICP-MS.

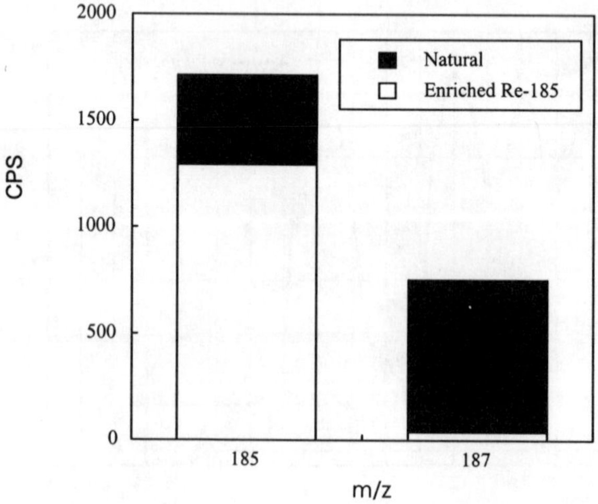

Figure 3 *Counts per second of Re isotopes by ICP-MS for determination of Re using enriched Re-185 as a yield tracer.*

4 CONCLUSION

The TEVA resin showed high selectivity for Re which was similar to that for Tc. When appropriate concentrations of nitric acid solutions, 1M and 8M, were used for washing and stripping, respectively, the elution behaviors of Tc and Re were the same. This meant that Re could be used as a yield tracer of Tc in environmental water samples and both elements could be concentrated in an 8-ml portion of 8M HNO_3 with more than 98% recovery. Also, because the resin could eliminate other matrix elements including Mo under the above mentioned combined use of nitric acid solutions, it should be applicable to determinations of Tc and Re concentrations by ICP-MS.

References

1 F. Luykx, in *Technetium in the Environment*, eds. G. Desmet and C. Myttenaere, Elsevier Appl. Sci. Pub, New York , 1986, pp. 21-27.
2 J.E. Till, in *Technetium in the Environment*, eds. G. Desmet and C. Myttenaere, Elsevier Appl. Sci. Pub, New York , 1986, pp. 1-20.
3 R.E. Wildung, T.R. Garland and D.A. Cataldo, *Health Phys*., 1977, **32**, 314.
4 P.J. Coughtrey, D. Jackson and M.C. Thorne, in *Radionuclide Distribution and Transport in Terrestrial and Aquatic Ecosystems, A Critical Review of Data*, A.A. Balkema, Rotterdam, 1983, Vol. 3, pp. 210-228.
5 E. Holm, J. Rioseco and M. García-León, *Nucl. Instr. Meth. Phys. Res.,* 1984, **223**, 204.
6 M. García-León, G. Manjón and C.I. Sánchez-Angulo, *J. Environ. Radioact.* 1993, **20**, 49.
7 S. Morita, K. Tobita and M. Kurabayashi, *Radiochim. Acta*, 1993, **63**, 63.
8 K. Tagami and S. Uchida, *Environ. Pollut.* 1997, **95**, 151.
9 S. Uchida, K. Tagami, E. Wirth and W. Rühm, *Environ. Pollut*., 1999, **105**, 75.
10 K. Tagami and S. Uchida, *J. Nucl. Radiochem. Sci.*, in press.
11 K. Tagami and S. Uchida, *Appl. Radiat. Isot*., 1996, **47**, 1057.
12 T. Sekine, M. Konishi, H. Kudo, K. Tagami and S. Uchida, *J. Radioanal. Nucl. Chem.* 1999, **239**, 483.
13 N. Matsuoka, T. Umata, M. Okamura, N. Shiraishi, N. Momoshima and Y. Takashima, *J. Radioanal. Nucl. Chem*., 1990, **140**, 57.
14 B.R. Harvey, K.J. Williams, M.B. Lovett and R.D. Ibbett, *J. Radioanal. Nucl. Chem.,* 1992, **158**, 417.
15 S.F. Fang and T.C. Chu, in *Proceedings of 10th International Congress of the International Radiation Protection Association*, Hiroshima, Japan, May 14-19, 2000; the International Radiation Protection Association, P-4a-237.
16 S. Uchida and K. Tagami, *J. Radioanal. Nucl. Chem*. 1997, **221**, 35.
17 K. Tagami and S. Uchida, *J. Radioanal. Nucl. Chem*. 1999, **239**, 643.
18 K. Tagami and S. Uchida, *Anal. Chim. Acta,* 2000, **405**, 227.
19 S. Uchida and K. Tagami, *Anal. Chim. Acta,* 1997, **357**, 1.
20 H.J.M. Bowen, in *Environmental Chemistry of the Elements*, Academic Press, London, 1979, pp. 22-23.
21 D.G. Brookins, in *Eh-pH Diagrams for Geochemistry*, Springer-Verlag, Berlin, 1988, pp. 97-101.
22 Eichrom Industries, Inc. in *Information sheet on TEVA resin (TCW01),* Eichrom Industries, Darien, IL, 1995.
23 C. Reimann and P. Caritat, in *Chemical Elements in the Environment*, Springer-Verlag,

Berlin, 1998, pp. 362-365.
24 D.C. Colodner, E.A. Boyle and J.M. Edmond, *Anal. Chem.,* 1993, **65**, 1419.
25 S. Uchida, K. Tagami and M. Saito, *J. Radioanal. Nucl. Chem*, in press.
26 R.J. Walker and J.D. Fassett, *Anal. Chem*., 1986, **58**, 2923.

THE USE OF *FUCUS VESICULOSUS* TO MONITOR THE TRANSPORT OF 129I IN THE ENGLISH CHANNEL

S.J. Parry and A.Davies

Imperial College of Science, Technology & Medicine, Silwood Park, Ascot, Berkshire SL5 7PY, UK

1. INTRODUCTION

Iodine-129 is the only naturally occurring radioisotope of iodine. It has a long half life, 1.57×10^{7} years, and although at present it does not pose any significant biological threat to humans, as levels are likely to continue rising steadily it will need to be monitored closely. Naturally ^{129}I is produced as a result of cosmic ray interactions with xenon isotopes in the upper atmosphere and in the marine and terrestrial environment it is produced as a result of the spontaneous fission of uranium. Since 1945 the ^{129}I concentration in the environment has been increasing as a result of anthroprogenic emissions. Nuclear weapons testing and the use of nuclear fuel to produce energy have contributed to the elevation in the level. At present however the largest anthroprogenic source of ^{129}I into the environment occurs as a result of the reprocessing of spent nuclear fuel.

Two sources dominate the ^{129}I emissions in Western Europe, namely the nuclear reprocessing facilities at COGEMA La Hague (France) and BNFL Sellafield (UK). La Hague, located 25 km from Cherbourg has two plants dedicated to the reprocessing of oxide fuels, the UP2-800 and the newer UPS plant. Sellafield has facilities for reprocessing Magnox fuel along with oxide fuel in the newer Thermal Oxide Reprocessing plant (THORP). Recent estimates suggest[1] that some 1400 kg of ^{129}I has been released into the environment, mainly the marine environment by the reprocessing plants at Sellafield and particularly Cap de la Hague in France over the last three decades with a gradual increase in discharges particularly over the past few years. This amount is not an insignificant quantity considering that the global ^{129}I reservoir is only in the region of 50000 kg [1]. The total global reservoir of stable iodine is estimated to be in the region of 8×10^{14} kg[2]. As a result of this it is necessary to discuss ^{129}I not only in terms of its concentration, but also in terms of the $^{129}I/^{127}I$ ratio. Isotopic ratios have now come into widespread use in this kind of work as it is assumed that there has been no recent addition of stable ^{127}I into the environment, thus the concentration is thought to be in a state of equilibrium. ^{129}I emissions continue and thus it can be assumed that it is only the ^{129}I that is in a dynamic state. For the determination of the $^{129}I/^{127}I$ ratio two distinct periods of time must be considered, namely the pre and the post bomb periods. The post bomb period is said to have begun in 1945 with the onset of nuclear testing leading to notable rises in the ratio. Pre 1945 $^{129}I/^{127}I$

ratios are now widely regarded to be in the region of 3×10^{-13} - 3×10^{-12} in the marine environment and slightly less at $10^{-15} - 10^{-14}$ in the lithosphere [1]. Post 1945 the marine ratio away from the direct influence of marine discharges has been shown to have increased to $2^{-11} - 2^{-10}$ and to $10^{-9} - 10^{-7}$ in the terrestrial environment [1].

The discharges from La Hague increased from 1990, and have now levelled out. Discharges from Sellafield were stable up to 1994, when the new THORP plant opened, leading to an increase in the level of reprocessing. It is now known that the Sellafield and La Hague discharges combine in the North Sea near the Norwegian coast[3]. Current estimates indicate that the emissions of ^{129}I into the sea at La Hague account for around 75% of the total yearly combined emissions, with discharges in the region of 220 kg/year [3]. The North Sea is thus treated as a large sink for ^{129}I before it is moved into the Atlantic where it is further diluted. Elevated marine $^{129}I/^{127}I$ ratios are consistently observed in the vicinity of reprocessing plants, especially the La Hague facility where ratios as high as 1.1×10^{-5} have been reported[4]. From work carried out by Raisbeck et al.[3], and others, the fate of conservative radionuclides leaving La Hague is relatively well understood. On leaving the plant, the radionuclides generally follow the coast as they pass up to Denmark where they are dispersed in the north Atlantic. However, some of radionuclides pass far enough out into the channel where they get caught up in the fast flowing mid channel current which flows west to east, where they pass quickly through the Straits of Dover and up to Scandinavia [5].

Modelling work carried out on the fate of radionuclides in the area suggests transit times for the radionuclides passing close to the coast from La Hague to the Straits of Dover to be in the region of 7 months [6-7]. The time decreases to 3 months for the radionuclides that are carried in the fast flowing central waters. These predictions are known to be dependent on a large range of factors with the most important being wind direction. If the wind is easterly the transit times are known to increase significantly [6].

A recent study by Frechou et al.[4] showed that the $^{129}I/^{127}I$ ratio decreased from 1.1×10^{-5} at the outlet pipe at La Hague to 0.27×10^{-5} at Wimereux, located about 200 km to the East. This shows a decrease in concentration of a factor of about 3.5 indicating that not much mixing occurs in the channel. Work by Raisbeck et al.[3] indicates that by the time the ^{129}I reaches Denmark the ratio has decreased quite considerably to 6.48×10^{-8}. These samples were however analysed in 1992 and can thus be expected to have increased since then. Values in Iceland were found to be as low as 5.97×10^{-10} in 1989 [1], close to the global background levels, indicating a large amount of mixing in the north Atlantic. Data obtained from past work [5] carried out in 1990 indicate that North Sea waters and Channel waters do not mix directly, with a region in the middle where mixing occurs. This work shows that no radionuclides released from La Hague would reach the East coast of Britain any further North than the Essex region. This was further confirmed by carrying out work on Sb^{125}, a radionuclide that is specifically known to be released from La Hague. No traces of this radionuclide were found on the East Anglian coastline. For that reason it was of interest to examine the movement of ^{129}I in this region. The aim of the work described in this paper was to establish $^{129}I/^{127}I$ ratios in the region of the South Coast of England, round to the east coast and North to the East coast of Scotland.

Algae are commonly used to monitor the presence of radionuclides in the marine environment. One of the most common and widely recognised seaweeds for this type of work is the brown seaweed bladder wrack *(Fucus vesiculosus).* It is ideal to collect and analyse as part of this study, since it is common and widespread throughout the UK, and parts of Europe. Iodine is taken up by the seaweed and mainly deposited in the thalus, although it does appear in the remainder of the plant to a lesser extent along with a vast range of other elements by adsorption. The mere presence of an element does not however indicate that the element is essential to the plants health. Iodine, however, does appear to be required by red and brown algae, although its metabolic role is not fully understood. As a result, the plant accumulates iodine and concentrations can build up, since the seaweed does not posses any mechanism to get rid of it. Concentration factors for some brown algae[8] have been observed to be as high as 32,310.

Many different techniques have been used to determine the ^{129}I content of a wide range of environmental samples. They include ICP-MS, laser resonance ionisation mass spectrometry (RIMS) and laser induced fluorescence spectrometry. These techniques although useful cannot be readily used for the analysis of environmental samples, since the detection limits of the techniques are not low enough. Losses are also introduced at the dissolution stage and during the actual analysis. ^{129}I can be detected using a direct method of gamma ray spectrometry with a low energy gamma ray at 40 keV. The detection limit, however, using this direct method of analysis is about 100 Bq/kg in vegetation samples. Separation techniques can be employed, such as ion exchange and precipitation, to preconcentrate the iodide as a silver salt. This improves the counting geometry, minimises attenuation effects in the sample and improves the detection limit, which can be as low as 100 mBq/kg. Accelerator mass spectrometry (AMS), can determine background ratios down to 10^{-15}, however, one of the most widely used technique is neutron activation analysis (NAA). Radiochemical neutron activation analysis (RNAA) designed to separate single elements or groups of elements from the matrix so that the ultimate selectivity and sensitivity for the analyte is obtained. Using this technique ^{127}I / ^{129}I ratios down to about 10^{-10} are detectable.

This paper describes a RNAA technique for the determination of ^{129}I/^{127}I ratios in *Fucus vesiculosus* collected from the English channel and the North Sea, to study the transport of ^{129}I in the UK marine environment and to establish the relative impacts of Sellafield and La Hague reprocessing plants.

2. EXPERIMENTAL

2.1 Sampling strategy

Nine sampling locations were chosen along the South and East English and Scottish coasts. A total of 12 samples were collected at locations thought to give a good indication of how the ^{129}I levels would change along the coast. The most westerly sampling location was at Keyhaven in Dorset. The site was chosen because it is due North of La Hague, and as the main current flows in the opposite direction, from West to East, concentrations of ^{129}I were expected to be low. Previous work, carried out on ^{125}Sb, showed that activity decreases rapidly with distance from he French coast. Clearly a large part of the south English coast is shielded by this drift effect and so any ^{129}I detected at this location can be treated as a background reading. It would be expected that, as a result of this drift effect, samples taken progressively further to the East would show increasingly more elevated values. The highest value would therefore be expected at the sampling site at St. Margaret's at Cliff in the Straits of Dover, where the channel is at its narrowest. Samples were also taken from locations in Essex and East Anglia, on the East coast, to give a more accurate indication of whether any iodine crossing the channel remains close to the coast or whether it mixes in the channel at the main crossover point. It proved difficult to obtain samples on the section of the East coast, from Mersea to Robin Hoods Bay, as the coastline was very sandy, and there were no rocky beaches and harbour walls, where seaweed tends to grow. Southwold Harbour in Suffolk did however provide a suitable, if not ideal, location to obtain some samples.

2.2 Sampling methodology

All the seaweed collected was firmly attached to rocks, and the whole plant was taken. The seaweed was sampled from locations that were at similar distances down the beach, as near to the low tide mark as possible to ensure that the plants were exposed to the water for similar lengths of time. Immediately after collection the seaweed was washed in seawater to ensure that it was clean of any sand and other materials, and then placed in standard resealable plastic food bags. The location and map grid reference of the site was noted and the samples placed in a cool box, kept as cold as possible by packing with ice. On returning to the laboratory the seaweed samples were transferred to a refrigerator for storage prior to analysis.

2.3 Sample preparation

About 500 g of fresh seaweed was weighed, placed in a large beaker and dried in an oven at 80^{0}C. The dried sample was placed in an electric blender and shredded. At this stage a 5 g sub-sample was taken for subsequent stable ^{127}I determination. The remaining sample was transferred to a pre-weighed round bottomed reaction vessel where 50 mg of KI was added to act as a carrier for the ^{129}I, thus minimising loss to the reaction vessel and the surroundings. One litre of 1M sodium hydroxide (NaOH) was added and the subsequent mixture was refluxed for 5-6 h on a heating mantle. The sample was transferred while still hot to a 2 L beaker and placed in an oven and kept at 95^{0}C until the sample was almost dry, at which point the temperature was increased to around 160^{0}C and left until the sample was totally dry. The sample, still in the beaker, was then transferred to a furnace and the temperature ramped up to 450^{0}C over 4 h and held for 20 h. The sample was then allowed to cool before being crushed and placed back in the furnace and ramped up to 600^{0}C over 2 h and held for 4 h before being allowed to cool.1M NaOH (250 ml) was added to the ashed sample and left on a hot plate for 2 h with occasional stirring. Deionised water was added as necessary to replace any evaporated water. The sample was filtered through a 9 cm GF/A glass filter paper and washed thoroughly with deionised water. The filtrate was transferred to a 2 L beaker and heated on a hotplate to reduce the volume to 500 ml.

2.4 Blanks, reference materials and standards

Procedural blanks were prepared as for the samples using the same procedure and reagents, with only the seaweed omitted. An 'in house' reference material was provided C. Frechou in which a large sample of *Fucus serratus*, closely related to *Fucus vesiculosus* was collected, dried and subsequently thoroughly homogenised. This 'in house' reference material contained a known ratio of ^{129}I/^{129}I at 1.85×10^{-5} so only a small, 10 g, sample was used at such a high concentration. This was refluxed in 100 ml of NaOH and the remaining procedure was identical to the one described above. Standard solutions were also prepared by omitting the reflux stage of the procedure and adding a known volume of standard ^{129}I of known activity to 250 ml of NaOH and KI carrier.

2.5 Anion Exchange

A Bio Rad AX1-X8 resin, 20-50 mesh in the hydroxide form with high selectivity towards iodide was conditioned with 1M NaOH. It was then added to the sample solution and stirred for 1 h to allow the resin to take up as much of the iodine from the filtrate as possible. The resin was allowed to settle and the solution decanted off careful ensuring none of the resin was lost. The resin was then thoroughly cleaned with deionised water to ensure all traces of the sample had gone. The resin was then loaded into an ion exchange column and washed in with cold deionised water. 100 ml of 1M NaOH was then run through the column at a rate of about 1 ml per min. Any chlorine present in the sample was eluted off at this stage by adding 200 ml of 0.5M Na_2SO_4. The resin was again washed with deionised water and removed from the column onto filter paper and allowed to dry under a heat lamp at a temperature not exceeding 40^0C.

2.6 Neutron activation

The resin was transferred to a polyethylene capsule and heat sealed in preparation for irradiation. Samples were irradiated in the Imperial College Consort Mark II reactor in a neutron flux of 1×10^{16} m^{-2} s^{-1} for 7.5 h. This length of time was considered appropriately long to get a good conversion of ^{129}I into ^{130}I. The samples were then left overnight in order for the short lived radionuclides to decay before the samples were processed the next day.

2.7 Post-irradiation separation

The resin was loaded into an ion exchange column and eluted with 70 ml of 20% sodium nitrate solution. On elution chlorine is seen to come off the column first followed by the bromine. A further 150 ml of sodium nitrate is then used to elute the column until all the solution containing the now separated ^{130}I has passed through the column. 3 ml of 100,000 mg per litre ammonium bromide and 60 ml of 880 ammonia solution was added to the beaker containing the eluent and stirred with a glass rod. 5 ml of 15% silver nitrate solution was then added while stirring. A heavy pale yellow precipitate initially forms, but rapidly disperses to a translucent white suspension. The beaker is covered with clingfilm and left to stand for 20-30 minutes with stirring after 10 minutes prior to filtration. The suspension was filtered and the beaker was washed with two successive 20 ml portions of ammonia. The filter is finally washed by two successive 20-ml portions of methanol. A vacuum was maintained for a further five minutes in order for the precipitate to dry. The filter paper was carefully removed using tweezers and transferred to a polythene container and the sample was transferred for gamma-ray spectrometry.

2.8 Gamma spectrometry

The samples were counted overnight for 15000 s. The activities of the ^{130}I and ^{126}I isotopes were measured to obtain the derived activity of the ^{129}I and ^{127}I isotopes. The samples were automatically decay corrected and converted into a ratio of $^{129}I/^{127}I$.

2.9 Determination of stable ^{127}I

The reactor epithermal large volume irradiation system was used to determine the stable ^{127}I content of the samples using the dried seaweed sample from earlier were used in this procedure. Approximate 1.5 g samples were placed into irradiation capsules in duplicate along with a standard containing 10 μg/ml of iodine solution in talc. The samples were placed into the reactor in sequence for a period of 10 min. After irradiation, the samples were counted on a gamma ray detector for 600 s.

Table 1 *Concentrations of ^{129}I and ^{127}I in seaweed and their ratios (±2 standard deviations)*

Location	^{129}I (mBq/kg)	^{127}I (mg/kg)	Ratio $^{129}I/^{127}I$ (x 10^{-7})
Keyhaven	58±6	143±8	0.62 ± 0.07
Peacehaven	149±13	121±23	1.9 ± 0.18
St Margaret's at Cliffe	133±11	70.4±5.3	2.9 ± 0.29
	145±12	70.2±2.3	3.2 ± 0.28
Mersea	826±67	585±12	2.2 ± 0.2
Southwold Harbour	306±27	561±27	0.84 ± 0.09
	228±20	469±23	0.74 ± 0.08
Robin Hood Bay	131±14	392±19	0.51 ± 0.06
Blackhall Rocks	163±17	458±22	0.54 ± 0.07
Berwick upon Tweed	104±10	283±12	0.56 ± 0.06
Peterhead	81±9	241±11	0.52 ± 0.06
	84±11	276±12	0.46 ± 0.06

4 RESULTS

The two $^{129}I/^{127}I$ ratios obtained for the reference seaweed material (1.04 and 1.12 ± 0.11 x 10^{-5}) showed good agreement with RNAA values quoted (range 1.06-1.36 x 10^{-5}, and mean 1.22 ±0.21), indicating that the analytical procedure was working satisfactorily.

Table 1 shows the results obtained for the ^{129}I levels detected on the South and East coasts of the UK mainland, the concentrations of the stable iodine and the resulting ratio of $^{129}I/^{127}I$. The range of activities of ^{129}I is large, from 58 to 825 mBq/kg. However, much of this variation simply reflects the variation in concentration of iodine in the seaweed, since stable iodine values range from 70 to 585 mg/kg. When the ratio of the ^{129}I to ^{127}I is calculated, the resulting values are remarkably consistent.

5 DISCUSSION

The bulk of the ^{129}I passing through the Straits of Dover is carried away quickly by the fast flowing current. This is confirmed by the fact that the levels fall off from $3.0 \pm 0.3 \times 10^{-7}$ in the Straits of Dover to $2.2 \pm 0.2 \times 10^{-7}$ at Mersea. The distance between the two locations is only in the region of about 80 km, which indicates that the ^{129}I does not follow the coast in this region. This agrees well with previous work by Guegueniat et al.[6] showing a region that includes Mersea where the North Sea and the channel waters mix and represent an area of relatively calm water. The small region in the South of Dover where the sample at St Margarets at Cliffe was taken represents the only region where direct contamination could potentially occur. It must be stressed however that the highest $^{129}I/^{127}I$ ratio observed of 4.9×10^{-7}, does not represent any significant health threat to the general public. The mean value for ^{129}I obtained was 139 mBq/kg in the Straits of Dover.

A direct comparison can be made between these data and the $^{129}I/^{127}I$ ratios found on the French side of the channel by Frechou et al.[4]. Wimereux is located about 200 km East of La Hague, almost directly opposite St Margarets at Cliffe, in the Straits of Dover. At Wimereux a $^{129}I/^{127}I$ ratio of 27×10^{-7} was detected in seaweed compared with 3.0×10^{-7} detected on the English side of the Channel. The results are directly comparable, as both samples tested were seaweed, even though the species used in the French study was *Fucus serratus*. The sampling and subsequent analyses were carried out within a 10 month period of each other, and as a result no significant increases in concentration could be expected. This represents a dilution factor of about 8 from France to England, a distance of about 50 km. Work carried out on ^{125}Sb by Guegueniat et al.[6] showed that this mixing band represents a margin between the discharges present in the Channel and those present in the North Sea namely from Sellafield. They showed that no ^{125}Sb, a radionuclide characteristic of La Hague discharges, was detected on the East Anglian coast. This seems to confirm that the La Hague and Sellafield discharges can be treated separately from this region of the coast: North as a result of Sellafield and South as a result of La Hague. If the background value of about 0.5 x 10^{-7} found in the North Sea is subtracted from the ratios found on the South coast, the contribution to the ratio at Dover from La Hague becomes 2.5, a dilution of 10. These results are very much in line with the work of Guegueniat et al.[6], who measured the decrease in another conservative element, ^{125}Sb activity, away from La Hague. They showed values of about 12 mBq/L at Dover, compared to about 80 mBq/L on the French coast opposite, a dilution of 7.

On leaving Sellafield the ^{129}I is known to travel along the West Scottish coast[9] following the main current that flows in the Irish Sea. On reaching the North of Scotland a large amount of the ^{129}I seems to be carried straight over to the Norwegian coast by currents in the Atlantic, while the remainder moves down the East coast of Scotland. This is confirmed by the sample taken from Peterhead showing a ratio of 0.5×10^{-7}. At the sampling locations further south the ^{129}I levels were observed to remain constant at 0.5×10^{-7} down to Robin Hood's Bay.

The data obtained along the East Coast has a background value of around 0.5×10^{-7}, compared to post-nuclear values obtained using seaweed in North America and Japan[10] of $0.2\text{-}2 \times 10^{-10}$. Measurements of seaweed from China[11] in 1996-1998 have also given values of 2×10^{-10}. This is the first value to be obtained for the ^{129}I concentration in this area, although work has been reported for areas across the English Channel in Europe. This background value is the same as that found at Bornholm in Denmark in 1999 by Hou et al.[12], also using *Fucus vesiculosus* as the monitor. However, studies of the temporal variation at Klint, Denmark has shown the increase in $^{129}I/^{127}I$ ratio from 0.4×10^{-7} in 1986 to 3.7×10^{-7} in 1999. It is interesting to note that concentrations of ^{129}I found in certain Danish waters is the same as the highest found on the South or East coast of England. Hou et al. proposed that La Hague contributes about 87% of the activity found at Klint, and Sellafield the remaining 13%. It would seem that there is a background signature $^{129}I/^{127}I$ ratio around the UK 0.5×10^{-7}. These values will thus act as a benchmark to compare with results in the future, as reprocessing plants strive to reduce marine discharges to satisfy the OSPAR treaty.

References

1 X. Hou, H.Dhalgaard, B.Rietz, U.Jacobson, S.P.Nielsen and A. Aarkroy, *Analyst*, 1999, **124**, 1109
2 A. Schmidt, Ch.Schnabel, J.Handl, D.Jakob, R.Michel, H-A.Synal, J.M.Lopez and M.Suter, *Sci. Tot. Environ.*, 1998, **223**, 131.
3 G.M.Raisbeck, F.Yiou, Z.Q.Zhou, L.R.Kilius and P.J.Kershaw, *Radioprotection-Colloques*, **32**, C2, 91.
4 C.Frechou, D.Calmet, P.Bouisset, D.Piccot, A.Gaudry, F. Yiou and G.Raisbeck, *J.Radioanal.Nucl. Chem.*, 2001, **249**, 133
5 P.Gueguenlat, P.bailly du Boos, R.Gandon, J.C.Salomon, Y.Baron and R.Leon, *Estuarine Coast. Shelf. Sci, 1994, 39, 59.*
6 P.Gueguenlat, J.C.Salomon, M.Wartel, L.Cabioch and A.Frazier, *Estuarine Coast. Shelf. Sci.*, 1993, **36**, 477.
7 P.Perianex, *J. Environ. Radioact.*, 2000, **49**, 259.
8 R.Fuge and C.C.Johnson, *Environ. Geochem. Health,* 1986, **8**, 31
9 G.M.Raisbeck and F.Yiou, *Sci. Tot. Environ.*, 1999, **237**, 31.
10 L.R.Kilius, A.E.Litherland, J.C.Rucklidge and N.Baba, *Appl. Radiat.Isotop.*, 1992, **43**, 279.
11 X.Hou, H.Dahlgaard, S.P.Nielsen and Wenjun Ding, *Sci. Tot. Environ.*, 2000, **246**, 285.
12 X.Hou, H.Dahlgaard and S.P.Nielsen, *Estuarine Coast. Shelf. Sci.*, 2000, **51**, 571.

DETERMINATION OF ^{237}Np BY NEUTRON ACTIVATION ANALYSIS AND ALPHA SPECTROMETRY

L. Benedik and U. Repinc

Department of Environmental Sciences, Jo ef Stefan Institute, Jamova 39, SI-1000 Ljubljana, Slovenia

1 INTRODUCTION

The determination of critical natural and man-made radionuclides in environmental samples is obviously important in view of the increasing energy production by nuclear reactors, the associated fuel cycle, reprocessing and waste disposal, the increased potential for environmental contamination and public concern over the potential hazards. The major working methods for determination of radionuclides in a wide variety of environmental samples include non-destructive gamma spectrometry and radiochemical and radioanalytical (or radiometric) techniques leading to measurements by alpha spectrometry, beta counting, beta spectrometry and alpha scintillation techniques, as well as mass spectrometry and neutron activation analysis for some nuclides.[1-10]

The determination of alpha emitters is an important topic in relation to health, nuclear waste management from nuclear reactors, recycling and final storage of radioactive waste, control of illicit nuclear activities, etc.

^{237}Np is one of the most important alpha-emitting radionuclides contributing to the collective dose commitment delivery in the long-term.[11,12] ^{237}Np is a long-lived ($t_{1/2} = 2.14 \times 10^6$ y) alpha emitting radionuclide and is present in the environment primarily as a result of atmospheric weapons testing and discharges from nuclear fuel reprocessing facilities. ^{237}Np is produced in the uranium fuel cycle via the reactions $^{235}U(n,\gamma)$ $^{236}U(n,\gamma)$ $^{237}U \rightarrow$ ^{237}Np and $^{238}U(n,2n)$ ^{237}U ($t_{1/2} = 6.75$ d) $\rightarrow$ ^{237}Np, as well as being the decay product of ^{241}Pu ($t_{1/2} = 14$ y) $\rightarrow$ ^{241}Am ($t_{1/2} = 433$ y)$\rightarrow$ ^{237}Np. ^{237}Np from the decay scheme of ^{241}Am is a minor source of neptunium present in the environment. The most important source of ^{237}Np are aqueous waste discharges from nuclear reprocessing plants. The determination of neptunium is necessary for controlling its content at different steps of fuel reprocessing. A fundamental understanding of ^{237}Np behaviour is important for the longer-term consideration of radioactive waste disposal to the environment because of its long half-life and production from the decay of its parent ^{241}Am and grandparent ^{241}Pu.

There are several existing studies on neptunium determination in intertidal coastal and estuarine sediments in the Irish Sea.[13-18] In the Irish Sea, neptunium may be present from nuclear weapons fallout, cooling water from fission reactors, industrial processing of ^{237}Np produced in fission reactors and decay of ^{241}Am and its parent ^{241}Pu. The Irish Sea receives controlled discharges of low-level radioactive waste from nuclear facilities including the BNFL Sellafield nuclear fuel reprocessing plant in Cumbria and the BNFL Springfield uranium ore concentrate processing facility near Preston in Lancashire. According to literature data,[19] ^{237}Np is discharged from the Sellafield reprocessing plant in Cumbria at about 1% of the $^{239/240}Pu$ concentrations and occurs in sediments in the vicinity of the discharge at a level of about 2 Bq kg^{-1}.

Data on concentrations of ^{237}Np in the literature of environmental radioactivity are sparse. The major reason why data on ^{237}Np are lacking, apart from a now disappearing lack of interest, is an analytical one. There exists a number of lengthy analytical methods for its determination by gamma spectrometry,[20] alpha spectrometry (AS),[21-28] neutron activation analysis (NAA) via the $^{237}Np(n,\gamma)^{238}Np$ reaction,[13,26,29-32] thermal ionisation mass spectrometry (TIMS),[33,34] inductively coupled plasma mass spectrometry (ICP-MS)[35,36] and liquid scintillation counting.[37-39]

The aim of this study was to determine the activity of ^{237}Np in environmental samples by direct gamma spectrometry, alpha spectrometry following radiochemical separation by ion exchange chromatography after wet acid dissolution and/or leaching, and radiochemical neutron activation analysis using pre-irradiation separation by ion exchange chromatography. The methods were applied to reference materials and two sediments, one from the Cumbrian Coastline, the other from the river Ribble.

2 EXPERIMENTAL

2.1 Samples

Reference materials IAEA-135 (Radionuclides in Sediment) and IAEA-368 (Pacific Ocean Sediment), Ribble Sediment (sediment from the tidal zone of the river Ribble, Lancashire, UK) and Intertidal Sediment from the Cumbrian Coastline.
Ribble Sediment (kindly supplied by Dr. P. J. Day, University of Manchester) is influenced both by tidal borne radionuclides originating from Sellafield and the manufacture of fuel elements at Springfields upstream. An Intertidal Sediments from the Cumbrian Coastline was collected for the international characterisation in 1997 and the Jo ef Stefan Institute participated in the international exercise.[20]

Samples for alpha spectrometry and radiochemical analysis were ashed in porcelain dishes in an electric muffle furnace at 500°C for 24 h.

2.2 Gamma spectrometry

Total activity concentrations of radionuclides in samples were determined by direct gamma spectrometry. Approximately 100 g dry sediment was packed into a cylindrical polythene vessel. The vessels were sealed hermetically using insulating tape and stored for at least 20 days to allow radioactive equilibrium of the ^{226}Ra series. After that time, the samples were measured using a coaxial high pure germanium (HPGe) detector. ^{238}U was assumed to be in radioactive equilibrium with ^{234}Th and ^{234m}Pa. Its activity was calculated from the 63.3 keV line of ^{234}Th and from the 1001 keV line of ^{234m}Pa. ^{230}Th was determined via its gamma line at 67.7 keV. ^{226}Ra was determined via the gamma lines of the daughter products ^{214}Pb (295.2 and 351.9 keV) and ^{214}Bi (609.3 keV). ^{210}Pb was measured from the 46.5 keV gamma line, while ^{231}Pa was measured from the 27.4 keV and 283.6 keV gamma lines, ^{241}Am was measured from the 59 keV and ^{237}Np via of its daughter ^{233}Pa at 311.8 keV.

The detector was calibrated using standard IAEA reference materials and radioisotopic solutions. Corrections for self-absorption were calculated after measurements with an external ^{241}Am gamma ray source (59.5 keV), according to the method of Cutshall.[40]

2.3 Alpha spectrometry

In the case of ^{237}Np measurements by alpha spectrometry (E_α = 4.77, 4.79 MeV), samples require extensive chemical separation prior to counting to remove peak interferences from other alpha emitters such as ^{234}U (E_α = 4.72, 4.77 MeV), ^{231}Pa (E_α = 4.67, 5.01 MeV) and ^{230}Th (E_α = 4.61, 4.68 MeV). The purification of Np from U is very important in the radiochemical separation procedure, because of the chemical similarity of these two elements.

2.3.1 Standard and tracer. ^{237}Np solution: A calibrated solution (8.377E-03 Bq g^{-1}) was purchased from Isotrak, AEA Technology, QSA. A working solution (3.7 Bq g^{-1}) was prepared in 1 M nitric acid.

Tracer ^{239}Np: This was prepared by irradiation in the reactor of μg quantities of a solution of uranium. The neptunium fraction was purified from uranium by ion exchange chromatography.

2.3.2 Radiochemical procedure. Up to 5 g of ashed sample was leached with 50 mL conc. HCl on a hot plate at 70°C overnight with ^{239}Np tracer. Leaching was performed in a covered teflon beaker on a hot plate with magnetic stiring. The leachate and the residue were separated by centrifugation. The residue was washed with 5 M HCl. The residue was checked by measuring the ^{239}Np gamma peak a 277.6 keV. The final solution was made up to 9 M HCl. About 1 g hydroxylamine hydrochloride was added, and the solution was warmed gently for about 15 min to ensure that Np was present as Np(IV). The solution was then transferred to the top of an anion exchange column of height 10 cm, diameter 1 cm (Dowex 1x8, 100-200 mesh in the Cl^- form) and allowed to flow through it at about 1 mL min^{-1}. The column was then washed with 50 mL of 9 M HCl and 50 mL 7 M nitric acid.[41] Plutonium isotopes were stripped with 50 mL of 0.01 M NH_4I in 9 M HCl.[42] Neptunium isotopes were finally eluted from the column with 50 mL of 4.5 M HCl.[43] The solution was evaporated to dryness, dissolved in 1mL of 6 M HCl, and evaporated to small volume. The radioactive sources for alpha spectrometry were prepared according to Puphal and Olsen[44] by electrodeposition on 19 mm diameter stainless steel disks, active diameter 17 mm, from 5.7% ammonium oxalate in 0.3 M HCl solution at pH 1 - 1.5. Neptunium isotopes were electroplated for 2 h at 300 mA, using a Pt anode.

2.3.3 Counting. Alpha activities of neptunium isotopes were measured with silicon surface barrier detectors (PIPS), connected to an EG&G ORTEC Maestro Gamavision MCA Emulator multichannel analysis system.

2.3.4 Chemical yield and counting efficiency. The chemical yield of separated neptunium radioisotopes was measured from the gamma activity at 277.6 keV of added ^{239}Np by measuring the stainless steel disk after electrodeposition on an HP Ge detector connected to Canberra MCA by Genie-2000 Software, in comparison to an evaporated aliquot of the ^{239}Np tracer.

The counting efficiency of the PIPS detector was measured from an electroplated disk of ^{230}Th prepared from a tracer solution of known radioactive concentration.

2.4 Neutron activation analysis

Neutron activation analysis (NAA) for ^{237}Np has previously been applied to relatively high concentrations in nuclear materials[29] and process streams.[30] When ^{237}Np is irradiated in a reactor the following capture reaction is induced: $^{237}Np(n,\gamma)^{238}Np$. ^{238}Np is formed by neutron capture with an exceptionally large thermal cross section of 170 barns and resonance integral of 600 barns and possesses a favourable half-life ($t_{1/2}$ = 2.2 days) and gamma energies at 984.4, 1026 and 1028.5 keV.

Byrne[13] described the method for determination of ^{237}Np in Cumbrian sediments by NAA. The basis of this method is leaching of the ashed sample with hydrochloric acid, removal of the bulk of the iron, sorption of Np(IV) on an anion exchanger and its subsequent elution after washing off most of the other impurities. Recovery was monitored with ^{239}Np tracer. Then the preconcentrated Np fraction was irradiated in the reactor and ^{238}Np was purified by solvent extraction, followed by gamma spectrometry on a Ge detector and multichanel analyser system. We adopted this method, but for separation we used anion exchange chromatography[41].

2.4.1 Standard and tracer. ^{237}Np solution: A calibrated solution (8.377E-03 Bq g^{-1}) was purchased from Isotrak, AEA Technology, QSA. A working solution (3.7 Bq g^{-1}) was prepared in 1 M nitric acid.

Tracer ^{239}Np: This was prepared by irradiation in the reactor of μg quantities of a solution of uranium. The neptunium fraction was purified from uranium by ion exchange chromatography.

2.4.2 Irradiations. Samples were irradiated in the Institute's 250 kW Triga Mark II reactor at a flux of 2×10^{12} n cm^{-2} s^{-1} for about 20 hours, together with ^{237}Np standards.

2.4.3 Counting. Gamma ray spectrometry was performed with a HP Ge well-type detector connected to a Canberra MCA by Genie-2000 Software.

2.4.4 Radiochemical procedure. Up to 5 g of ashed sample was leached with 50 mL conc. HCl on a hot plate at 70°C overnight with ^{239}Np tracer. Leaching was performed in a covered teflon beaker on a hot plate with magnetic stirring. The leachate and the residue were separated by centrifugation. The residue was washed with 5 M HCl. The residue was checked by measuring the ^{239}Np gamma peak at 277.6 keV. The final solution was made up to 9 M HCl. About 1 g hydroxylamine hydrochloride was added, and the solution was warmed gently for about 15 min to ensure that Np was present at Np(IV). The solution was then transferred to the top of the anion exchange column of height 10 cm, diameter 1 cm (Dowex 1x8, 100-200 mesh in the Cl^- form) and allowed to flow through it at about 1 mL min^{-1}. The column was then washed with 50 mL of 9 M HCl and 50 mL 7 M nitric acid.[41] The neptunium isotopes were finally eluted from the column with 50 mL of 4.5 M HCl.[43] The solution was evaporated to dryness, disolved in 8 M nitric acid and sealed in a plastic vial. Measurement of ^{239}Np allows determination of the recovery. Then the sample was irradiated in the reactor with a ^{237}Np standard. After 2-3 days the ampoule was opened and the solution transferred to a 50 mL baker with concentrated nitric acid. The solution was evaporated to dryness and converted to Cl^- form. The procedure for purification of Np was the same as in the first stage. Neptunium isotopes were measured on a Ge detector at 277.7 keV (^{239}Np) and at 984.4, 1026, 1028.5 keV (^{238}Np).

3 RESULTS AND DISCUSSION

The results obtained by gamma spectrometry for the reference materials and sediments are shown in Table 1.

Table 1 *Activity concentration of gamma emitters in samples*

Sample	*Nuclide*	*Activity concentration ($Bq\ kg^{-1}$)* This work	% uncertainty	Reference or literature value
IAEA-368 Pacific Ocean Sediment	^{210}Pb	28	4.6	23.2
	^{241}Am	< 5		1.3
	$^{238}U/^{234}Th$	27	2.8	31
	^{226}Ra	22	2.2	21.4
	^{137}Cs	< 1		0.34
	$^{237}Np/^{233}Pa$	< 1.5		0.25
IAEA-135 Radionuclides in Sediment	^{210}Pb	45	7.6	48
	^{241}Am	325	3.0	318
	$^{238}U/^{234}Th$	30	3.0	30
	^{230}Th	71	5.6	69.1
	^{226}Ra	27	4.3	23.9
	^{137}Cs	1151	1.2	1108
	$^{237}Np/^{233}Pa$	< 1.5		0.87
Ribble Sediment	^{210}Pb	78	3.6	
	^{241}Am	351	3.0	
	$^{238}U/^{234}Th$	145	1.9	
	^{230}Th	820	4.3	
	^{226}Ra	73	2.2	
	^{231}Pa	131	5.9	
	^{137}Cs	574	0.5	
	$^{237}Np/^{233}Pa$	1.8	16.7	
Cumbrian Sediment	^{210}Pb	77	13.0	70.4
	^{241}Am	1460	1.0	1590
	$^{238}U/^{234}Th$	46	2.6	44.5
	^{230}Th	54	16.8	62.9
	^{226}Ra	42	2.3	43.2
	^{231}Pa	< 10		
	^{137}Cs	683	1.0	726
	$^{237}Np/^{233}Pa$	6.2	8.1	6.33

Direct gamma spectrometry is a useful method for determination of neptunium activities higher than 1.5 Bq kg^{-1} with an appropriate measuring time. The results for Ribble Sediment and Cumbrian Sediment obtained by gamma spectrometry for other radionuclides show that the former sediment is more contaminated from Springfields manufacture of fuel elements with ^{230}Th and ^{231}Pa as daughters of the uranium decay chain. Cumbrian sediment contains less ^{230}Th but a lot of ^{241}Am. Figure 1 shows a partial gamma spectrum of Ribble Sediment in the range of 100 to 350 keV. The peak of ^{233}Pa was identified and its activity was determined as 1.8 Bq kg^{-1}.

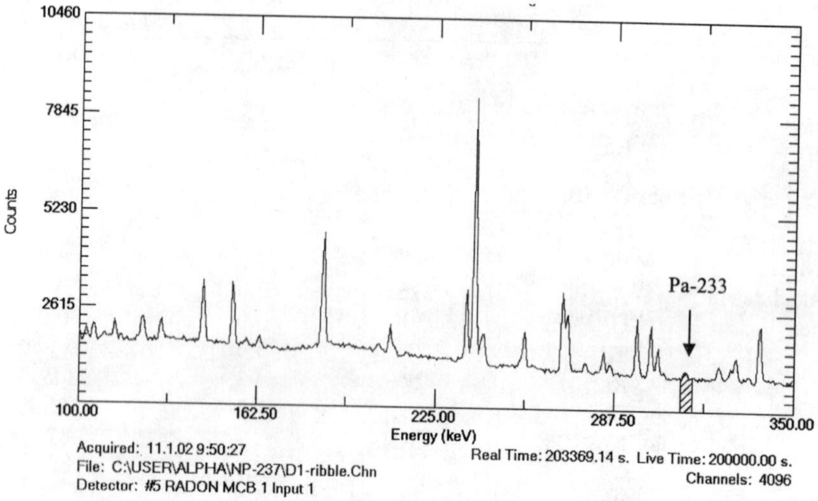

Figure 1 *Gamma spectrum of Ribble Sediment, counted for 200000 s*

Figure 2 shows an alpha spectrum of neptunium radioisotopes, separated by the above method. The separation procedure using ion exchange chromatography is very effective. The chemical yield of neptunium was measured from the gamma activity at 277.6 keV of added ^{239}Np. The yield was very high (90%) and the radiochemical purity good. As is evident from Figure 2, the alpha spectrum shows some impurities of ^{239}Pu at 5155 keV, which however do not interfere with the ^{237}Np peak.

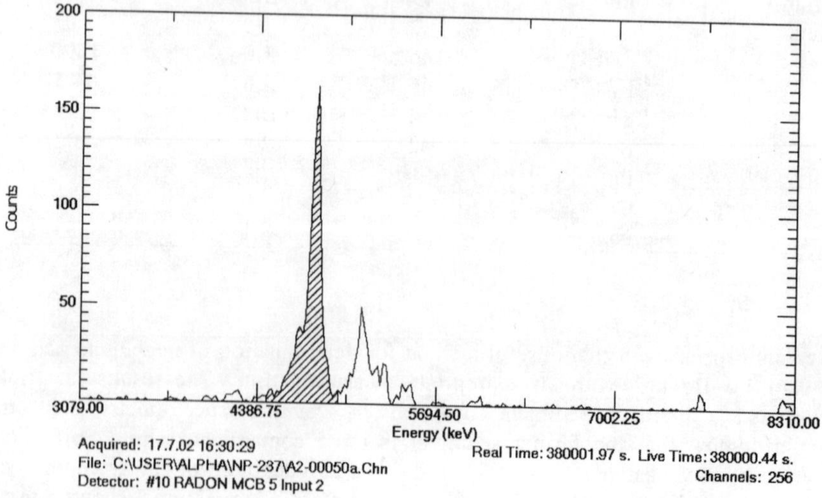

Figure 2 *Alpha spectrum of neptunium fraction isolated from Ribble Sediment*

Figure 3 shows a partial gamma spectrum of the ^{238}Np fraction isolated from the Ribble Sediment. ^{238}Np has its major peak at 984.4 keV and a characteristic doublet at 1206 and 1028.5 keV.

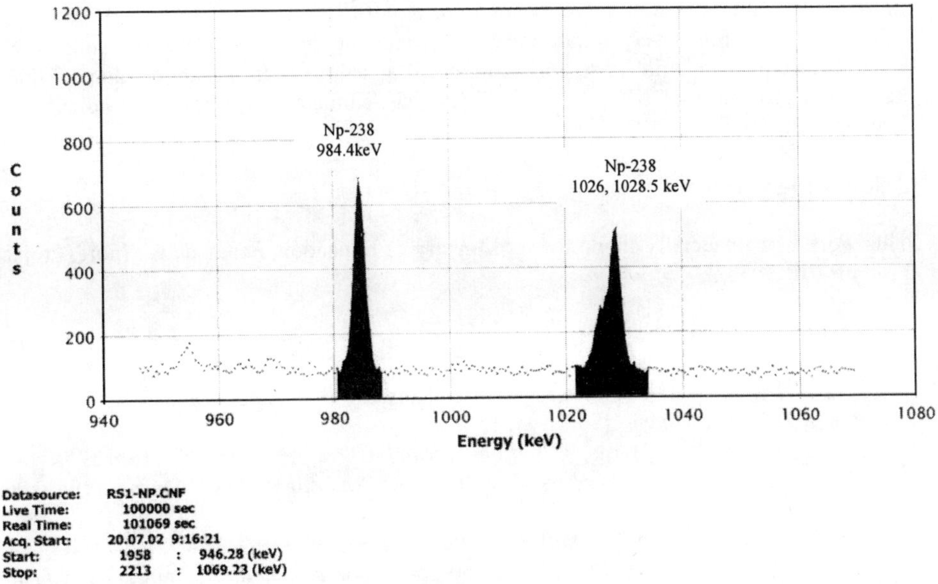

Figure 3 *Partial gamma spectrum of the ^{238}Np fraction isolated from Ribble Sediment by preseparation RNAA*

Table 2 shows results for the activity concentration of ^{237}Np found in reference materials and sediments. It is evident that the activity concentration of ^{237}Np found by direct gamma spectrometry via its daughter ^{233}Pa, alpha spectrometry and preseparation radiochemical neutron activation analysis are in a good agreement.

Table 2 *Activity of ^{237}Np found in sediment with various methods*

Sample	*This work ($Bq\ kg^{-1}$)*			*Literature value ($Bq\ kg^{-1}$)*
	Gamma	*Alpha*	*RNAA*	
IAEA-135	< 1.5	0.94 ± 0.1	1.05 ± 0.2	0.87[28]
IAEA-368	< 1.5	0.25 ± 0.08	0.33 ± 0.11	0.25[28]
Ribble Sediment*	1.83 ± 0.3	1.66 ± 0.2	1.72 ± 0.2	-
Cumbrian Sediment**	6.2 ± 0.5	5.9 ± 0.2	6.0 ± 0.3	6.33 ± 0.53[20]

* Sediment from the tidal zone of the river Ribble, Lansashire, UK

** Interdial sediment from the Cumbrian Coastline, UK

It has be been shown that direct gamma spectrometry allows determination of ^{237}Np via ^{233}Pa at activities higher than 1.5 Bq kg^{-1}. For lower activities it is possible to use two approaches: alpha spectrometry and/or preseparation RNAA. In both methods a good radiochemical separation of neptunium from uranium is important. In alpha spectrometry ^{234}U overlaps with ^{237}Np. In the case of preseparation of neptunium and then irradiation, impurities of uranium result in production of origin of ^{239}Np.

Quality control is very important in determination of man-made radionuclides. No certified reference material exists for ^{237}Np in soil or sediment. Information value 0.007 Bq kg^{-1} is reported only for NIST-SRM 4357 (Ocean Sediment Environmental Radioactivity Standard).

Acknowledgements

This work was financially supported by Ministry of Education, Science and Sport (Project group P-0106-0532).

References

1 L.L. Smith, J.S. Crain, J.S. Yaeger, E.P. Horwitz, H. Diamond and R. Chiarizia, *J. Radioanal. Nucl. Chem.*, 1995, **194,** 151.
2 L. Benedik and A. R. Byrne, *J. Radioanal. Nucl. Chem. Articles*, 1995, **189,** 325.
3 M.J. Vargas, A.M. Sanchez, and F.V. Tome, *J. Radioanal. Nucl. Chem.*, 1995, **196,** 345.
4 Y.J. Zhu and D.Z. Yang, *J. Radioanal. Nucl. Chem.*, 1995, **194,** 173.
5 J.S. Crain, L. L. Smith, J.S. Yaeger and J.A. Alvarado, *J. Radioanal. Nucl. Chem.*, 1995, **194,** 133.
6 C. Testa, D. Disideri, M.A. Meli and C. Rosseli, *J. Radioanal. Nucl. Chem.*, 1995, **194,** 141.
7 J.I. Garcia Alonso, D. Thoby-Schultzendorff and B. Giovannone, L. Koch, *J. Radioanal. Nucl. Chem.*, 1996, **203,** 19.
8 L. Benedik and A.R. Byrne, *Acta Chim. Slov.*, 1994, **41**, 1.
9 A.R. Byrne and A. Komosa, *Sci. Total Environ.*, 1993, **130/131**, 197.
10 J.H. Kaye, R.S. Strebin and R.D. Orr, *J. Radioanal. Nucl. Chem.*, 1995, **194,** 191.
11 M.D. Hill and I.F. White, *Nucl. Energy,* 1982, **21**, 225.
12 J.H. Rees and C.M. Shipp, *Nucl. Energy,* 1983, **22**, 423.
13 A.R. Byrne, *J. Environ. Radioactivity*, 1986, **4**, 133.
14 A.S. Hursthouse, M.S. Baxter, F.R. Livens and H.J. Duncan, *J. Environ. Radioactivity*, 1991, **14,** 147.
15 D.J. Assinder, M.Yamamoto, C.K. Kim, R. Seki, Y. Takaku, Y. Yamauchi, K. Komura, K. Ueno and G.S. Bourne, *J. Environ. Radioactivity,* 1991, **14**, 135.
16 C.K. Kim, S. Morita, R. seki, Y. Takaku, N. Ikeda and D.J. Assinder, *J. Radioanal. Nucl. Chem., Articles,* 1992, **156**, 201.
17 S.R. Jones, S.M. Willans, A.D. Smith, P.A. Cawse and S.J. Baker, *Sci. Total Environ.*, 1996, **183,** 213.
18 D.J. Assinder, *J. Environ. Radioactivity,* 1999, **44**, 335.
19 B.R. Harvey, *Potential for post depositional migration of neptunium in Irish Sea sediments*, in *Impact of Radionuclide Releases into the Marine Environment*, IAEA, Vienna, IAEA-SM-248/104, 1981, pp. 93-103.
20 I. Adsley, D. Andrew, D. Arnold, R. Bojanovski, Y Bourlat, A.R. Byrne, M-T. Crespo, J. Desmond, P. De Felice, A. Fazio, J.L. Gascon, R.S. Grieve, A.S. Holmes,

S.M. Jerome, M. Korun, M. Magnoni, K.J. Odell, D.S. Popplewell, I. Poupaki, G. Sutton, J. Toole, M.W. Wakerley, H. Wershofen, M.J. Woods and M.J. Youngman, *Appl. Radiat. Isot.*, 1998, **49**, 1295.

21 E. Holm and M. Nilsson, *Methods for determination of ^{237}Np in low level environmental samples,* in *Radioelement Analysis – Progress and Problems,* Proc. 23rd Conf. On Anal. Chem. In Energy Technol., Gaatlingburg, 1979, pp. 231-236.

22 E. Holm, *Release of ^{237}Np to the environment: measurements of marine samples contaminated by different sources,* in *Impact of Radionuclide Releases into the Marine Environment*, IAEA, Vienna, IAEA-SM-248/104, 1981, pp. 155-160.

23 R.K. Schulz, G.T. Wink and L.M. Fujii, *Soil Science,* 1981, **132**, 71.

24 R.J. Pentreath, *The biological availability to marine organisms of transuranium and other long-lived nuclides,* in *Impact of Radionuclide Releases into the Marine Environment*, IAEA, Vienna, IAEA-SM-248/104, 1981, pp. 241-272.

25 M. Yamamoto, S. Igarashi, K. Chatani, K. Komura and K. Ueno, *J. Radioanal. Nucl. Chem., Articles,* 1990, **138**, 365.

26 P. Germain, P.M. Gueguenait, S. May and G. Pinte, *J. Environ. Radioactivity,* 1987, **5**, 319.

27 A.S. Hursthouse, M.S. Baxter, K. McKay and F.R. Livens, *J. Radioanal. Nucl. Chem., Articles,* 1992, **157,** 281.

28 J. Moreno-Bermudez, PhD. Thesis, Technical University of Budapest, 2001.

29 P. Guay and F. Pinelli, *J. Radioanal. Chem.,* 1973, **16**, 89.

30 R. Raghavan, V.V. Ramakrishna, S.K. Patil and M.V. Ramaniah, *J. Radioanal. Chem.,* 1976, **33,** 31.

31 H. Ruf and M. Friedrich, *Nuclear Technology,* 1977, **37,** 79.

32 P. Germain and G. Pinte, *J. Radioanal. Nucl. Chem.,* 1990, **138,** 49.

33 J.H. Landrum, M. Linder and N. Jones, *Anal. Chem.,* 1969, **41**, 840.

34 L.W. Cooper, J.M. Kelley, L.A. Bond, K.A. Orlandini and J.M. Grebmeier, *Marine Chemistry,* 2000, **69,** 253.

35 M. Gastel, J.S. Becker, G. Küppers and H.-J. Dietze, *Spectrochimica Acta,* 1997, **52**, 2051.

36 J.B. Truscott, P. Jones, B.E. Fairman and E.H. Evans, *Analytica Chimica Acta,* 2001, **433**, 245.

37 D.Y. Yang, Y. Zhu and R. Jiao, *J. Radioanal. Nucl. Chem.,* 1994, **183**, 245.

38 J.R. Cadieux and S.H. Reboul, *Radioactivity and Radiochemistry,* 1996, **7,** 30.

39 J. Aupiais, N. Dacheux, A.C. Thomas and S. Matton, *Analytica Chimica Acta,* 1999, **398**, 205.

40 N.H. Cutshall, I.L. Larsen and C.R. Olsen, *Nucl. Instrum. Meth.,* 1983, **206**, 309.

41 A.R. Byrne and L. Benedik, *Talanta,* 1988, **35**, 161.

42 H. Pintar and L.Benedik, *A sequential separation of man-made radionuclides,* in *YISAC 2000, young investigators seminar on analytical chemistry, Proceedings,* Graz, 2001, pp. 53-58.

43 D.N. Edgington, *Int. J. Appl. Radiat. Isot.,* 1967, **18**, 11.

44 K.W. Puphal and D.R. Olsen, *Anal. Chem.,* 1972, **44,** 284.

MAN-MADE RADIONUCLIDES IN AGE-DATED SEDIMENT CORES FROM THE BLACK SEA

R.A. Aliev, St.N. Kalmykov and Yu.A. Sapozhnikov

Department of Chemistry, Lomonosov Moscow State University, Moscow 119992, Russia

1 INTRODUCTION

The Black Sea is a closed water reservoir that obtained man-made radionuclides both from global fallouts and the Chernobyl accident.

The global man-made radionuclide fallout as a result of the nuclear weapon tests in the atmosphere, could be observed everywhere in bottom sediments starting from 1952-1953. Since the maximum fallout latitude (45° N) is crossing the Black Sea it should obtain rather high levels of fallout radionuclides[1]. Juzdan[2] estimated the deposition of ^{90}Sr and ^{137}Cs at this latitude by the end of 1985 to be 1.84 GBq/km^2 and 2.77 GBq/km^2 respectively. This value for $^{239,240}Pu$ was 81.4 MBq/km^2 as estimated by Hardy et al.[3]. The maximum activity of man-made radionuclides in vertical profiles of bottom sediments caused by intensification of the nuclear weapon tests is usually observed in the beginning of 60s with the smooth recession after the nuclear weapon tests (except for underground tests) were prohibited.

The Chernobyl accident has resulted in the significant pollution of the Black Sea by long-lived radionuclides, in particular ^{137}Cs and ^{90}Sr. During the first month, after the accident, the Black Sea received the majority of radionuclides from atmospheric fallout. Later, the radionuclides were mostly transported by the rivers, mainly the Dnieper and the Danube. River transport is most important for ^{90}Sr. Caesium was strongly bound with sediments of the Dnieper system and relatively large amounts (more than 100 Bq/kg) of ^{137}Cs were found in the sediments of Dnieper – Bug Gulf[4].

The primary ^{137}Cs atmospheric fallout was governed by the meteorological conditions and occurred extremely non-uniformly on the sea surface. During the first days (April 26-28) the main direction of a radioactive cloud transfer was northwest and, as a result, Northern and Western Europe has undergone contamination. During the next days, the wind changed to a westerly direction. In the first days of May 1986, local deposition of radionuclides was observed in the area of the Caucasus Coast. The Crimean coast was practically not affected, however, high ^{90}Sr activities were reported (up to 2.52 Bq/l) in seawater to the west of Cape Tarkhankut in the summer of 1991[5].

2 METHODS AND RESULTS

2.1 Sampling

Sediment samples BS-5, BS-11, BS-23, BS-27, BS-30, BS-33 were collected by multicorer during the 55th Cruise of RV "Professor Vodyanitsky". Cores AK-23, AK-21, #4 were supplied by Kazimeras Shimkus, SD SIO RAS. Location of sampling sites is shown in Table 1 and in Figure 1.

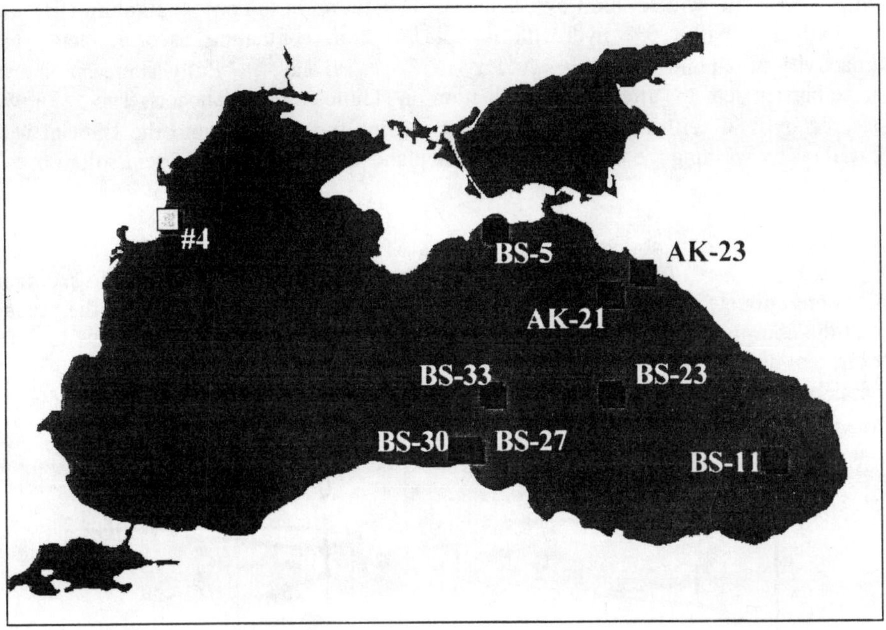

Figure 1 *Location of the sediment samplings from the Black Sea*

Table 1 *Sampling of sediment cores*

Region	*Core*	*Coordinates*	*Depth, m*	*Year*
Caucasus coast	AK-23	44°34,7' N. 38°02,4' E	8	1997
	AK-21	44°15,1' N 37°32,2' E	1890	1997
Coast of Crimea	BS-5	44°47,8' N. 35°32,3' E	62	2000
Danube Avandelta	# 4	45°05' N 29°45' E	20	1998
Eastern Sector	BS-11	41°59,9' N 40°11,4' E	1892	2000
	BS-23	43°00,1' N 37°19,9' E	2168	2000
	BS-33	42°59,96' N 35°40' E	2199	2000
Turkish Coast (Sinop Peninsula)	BS-27	42°02,8' N 35°07,2' E	57	2000
	BS-30	42°04,64' N 34°54,07' E	54	2000

2.2 Radionuclide determination

Gamma-spectrometry was performed for ^{210}Pb and ^{137}Cs determination using HPGe detector.

The subsamples for plutonium determination were dried at $60^{0}C$ for 24 hours and weighed afterwards. ^{236}Pu or ^{242}Pu was added as a chemical yield monitor. In order to remove any organic substances present, samples were placed in furnace for at least 10 hours at a temperature of 400-450 ^{0}C. The digestion procedure included treatment with concentrated nitric acid and/or aqua regia. Plutonium was extracted in the tetravalent state from 3–5 M nitric acid solution by 0.1 M TOPO solution in toluene. Plutonium (III) was back extracted with 10% hydrochloric acid solution containing ascorbic acid. The radioactivity of α-emitting isotopes (^{236}Pu or ^{242}Pu, ^{238}Pu and $^{239,240}Pu$) was measured on a surface barrier detector after electrodeposition on stainless steel planchets. The ^{241}Pu was measured by LSC, either in 2π-geometry by placing the planchet into the LS vial with cocktail or, by washing the plutonium from the planchets using 1M nitric acid solution.

2.3 Results

2.3.1 The Caucasus Coast and The South Crimea Coast. Two bottom sediment cores were collected near the Caucasus coast - one in the Gelendzik gulf and the second one at the continental slope at a depth of 1890 m. The ^{137}Cs vertical distributions in these bottom sediment cores are presented in Figure 2.

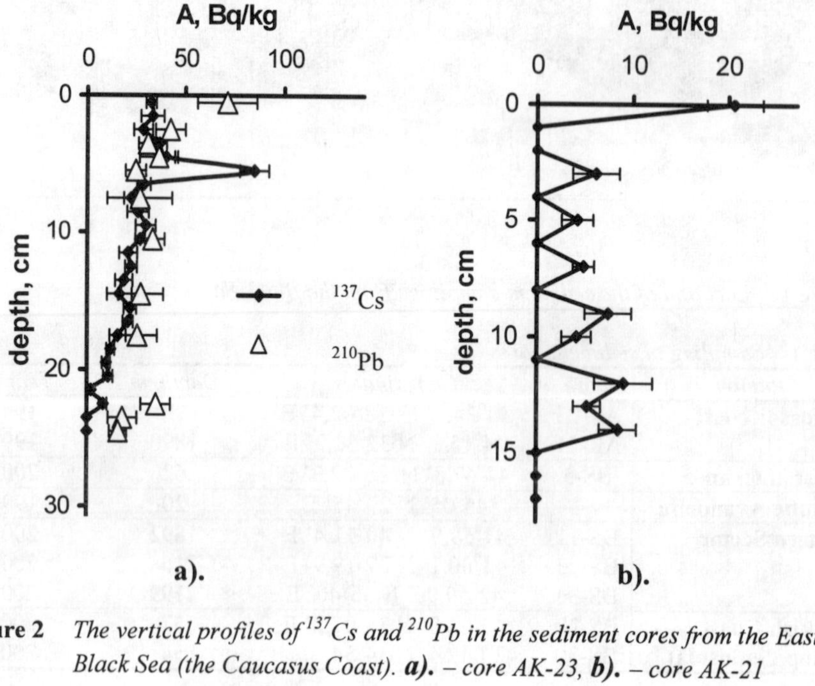

Figure 2 *The vertical profiles of ^{137}Cs and ^{210}Pb in the sediment cores from the Eastern Black Sea (the Caucasus Coast).* ***a).*** *– core AK-23,* ***b).*** *– core AK-21*

For the samples collected in the Gelendzik Gulf the maximum, possibly corresponding to the Chernobyl accident, is well observed in the 4-6 cm layer. To provide information on the history and origin of the man-made radionuclides in the bottom sediments ^{210}Pb age-dating was performed. The average sedimentation rate was 0.4±0.1 cm/year that correlates with data[6], which indicated rates of 0.1 – 1.0 cm/year in the shelf and margin regions. The plotted ^{210}Pb activity (Bq) vs. depth (cm) is presented in Figure 2 for the sediment core collected in the Gelendzik gulf. The sedimentation rate data confirm the Chernobyl origin of the ^{137}Cs in the 4-5 cm horizon. The decrease of ^{137}Cs activity with depth corresponds to the global fallout due to atmospheric nuclear tests. The lowest horizon containing measurable ^{137}Cs activity corresponds to the beginning of the 50s.

For the samples collected at the continental slope the vertical profile of ^{137}Cs could be explained by the post-sedimentation mixing.

One sediment core (BS-5) was collected near the Crimea Coast at a depth of 62 m. The vertical profile of ^{137}Cs activity in this core is presented at the Figure 3. There is a sharp maximum of ^{137}Cs activity corresponding to horizon 0.5-1 cm, caused by the Chernobyl accident.

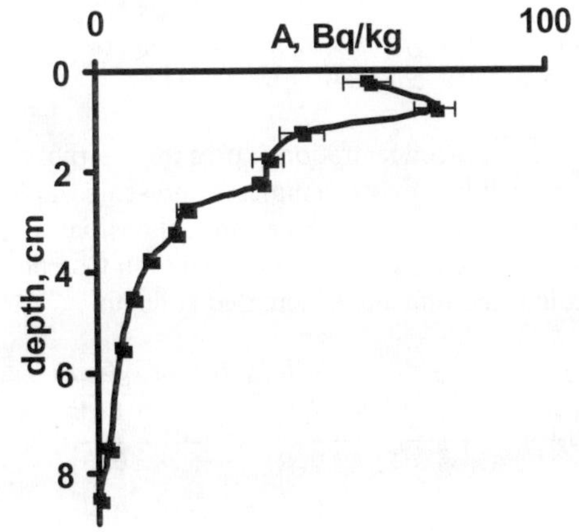

Figure 3 *The vertical profile of ^{137}Cs in the sediment core BS-5 from the Eastern Black Sea (the Crimea coast)*

2.3.2 *Western Sector – Danube Avandelta.* The ^{137}Cs vertical distribution in the sediment core collected in the Danube river avandelta is presented in Figure 4.

As in the case of the samples collected at the continental slope near the Caucasus coast the post-sidementational mixing of the samples could be the possible reason for the observed ^{137}Cs vertical profile. Annual variations in Danube runoff may also contribute to such a profile of ^{137}Cs.

In order to study the origin of the radionuclides, plutonium isotope ratios may be used. ^{241}Pu having a half-life period of 14.5 years, can serve as sedimentation monitor together with $^{238}Pu/^{239,240}Pu$ and $^{241}Pu/^{239,240}Pu$ ratios. As observed in Figure 5 the isotope activity ratio varies significantly depending on origin of plutonium[7].

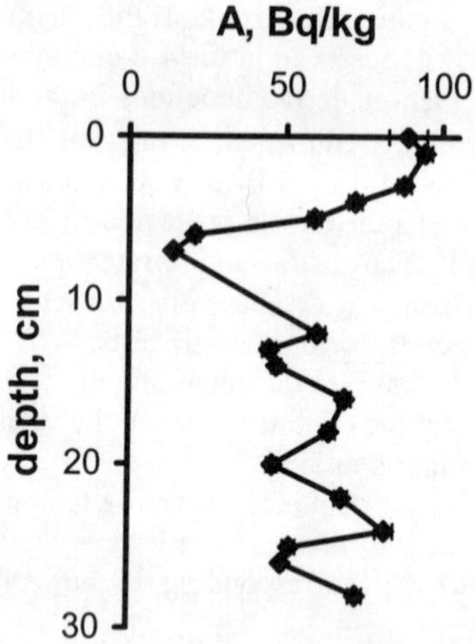

Figure 4 *The vertical profile of ^{137}Cs in the sediment core from the Western Black Sea – Danube Avandelta, core #4*

^{238}Pu, ^{239,240}Pu and ^{241}Pu concentrations, plus their activity ratios were determined for several horizons of the sediment core samples as presented in Table 2. The obtained data indicate that sediments are mixed and the radionuclides have both Chernobyl and fallout origin. The upper layers were mostly contaminated with Chernobyl plutonium whereas the relative content of fallout radionuclides increased at depth.

Table 2 *Plutonium isotopes in sediment core #4.*

H, cm	^{238}Pu, Bq/kg	^{239,240}Pu, Bq/kg	^{241}Pu, Bq/kg
0-1	2.72 ± 0.59	5.28 ± 0.83	230 ± 90
2-3	2.32 ± 0.49	5.16 ± 0.74	240 ± 100
4-5	2.16 ± 0.17	6.01 ± 0.69	280 ± 50
24-25	1.95 ± 0.23	5.09 ± 0.75	170 ± 80

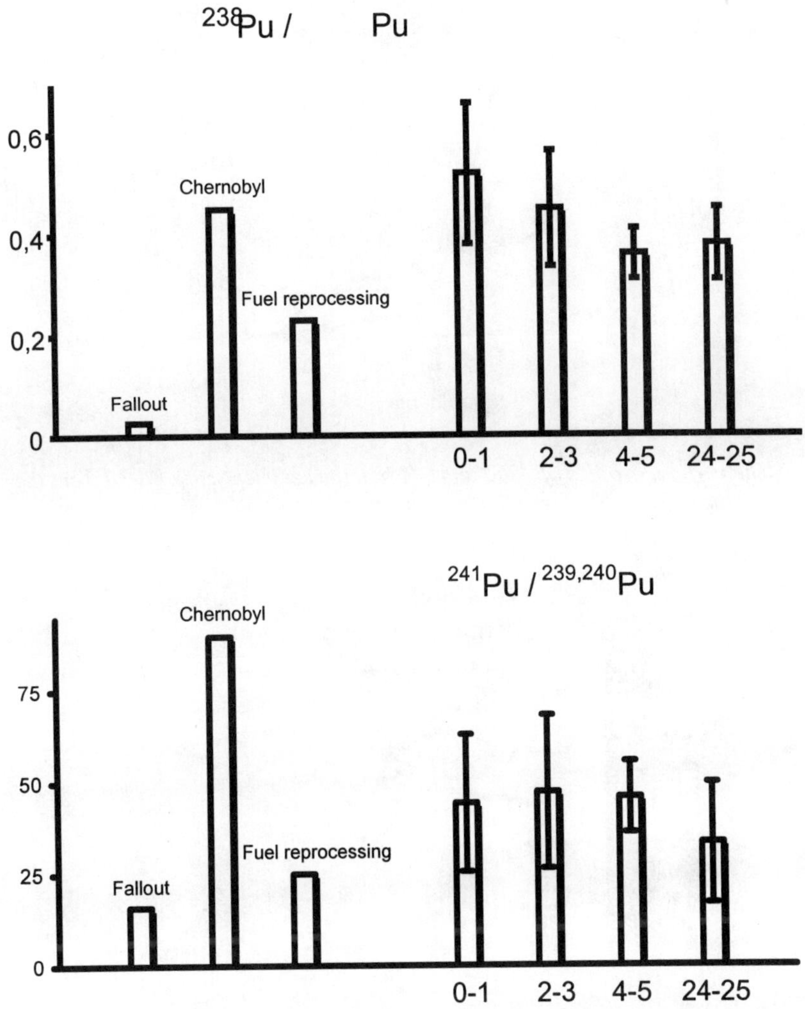

Figure 5 *Plutonium isotopic ratio from different origin vs. Danube Avandelta, core #4*

2.3.3 The Turkish Coast – The Sinop Peninsula. Two sediment cores were collected near the Sinop Peninsula. The vertical profiles of ^{137}Cs in these cores are presented in Figures 6 and 7. Non-regular distribution of ^{137}Cs in BS-27 can be explained by bioturbation.

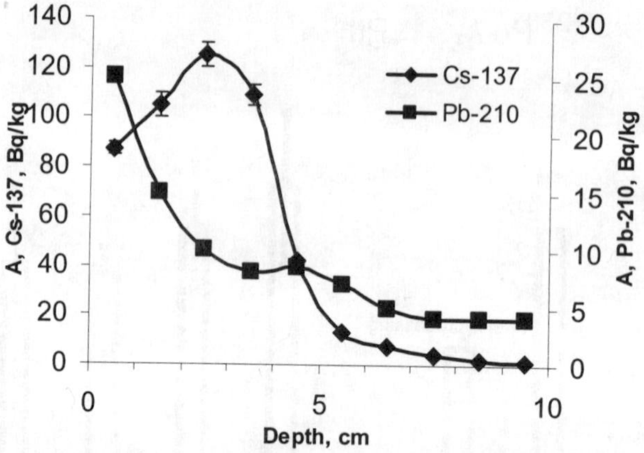

Figure 6 *The vertical profiles of ^{137}Cs and ^{210}Pb in the sediment core BS-30 from the Southern Black Sea – the Sinop Peninsula*

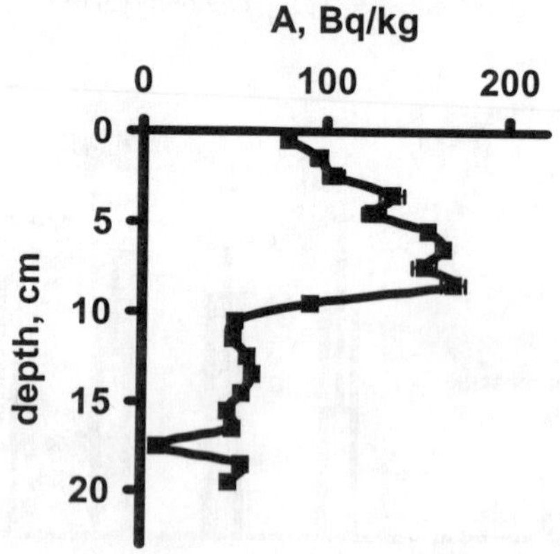

Figure 7 *The vertical profile of ^{137}Cs in the sediment core BS-27 from the Southern Black Sea – the Sinop Peninsula*

2.3.4 Eastern Sector. Three sediment cores were collected in the deep-water part of the Eastern Black Sea (BS-11, BS-23, BS-33). The vertical distribution of ^{137}Cs in BS-11 has two sharp maximums, corresponding to the Chernobyl accident and the atmospheric nuclear tests (Figure 8). In the BS-23 the maximum of ^{137}Cs was found in the upper layer. We believe that sedimentation rate in BS-23 is relatively slow (Figure 8).

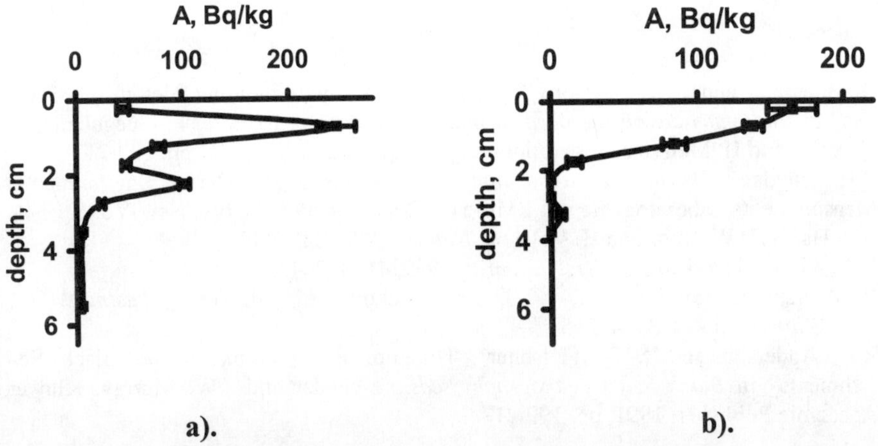

Figure 8 *The vertical profiles of ^{137}Cs and ^{210}Pb in the sediment cores from the Eastern Black Sea. a).BS -11, b). BS-23*

No radiocaesium was found in the BS-33 core by gamma-spectrometric analysis. We believe that, in this part of the Black Sea, the peculiarities of water mass circulation don't allow ^{137}Cs accumulation immediately in underlying bottom sediments. The water mass transport and sedimentation processes in this region should be the subject of further investigations.

3 CONCLUSION

1. The Chernobyl accident is the main source of the Black Sea contamination by man-made radionuclides.
2. The specific activity of ^{137}Cs in the bottom sediments of the Black Sea reaches the level of 200 Bq/kg (dry weight).
3. The vertical distribution of ^{137}Cs in the Black Sea sediments in most cases has a sharp maximum, corresponding to the Chernobyl accident. Sometimes a second maximum caused by the atmospheric nuclear tests also occurs (corresponds to early 60s).
4. Sometimes the vertical distribution of ^{137}Cs in the sediments is irregular. There are three main possible reasons for this fact:
   - mechanical disturbance (e.g. AK-21);
   - bioturbation, (e.g. BS-27);
   - annual changes of river runoff, like in Danube Avandelta (e.g. #4).
5. Recently ^{137}Cs fluxes to marine sediments dramatically decreased. It means, that self-cleaning of water column takes part during the last years.

References

1 K. Buesseler and H. Livingston, "Natural and man-made radionuclides in the Black Sea" in *Radionuclides in the oceans. Inputs and inventories,* eds., P. Guegueniat, P. Germain and H. Metivier, Les editions de physique, Les Ulis, 1996, pp. 199-217.

2 Z.R. Juzdan, *Worldwide deposition of* ^{90}*Sr through 1985.* Environmental Measurements Laboratory Report EML-515, Department of Energy, New-York, 1998.

3 E.P. Hardy, P.W. Krey and H. Volchok, *Nature,* 1973, **241**, 444

4 R.A. Aliev, *Vestnik MGU, Ser.2 Khimiya*, 1999, **41**, 4, 264.

5 Yu.A. Sapozhnikov, I.P. Efimov, L.D. Sapozhnikova and V.P. Remez, *Vestnik MGU, Ser.2 Khimiya*, 1992, **33**, 4, 395.

6 R.F. Anderson and M.Q. Fleishner, "Uranium Precipitation in the Black Sea sediments" in *Black Sea Oceanography, eds.,* E. Ezdar and J.W. Murray, Kluwer Academic Publisher, 1991, pp. 199-217.

DETERMINATION OF PLUTONIUM IN THE AIR BY ISOTOPIC DILUTION ICP-MS AFTER TNOA EXTRACTION CHROMATOGRAPHIC SEPARATION

Yu-Ren Jin, Jin-Feng Gao, Jun-Feng Bai, Guo-Qin Zhou, Lin Li, Jin-Chuan Xie
Jian-Feng Wu, Xu-Hui Wang , Feng-Rong Zhu

Northwest Institute of Nuclear Technology, Xi'an, 710024, P.R.China

1 INTRODUCTION

Environmental plutonium is primarily due to the fallout of atomic weapon testing [1] which deposited around 10^4 TBq of plutonium on the Earth's surface [2]. More localized sources of plutonium in the environment are from the leakage from nuclear-powered generating stations, reprocessing plants and other nuclear facilities. The interest of the determination of environmental plutonium arises from its radioecology and radiotoxicity. The plutonium isotopes in the environment are mainly ^{238}Pu, ^{239}Pu, ^{240}Pu, ^{241}Pu and ^{242}Pu, of which ^{239}Pu is the most abundant isotope. It is known that the isotopic composition of plutonium varies with its source term, so, the isotope ratio, ^{240}Pu/^{239}Pu, can be used to identify different sources of plutonium released. The measurement of aerosol samples provides valuable input data for the estimation of the radiation exposure of the population. Plutonium aerosol monitoring should be conducted in the cases such as during the process of sorting the soil contaminated by plutonium [3].

Alpha spectrometry is, hitherto, widely utilized to determine environmental plutonium, but it fails to offer isotopic ratio of ^{240}Pu to ^{239}Pu because of overlap of spectra. Moreover, it is time-consuming because of the low specific radioactivity in uncontaminated environmental sample. Mass spectrometry, such as thermal ionization mass spectrometry, resonance ionization mass spectrometry and accelerator mass spectrometry, can provide full isotopic information, but their low sample throughput as well as tedious pre-treatment procedure restrain their extensive applications in environmental plutonium analysis. The advent of inductively coupled plasma mass spectrometry (ICP-MS), which combined with high sensitivity, high throughput and ability to measure isotopic ratios, make the routine analysis of environmental plutonium practicable, and it has been successfully used in the determination of long-lived radioactive nuclides [4,5].

When ICP-MS is used to measure ultra-trace plutonium, one should pay attention to the existence of interfaces . The main interferences may be polyatomic ions, of which $^{238}U^1H^+$ is the most conspicuous [6-10]. But we have noticed other polyatomic ions, derived from Hg, Pb and Tl, may also affect the measurement of ultra-trace plutonium if unsatisfactory separation was encountered or there were improper constituents existed in the solution for measurement; therefore chemical separation is absolutely necessary in order to deplete these interferences. Anion exchange [5,11,12,13] and TEVA [11] or UTEVA [14]-extraction chromatography have been used to separate plutonium from matrix. In this paper

TNOA-extraction chromatography was used to separate plutonium from aerosol filter material. Particular attention was given to the chromatographic behaviour of precursors of interfering polyatomic ions. The quantification of ^{239}Pu was carried out by isotopic dilution using ^{242}Pu as spike.

2 EXPERIMENTAL

2.1 Reagents And Standard Solutions

Ultra pure reagents of nitric acid, hydrochloric acid, oxalic acid and hydrogen peroxide were used in the experiments. The standard solutions were diluted with 2% (m/m) nitric acid in 18.2MΩ·cm deionized water obtained from Ultra-Pure Water System (Millipore, United States). ^{242}Pu was produced by the State Scientific Center of Russia, in which the atomic ratios of $^{239}Pu/^{242}Pu$ and $^{240}Pu/^{242}Pu$ are 0.0012 and 0.017, respectively.

2.2 Sampling and Pre-Treatment

Aerosol samples were collected at a height of 1m above ground on polycarbonate filters (20×30cm) at an air flow of about 100 $m^3 \cdot h^{-1}$. A known amount of ^{242}Pu stock solution (~0.2ml of 0.15 $ng \cdot ml^{-1}$ in 2 $mol \cdot l^{-1}$ nitric acid) was added to the filter. The filter was ashed at 550□ for 8 h in a muffle furnace. The ashed residue was further treated with mixed acid of hydrofluoric acid and nitric acid for several times, and 30% hydrogen peroxide was added to obtain Pu(IV), and finally, it is converted to a solution of 8 $mol \cdot l^{-1}$ nitric acid.

2.3 Separation Procedure

Extraction chromatography was applied to separate plutonium from aerosol using TNOA as stationary phase (40% of TNOA to Kel-F, m/m, 100-200 mesh) in a mini-column (diameter: 2mm, length: 5mm). The column was firstly conditioned with 8 $mol \cdot l^{-1}$ nitric acid, and then the resultant solution (about 1.5 ml) was loaded onto the column. The column was washed with 1ml 8 $mol \cdot l^{-1}$ nitric acid, and rinsed with 1.5 ml 9 $mol \cdot l^{-1}$ chloric acid, followed by 1 ml 3 $mol \cdot l^{-1}$ nitric acid. Plutonium was stripped with 1ml 0.15 $mol \cdot l^{-1}$ nitric acid-0.025 $mol \cdot l^{-1}$ oxalic acid.

2.4 Measurement

All measurements were performed on a Finnigan MAT ELEMENT high resolution ICP-MS with Cetac Technologies concentric nebulizer and an Aridus membrane desolvator. A mass resolution of 300 was used. The instrumental settings are summarized in Table 1. Before a measurement sequence was performed, the instrument was carefully mass calibrated. The argon gas flow rate and the ion lenses were adjusted to get maximum sensitivity at m/z238 by monitoring 10 $ng \cdot ml^{-1}$ uranium standard solution. The sensitivity can be tuned to approximately 2×10^5cps per ppb ^{238}U.

A blank (2% nitric acid) procedure was performed in advance of sample analysis. The formation rate of m/z239 originated from ^{238}U and the mass bias coefficient were determined before or after the measurement of the sample. The data collecting parameters are also listed in Table 1. The signal intensity at m/z 238 was monitored, in case unsatisfactory separation of uranium was encountered. The atomic ratio, $^{239}Pu/^{242}Pu$, was corrected according to equation (1).

Table 1 *Instrumental operating conditions and measurement parameters*

Instrumental operating conditions		Measurement parameters	
Plasma power	1350 W	Isotopes	^{238}U, ^{239}Pu, ^{240}Pu, ^{242}Pu
Rf power	<7 W	Acquisition mode	E-scan
Sample uptake rate	0.06 ml·min^{-1}	No. of scans	30
Plasma gas flow rate	13 l·min^{-1}	Acquisition points	10
Auxiliary gas flow rate	1.2 l·min^{-1}	Search window	50%
Nebulizer gas flow rate	0.85 l·min^{-1}	Integration window	80%
Ion lens settings	Optimized daily	Dwell time per sample (ms)	^{238}U,10; ^{239}Pu, 200; ^{240}Pu, 500; ^{242}Pu, 50
Ion sampling depth	Optimized daily		
Sampler skimmer cone	Nickel	No. of sampling per nuclide	20

$$^{239/242}R_{\mathrm{m}} = (^{239/242}R - \frac{^{239}I_{\mathrm{b}} + ^{238}I \times K_U}{^{242}I}) \times \frac{^{242}I}{(^{242}I - ^{242}I_{\mathrm{b}})\eta} \qquad (1)$$

Where $^{239/242}R_m$ is the actual atomic ratio, $^{239}Pu/^{242}Pu$, in the solution measured, $^{239/242}R$ refers to the atomic ratio reported by the instrument. K_U is the formation rate at m/z239 by uranium. I denotes signal intensity, cps. η is mass bias calibration coefficient, which was determined by using standard plutonium solution before or after ICP-MS measurement. The subscripts, b and d, denote blank and dilute, respectively.

The amount of ^{239}Pu was calculated according to Equation (2).

$$^{239}Pu = ^{242}Pu_{\mathrm{d}} \times \left(^{239/242}R_{\mathrm{m}} - ^{239/242}R_{\mathrm{d}}\right) \times \frac{239}{242} \qquad (2)$$

Where $^{242}Pu_{\mathrm{d}}$ is the amount of ^{242}Pu added, g. $^{239/242}R_d$ is the atomic ratio in dilute.

3 RESULTS AND DISCUSSIONS

3.1 Polyatomic Ions

The interferences involved in ICP-MS determination of environmental plutonium are mainly matrix effect and spectral interferences. The matrix effect may suppress the signal intensity, but it could be eliminated as long as the sample was subjected to pre-separation. The latter will, therefore, be particularly investigated in the present work. When the solution of 10ppb uranium was injected to the plasma, high signal intensities at m/z 239, m/z240 and m/z 237 were observed, as illustrated in Figure 1, and the counting rate at m/z239 is much larger than that at m/z237, due to the existence of polyatomic ion, $^{238}U^{1}H^{+}$. To reveal the dependency of signal intensity of m/z239, m/z237and m/z240 with m/z238, solutions with different concentration of uranium were measured. The intensities at m/z 239, m/z240 and m/z237 are linearly proportional to that at m/z238, as illustrated in Figure 2. These results mean that the calibration of the interference from ^{238}U could be fulfilled by the ratio of m/z239 to m/z238, K_U. It should be stressed that this ratio is dependent on the instrumental settings; therefore, it should be determined before or after ICP-MS measurement. K_U at m/z 239 is 4.4×10^{-5}, when the pneumatic nebulizer was used, is in agreement with that reported (Table 2) by other authors [6-9]. The amounts of uranium is around μg·g^{-1}, and the ^{239}Pu equivalent, accordingly, caused by tailing and polyatomic ion from uranium in one gram of soil is 4.4×10^{-11}g, which is much larger than the amount of ^{239}Pu in one gram of soil, the latter is around 10^{-13}g in one gram of surface soil [15]. Therefore, the decontamination of uranium from plutonium before ICP-MS measurement remains essential.

In our preparative experiment, a reference soil sample (GBW07403, National Center of Standard Substances of China), with ^{242}Pu added, was used to examine the validity of separation procedures. The procedure used was almost the same as described above, but plutonium was stripped using 0.36 $mol \cdot l^{-1}$ chloric acid, and the resultant eluate was injected directly to ICP in order to promote sample throughput and the recovery (From our experiences, an additional dryness could lead to 30~40% loss of ultra-trace plutonium). We

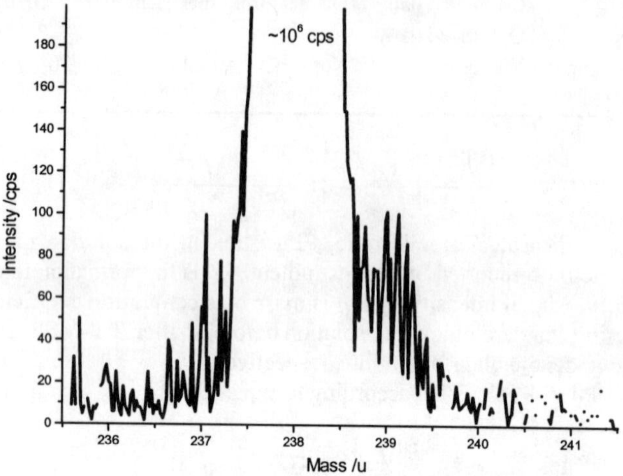

Figure 1 *Inductively coupled plasma mass spectrum of uranium, 10 ppb*

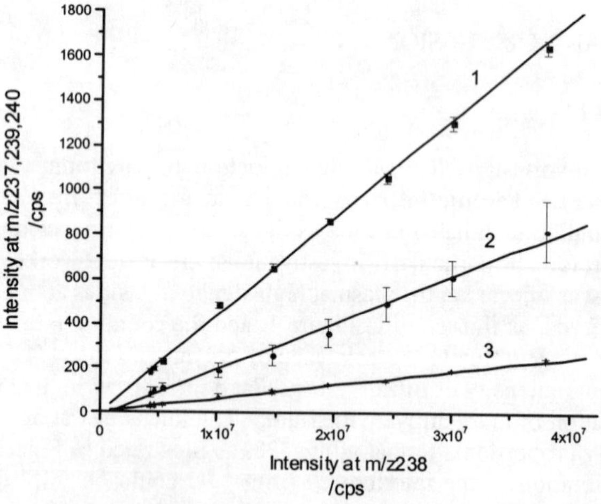

Figure 2 *The dependencies of signal intensity at m/z239, 237, 240 with signal intensity at m/z238; Sample: Uranium solution of different concentration. Nebulizer: pneumatic nebulizer. 1, $^{239}I/^{238}I$; 2, $^{237}I/^{238}I$; 3, $^{240}I/^{238}I$*

noticed a surprisingly high intensity at m/z 242; and therefore amass scan, as illustrated in Figure 3, was conducted. Figure 3 shows that, when a concentric nebulizer was used, the count rate at m/z 238 is only ~100 cps after the sample was subjected to chromatographic separation, so uranium was nearly completely eliminated. But the count rate in the mass range from m/z 241 to m/z 248 is abnormity; we designate them as polyatomic ions that was illustrated in the figure. Therefore, the spectral interferences come from not only the tailing and polyatomic ion of neighboring actinide, ^{238}U, but also heavy metals such as Hg, Pb and Tl. We could not ascertain that other polyatomic ions did not exist, so the formation rate of polyatomic ion and the detailed chromatographic behaviour of the precursors should be investigated. The main polyatomic ions interfering in the determination of ^{239}Pu, ^{240}Pu and ^{242}Pu are summarized in Table 3. Table 3 shows that nitric acid is a more appropriate medium than chloric acid, since the former produces few kinds of interfering polyatomic ions. In the subsequent experiments, dilute nitric acid containing small amount of oxalic acid was used to strip plutonium, this eluant could be injected directly to the nebulizer and the deposition of carbon on the sampling corn was negligible. Moreover it is beneficial for the elimination of memory effect; this is discussed later (Section 3.2). The polyatomic yields measured are listed in Table 4, and the results in the table show that pre-separation is required in order to avoid possible interferences. Fortunately, TNOA-extraction chromatography can effectively remove these precursors. Figure 4 shows the elution curves of metal ions using TNOA extraction chromatography. As illustrated in the figure, most of uranium, lead and mercury could be washed off with 8 mol·l^{-1} nitric acid, and the adsorbed thallium could be rinsed with 9 mol·l^{-1} chloric acid, and the remaining Hg and Tl can further be eluted with 3 mol·l^{-1} nitric acid.

Table 2 *The tailing and uranium hydride in ICP-MS*

K_U $^{239}I/^{238}I$	$^{240}I/^{238}I$	Instrument, Sample injection*	Refs
7.8×10^{-5}		Quad, PN	6
7.1×10^{-5}		Quad,USN	
0.95×10^{-5}		Quad,USN/MD	
$(3.84\pm0.15)\times10^{-5}$	$(9.7\pm2.5)\times10^{-7}$	HRICP-MS, PN	7
6.7×10^{-5}		HRICP-MS, PN	8
$(3.2\pm0.2)\times10^{-5}$		Quad II, PN	9
$(1.8\pm0.1)\times10^{-5}$		Quad II, USN	
5.8×10^{-6}		Multi-collector,HRICP-MS, MD	10
$(4.4\pm0.08)\times10^{-5}$	$(6.17\pm0.07)\times10^{-6}$	HRICP-MS, PN	Present work
$(1.6\pm0.09)\times10^{-5}$		HRICP-MS, MD	

*PN, Pneumatic Nebulizer. MD, Membrane Desolvator, USN, UltraSonic Nebulizer

Table 3 *Polyatomic ions interfere the ICP-MS determination of ultra-trace plutonium*

M/Z	Polyatomic ions							
239	$^{199}Hg^{40}Ar^+$	$^{202}Hg^{37}Cl^+$	$^{203}Tl^{36}Ar^+$	$^{204}Pb^{35}Cl^+$	$^{205}Tl^{35}Cl^+$	$^{207}Pb^{32}S^+$	$^{208}Pb^{31}P^+$	$^{238}UH^+$
240	$^{200}Hg^{40}Ar^+$	$^{203}Tl^{37}Cl^+$	$^{204}Pb^{36}Ar^+$	$^{208}Pb^{32}S^+$	$^{209}Bi^{31}P^+$			
242	$^{202}Hg^{40}Ar^+$	$^{205}Tl^{37}Cl^+$	$^{206}Pb^{36}Ar^+$	$^{207}Pb^{35}Cl^+$				

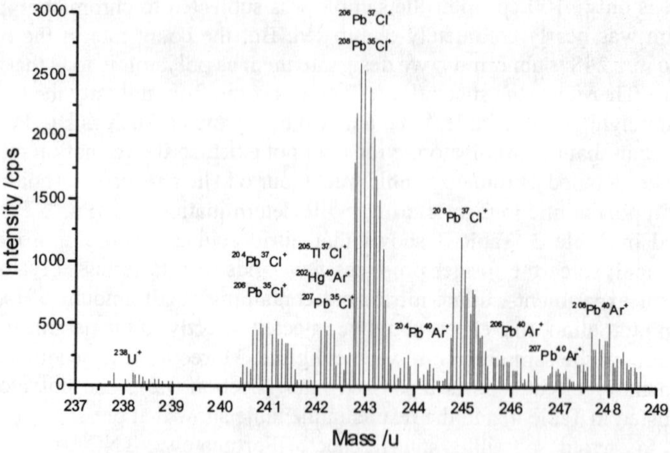

Figure 3 *Mass spectrum of plutonium eluant; Medium, 0.36 mol· l^{-1} chloric acid*

Table 4 *Yields of polyatomic ion*

	Medium	M/Z	Yield	RSD
UH/U	2% nitric acid	239	4.1×10^{-5}	1.6
PbCl/Pb	0.36 mol·l^{-1} chloric acid	239	5.9×10^{-6}	7.6
PbAr/Pb	0.36 mol·l^{-1} chloric acid	240	7.5×10^{-7}	8.2
HgAr/Hg	2% nitric acid	239,240,242	7.5×10^{-7}	6.5
TlCl/Tl	0.36 mol·l^{-1} chloric acid	242, 240	2.9×10^{-7}	8.0
TlAr/Tl	0.36 mol·l^{-1} chloric acid	239	3.1×10^{-7}	8.5

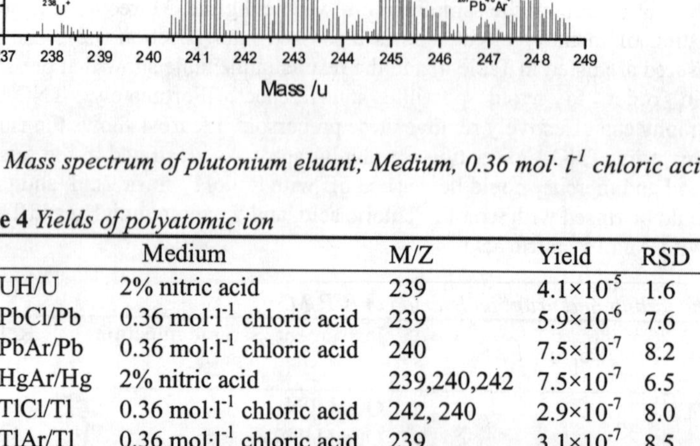

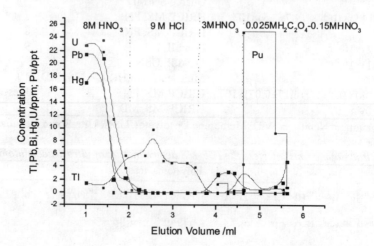

Figure 4 *Step elution of U, Pb, Hg, Tl and Pu;Load: 1 ml mixed metal solution in 8 mol·l^{-1} nitric acid; Column: Mini-column with inner diameter 2 mm, length 50mm, filled with 40% TNOA in Kel-F*

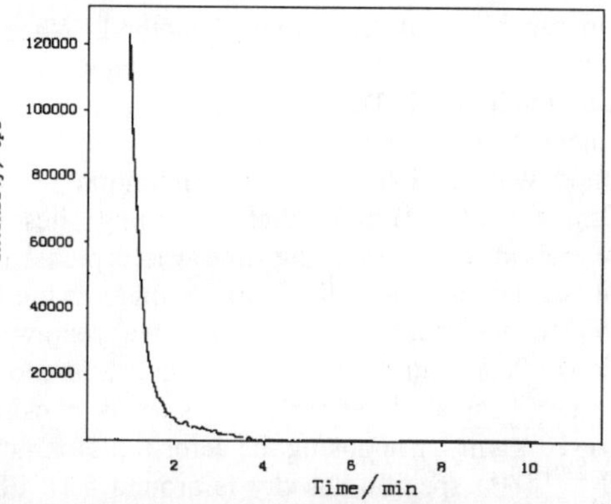

Figure 5 *The elimination of the memory effect of plutonium by rinsing with 0.025 mol·l^{-1} oxalic acid in 0.15 mol·l^{-1} nitric acid; Pre-injection with 1 ng·ml^{-1} ^{242}Pu solution in 2% nitric acid for 5 min followed by 2% nitric acid for 2 min*

3.2 Memory Effect

The retention of analyte in the sample injection system or the torch may induce additional signal intensity of the following sample to be analysed. The memory effect has been previously reported for the more volatile elements such as cadmium, iodine, lead, lithium and mercury [13]. Moreover, we observed this phenomenon during the measurement of environmental plutonium.

When a solution of ^{242}Pu (1 ng·ml^{-1} in 2% nitric acid) was introduced into ICP for 5 min and then injected with 2% nitric acid. The signal intensity at m/z242 is still much larger than that of the blank. This result showed that the memory effect of plutonium did exist. In order to find the location of plutonium retention and the method to eliminate memory effect, two experiments were performed. First, the solution of ^{242}Pu was injected for 5 min, followed by rinsing with 2% nitric acid for 2 min, and then a new plastic tubing was used to transmit 2% nitric acid. The signal intensity at m/z242 was observed to return to the background promptly. This result suggests that the memory effect was caused by the retention of plutonium in plastic tubing. Secondly, after ^{242}Pu was injected for 5 min and rinsed with 2% nitric acid for 2 min, 0.025 mol·l^{-1} oxalic acid in 0.15 mol·l^{-1} nitric acid was injected, the counting rate at m/z242 decreased dramatically, as shown in Figure 5. Therefore, this protocol is most appropriate for the elimination of memory effect. The same solution, as mentioned above, was also used to strip plutonium from the column; therefore it is recommended that the plutonium eluate is introduced to ICP directly.

3.3 Analysis of ^{239}Pu in Aerosol

The detection limit (DL) of ^{239}Pu was calculated as the equivalent concentration to three times the standard deviation of the 2 % nitric acid blank. Minor background signals were found at the plutonium mass range from m/z239 to m/z242, with <1 cps for m/z239, m/z240 and m/z242 when a micro-concentric nebulizer was used for sample introduction. The DL is 6 femtogram for ^{239}Pu, and it is 0.5 femtogram when membrane desolvating nebulizer was used to introduce sample. This DL is rather a good one in comparison with the reported DLs [4,5,7,16,17]. Background is the concomitant of sample treatment; therefore, the limit of determination, LOD, which is based on 3σ criterion of the background, is more useful to characterize the detection capability of an analytical method. When a micro concentric nebulizer was used, the limit of determination is 16.3 fg for aerosol sample via mini-column TNOA extraction chromatography [18]. The front 0.8 ml plutonium eluant was submitted for the ICP-MS measurement. The recovery of plutonium in this aliquot is 75%.

The established method was used to determine plutonium in aerosol. Seventy-two aerosol samples (>2000m^3 air filtered) from three sampling sites were analyzed. The relationship of the concentration to the sampling time was depicted in Figure 6-A, B, C. Some experimental points deviate dramatically from the mean value (in dotted lines), we impute this phenomenon to weather conditions. As we redraw them with weight concentration, the variations of concentration are much smaller, as shown in Figure 7-A, B, C. The amount of airborne ^{239}Pu at three sampling sites is consistent, and the mean concentration is $(4.8\pm2.9)\times10^{-17}$g·m^{-3}, Supposing the aerosol plutonium comes from global fallout, then the derived $^{239+240}$Pu specific activity is around 0.18 μBq·m^{-3} by using the average isotopic ratio, ^{240}Pu/^{239}Pu (0.176) of the global fallout[17]. This concentration is very close to world-wide background reported in the early 1990's [3].

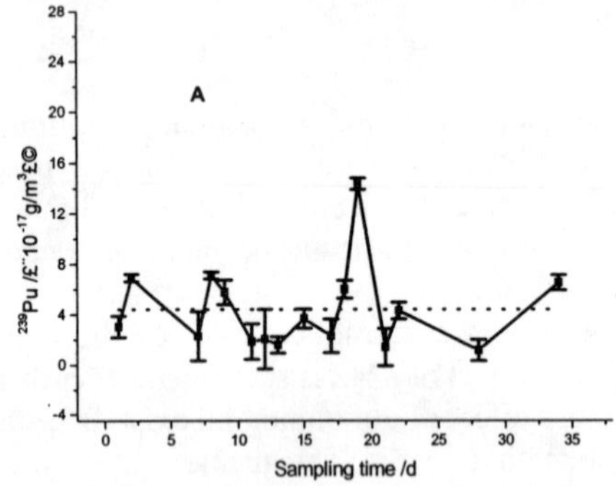

Figure 6 *Volumetric concentration of airborne ^{239}Pu at different sampling time; At site: A*

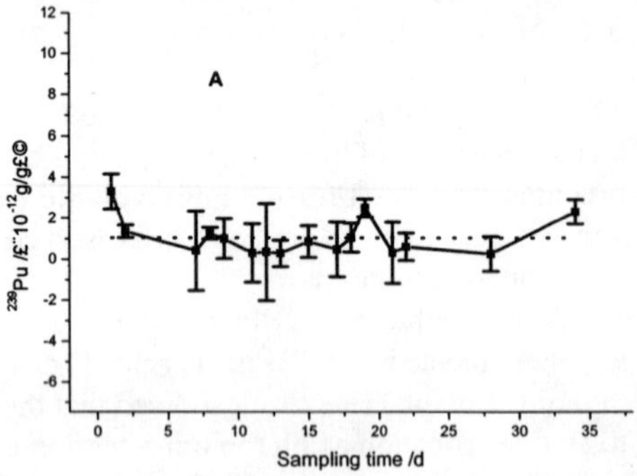

Figure 7 *Weight concentration of ^{239}Pu in dry aerosol corresponding to that of Figure 6; At site:A*

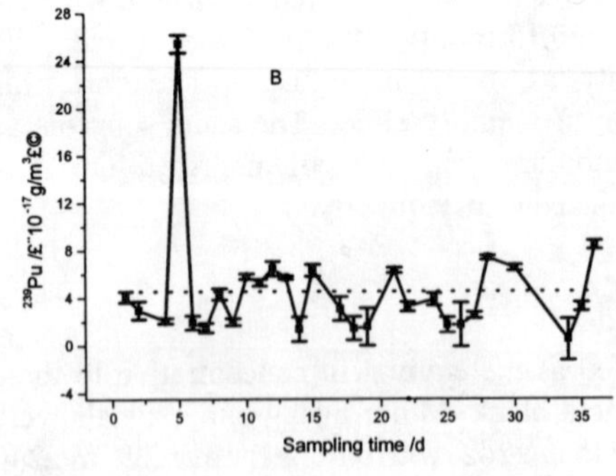

Figure 6 *Volumetric concentration of airborne ^{239}Pu at different sampling time; At site: B*

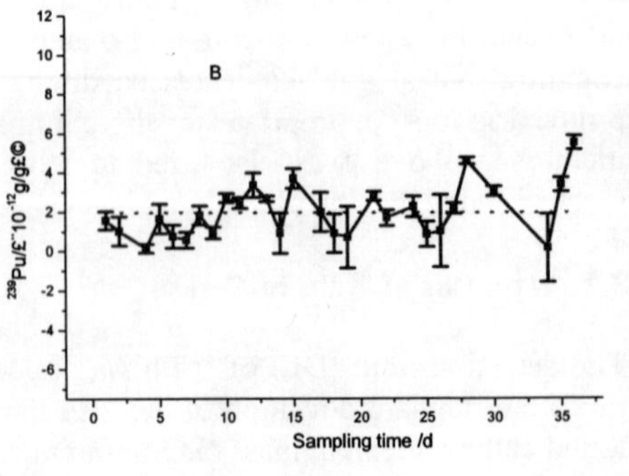

Figure 7 *Weight concentration of ^{239}Pu in dry aerosol corresponding to that of Figure 6; At site:B*

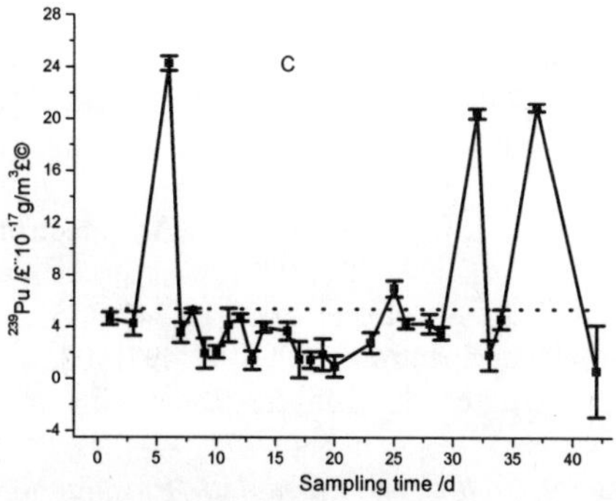

Figure 6 *Volumetric concentration of airborne ^{239}Pu at different sampling time; At site: C*

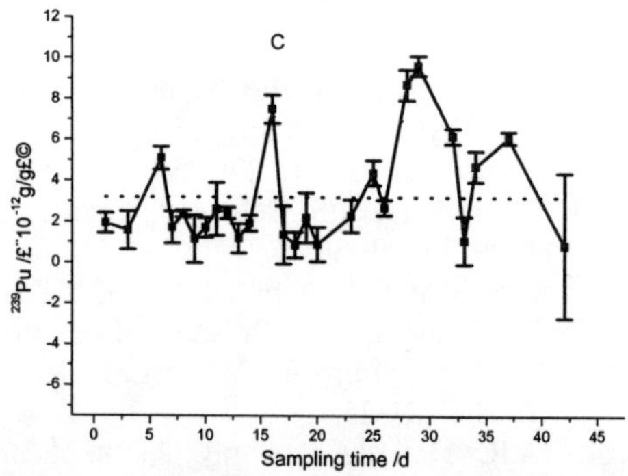

Figure 7 *Weight concentration of ^{239}Pu in dry aerosol corresponding to that of Figure 6; At site:C*

4 SUMMARY

Precursors U, Pb, Hg, Tl and others can form polyatomic ions which interfere with inductively coupled plasma mass spectrometric determination of environmental plutonium. The formation rates of polyatomic ions from these precursors were measured, and their chromatographic behaviour were investigated. The precursors can be rinsed out using 8 $mol{\cdot}l^{-1}$ nitric acid, 9 $mol{\cdot}l^{-1}$ hydrochloric acid and 3 $mol{\cdot}l^{-1}$ nitric acid, respectively. The plutonium was stripped from tri-n-octylamine (TNOA) extraction chromatographic column by using 0.025 $mol{\cdot}l^{-1}$ oxalic acid-0.15 $mol{\cdot}l^{-1}$ nitric acid. Memory effect was also observed during the measurement, which was ascribed to the retention of the analyte on the soft plastic sampling tubing. Oxalic acid is effective in eliminating the effect; therefore the plutonium eluant, with the addition of oxalic acid is recommended for direct injection. Isotope dilution was used for the measurement of ultra-trace ^{239}Pu using ^{242}Pu as the spike. The amount of ^{239}Pu in regional aerosol samples were analyzed with detection limit, 6 fg, and the limit of determination, 16fg. The amount of ^{239}Pu in three sampling sites is consistent. The mean concentration of ^{239}Pu was $(4.8{\pm}2.9){\times}10^{-17}g{\cdot}m^{-3}$.

References

1 L.W. Green, F.C. Miller, J.A. Sparling, S.R. Joshi, J. Am. Soc. Mass Spectrom., 1991,**2**,240-244.

2 R. Pentreath, J. Appl. Radiat. Isot., 1995,**46**,1279.

3 J.H. Shinn, C.F. Fry, J.S. Johnson, UCRL-ID-116495, February 1, 1994.

4 C-K. Kim, R. Seki, S. Morita, S-I. Yamasaki, A. A. Tsumura, K.Y. Takaku, Y. Igarashi, M. Yamamoto, J. Anal. Atom. Spectrom., 1991,**6**,205-209.

5 S. Stürup, H. Dahlgaard, S.C. Nielsen, J. Anal. Atom. Spectrom., 1998,**13**,1321-1326.

6 I. Rodushkin, P. Lindahl, E. Holm, P. Roos, Nuclear Instruments and Methods in Physics Research 1999,**A423**, 472-479.

7 J.S. Becker, H-J. Dietze, Determination of long-lived radionuclides by double-focusing sector field ICP mass spectrometry, In: Advances in Mass Spectrometry, Vol.14, Edn, E.J. Karjalainen, A.E. Hesso, J.E. Jalonen, U.P. Karjalainen, Elsevier Science Pulishers

B.V., Amsterdam, 1998, pp. 687.
8 J.S. Crain, J. Alvarado, J. Anal. Atomic Spectrometry. 1994,**9**,1223-1227.
9 V.P. Perelygin, Yu.T. Chuburkov, Nucl. Tracks Radiat. Means., 1993,**22**,869-872.
10 R.N.Taylor, T. Warneke, J.A. Milton, P.E. Warwick, R.W. Nesbitt, J. Anal. At. Spectrom., 2001,**16**,279-284.
11 Y. Muramatsu, S. Uchida, K. Tagami, S. Yoshida, T. Fujikawa, J. Anal. At. Spectrom., 1999, **14**,859-865.
12 C-S. Kim, C-K. Kim, J-I. Lee, K-J. Lee, J. Anal. At. Spectrom., 2000,**15**,247-255.
13 R. Chiappini, J-M. Tailade, S. Brebion, J. Anal. At. Spectrom., 1996, **11**,497-503.
14 C. Apostolidis, R. Molinet, P. Richir, M. Ougier, K. Mayer, Radiochim. Acta, 1998,**83**,21-25.
15 A.R. Date, A.L. Gray, in *Applications of Inductively Coupled Plasma Mass Spectrometry* .Edn., A.R.Date and A.L.Gray, Chapman and Hall, New York, 1989.
16 W.C. Inkret, D.W. Efurd, G. Miller, D.J. Rokop, T.M. Benjamin, International Journal Of Mass Spectrometry 1998,**178**,113-120.
17 P.W. Krey, E.P. Herdy, C. Pachucki, F. Rourke, J. Coluzza, W.K. Benson, Proceedings of a Symposium on Transuranium Nuclides in the Environment. 1976, **IAEA-SM-199-39.** 671-678, IAEA, Vienna.
18 Y-R. Jin, L-Z. Zhang, S-J. Han, F-R. Zhu, J-F. Bai, G-Q. Zhou, F. Ma, L-X. Zhang, Acta Chimica Sinica, 2000, **58**, 1291-1295(in Chinese).

SIMPLE METHOD FOR C-14 ANALYSIS IN ORGANIC MATERIAL AND ITS DISTRIBUTION IN FOREST AND CULTIVATED FIELD

M. Atarashi-Andoh and H. Amano

Department of Environmental Sciences, Japan Atomic Energy Research Institute, Tokai-mura, Naka-gun, Ibaraki 319-1195, Japan

1 INTRODUCTION

C-14 is one of the naturally occurring radionuclides, produced in the upper atmosphere by cosmic ray neutrons in the reaction $^{14}N(n, p)^{14}C$. The specific activity of C-14 in trees that grew in the pre-industrial atmosphere (19th century) was 226 mBq(g carbon)$^{-1}$.[1] However, as a result of atmospheric nuclear-weapon tests in the 1950s to early 1960s, the specific activity of C-14 in the environment increased dramatically to almost double that from natural sources in the high latitude troposphere in 1963.[2] Since the nuclear test ban treaty in 1962, C-14 in the atmosphere has gradually decreased. However, because of the long radioactive half-life of C-14 (5730 yr), fallout C-14 from former nuclear-weapon tests still remains in soil and C-14 flux from the soil surface to the atmosphere was observed in a forest.[3] Patterns of C-14 enrichment in soil profiles will provide important information for estimating carbon turnover and carbon flux from soil. C-14 is also released from nuclear power plants and nuclear reprocessing plants. The first commercial spent fuel reprocessing plant in Japan is now under construction. Recent increase in the peaceful use of nuclear energy has made monitoring of C-14 in the environment more important.

Many studies have been performed to establish the low-level C-14 specific activity in organic materials using liquid scintillation counting (LSC) and accelerator mass spectrometry (AMS). For LSC, the benzene synthesis method is usually adopted because of its high counting efficiency. However, using benzene synthesis-LSC requires relatively large samples (several grams of carbon) and complex preparation procedures, and using AMS is relatively costly. For these reasons, direct CO_2 absorption and LSC method have an advantage when many samples must be measured. We have developed a simple analytical method of C-14 measurement using fast bomb combustion and LSC. C-14 specific activities in plants and soils in both a forest and a cultivated field were measured using this method.

2 MATERIAL AND METHODS

2.1 Preparation for C-14 analysis

Figure 1 shows a schematic diagram of the set-up. A sample containing organic material was burned under high oxygen pressure (300 psi) in a fast oxygen combustion bomb apparatus (DELFI SCIENCE INTERNATIONAL, Inc). About 5 g samples were burned at a time. A cold trap using liquid nitrogen caught the generated gas mixture, including CO_2. The cold trap was then warmed to room temperature and the trapped gas was purged with nitrogen gas into a $KMnO_4$ solution in order to remove nitrogen oxide and sulfur oxide. The gas was then passed through a water trap (Drierite, W. A. Hammond Drierite Co. Ltd.) and a CO_2 trap (Molecular Sieve (M.A.) 4A, Union Carbide Co.). The CO_2 trapped by M.S. 4A was released using a furnace (420 °C) and purged into a CO_2 absorbent material (CARBO-SORB E, Packard Ltd.) mixed with a scintillation cocktail (PERMA FLUORE E^+, Packard Ltd.) with nitrogen gas after passing through a Drierite column to ensure the gas was absolutely water vapor free. CARBO-SORB E and PERMA FLUORE E^+ were used for environmental samples by Momoshima et al..[4] CORBO-SORB E can absorb 4.8 mmol ml^{-1} of CO_2, which corresponds to 57.6 mg of carbon. To prevent CARBO-SORB E from becoming solid when it absorbs CO_2 up to its limit, PERMA FLUORE E^+ was mixed with it before absorption. The temperature of the reagent mixture increases as CO_2 is trapped, so the tube of reagents should be put in an ice bath during the process. After absorption of CO_2, the 20 ml mixture of reagents was pipetted into a glass vial and weighed. The increase in weight per 20 ml of reagent mixture was the mass of absorbed CO_2.

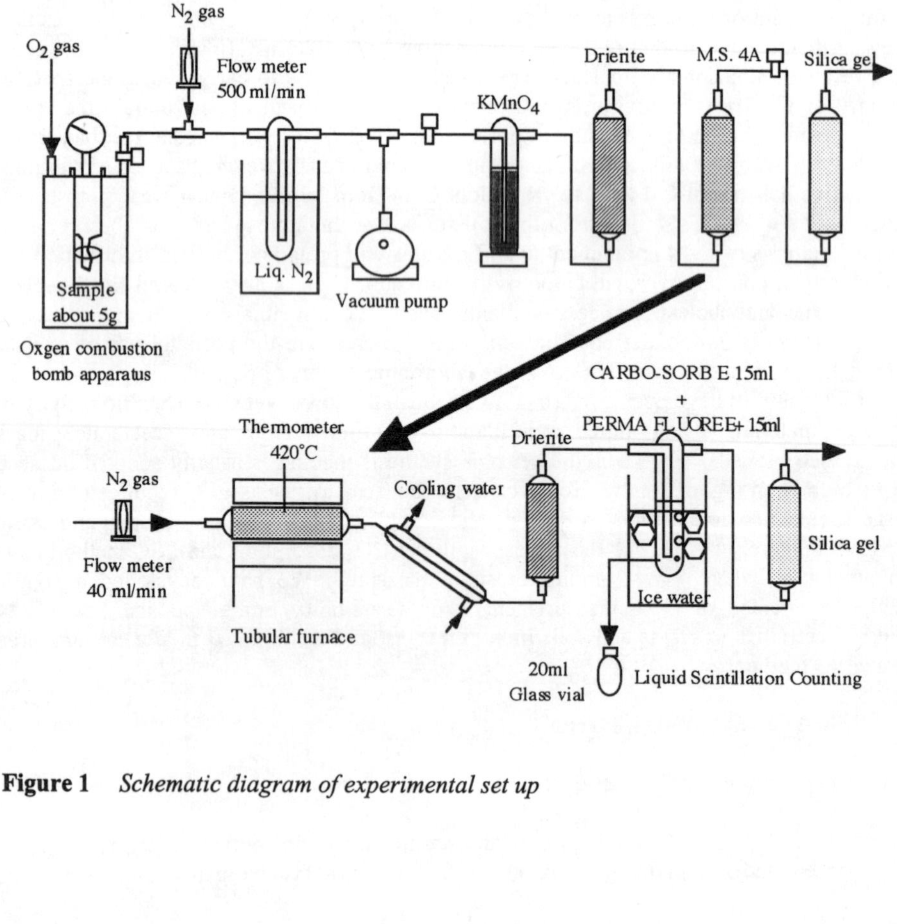

Figure 1 *Schematic diagram of experimental set up*

2.2 Liquid scintillation counting

NIST reference material SRM-4990C (oxalic acid with 18.41dpm (g carbon)$^{-1}$) was used to determine the counting efficiency. Figure 2 shows a schematic diagram of the sample preparation for counting efficiency measurement. Oxalic acid was dissolved in purified water with H_2SO_4, and $KMnO_4$ was added slowly to evolve CO_2. The CO_2 was purged into a water trap (Drierite) followed by a mixture of CARBO-SORB E and PERMA FLUORE E^+ by nitrogen gas with 40 ml min^{-1} flow rate. After CO_2 absorption, the 20 ml reagent mixture was pipetted into a glass vial. The mass of absorbed CO_2 was evaluated by the same gravimetric method as for C-14 measurement in organic material.

The C-14 radioactivity was measured with a low background liquid scintillation counter (LSC-LB III, Aloka) by 500-minute counting.

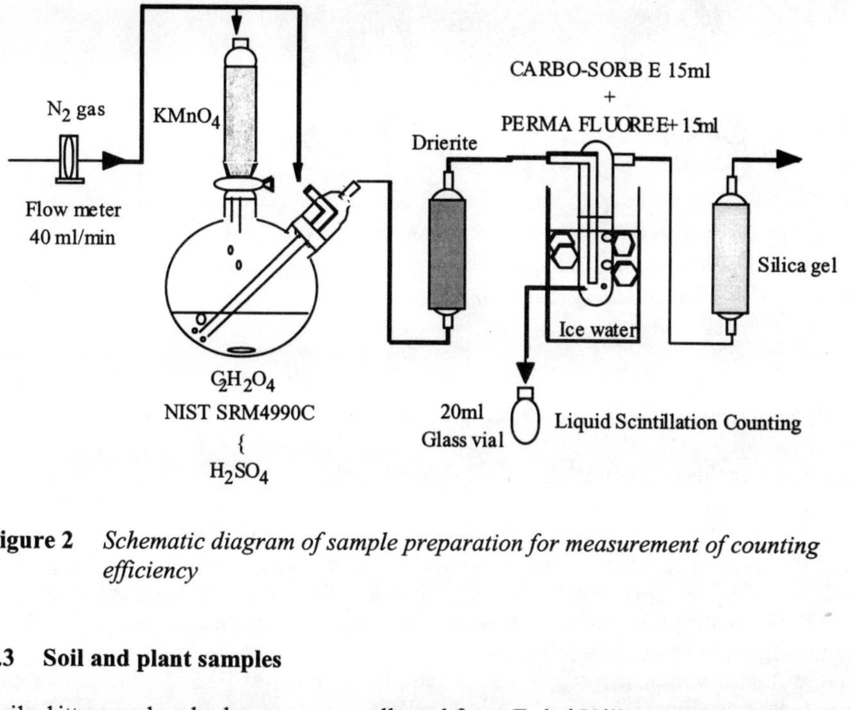

Figure 2 *Schematic diagram of sample preparation for measurement of counting efficiency*

2.3 Soil and plant samples

Soils, litters and cedar leaves were collected from Tokai Village and Satomi Village in 2001 and 2002, and rice samples were collected from Mito City in 2001 in Japan. Figure 3 shows the location of the sampling points. There are nuclear facilities in Tokai Village, including a nuclear power plant and a spent fuel reprocessing plant. Satomi Village and Mito city are control sites. Litter layers were divided into 2 or 3 layers. After removal of litters, soil samples were collected at depths of 0~2, 2~5 cm from the soil surface. At the sampling points in Satomi Village, as the cedar trees were too high to take their leaves directly, newly fallen leaves were collected. At the sampling points in Tokai Village, tips of cedar leaves were collected. Rice sample were collected in July and September in 2001. The rice plants were 20 days before flowering in July, and collected rice plants were divided into four parts: root, lower leaf (from soil surface to 20cm high), middle leaf

(20cm~40cm) and upper leaf (from 40cm to top of leaf (about 70cm)). Rice ears were collected just before harvest time in September.

The samples were dried in an oven (50 °C) to attain constant weight. Litters, cedar leaves and rice plants were ground into very fine powder. Soil samples were ground and sieved through a 2mm mesh after removal of stone, roots and plant debris by hand picking. Then C-14 specific activity was measured by the aforementioned method.

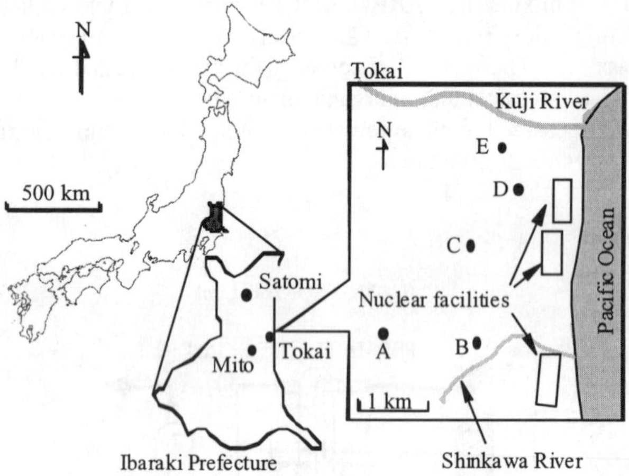

Figure 3 *Location of soil and plant samples. Points A to E indicate sampling locations in Tokai village.*

3 RESULTS AND DISCUSSION

3.1 Liquid scintillation counting

The external standard channel ratio (ESCR) was examined using NIST reference material to check the quenching level and the counting efficiency. The sample's quenching level depended on the mass of absorbed CO_2.[4, 5] In this experiment, ESCR increased with increase in CO_2 absorption, but the counting efficiency did not change much with ESCR and the average value was 60 % (Figure 4).

Figure 5 shows the change in counting rate, background and ESCR for leaf sample with time. As a background, a mixture of CARBO-SORB E and PERMA FLUORE E$^+$ without any absorption was used, because background counting does not depend on the mass of absorbed CO_2.[5] The background counting rate was about 5 cpm. The counting rate for leaf sample and background did not change much for 12 days after CO_2 was fixed. However, ESCR for leaf sample slightly decreased after 4 days. Therefore, we used the average counting efficiency (60 %) following plant and soil sample analysis. The detection limit in this experimental condition was 20 mBq (g carbon)$^{-1}$, so environmental C-14 levels (about 250 mBq (g carbon)$^{-1}$) could be evaluated by this method.

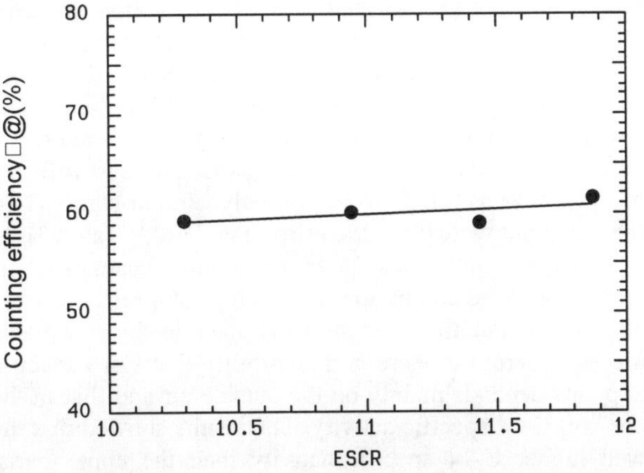

Figure 4 *Change in counting efficiency with ESCR*

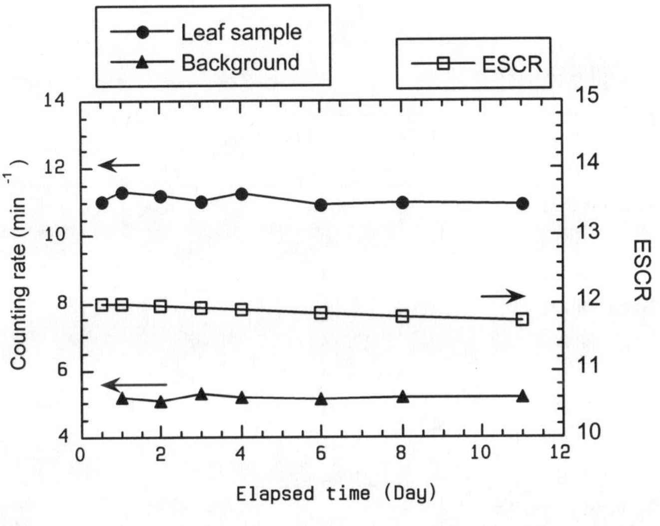

Figure 5 *Change in counting rate and ESCR with time*

3.2 Application to soil and plant samples

C-14 specific activities in cedar leaves were 249 and 247 mBq (g carbon)$^{-1}$ in Satomi Village and 256~278 mBq (g carbon)$^{-1}$ in Tokai Village respectively (Table 1). Momoshima et al. reported that about 20 mBq (g carbon)$^{-1}$ excess was observed around the nuclear power plant in vegetation samples in comparison with the areas located far from nuclear facilities.[6] In this study, C-14 specific activity in cedar leaves around nuclear facilities exceeded that in the control areas by 8~30 mBq (g carbon)$^{-1}$.

C-14 specific activities in litters and soils in forests were higher than those in live cedar leaves for all sampling points (Table 2). C-14 concentration in the atmosphere showed the highest level in the 1960s and continuously decreased after that.[2] Therefore, higher C-14 specific activities in litter and soil are ascribed to C-14 fallout from nuclear-weapon tests in the 1950s and early 1960s. Table 2 also shows that C-14 specific activities in litter and soil collected in Tokai Village were higher than those collected in Satomi Village. This is because of the effects of emission from nuclear facilities and also environmental factors such as decomposition rate of litters and amount of fallen leaves.

It was reported that C-14 specific activity in rice in Japan was 256~280 mBq (g carbon)$^{-1}$ in 1990 and 235~269 mBq (g carbon)$^{-1}$ in 1994 respectively.[7] In our study, C-14 specific activities in all parts of the rice plant were 228~254 mBq (g carbon)$^{-1}$ (Table 3). In paddy fields, rice plants are planted close together and there are small exchanges of air between the areas over the rice plants and those among the rice plants. CO_2 released from paddy soil by decomposition of the soil organic matter is probably absorbed by rice plants before it is released to the outside air. Therefore, there is a possibility that C-14 specific activity in the upper parts of rice plants depends mainly on the outside air and that in the lower parts depends mainly on the soil C-14 specific activity. The results showed that the lower parts of rice plants showed higher C-14 specific activity than the upper parts. However, further study is needed to determine the origin of the C-14 in different parts of the rice plant.

Table 1

C-14 specific activity in Japanese cedar leaves

Location	C-14 specific activity mBq/g carbon
Satomi Village	
Site 1	249±4[a]
Site 2	247±4
Tokai Village	
A	278±4
B	256±4
C	262±4
D	265±4
E	259±4

[a] Counting error 1σ

Table 2

Distribution of C-14 specific activity in forests

(mBq (g carbon)$^{-1}$)

Location	Cedar leaf	Litter			Soil	
		Upper	Middle	Lower	0□2cm	2□5cm
Satomi Village						
Site 2	247±4	247±4		263±4	261±4	251±5
Tokai Village						
A	278±4	294±4		306±4	283±4	298±6
C	262±4	293±4	271±3	301±4	298±6	292±9

Table 3

C-14 specific activity in each part of the rice plant

Part of rice plant	C-14 specific activity mBq/g carbon
Ear	234±4
Upper leaf	228±4
Middle leaf	245±5
Lower leaf	254±5
Root	242±4

4 CONCLUSION

A simple and quick method for C-14 measurement was developed. With this method, about 0.6g of carbon was absorbed in 20 ml of mixture of CO_2 absorbent and scintillation cocktail, and C-14 specific activity was measured by a liquid-scintillation counter by 500-minute counting. The resulting counting efficiency was 60 % and the detection limit was 20mBq(g carbon)$^{-1}$. The environmental C-14 levels (about 250 mBq(g carbon)$^{-1}$) could be evaluated by this method.

Using this method, C-14 in soil and plant samples was measured in an area around nuclear facilities and control areas. C-14 specific activity in Japanese cedar leaves around nuclear facilities exceeded that in the control areas by 8 to 30 mBq (g carbon)$^{-1}$. In forests, C-14 specific activities in litters and soils were higher than that in live cedar leaves. This is because the main parts of bomb-originated C-14 is still contained in litters and soil surfaces.

References

1 Telegadas, K., Health and Safety Laboratory Fallout Program Quarterly Summary report HASL-243. New York, 1971, p. I-2
2 United Nations, Report of the United Nations Scientific Committee on the Effects of Atomic Radiation. Supplement No.16 (A/5216), United Nations, New York, 1962, p. 262
3 J. Koarashi, H. Amano, M. Andoh, T. Iida and J. Moriizumi, *J. Environ. Radioactivity*, 2002, **60**, 249
4 N. Momoshima, H. Kawamura and K. Takashima, *J. Radioanal. Nucl. Chem.*, 1993, **173**, 323.
5 T. Koma, Master's thesis, Nagoya Univ., 1997
6 N. Momoshima, H. Kawamura, Y. Takashima and S. Ezumi, *Hokenbutsuri*, 1995, **30**, 121.
7 H. Watanabe, PNC TN1410 97-006, 1997

^{18}F-LABELLING OF HUMIC SUBSTANCES

K. Franke[1], J. T. Patt[1], M. Patt[2], D. Rößler[2], H. Kupsch[2]

[1]Institut für Interdisziplinäre Isotopenforschung, Leipzig, Germany
[2]Klinik und Poliklinik für Nuklearmedizin, Universität Leipzig, Germany

1 INTRODUCTION

Humic substances are ubiquitous in water, sediment, and soil. They are the major components of materials that comprise the natural organic matter. These complex substances are a heterogeneous mixture of non-living organic matter and its decomposition products. Due to their large number of different functional groups and amphipathic properties humic substances contribute to many geochemical and environmental processes.[1,2,3,4,5] The importance of humic substances for many phenomenon's like nutrient cycles, acid-base buffering, sequestration, transport, and retaining of metals and binding of organic molecules is well recognized. Especially the capability of humic substances to resist microbial degradation over long time induced the idea of pseudostructures of humic substances. Under c ertain circumstances, humic s ubstances can form refractory colloids, precipitate via aggregation or remain in solution as negatively charged complexes. Many studies have investigated the conformational nature of the humic substances for a better understanding and interpretation of these data. However, investigations at natural concentration levels of humic substances are required concerning the inter- and intra molecular interaction mechanisms. We developed a new radiolabelling method for humic substances to enhance the possibilities of classical investigations at the molecular level. ^{18}F was used for radiolabelling of humic acids and fulvic acids. The applicability of this method was demonstrated in the high performance size exclusion chromatography (HPSEC) and the isoelectric focusing (IEF), widely used tools for characterization of molecular properties of humic substances.

2 MATERIALS AND METHODS

2.1 Humic Substances

For our studies, water samples were collected at a bog ("Kleiner Kranichsee") located near Carlsfeld, Saxony, Germany. A detailed description of the sampling area can be found in former works.[6] Humic substances were extracted from the bog water according to a procedure of the International Humic Substance Society.[7] For that purpose the bog water was first filtered (0.45 µm) and acified with HCl to pH 2. After passing a XAD-8 resin the humic acid (HA) and fulvic acid (FA) were obtained by back elution with dilute NaOH.

HA were separated from FA by precipitation at pH 1. Before final lyophilization the humic substances were purified.

2.2 Labelling

^{18}F was used for radiolabelling of HA and FA. A couple of ^{18}F-labelling methods are performed by nucleophilic substitution reactions by the use of non carrier added [^{18}F]fluoride in dipolar aprotic mediums. For this type of reaction, the unprotonated [^{18}F]F^{-} ion with it's nucleophilic reactivity is necessary. This is not given in the presence of proton donating groups as at HA or protic solutions. Additionally, in spite of their amphipathic properties HA and FA have often poor solubility in non-aqueous solvents. On the other hand are HA and FA possibly denaturised by the use of non-aqueous solvents and the geo-chemical properties of the HA and FA are changed. For that reason an alternative procedure was required for ^{18}F-labelling. Based on a known method to synthesize [^{18}F]4-fluoro-benzenediazonium ions a suitable labelling procedure was developed for humic acid and fulvic acid. The labelling was performed via azo-coupling of the 4-fluoro-benzenediazonium ions at the electron rich aromatic constituents of the HA and FA (Scheme 1).

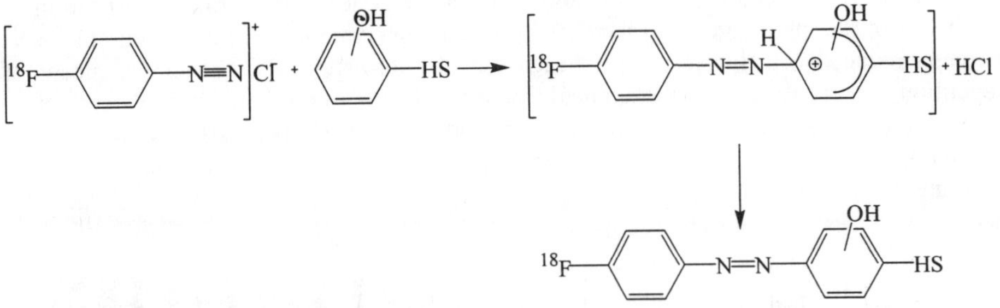

Scheme 1

2.3 High Performance Size Exclusion Chromatography

HPSEC was used to evaluate the labelling method. The instrument was equipped with a TSK gel column TSK-G3000PWXL (300 mm x 7.8 mm). The size fraction range is between 0.5 kD – 800 kD. The mobile phase consists of 0.02 M KCl in MilliQ water with a 0.05 M Tris-buffer (pH 8). Sodium polystyrene sulfonates (M_n: 4.3 kD, 8.6 kD, 17.4 kD, 33.8 kD), blue dextran (M_n: 100 kD), and acetone were used as standards. A NaI(Tl) detector was utilized to measure the radiochromatograms.

2.4 Isoelectric focusing

A Multiphor II (Pharmacia) was utilized for IEF. The used gel consists of polyacrylamide (T = 40 %, C = 3,33 %). The applied pH-range is 3 to 10. The parameters of the pre-focussing (0:00 – 0:30) and main-focussing (0:30 – 2:21) are given in Table 1. The temperature was adjusted at 4°C, the maximal current at 50 mA, the maximal electric power at 20 W.

Table 1 *Parameters of the isoelectrical focussing* (g: gradient; c: constant).

Time [min]	Voltage [V]
0:00 - 0:30	800 (c)
0:30 – 0:41	500 (g)
0:41 – 0:51	500 (c)
0:51 – 1:01	800 (g)
1:01 – 1:31	800 (c)
1:31 – 1:51	1200 (g)
1:51 – 2:21	1200 (c)

3 RESULTS

The radiochemical yield was optimised by adapting the pH in the range 5.5 to 10. Figure 1 represents the achieved labelling yields. The yields were estimated via ultra filtration. The fraction < 3000 D corresponds to the exchangeable ^{18}F, the fraction > 3000 D to the HS species. In the alkaline medium (pH 8.5 - 10), one finds the highest rates of the azo-coupling reaction (70 %). The ability of phenols for electrophilic substitution is increased in this pH-range. On the other hand, the concentration of the diazonium ion is strongly reduced with increasing pH because of the formation of diazoate. By means of repetition of ultra filtration we examined the results of the first ultra filtration step. About 1% - 2% of total activity is present in the fraction < 3000 D. This share is possibly not caused by exchangeable ^{18}F, but corresponds to the known properties of humic substances to tunnel filters and membranes.

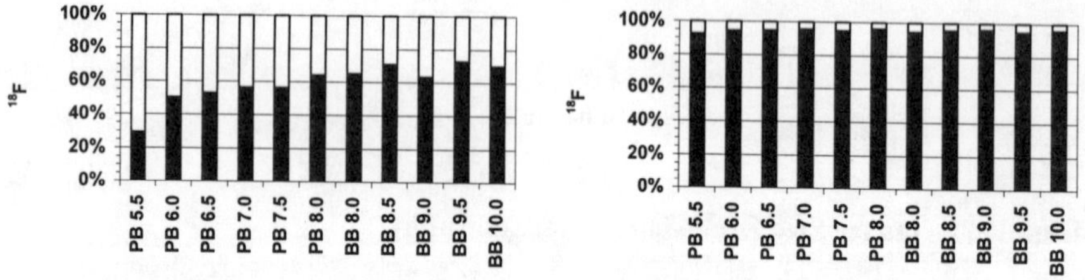

Figure 1 *Radiochemical yield of the [^{18}F]HA-labelling in dependence on the labelling pH evaluated via ultra filtration (cut off 3000 D);*
left: 1st ultra filtration / right: 2nd ultra filtration
> 3000 D – back / < 3000 D - white
pH 5.5 – pH 8 phosphate buffer (PB), pH 8 - pH 10 borate buffer (BB).

A first evaluation of the behaviour of the ^{18}F-labelled HA was performed with the HPSEC. The [^{18}F]HA radiochromatograms (Figure 2/left) are put side by side with known [^{131}I]HA radiochromatograms and their UV absorption signal (Figure 2/right).[8,9,10] Obviously, we contained a larger peak width for [^{18}F]HA then for [^{131}I]HA. A slight change in the experimental set-up causes these differences. The size of the sample loop in front of the detector had to be increased to achieve a better minimal detectable activity of ^{18}F. Neglecting this technically caused differences, we found similar molecular weight distributions of the investigated HA.

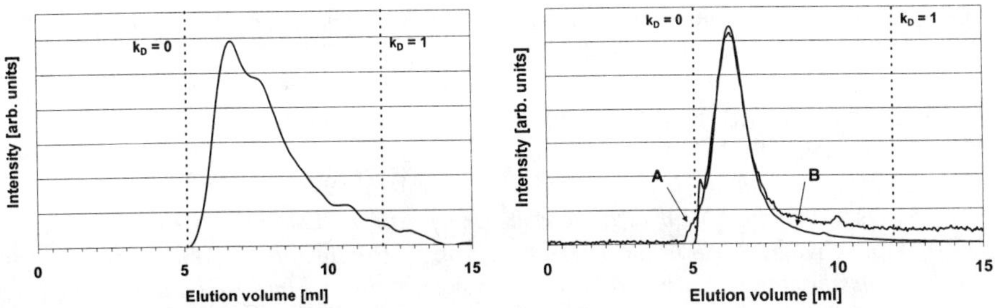

Figure 2 *Comparison of HPSEC chromatograms; left: [^{18}F]HA; right: [^{131}I]HA (A) and the UV absorption signal (B)*

Further evaluation shows the applicability of the ^{18}F-labelled HA in IEF investigations (Figure 3). The two-dimensional image of the [^{18}F]HA distribution is shown at the left side. Exemplarily the electropherograms of the at pH 9.5 and pH 10 labelled [^{18}F]HA are shown in comparison to the electropherogram of [^{18}F]4-fluoro-benzenediazonium ions and its decomposition products (not characterized) on the right side. The found isoelectric points are characteristic for these humic acids.[6]

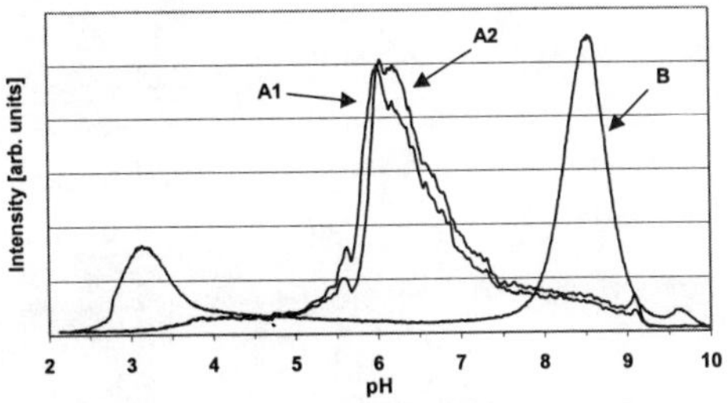

Figure 3 *Radioelectropherograms; A1: [^{18}F]HA pH 9.5; A2: [^{18}F]HA pH 10; B: [^{18}F]4-fluoro-benzenediazonium ion signal*

4 CONCLUSION

The reaction of the ^{18}F-labelled diazonium ion with HA is a versatile reaction for the production of ^{18}F-labelled HA in good radiochemical yield. The advantage of this method are the smooth reaction course and the aqueous reaction conditions. This preparation complete other methods for a covalent labelling of the carbon backbone of the HA. The extremely low detection limits (online/offline) of the radiomarkers and their easy measurement offers alternatives for experiments focused on natural concentration. Especially investigations of inter and intra molecular interactions and competition reactions of HA are feasible and much more comfortable, respectively. Furthermore, the possible use of PET offers new capabilities of representation the tridimensional distribution of HA in geomatrices.

References

1 G.R. Aiken, P. MacCarthy, R.L. Malcolm and R.S. Swift, *Humic Substances in Soil, Sediment, and Water*, Wiley, New York, 1985.

2 G. Davies and E.A. Ghabbour, *Humic Substances: Structures, Properties and Uses*, Royal Society of Chemistry, Cambridge, 1998.
3 M.H.B. Hayes and R.S. Swift, in *The chemistry of soil constitutants*, ed. D.J. Greenland and M.H.B. Hayes, Wiley, Chichester, UK, 1978, p.179.
4 N. Senesi and T. M. Miano, *Humic Substances in the Global Environment: Implications for Human Health*, Elsevier, Amsterdam, 1994.
5 F. Stevenson, *Humus Chemistry: Genesis, Composition, Reactions*, Wiley, 1982.
6 U. Gottschalch, *Charakterisierung von Huminstoffen mit Hilfe der Ionenfokussierenden Elektrophorese sowie die Untersuchung komplexer Wechselwirkungen zwischen Huminstoffen mit Aluminium-, Mangan- und Uranionen*, Dissertation, Universität Leipzig, 1997.
7 E.M. Thurman and R.L. Malcolm, *Environ. Sci. Technol.*, 1981, **15,** 463.
8 P. Warwick, I. Mason, A. Hall and R. Holmes, *Radiochim. Acta*, 1994, **66/67**, 427.
9 P. Warwick, L. Carlsen, A. Randall and P. Lassen, *Chem. Ecol.*, 1993, **8**, 65.
10 K. Franke, D. Rößler and H. Kupsch, *Proceedings 20th Anniversary Conference,* International Humic Substances Society, Boston, USA, 21.07.2002 - 26.07.2002, p. 22.

CHEMICAL AND RADIOLOGICAL CHARACTERISATION OF SANTOS ESTUARY SEDIMENTS

P.S.C. Silva[1], B.P. Mazilli[1] and D.I.T. Favaro [2]

[1]Departamento de Radioproteção Ambiental
[2]Laboratório de Análise Por Ativação
Instituto de Pesquisas Energéticas e Nucleares
Caixa Postal 11049, São Paulo, Brasil

1 INTRODUCTION

Estuaries may accumulate material transported by rivers, including substances of natural or industrial origin, such as major, minor or trace elements. The fate of these substances is a major concern in environment studies. It is also known that phosphate deposits are generally characterised by enhanced radionuclides concentration compared to average natural levels. Mining and processing of phosphate ore redistribute radionuclides into final products, by-products and solid waste (phosphogypsum). Santos estuary, located in São Paulo State, Southwest Brazil, and comprising three main counties (Santos, São Vicente and Cubatão), is considered one of the most important industrial regions of Brazil. The phosphate fertiliser complex located in Santos estuary has produced approximately 69 million tonnes of phosphogypsum, which is stockpiled and presents a potential threat to the environment[1,2]. Sediment samples from Santos estuary were analysed by neutron activation analysis and gamma spectrometry for elements As, Ba, Br, Ce, Co, Cr, Cs, Eu, Fe, Hf, K, La, Lu, Na, Nd, Rb, Sb, Sc, Se, Sm, Ta, Tb, Th, U, Yb and Zr, and radionuclides ^{226}Ra, ^{228}Ra, ^{210}Pb and ^{40}K. The main aim of this characterisation is to identify contaminated areas due to industrial activities. Statistical analysis of the results showed that samples close to the industries present anomalous activity concentrations, above average values of the region, as well as, relatively high concentrations for rare earth elements and other trace metals. The correlation analysis showed that activity concentration and trace elements concentration correlate well with total organic carbon content. The radiological environmental impact is mainly due to the release of ^{226}Ra, ^{228}Ra, Th isotopes and ^{210}Pb from the phosphogypsum piles.

2 METHODOLOGY

Eighteen sediment samples were collected by Companhia Brasileira de Saneamento Básico (CETESB), in points showed in figure 1. All samples were analysed by neutron activation analysis (NAA), for determination of the following elements: As, Ba, Br, Ce, Co, Cr, Cs, Eu, Fe, Hf, K, La, Lu, Na, Nd, Rb, Sb, Sc, Se, Sm, Ta, Tb, Th, U, Yb and Zr. The elements' determination was made by irradiation of approximately 150mg of each sample, during 16 hours at a neutron flux of 10^{12} n.cm^{-2}s^{-1}, at Instituto de Pesquisas Energéticas e Nucleares (IPEN) research reactor IEA-R1. The induced radioactivity was measured with a

Ge-hyperpure detector, Intertechnique, with 2.1 keV resolution for the 1332 keV ^{60}Co photo peak. The concentration of the analysed elements was determined by comparing activities obtained in the sediment samples with standard materials Buffalo River Sediment (NIST-2704) and Soil-7 (IAEA).

Among the collected samples, 13 were analysed by gamma spectrometry for radionuclides determination (^{226}Ra, ^{228}Ra, ^{210}Pb and ^{40}K). For this determination samples were sealed in polyethylene containers for about four weeks prior to the measurement, in order to ensure that equilibrium has been reached between ^{226}Ra and its decay products of short half-life, ^{214}Bi and ^{214}Pb. The ^{228}Ra content of the samples was determined by measurement of ^{228}Ac photo peaks and the concentrations of ^{210}Pb and ^{40}K were carried out by measuring the activity of their own photo peaks. Samples were measured by using a Ge-hyperpure detector, EGNC 15-190-R from Eurisys, with 15% efficiency, during 60.000s. The detector was calibrated using natural soil and rock spiked with radionuclides certified by Amersham. The system background was obtained by measuring the same sample counting geometry for 300.000s. The gamma spectra obtained were analysed by using WinnerGamma[3] program.

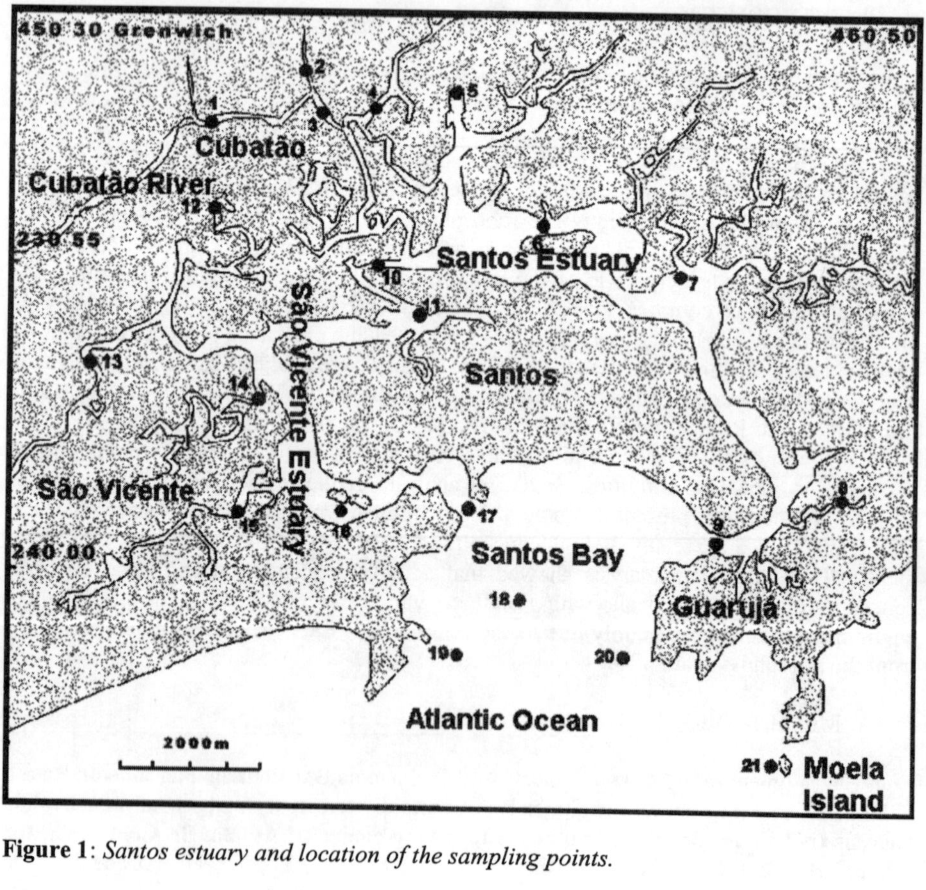

Figure 1: *Santos estuary and location of the sampling points.*

Self-absorption correction for ^{210}Pb was applied since the attenuation for its low energy gamma rays is highly dependent upon sample composition. The approach used was modified from that suggested by Cutshall *et al*.[4]

As organic matter is a good absorber of metals and radionuclides, its amount was also determined in the samples by heating at 360ºC – 370ºC for 2 hours[5].

All sediment samples studied were prepared by drying at a temperature of 110°C to constant mass, ground to a grain-size of less than 250μm for gamma spectrometry, 105 μm for NAA and finally homogenised.

3 RESULTS AND CONCLUSIONS

Table 1 shows the results obtained in the samples measurement by NAA, the mean values, the relative standard deviation and the range of values. It can be seen that the elements Br and U presented higher variations and Ba the smallest one, although for the majority of the analysed elements the variations obtained were bigger than 50%, reflecting the complexity of this kind of environment.

Table 1:*Mean value, relative standard deviation and range obtained for sediments by NAA.*

	mean (μg/g)	Rsd (%)	Range (μg/g)
As	6	61	2 - 11
Ba	329	27	116 - 488
Br	50	84	2 - 133
Ce	64	56	8 - 136
Co	6	50	1,3 - 12
Cr	35	55	7 - 78
Cs	3	56	0,5 - 8
Eu	0,9	39	0,2 - 1,5
Fe(%)	2	49	0,4 - 4
Hf	6	62	1,2 - 13
K(%)	1,2	40	0,4 - 1,8
La	28	61	6 - 65
Lu	0,3	54	0,04 - 0,6
Na	10443	78	1674 - 33037
Nd	31	61	4 - 73
Rb	69	51	17 - 180
Sb	0,4	52	0,2 - 0,8
Sc	7	54	0,8 - 13
Se	0,5	60	0,2 - 1,4
Sm	5	65	1,0 - 12
Ta	0,8	44	0,1 - 2
Tb	0,4	57	0,1 - 0,9
Th	10	63	1 - 24
U	3	86	0,9 - 14
Yb	2	54	0,5 - 4
Zr	188	50	43 - 385

Elbaz-Poulichet and Dupuy[6] showed that phosphate deposits contribute to the enhancement of rare earth elements in the environment. In a previous paper, the authors presented the concentration of rare earth elements in the same sediment samples[7]. An increase in rare earth elements was found in points 1, 2 and 3, probably due to their release together with Th and Ra isotopes from phosphogypsum stockpiled in the surrounding area.

The correlation coefficient was also determined indicating that: a) the rare earth elements are strongly correlated between themselves and also with trace metals, Fe, Zr and Th; b) the elements U and Th are well correlated between themselves, Th is strongly correlated with all rare earth elements, trace metals and Fe; c) U is well correlated only with Cs and Ta, among the trace metals.

Fe is generally well correlated with all trace metals. Therefore, it can be used to normalise experimental results and to identify possible enrichments among the analysed elements. Since aluminium was not detected by NAA, iron was used to normalise the studied elements.

Although there are a lot of possible external sources of iron, the results obtained by using Fe as a normaliser showed good agreement for Co, Ce, Cr, Cs, Eu, La, Lu, Nd, Sc, Sm, Ta, Th, Yb and Zr. Slight enrichments were observed for As in points 17 and 19; for Se in points 7, 9, 16, and 17 and for U in points 12 and 13. An explanation for U enrichment may be the great concentration of organic matter.

Table 2 shows activity concentrations for U, Th, ^{226}Ra, ^{228}Ra, ^{210}Pb and ^{40}K and activity ratios ^{226}Ra/^{238}U and ^{228}Ra/^{232}Th.

Points 1, 2 and 3 presented concentrations for ^{226}Ra and ^{228}Ra above the mean value of the region, probably due to releases from stockpiled phosphogypsum originated in the phosphoric acid industries. This contamination is supported by the results obtained for the activity ratio ^{226}Ra/^{238}U and ^{228}Ra/^{232}Th in the same points, which were higher than usually expected due to the solubility of radium isotopes in estuarine systems[8, 9]. This enrichment can also be seen when normalising the activity concentrations of Ra isotopes with iron concentration. The correlation coefficient analysis shows that Ra isotopes have a strong correlation with rare earth elements, Fe, Zr and trace metals.

For U distribution, points located in the vicinity of the industrial area presented mean activity concentration of 117Bk/kg, whereas more distant points (9, 16, 17, 18, 19 and 20) presented concentrations below 50Bq/kg, with mean value of 32Bq/kg.

The results obtained for the correlation coefficients between Th and Ra-isotopes, U and Ra-isotopes and organic matter and Ra-isotopes show that points 1, 2 and 3 present anomalous concentrations of Ra-isotopes and Th. The correlation analysis for ^{210}Pb showed that this element correlates with rare earth elements, trace metals and Fe and it seems to be enriched just in the areas with high organic matter content. ^{40}K showed the same trend as ^{210}Pb.

4 CONCLUSIONS

The results obtained reflect the complexity of the region, however good correlation was observed between Fe concentration and Co, Ce, Cr, Cs, Eu, La, Lu, Nd, Sc, Sm, Ta, Th, Yb, Zr, and Ra-isotopes content. These results corroborate the already observed enrichment of ^{226}Ra, ^{228}Ra and Th in the surroundings of the phosphoric acid industries located in the region, both by the correlation analysis and by the iron normalisation. It can also be seen, that the organic matter content, in the estuary sediment, is a powerful trap for nuclides U and ^{210}Pb.

Table 2: *Activity concentrations (Bq/kg) for U, Th, ^{226}Ra, ^{228}Ra, ^{210}Pb and ^{40}K and activity ratios $^{226}Ra/^{238}U$ and $^{228}Ra/^{232}Th$.*

	U	Th	^{226}Ra	^{228}Ra	^{210}Pb	^{40}K	^{226}Ra/^{238}U	^{228}Ra/^{232}Th
A 1	69 ±6	79 ±4	44 ±5	77 ±6	49 ±9	379 ±30	1,3 ±0,1	0,98 ±0,05
A 2	109 ±8	96 ±5	43 ±5	73 ±6	44 ±9	392 ±32	0,81 ±0,07	0,76 ±0,04
A 3	86 ±7	60 ±3	40 ±5	59 ±6	41 ±8	546 ±39	0,96 ±0,09	0,99 ±0,06
A 6	87 1±8	37 ±2	- -	- -	- -	- -	- -	- -
A 7	48 ±13	15 ±1	- -	- -	- -	- -	- -	- -
A 8	95 ±27	44 ±4	19 ±4	25 ±5	60 ±11	430 ±38	0,4 ±0,2	0,6 ±0,1
A 9	43 ±4	32 ±2	13 ±±3	15 ±3	37 ±7	341 ±29	0,6 ±0,1	0,47 ±0,06
A 10	103 ±9	44 ±2	20 ±4	29 ±4	47 ±9	418 ±34	0,40 ±0,07	0,66 ±0,05
A 12	348 ±27	77 ±4	31 ±5	36 ±5	114 ±21	486 ±39	0,18 ±0,04	0,47 ±0,04
A 13	145 ±20	41 ±2	20 ±4	24 ±4	47 ±10	428 ±31	0,28 ±0,08	0,58 ±0,06
A 14	116 ±16	42 ±2	15 ±3	20 ±3	95 ±18	370 ±30	0,27 ±0,08	0,48 ±0,05
A 15	84 ±13	34 ±2	15 ±3	16 ±4	69 ±14	315 ±31	0,4 ±0,1	0,47 ±0,08
A 16	26 ±13	10,2 ±0,4	8 ±2	10 ±3	27 ±4	189 ±21	0,6 ±0,4	1,0 ±0,1
A 17	23 ±5	4,5 ±0,2	6 ±2	8 ±3	13 ±0	186 ±20	0,5 ±0,2	1,8 ±0,2
A 18	43 ±10	29 ±1	- -	- -	- -	- -	- -	- -
A 19	26 ±8	11,0 ±0,4	- -	- -	- -	- -	- -	- -
A 20	31 ±8	21 ±1	- -	- -	- -	- -	- -	- -
A 21	74 ±18	41 ±2	13 ±5	23 ±6	73 ±8	377 ±40	0,4 ±0,2	0,6 ±0,1

- not measured yet

ACKNOWLEDGMENTS

This work was supported by Fundação de Amparo à Pesquisa do Estado de São Paulo – FAPESP, under fellowship contract 99/06952-4 and by Conselho Nacional de Desenvolvimento Científico e Tecnológico – CNPq, grant 300835/95-7. The samples analysed were provided by CETESB (Companhia de Tecnologia de Saneamento Ambiental).

BIBLIOGRAPHY

1 G.J. McNabb, J.A. Kirk and J.L. Thompson, Radionuclides from phosphate-ore-processing plants: the environmental impact after 30 years of operation. *Health Physics*, 1979, **37**, 585.

2 R. Periáñez, R. and M. Garcia-León, Ra-isotopes around a phosphate fertilizer complex in an estuarine system at the southwest of Spain. *Jour. Of Radioanal. And nucl. Chem.*, 1993, **172**, 71.

3 InterWinner Spectroscopy Program Family Version 4.1 (release 1st of June 1998) Eurisys Mesures.

4 N. H. Custhall, I.L. Larser and C.R. Olsen, Direct analysis of ^{210}Pb in sediment samples: self-absorption corrections. *Nuclear Instruments and methods*, 1983, **206**, 309.

5 M. Kralik, A rapid procedure for environmental sampling and evaluation of polluted sediments. Appl. Geochem, 1999, **14**, 807.

6 F. Elbaz-Poulichet and C. Dupuy, Behaviour of rare earth elements at the freshwater – seawater interface of two acid mine rivers: the Tinto and Odiel (Andalucia, Spain). *Appl. Geochem*. 1999, **14**, 1063.

7. P.S.C. Silva, B.P. Mazzilli, and D.I.T. Favaro, Estudo dos Elementos Terras Raras em Sedimentos do Estuário de Santos e São Vicente. Proceedings of VIII Congresso Brasileiro de Geoquímica, Curitiba, 2001.

8 M. Gascoyne, 'Geochemistry of the Actinides' in Ivanovich, M and Harmon, R. S. *Uranium Séries Disequilibrium: Applications to Environmental Problems*. Oxford, Clarendom Press, 1982, Chapter 2, pp. 33 – 55.

9 M.S. Moore, Radium isotopes in the Chesapeake Bay. *Estuar. Coast Shelf. Sci*. 1981, **12**, 713.

THERMOCHROMATOGRAPHIC STUDIES ON THE SEPARATION OF PLUTONIUM FROM SOIL SAMPLES

U.G. Schneider[1], U. Krähenbühl[1] and S. Röllin[2]

[1] Laboratory for Radio- and Environmental Chemistry, Department of Chemistry and Biochemistry, University of Bern, Freiestrasse 3, CH-3012 Bern, Switzerland
[2] Spiez Laboratory, CH-3700 Spiez, Switzerland

1 INTRODUCTION

Thermochromatographic separation of Pu from soil samples and measurement by ICP-MS becomes a fast alternative to other techniques of determination.

Experiments with a temperature gradient and $N_2/SOCl_2$ as reactive gas resulted in incomplete separation of Pu from U.[1] Thus the formation of UH^+ in the plasma of the main isotope ^{238}U leads to masking the isotope ^{239}Pu. Therefore, an additional U extraction step from the solution was necessary.

The intention of these studies is to optimise the separation of Pu from the naturally occurring U in the soil. In addition to $N_2/SOCl_2$, gas mixtures of $Cl_2/SOCl_2$ and of Cl_2/CCl_4 were tested with the idea to form more volatile compounds like UCl_5 and UCl_6, thus transporting more U over the Pu adsorption position to far lower temperatures.

2 EXPERIMENTAL

2.1 Preparation of the Chromatography Column and the Soil Sample

The chromatography column (Figure 1, *4*) consists of a quartz glass tube (10 mm i.d., length 60 cm) with a filling of quartz glass sand (0,3 – 0,6 mm) held in position by quartz glass wool. This column is placed into a vertical tube furnace (*3*) (length 50 cm) and heated up in a stream of Cl_2 for pre-chlorination. After one hour at 950°C this furnace is switched off and the column cools down. At around 50°C the stream of Cl_2 is stopped and the column is flushed for some minutes with Ar dried by molecular sieve 5 Å (*1*).

The soil sample is spiked, mixed with 30% H_2O_2 and 36% HCl, respectively and evaporated to dryness (150°C). The pre-chlorinated chromatography column is moved out of the furnace and the pretreated sample material filled into sample position (*5*).

2.2 Thermochromatography

The column is fixed into the chromatography setup (length 36 cm), consisting of two vertical tube furnaces (*6*,*7*). Experiments are carried out with a reactive gas flow of 40 ml/min. Chlorine (or Nitrogen), after being saturated at room temperature with $SOCl_2$

or CCl_4 (*2*), passes through the column from the bottom to the top. Within one hour the main furnace (*6*) is heated up to 1120°C in the sample position. Both $SOCl_2$ and CCl_4 as reactive gas components decompose by a pyrolysis reaction, forming free Cl_2 and a wide range of sulphur chlorides (+ SO_2) or carbon chlorides, respectively.[2,3] The oxides of the sample react with this gas mixture, forming gaseous SO_2 or $COCl_2$ and chlorides, which are volatilised, transported upwards and depose on the quartz glass filling. The upper furnace (*7*) is a gradient furnace with a logarithmic heating coil. Within one hour, this furnace is slowly heated up and a stable temperature gradient is achieved. Then the chromatography is run for two more hours.

The different metal chlorides are transported to different positions in the column, depending on their amount and their thermodynamic properties. The deposition mechanism changes abruptly at about one monolayer coverage of the surface.[4] Element chlorides in trace amounts which do not form a monolayer interact with the surface and the transport depends strongly on their adsorption properties. For main elements of soil samples occurring in macro amounts, the deposition process is predominantly a desublimation, depending barely on the adsorption of the chloride on a surface.

Very volatile compounds can be collected with a Liebig condenser (*8*), which is put at the top of the column. It is filled with quartz glass sand like the column and cooled with 2-propanol kept at –30°C by dry ice. The reactive gas is neutralised with 5 M NaOH (*9*), where a magnetic stir bar with an upright spike cleans solid reaction products from the end of the exhaust hose, thus preventing clogging.

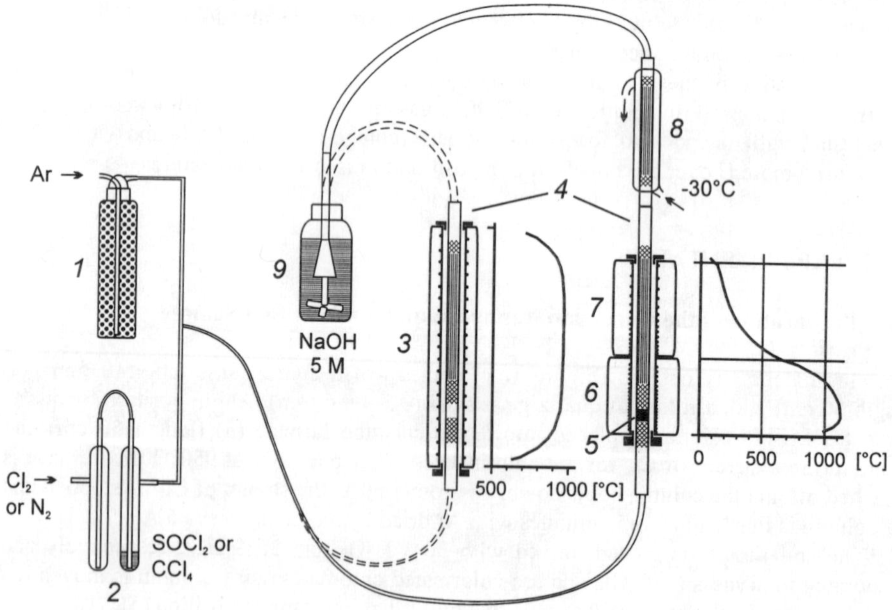

Figure 1 *experimental setup: (1) molecular sieve 5Å, (2) saturation with $SOCl_2$ or CCl_4, (3) furnace for pre-chlorination, (4) filled quartz glass tube, (5) sample, (6) main furnace, (7) gradient furnace, (8) filled Liebig condenser, (9) 5 M NaOH*

After the experiment a fine glass tube connected with a vacuum pump is used to sample well defined fractions of the quartz sand filling covered with chlorides out of the upright standing chromatography column. The fractions are leached with 3% HNO_3, filtered, diluted and measured with ICP-MS and ICP-OES, respectively.

For these studies all experiments were performed with different reactive gas mixtures on 1 g of IAEA-Soil-6 in order to keep the matrix chemistry constant and to get comparable element distributions.[5] The soil was spiked with ^{242}Pu. This isotope is not influenced by UH^+ formation during the ICP-MS measurements. For comparison, these experiments were also made with ^{242}Pu tracer only.

U, Pu and K have been measured quantitatively and some 50 elements semi-quantitatively for getting element distributions along the column.

3 RESULTS AND DISCUSSION

3.1 Experiments with ^{242}Pu Tracer

The thermochromatographic experiments with ^{242}Pu tracer and different reactive gas mixtures show only one dominant Pu peak (Figure 2). Considering the different sampling steps, it seems to be the same position in the temperature range around 300° - 360°C and accordingly the same compound adsorbed. The formation of the transported compound is independent of the gas composition. The volatility of $PuCl_3$ is too low for being transported at temperatures lower than 920°C.[6] In the presence of Cl_2, however, $PuCl_3$ is transformed to the more volatile $PuCl_4$, which exists only in the gas phase.

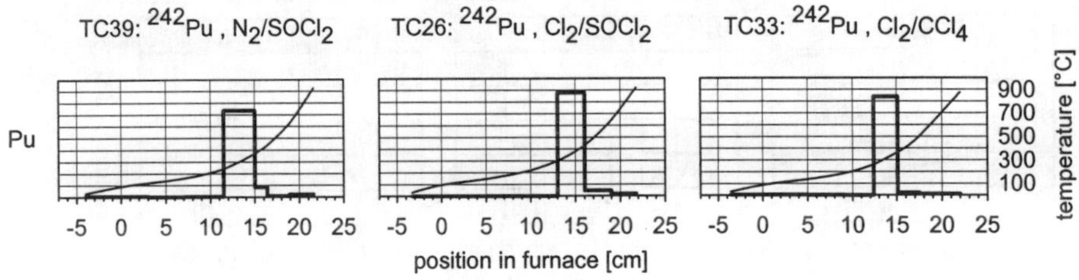

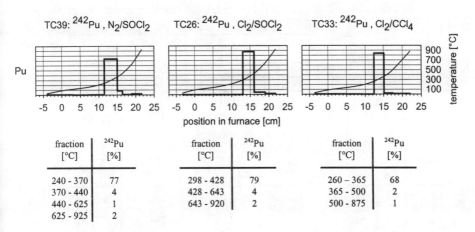

fraction [°C]	^{242}Pu [%]
240 - 370	77
370 - 440	4
440 - 625	1
625 - 925	2

fraction [°C]	^{242}Pu [%]
298 - 428	79
428 - 643	4
643 - 920	2

fraction [°C]	^{242}Pu [%]
260 – 365	68
365 - 500	2
500 - 875	1

Figure 2 . *Distributions of Pu obtained by thermochromatographic experiments with different reactive gas mixtures on ^{242}Pu tracer, ordinate arbitrary units, the tables below show yields of the Pu containing fractions*

3.2 Experiments with ^{242}Pu Tracer and IAEA-Soil-6

3.2.1 Behaviour of some Main Elements of Soils. The distributions of most of the main and trace elements show more than one peak. Sodium shows up to four peaks (Figure 3), depending on the number of sampling steps. Potassium shows two peaks, one around the melting temperature of KCl (772°C) and one around 400°C. The low-temperature K peak is associated with a small peak of Al, probably because of the formation of the double chloride $KAlCl_4$. The main part of Al is located from 170°C downward in presence of $SOCl_2$ in the gas mixture. Whithout $SOCl_2$ in the gas mixture, Al leaves the furnace and the deposition starts from 108°C downward. Most of the Fe and Mn are concentrated in single main peaks which do not shift the position by changing the gas mixture.

3.2.2 Behaviour of Pu and U. When using $N_2/SOCl_2$ as the reactive gas, the element distribution diagram (Figure 3, TC36) shows a major Pu peak around 400°C. The Cl_2/CCl_4 experiment (Figure 3, TC48) with its very narrow sampling steps resolves clearly two separate Pu deposition zones at 250°C – 292°C and 383°C – 584°C. The three Pu containing samples of the $Cl_2/SOCl_2$ experiment (Figure 3, TC30) form a broad distribution in a similar temperature range of 210°C – 680°C. In comparison with the analogous experiments with ^{242}Pu tracer only (Figure 2), the greater part of the Pu remains at the same or higher temperatures, while a smaller part is transported to even lower temperatures than without a soil matrix. An influence of the matrix elements of the soil sample is obvious.

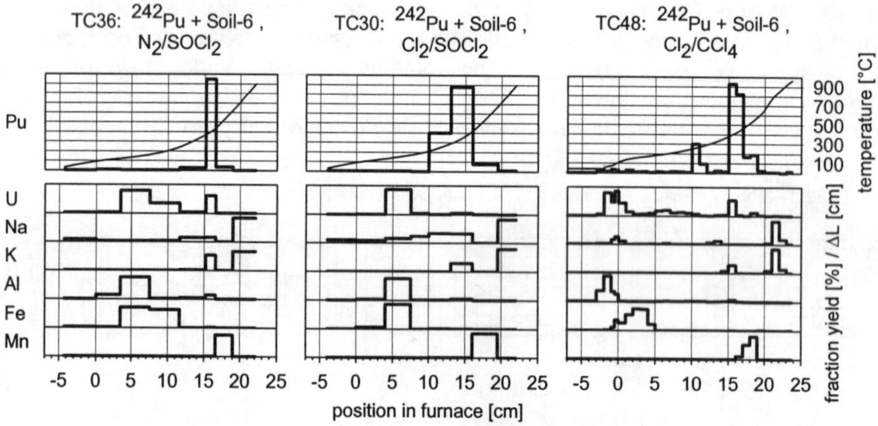

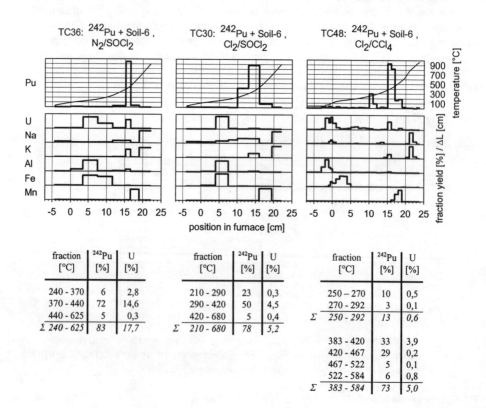

fraction [°C]	^{242}Pu [%]	U [%]
240 - 370	6	2,8
370 - 440	72	14,6
440 - 625	5	0,3
Σ 240 - 625	*83*	*17,7*

fraction [°C]	^{242}Pu [%]	U [%]
210 - 290	23	0,3
290 - 420	50	4,5
420 - 680	5	0,4
Σ 210 - 680	*78*	*5,2*

fraction [°C]	^{242}Pu [%]	U [%]
250 – 270	10	0,5
270 - 292	3	0,1
Σ 250 - 292	*13*	*0,6*
383 - 420	33	3,9
420 - 467	29	0,2
467 - 522	5	0,1
522 - 584	6	0,8
Σ 383 - 584	*73*	*5,0*

Figure 3 *Distributions of Pu, U and some matrix elements obtained by thermo-chromatographic experiments with different reactive gas mixtures on 1 g of IAEA-Soil-6 spiked with ^{242}Pu, ordinate arbitrary units, the tables below show the Pu containing fractions with yields for Pu and U*

In all the reactive gas experiments, the highest Pu content seems to be associated with $KAlCl_4$ around 400°C. In all cases a minor influence of the main peak of Mn ($MnCl_2$) around 450°C – 600°C can be seen, collecting some of the Pu at this temperature. So far, the Pu contents in the low temperature positions and between the $KAlCl_4$- and the $MnCl_2$ peak (TC48) cannot be related to any matrix element compound.

U and Al are the only elements, which show identical behaviour, when the reactive gas mixture is changed. With $N_2/SOCl_2$ and $Cl_2/SOCl_2$ (Figure 3, TC36 + TC30) the main part of U adsorbs at around 150°C, with Cl_2/CCl_4 (Figure 3, TC48) it is transported to lower than 135°C and 72% of the element are found in the Liebig condenser (not shown). In all experiments, small peaks of U are located at higher temperatures, especially at the $KAlCl_4$ position and less intense at the $MnCl_2$ position.

As expected, the separation of Pu from U is deteriorated with $N_2/SOCl_2$. The use of Cl_2 leads to a better separation at the main peak of Pu around 400°C ($KAlCl_4$). In $KAlCl_4$- or $MnCl_2$ position, however, both Pu and U are present in some quantities. Far better separations were achieved for the Pu fractions, which could not be associated with any matrix element. Almost no U is present in these positions. This is the further transported Pu at 210°C – 290°C (TC30) and 250°C – 292°C (TC48). The fraction found in the separation column TC48 between $KAlCl_4$ and $MnCl_2$ at 420°C – 467°C is the most interesting, since it contains 29% of the Pu and 0,2% of the U.

4 CONCLUSIONS

In thermochromatographic experiments with different chlorinating reactive gas mixtures, pure Pu is transported independently of the gas mixture, probably as $PuCl_4$. In presence of soil, Pu is predominantly associated with $KAlCl_4$. Furthermore, with $MnCl_2$ two compounds containing Pu are recognised which are not identified so far. The U/Pu separation at the position of $KAlCl_4$ is better by using Cl_2. The best U/Pu separation is achieved in the fractions with the compounds not identified. Further experiments with better resolution are necessary to identify either specific Pu carriers or U carriers.

References

1 L. Wacker, U. Krähenbühl and B. Eichler, *Radiochimica Acta*, 2002, **90**, 133.
2 *Gmelins Handbuch der Anorganischen Chemie*, 8th Edn., Verlag Chemie GmbH, Weinheim/Bergstr., 1963, Vol. Schwefel B3, p. 1796.
3 *Gmelin Handbuch der Anorganischen Chemie*, 8th Edn., Springer Verlag, Berlin, Heidelberg, New York, 1974, Vol. Kohlenstoff D2, p. 219.
4 B. Eichler, F. Zude, W. Fan, N. Trautmann and G. Herrmann, *Radiochimica Acta*, 1993, **61**, 81.
5 L. Pszonicki, A.N. Hanna and O. Suschny, Report on the Intercomparison Run Soil-6, International Atomic Energy Agency, Vienna, 1984, Report IAEA/RL/111.
6 F. Dienstbach and K. Bächmann, *Analytical Chemistry*, 1980, **52**, 620.

Acknowledgements

Spiez Laboratory, CH-3700 Spiez, Switzerland for financial and analytical support

PARTITIONING OF PLUTONIUM ISOTOPES IN LAKE BOTTOM SEDIMENTS

R. Gvozdaitė, R. Druteikienė, B. Lukšienė, N. Tarasiuk, N. Špirkauskaitė

Nuclear and Environmental Radioactivity Research Laboratory of the Institute of Physics, A. Goštauto 12, 2600, Vilnius, Lithuania

1 INTRODUCTION

The investigations in the field of radionuclide behavior in the environment are of particular concern in several aspects: the first one is radioecological because of the necessity to evaluate the influence of ionizing radiation caused by radioelements on the animate organisms and among them on man and the consequences resulting from it. The second important aspect is geophysical when peculiarities of the distribution of radionuclides are applied to the investigations of geophysical processes occuring in the biosphere. The problem of the radionuclide behavior in the water ecosystems is in the center of interest of many radioecological studies all over the world[1-6]. Natural water basins are important both in recreative and nutrition aspects therefore the determined activity ratios of radionuclides and analysis of migration processes allow to define the present situation and to make prognosis on the possibilities of water basins selfcleaning and its rates. It is natural because lakes in many countries are one of the most important and easily accessible repositories of freshwater.

Plutonium deposition to the bottom sediments is one of the water basins selfcleaning mechanisms. Plutonium behavior in lakes ecosystem, the physico-chemical features of this radionuclide and their changes in time after they were introduced into water depends on the radionuclide oxidation state and hydrochemical parameters (pH, salinity, redox conditions, organic matter) as well as. It is evident that plutonium fixation in bottom sediments is not irreversible process, and its transfer back to the water layer is not clear yet.

2 MATERIALS AND METHODS

2.1 Characteristics of the lakes

3 shallow lakes from different parts of Lithuania, Fig. 1., were selected to study the accumulation abilities of ^{238}Pu, ^{239,240}Pu in their sediments. Lakes were selected paying attention to their different origin and some morphometric data of these lakes are presented in Table 1.

Limnoglacier type - Lake uvintas is very shallow and looks like a large marsh. Bottom sediments are of the peat and sapropelic type. A natural reservation created for birds protection Lake uvintas for many years was in the center of interest of many scientific institutions of Lithuania.

Figure 1 *Disposition of studied lakes: 1 – uvintas, 2 – Juodis, 3 – Asavas-Asavėlis*

Table 1 *Morphometric data of the lakes selected for the radioecological study*

Lake	Area, km^2	Drainage basin, km^2	Hydraulic retention time, year	Precipitation, mm/year	Max. depth, m	Mean depth, m
Asavas-Asavėlis	2,23		0,20	630	6,7	4,20
Juodis	0,10	3,5	0,3-0,9	570	3,5	2,00
uvintas	10,30	345	0,11	600	2,5	0,67

A groove type - Lake Juodis was created by the melting of separate pieces of ice left from the glacier. Lake Juodis is located 17 km to the north-east from Vilnius city, in a woody region. It is a small running lake to a lake chain connected by a brook inflowing to the Neris River. Lake Juodis consists of two parts. The southern part of the lake is wider and deeper (~3,5 m); the northern one is more shallow (1-1,5 m). The accumulation zone consists of sapropelic type sediments.

Lake Asavas-Asavėlis is located in the upper reaches of the Šventoji River at 20 km distance from the Ignalina NPP. It consists of two parts. The largest part (Lake Asavas) is connected with the smaller one (Lake Asavėlis) by a shallow strait (~2.0 m depth). The woodland adjoining the strait is very wet and looks like a swamp. The accumulation zone consists of sapropelic type sediments.

The Pu content was measured in the upper sediment layer of the 20 cm thickness, slicing it into ~ 2 cm horizons. Sediment samples in Lake Asavėlis were taken seasonally. They

were analyzed for vertical profiles of $^{239,240}Pu$ and ^{238}Pu specific activities. Pu isotopes physico-chemical forms were studied in the upper sediment layer (~ 2 cm) of this lake. Sediment samples in Lakes uvintas and Juodis were taken in winter and summer.
Temperature, oxygen and H_2S concentrations in the bottom water were measured in Lakes Asavėlis, Žuvintas and Juodis. Methane concentrations in the bottom water were additionally measured in Lakes Juodis and uvintas.

2.2 Plutonium radiochemical procedure and measurement

For determination of plutonium specific activity, a dry bottom sediment sample was heated at 700 °C 2 hours in a muffle furnace. After adding ^{242}Pu as a radiochemical yield tracer, the soil sample was digested repeatedly with 8 mol/L HNO_3. Pu was separated and purified due to radiochemical separation procedures by means of strong basic anion exchange resin DOWEX 1×8 and then electrodeposited on a stainless-steel disk from Na_2SO_4 \ H_2SO_4 electrolyte solution for 1 hour using a current density 0,6 $A \cdot cm^{-2}$. The alpha spectrometric measurement chain is composed of a Canberra PD detector (area-450 mm^2, resolution 17 keV (FWHM) at 4-6 keV) coupled to a SES-13 spectrometer. Alpha efficiency 25 %, the detection limit for a counting time of 86400 s is about 10^{-3} Bq of $^{239,240}Pu$.
The plutonium radiochemical calibration was carried out during the ITWG inter-laboratory comparison exercise "Plutonium containing material" organized by EURATOM.

2.3 Sequential extraction method.

Sequential extraction, as a tool to identify physico-chemical associations of plutonium isotopes with macroelements of matrix, which they penetrated, has been chosen. Selective extractants were chosen basically according to Tessier et al.[7] The principal scheme and chemical phases leached by extractants are presented in Table 2.

Table 2 *Scheme of Pu sequential extraction*

Form	Sequential extractant	Chemical phase
F_1	Double-distilled water	Soluble solids, trace elements in ionic phase and soluble complex compounds
F_2	1 mol/L ammonium acetate (pH7)	Exchangeable ions
F_3	30 % hydrogen peroxide, pH2 with HNO_3	Organics, sulfides
F_4	1 mol/L acetic acid	Carbonates, some Fe and Mn oxides
F_5	0,04 mol/L hydroxylamine chloride in 25% acetic acid (1:1)	Iron and manganese oxides (moderately reducible phase)
F_6	HNO_3 and HF	Residual fraction

The extraction procedures were performed at 60 °C (F_3, organics), 96 °C (F_5, oxides) and room temperature (F_1, F_2, F_4, F_6) by shaking the whole sample of sediments with selective reagent for a definite duration (from 1 to 12 hours) on a water bath shaker (type 357). The ratio of solids to the extractant was kept at 1:10 (100 g dry weight of sediments and 1 L of the extragent). The extractant was separated by filtration, and the residue from the previous step was washed twice with distilled water. After the filtration and washing steps, filters with residual solids were reintroduced into the reaction vessel for the next extraction. For

Pu analysis each fraction of sequential extraction was spiked with the ^{242}Pu yield-tracer and evaporated. The further radiochemical analysis and spectrometric measurement has been performed as mentioned above.

3 RESULTS AND DISCUSION

3.1 ^{239,240}Pu and ^{238}Pu specific activities in the bottom sediments of studied lakes

The seasonal partitioning of ^{239,240}Pu and ^{238}Pu in bottom sediments of Lake Asavėlis is shown in Figs 2 and 3. Data on vertical profiles of ^{239,240}Pu and ^{238}Pu activity concentrations in the upper layer of sediments have shown that specific activity varies from 1,5±0,2 Bq·kg^{-1} to 7,4±0,7 Bq·kg^{-1} for ^{239,240}Pu and from 0,07±0,02 Bq·kg^{-1} to 0,22±0,06 Bq·kg^{-1} for ^{238}Pu in winter. The range of activity concentration was obtained in 3,6±0,3-9,9±0,9 Bq·kg^{-1} for ^{239,240}Pu, 0,11±0,03-0,29±0,06 Bq·kg^{-1} for ^{238}Pu; 6,0±0,7-9,1±0,9 Bq·kg^{-1} for ^{239,240}Pu, 0,15±0,04-0,46±0,11 Bq·kg^{-1} for ^{238}Pu; 1,4±0,1-7,9±0,6 Bq·kg^{-1} for ^{239,240}Pu, 0,11±0,02-0,24±0,05 Bq·kg^{-1} for ^{238}Pu in spring, summer and autumn respectively. The ^{239,240}Pu and ^{238}Pu load in the upper sediment layer of the 10 cm thickness in winter came up to 58,6 Bq·m^{-2} and 1,7 Bq·m^{-2} respectively. The approximate load values for ^{239,240}Pu and ^{238}Pu (95,3 and 2,9 Bq·m^{-2}; 50,1 and 2,0 Bq·m^{-2}; 62,9 and 2,2 Bq·m^{-2} respectively) were obtained in spring, summer and autumn.

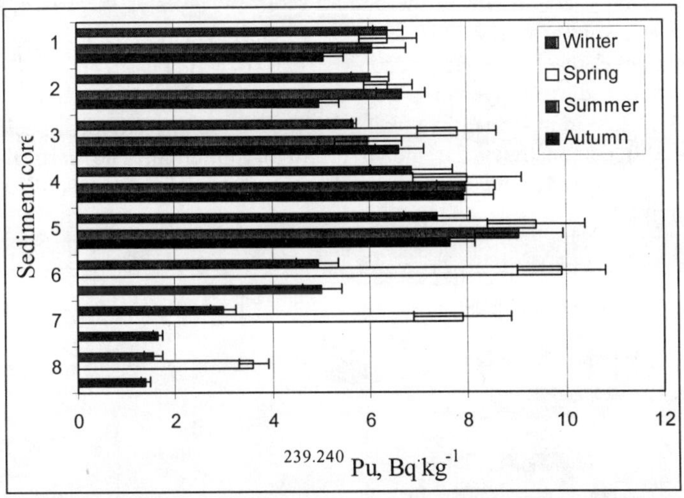

Figure 2 *Vertical profiles of the ^{239,240}Pu specific activity in the bottom sediments of Lake Asavėlis*

Ratios of ^{238}Pu and ^{239,240}Pu specific activities (0,02-0,09) in the sediments of Lake Asavėlis show that the plutonium components of the global origin are dominant in the Pu load.

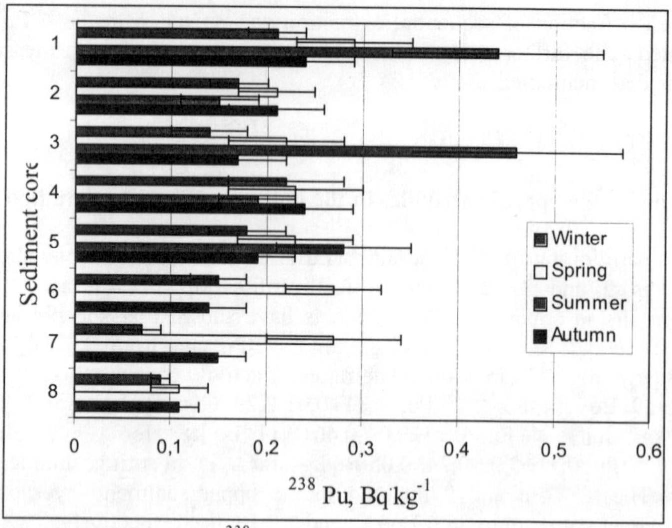

Figure 3 *Vertical profiles of the ^{238}Pu specific activity in the bottom sediments of Lake Asavėlis*

The specific activity of ^{239,240}Pu in the bottom sediment layer of the 10 cm thickness in Lake uvintas varied from 2,21±0,19 to 7,40±0,61 Bq·kg^{-1}. ^{238}Pu specific activity concentrations level reached 0,031±0,020 – 0,277±0,045 Bq·kg^{-1}. The ^{239,240}Pu and ^{238}Pu load in the upper sediment layer in Lake uvintas is significantly lower (^{239,240}Pu – 37,8 Bq·m^{-2}, ^{238}Pu – 1,3 Bq·m^{-2}) than in Lake Asavėlis. Plutonium isotopes specific activities ratio (0,01-0,05) demonstrates the global origin of pollution. The vertical profiles of ^{239,240}Pu and ^{238}Pu activity concentrations in the upper layer of sediments of Lake uvintas are shown in Fig 4.

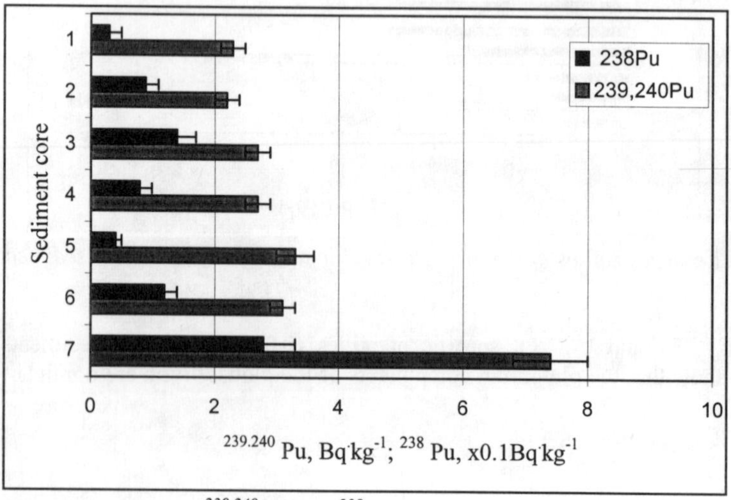

Figure 4 *Vertical profiles of ^{239,240}Pu and ^{238}Pu specific activities in the bottom sediments of Lake uvintas*

The analysis of $^{239,240}Pu$ and ^{238}Pu in sediments taken in Lake Juodis in winter 1999/2000 has shown that specific activity were markedly higher than in Lake Asavėlis and Žuvintas. Values of specific activity ranged from 8,01±0,56 Bq·kg^{-1}to 19,8±1,11 Bq·kg^{-1} for $^{239,240}Pu$ and from 0,100±0,030 Bq·kg^{-1} to 1,120±0,180 Bq·kg^{-1} for ^{238}Pu. Ratios of ^{238}Pu and $^{239,240}Pu$ specific activities (0,01-0,06) in the sediments of Lake Juodis show that the plutonium components are of the global origin. The vertical profiles of $^{239,240}Pu$ and ^{238}Pu activity concentrations in the upper layer of sediments of Lake Juodis are shown in Fig 5.

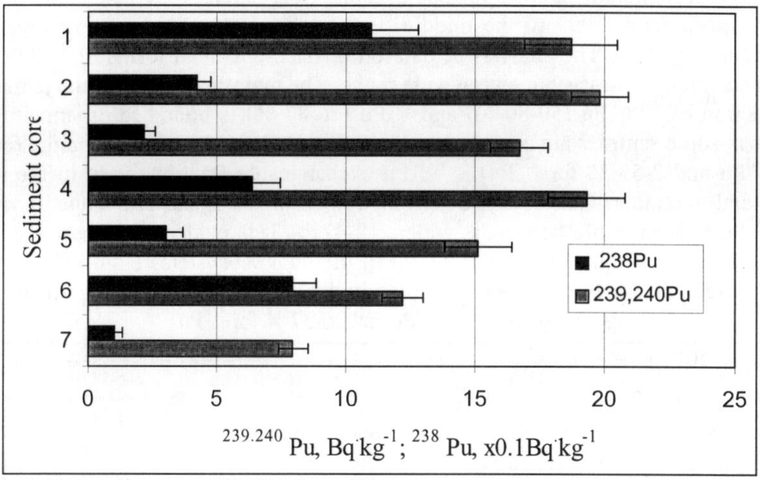

Figure 5 *Vertical profiles of $^{239,240}Pu$ and ^{238}Pu specific activity in the bottom sediments of Lake Juodis*

As can be seen (Figs 2-5), the vertical distribution of plutonium isotopes specific activities in the upper layer of bottom sediments is very complicated in all studied lakes. The analysis of the sediments samples indicated the fluctuation of plutonium activity in whole vertical profile. In consideration of plutonium origin (global fallout) in all studied lakes, it can be maintained that it is strongly fixed in upper sediments layer due to their chemical properties. Significant role for accumulation of Pu can be attributed to the presence of larger amount of organic matter in sediment layer. As mentioned above, bottom sediments are of the peat and sapropelic type in all 3 lakes. Analysis of the organic content of the lake sediments (24,9-70,8 %), show that this component may be very significant. Therefore uneven distribution of $^{239,240}Pu$ and ^{238}Pu throughout all sediment horizons may be related to the diffusive mixing processes. The effects related to the organics decomposition when methane and CO_2 have been produced and their release from the sediments followed by the surface sediment mixing.

Therefore, mobilities of radionuclides in sediments rich in organics must be influenced not only by their chemical properties but also must be related to the decomposition rate of the organic component of the sediments. Methane and CO_2 releases can turn surface sediments into a somewhat boiling layer with the bottom water filled in the liberated spaces. This mixing effect must also be proportional to the decomposition rate of organics. According to these considerations, different kind of vertical profiles of radionuclide specific activities must represent different rates of the organics decomposition as well as.

3.2 ^{239,240}Pu and ^{238}Pu physico-chemical forms in lake bottom sediments

The definition of radionuclide species in bottom sediments provides very valuable information on the radionuclide mobility in the aquatic ecosystems. Investigations were focused on the study of radionuclide physico-chemical forms in the sediment surface layers.

Physico-chemical forms of ^{238}Pu, ^{239,240}Pu were mainly studied in the surface sediments of Lake Asavėlis. Sediment samples for this analysis were taken in the lake center in spring, summer and autumn, 1999 in the middle of the lake. Data of these measurements are presented in Figs.6, 7. The analysis of data on physico-chemical forms of ^{238}Pu, ^{239,240}Pu shows their fractions to vary in a very wide range. The main feature of results is that a very large fraction of ^{239,240}Pu (50-80 %) and ^{238}Pu (26-37 %) is bound to organics. Residual fractions in some samples are comparatively small and also vary in a wide range (6,5-34 % for ^{239,240}Pu and 2-55 % for ^{238}Pu). ^{239,240}Pu exchangeable fractions vary in the range of 1,0-9 % and maximum values were determined in spring samples. The range of variations of the ^{238}Pu exchangeable fraction is wider – 8-42 %. It is evident that these data are the sign of the high mobility of ^{238}Pu, ^{239,240}Pu in the lake ecosystem. High values of ^{238}Pu, ^{239,240}Pu potentially mobile fractions are mainly due to their associations with the organic substances in sediments (50-80 % for ^{239,240}Pu and 26-37 % for ^{238}Pu).

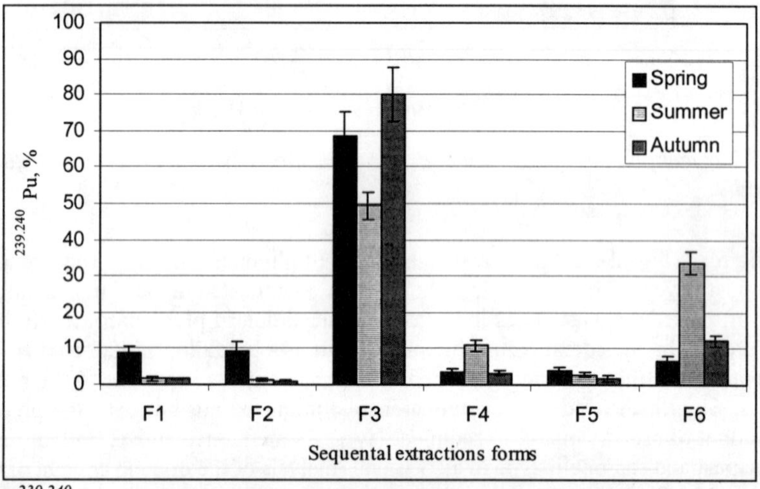

Figure 6 *$^{239.240}$Pu physico-chemical forms in the sediments of Lake Asavėlis*

The other feature of data is a rather small fraction of ^{239,240}Pu specific activity associated with oxides – 1,8-3,8 % (6,1-9,0 % for ^{238}Pu). A range of variations of the fraction associated with carbonates is equal to 3,3-11 % for ^{239,240}Pu and for ^{238}Pu (1,6-13 %).

Physico-chemical forms of ^{238}Pu, ^{239,240}Pu were also measured in the surface layer of the sediment sample taken in Lake uvintas in winter 1999/2000. Data are presented in Fig 8. The main features are rather low values of the exchangeable fraction (1,3 %) and the fraction associated with the organics (11,2 % for ^{239,240}Pu; 6,6 % for ^{238}Pu). Nevertheless the total exchangeable fraction is large (17,8 % ^{239,240}Pu and 38,2 % for ^{238}Pu) and, apparently, its rise is due to the processes of the redistribution of Pu in sediments and lake water during the decomposition of the organic substances. Besides a large fraction of ^{239,240}Pu associated with the oxides must be mentioned (15,3 % for ^{239,240}Pu; 15,8 % for

^{238}Pu). Potentially mobile fractions of ^{238}Pu and $^{239,240}Pu$ are equal to 48,7 and 46,1 %, respectively.

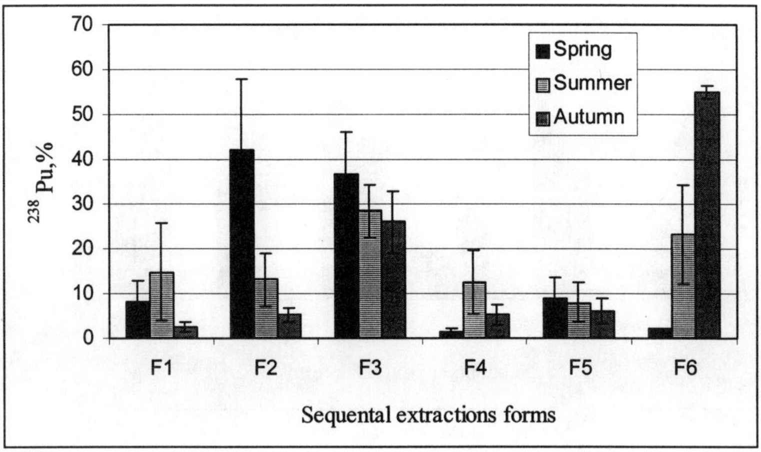

Figure 7 *^{238}Pu physico-chemical forms in the sediments of lake Asavėlis*

The results of the investigation of physico-chemical forms show a complexity of the ^{238}Pu, $^{239,240}Pu$ behavior in the lake ecosystem. A high mobility and a wide range of reduction states allow this radionuclide to participate in a large number of remobilization – fixation cycles in the lake ecosystem. Apparently often changing conditions during warm seasons impose some restrictions on the plutonium transfer from the lake ecosystem. A rise of the CO_2 concentration in lake water under the ice cover in winter and the respective decrease in pH promote the solubility of Pu and its removal from the lake ecosystem.

Data in Tables 3 show that pH of the pore water of sediments is, as a rule, lower than the lake surface water ones. The largest differences of pH were measured in the sediment and water samples taken in Lake Asavėlis (0,39).

It was discovered[8] that sometimes anoxic conditions in the bottom water in separate parts of Lake uvintas are formed at the end of long and cold winter.

Therefore investigations on the formation of anoxic conditions in Lakes Asavėlis, Juodis and uvintas were focused on measurement of oxygen, H_2O and methane concentrations in winter 2000/2001. Obtained data and additional parameters (surface and bottom water temperatures, the measurement depth) are presented in Table 3. Oxygen concentration in the middle of the southern part of Lake Juodis was very low (~2 mg·L^{-1}).

On the contrary, oxygen concentrations in the bottom water of Lake Žuvintas and Asavėlis were were significantly higher. The lack of oxygen compared with the saturation level did not reach 45 %. (For the simplicity in Table 3 saturation level of oxygen in the double-distilled water is used (in brackets)). Measurement sites in Lake uvintas were chosen in the southern part of the lake at rather large distances from the inflow of the Bambena River.

Measurements were carried out during a short floody period In Lake Asavėlis in winter 2001. A large puddle of thaw water was formed over the ice cover at the first sampling site (A). Therefore inverse temperature stratification was measured. Data on oxygen concentrations in the bottom water show, Table 3, that the highest oxygen consumption is in the central part of Lake Asavėlis, promoting the formation of the anoxic conditions at the some sediment depth near the sediment surface.

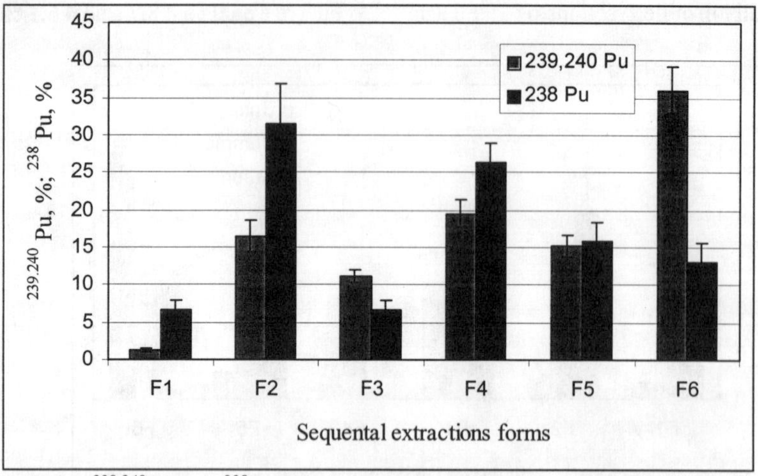

Figure 8 *$^{239.240}$Pu and ^{238}Pu physico-chemical forms in winter 1999/2000 in lake uvintas*

Table 3 *.Data of the measurements of the temperature stratification, pH, conductivity (C, μS cm^{-1}) and the concentrations of oxygen and methane in the bottom water in studied lakes in winter 2001. In brackets (for comparison) - saturated oxygen concentrations in the double-distilled water at the temperature of interest (nm – not measured)*

Lake name, date	Surface			Bottom					
	T, °C	pH	C	Depth, m	T, °C	pH	C	Oxygen, mg·L^{-1}	Methane, mL· L^{-1}
Asavėlis, February 8, 2001									
A (near the outflow)	3,0	8,00	300	3,5	2,4	7,61	372	10,24 (13,71)	nm
B (Lake center)	0	-		5,5	4,4			3,58 (13,00)	nm
C (strait)	0			1,5	1,6			12,48 (14,02)	nm
uvintas, February 14, 2001									
A (Southern part)	2,3	7,62	381	2,5	4,5	7,45	460	8,48 (12,96)	0,40
B (Southern part)	2,3			2,5	4,5			7,2 (12,96)	0,09
C (Southern part)	2,3			2,5	4,5			8,06 (12,96)	0,27
Juodis, February 20, 2001									
A (Center of the southern part)	-	7,85	269	3,5	5,4	7,65	430	2,24 (12,66)	5,17
B (Center of the northern part)	-	-	-	1,5	3,8	-	-	11,28 (13,20)	1,94

H_2S concentrations were at the undetectable level in the bottom water in all cases.

The other way of the organics decomposition under anoxic conditions in sediments – methane production – was recorded in Lakes uvintas and Juodis. It is known[9] that the surface layer of sediments of almost 20 cm thickness is most productive of methane in anoxic conditions.

Data on the methane concentrations in the bottom water in Lake uvintas show them to be rather low but not zero ones. It means that anoxic conditions in sediments are formed below some depth.

Significant methane concentrations were measured in the bottom water in the middle of the southern part of Lake Juodis. Apparently, it means that anoxic conditions in this case were formed at a smaller sediment depth than that in Lake uvintas.

As it was mentioned above, production of methane and the spontaneous release of its large amount from the sediments rich in organics must be related to the processes of the surface sediment mixing (methane mixing). It can be suggested that the methane mixing coefficient of the surface sediments must be proportional to the organics decomposition rate of the sediments.

4 CONCLUSIONS

$^{239,240}Pu$ and ^{238}Pu activity concentrations can reach their maximum values (~20 $Bq \cdot kg^{-1}$ and ~ 1,2 $Bq \cdot kg^{-1}$, respectively) in sediments of the sapropelic type. Pu radionuclides of the global origin are mainly responsible for the Pu load on the lake sediments in Lithuania.

Plutonium isotopes behavior in shallow lake ecosystems in Lithuania is tightly bound with the sediment organics features. Sediments of Lithuanian lakes are distinguished for the large organics content (25-71 %). Potentially mobile fraction of the plutonium load on the lake surface sediments can be very high (56-94 % for $^{239,240}Pu$). The processes of the radionuclide turnover in lake ecosystems to be tightly bound with processes of the sediment organics decomposition. Decomposition of sediment organics under anaerobic conditions is followed by methane production, its spontaneous release to the bottom water and, as a result, by the sediment surface layer mixing promoting high mobilities of the radionuclides in sediments.

ACKNOWLEDGEMENT

The work has been partially funded by IAEA grant (Research Project No. 10544/RO) in 1999/2001.

References

1 P. Spezzano, J. Hilton, J.P. Lishman, T.R. Carrick, *J. Eviron. Radioactivity*, 1993, **19**, 3, 213.
2 D. Hongve, I.A. Blakar, J.E.Brittain, *J. Eviron. Radioactivity*, 1995, **26**, 2, 157.
3 J.A. Robins, A.W.Jasinski, *J. Eviron. Radioactivity*, 1995, **27,** 1, 13.
4 I.I.Kryshev, G.A.Romanov, L.N.Isaeva, Yu.B.Kholina, *J. Eviron. Radioactivity*, 1997, **34**, 3, 223.
5 L.Hakanson, *J. Eviron. Radioactivity*, 1999, **44,** 1, 21.
6 J.T.Smith, A.V.Kudelsky, I.N.Ryabov, R.H.Hadderingh, *J. Eviron. Radioactivity*, 2000, **48**, 3, 359.
7 A. Tessier, P. Campbell, M. Bisson, *Anal. Chem.* 1979, **51**, 7, 844.

8. I. Baltrusaitiene, in *The uvintas reservation (The data of the complex expeditions in 1979-1985),* 1st Edn., ed. V. Kontrimavicius, Vilnius, Academia, 1993, ch. 1, p. 103(in Russian).
9. S.I.Kusnetsov, A.I.Saralov, T.N.Nazina in *Microbyological processes of the carbon and nitrogen turnover in lakes.),* 1st Edn., ed. S.I.Kusnetsov, Moscow, Nauka, 1985, ch. 3, p. 29 (in Russian).

RADIOANALYTICAL METHOD FOR THE DETERMINATION OF ^{90}Sr IN SOIL SAMPLES BY YTTRIUM SOLVENT EXTRACTION AND CERENKOV COUNTING

D. Pérez Sánchez [1,2], A. Martín Sánchez [1], M. R. García Sanz [3] and A. Fernández Timón [1].

[1] Departamento de Física, Universidad de Extremadura, 06071, Badajoz, Spain.
[2] Centro de Protección e Higiene de las Radiaciones, A.P.6195, La Habana, Cuba.
[3] CIEMAT, 28040, Madrid, Spain.

1 INTRODUCTION

The isotope ^{90}Sr is one of the most hazardous man-made radionuclides. It has been deposited in the environment as a result of the global fallout from nuclear weapons testing and accidental releases such as Chernobyl. This artificial radionuclide has high dosimetric importance because it accumulates in bone tissues and has a relatively long half-life ($T_{1/2}$=28.7 y). For these reasons it has become, together with ^{137}Cs, the principal radionuclide in environmental monitoring and radioecological research. Fast and reliable methods for the analysis of ^{90}Sr environmental samples must be developed.

The method commonly used for ^{90}Sr in soil is the classic fuming acid method which is hazardous, and the in-growth of ^{90}Y is time consuming in the analysis[1]. Therefore, it has been substituted by yttrium extraction with HDHEP. Among several published methods, yttrium extraction is becoming widespread for the routine analysis of samples. This method simplifies the chemical treatment and the measurement [2]. A simple sensitive method utilising Cerenkov radiation and a low level liquid scintillation analyser has been developed to estimate ^{90}Sr in soil samples through the measurement of ^{90}Y.

The aim of this work is to describe an accurate rapid and relatively simple procedure for monitoring and determining ^{90}Sr in soil samples. The procedure has been tested and validated with certified reference material and with an inter-comparison soil sample exercise.

2 EXPERIMENTAL

Figure 1 shows a flow chart of the proposed chemical procedure. All reagents used were of analytical grade. The standard reference solution of ^{90}Sr(^{90}Y) used for calibration was obtained from CIEMAT. A QUANTULUS 1220 Liquid Scintillation Spectrometer was used to measure the Cerenkov radiation.

2.1 Radiochemical Procedure

2.1.1 Sample Preparation and Treatment. Soil samples were dried at 105 ºC and ashed at 550 to 600 ºC. A mass of dried sample ranging between 10 and 30 g was taken for the analyses (the quantity depending on the expected activity concentration). The sample was transferred to a glass beaker and 50 mg of Sr^{2+} and 10 mg of Y^{2+} carriers were added.

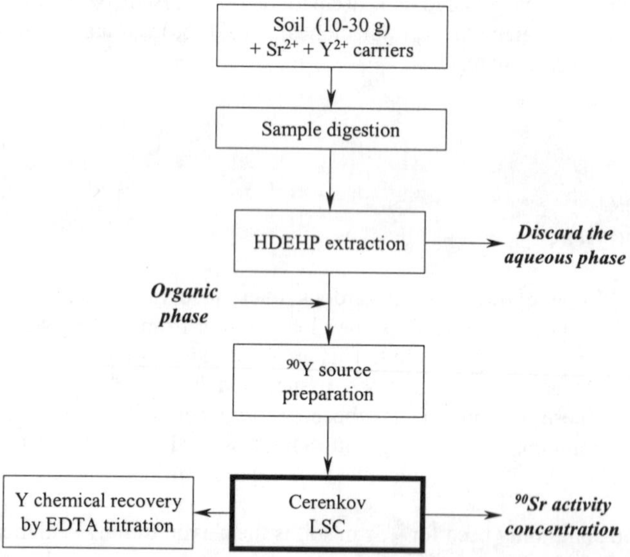

Figure 1 *Scheme of the ^{90}Sr analytical procedure used in soil samples.*

The samples were dissolved with 80 mL 8M HNO_3, boiling for 1 hour with magnetic stirring. The solution was cooled and 60 mL of aqua regia were added. The sample was digested by boiling again with magnetic stirring for 3 h. When necessary, further acid mixture was added during the digestion. Finally, 2 mL of 30% H_2O_2 were added and the suspension was heated to 90°C, cooled, and vacuum filtered through a glass fiber filter. The residue was throughly washed in 60% HNO_3 and then discarded. Finally the total aqueous part of the sample was heated to dryness and the residue dissolved in 40 mL of 0.01M HCl.

2.1.2 Extraction of ^{90}Y with HDEHP[3]. Before the extraction of yttrium, 2g of citric acid were added to the sample and the pH adjusted to 1.0 to 1.2 using 25% ammonium solution with stirring. The solution was transferred to a separating funnel and extracted with 50 mL of 10% HDEHP in toluene by shaking. The time was recorded because the ^{90}Y starts to decay from this moment in the organic phase. The aqueous phase was discarded. The organic phase was washed by shaking with 0.1M HCl several times and the phases were again left to separate. The yttrium was back separated into the aqueous phase by shaking with 50 mL 3M HNO_3. Nevertheless, as a check, the aqueous solution stripped in all the extraction processes should be retained, because after two weeks ^{90}Sr and ^{90}Y must reach equilibrium.

2.1.3. Source Preparation for Counting. The aqueous phase from the extraction was transferred to a centrifuge tube and concentrated ammonia solution was added to produce a pH of between 9-10. The solution was centrifuged and the supernatant discarded. Finally the hydroxide precipitate was dissolved with 1 mL of HNO_3 and the resulting solution transferred to a 20 mL polyethylene vial, washing and diluting with 20 mL of distilled water. Then the sample was Cerenkov counted in the liquid scintillation spectrometer. The chemical recovery was determined by complexometric tritration with EDTA [4].

2.2 Counting Procedure

2.2.1 Sample Counting. The vial was measured for 200 minutes using the Cerenkov protocol defined in the QUANTULUS 1220 liquid scintillation spectrometer. Detection of Cerenkov photons is similar to that for the ^{3}H beta emissions in a scintillation liquid. Hence, devices designed for measuring beta emissions by liquid scintillation can be used to measure Cerenkov radiation for a sample[5]. The counting window defined for ^{90}Y determinations was chosen by taking into consideration the quenching level of each sample. The ^{90}Y Cerenkov counting protocol was previously calibrated by a method of channel ratios using the standard solution of ^{90}Sr and ^{90}Y in equilibrium. A spectrum for Cerenkov ^{90}Y in a soil sample is shown in Fig. 2. The quenching parameter for each sample was determined to take a value for the corrected efficiency as described below. A second measurement of the sample can be made for the quality control of the chemical separation of yttrium by determining the ^{90}Y decay after 48 hours.

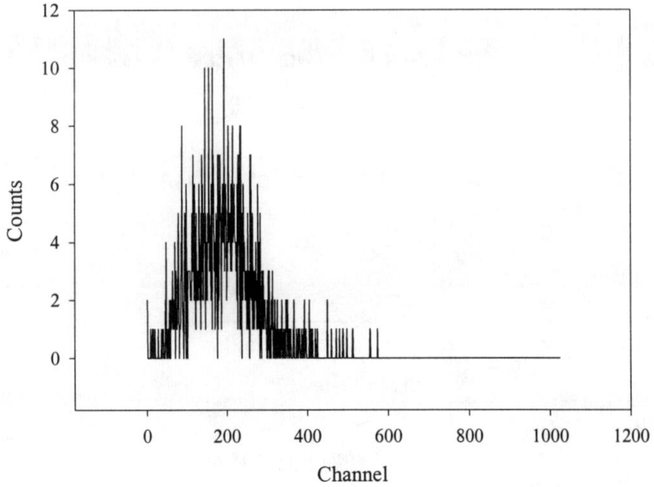

Figure 2 *A spectrum for Cerenkov counting of the soil intercomparison sample LMR-1.*

2.2.2 Calibration and Counting Conditions Optimization. The measurement conditions of the spectrometer, including type of vial and sample volume, were optimised before the measurements. Polyethylene vials gave a lower background than glass vials [6]. The greatest sample volume (20 mL) gave the lowest value for the Factor of Merit (Table 1). Nevertheless, the efficiency was similar for all the volumes, so that, in order to minimize colour effects and maximize the useful volume for the sample solution, a volume of 20 mL was chosen for all determinations.

Table 1 *Background, efficiency, Factor of Merit (FOM) and Minimum Detectable Activity (MDA), obtained for several volumes of sample.*

V (ml)	Background (cpm)	Efficiency	FOM (cpm^{-1})	MDA (10^{-3} Bq)
5	0.678 ± 0.08	0.56	0.46	2.44
10	0.769 ± 0.09	0.58	0.43	2.46
15	0.901 ± 0.10	0.58	0.38	2.59
20	0.972 ± 0.10	0.60	0.37	2.59

For the Cerenkov efficiency calibrations, a set of vials containing known activities of the ^{90}Sr-^{90}Y standard solution with different quenching levels (using $K_3[Fe(CN)_6]$ as colour quenching agent [7]) were prepared. The measurements of these standards gave the quenching colour factor and efficiency curves shown in Fig.3. Another set of vials without adding the standard solution was prepared in the same way for the background measurements.

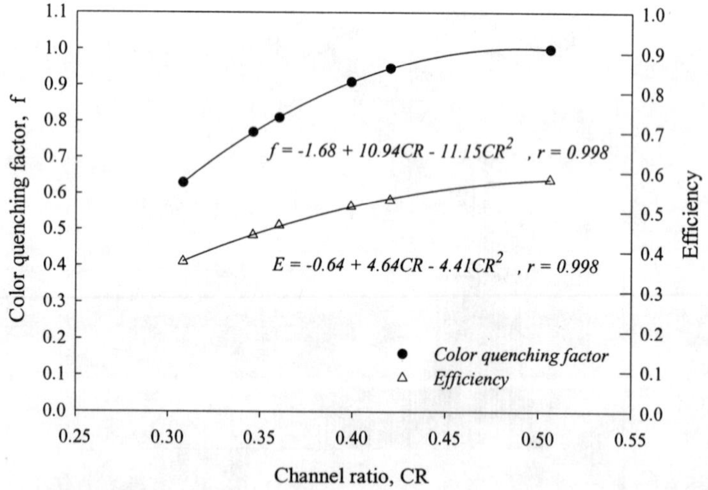

Figure 3 *Colour quenching factor and efficiency versus channel ratio (ratio between the counts in channels above that of the maximum and the total number of counts in the Cerenkov spectrum).*

After the determination of the ^{90}Y Cerenkov efficiency and background due to colour quenching level, the ^{90}Sr activity concentration in the sample was calculated according to

$$A(^{90}Sr) = \frac{(S_{Sample} - B_{Background}) \cdot \exp(\lambda t)}{W \cdot R_{QY} \cdot E(^{90}Y)},$$

where $A(^{90}Sr)$ is the specific activity of ^{90}Sr, S_{Sample}, the counting rate for the sample, $B_{Background}$, the background counting rate, $E(^{90}Y)$, the Cerenkov counting efficiency for ^{90}Y, R_{QY}, the chemical recovery of Y, and W, the weight of the soil sample. The factor $\exp(\lambda t)$ takes the decay into account.

3 RESULTS AND DISCUSSION

The procedure was tested with two IAEA reference material soil samples (IAEA-375, SOIL-6), and with the sample coming from an intercomparison exercise for the determination of radionuclides in soil samples (LMR-1) organized by the Consejo de Seguridad Nuclear (Spanish Regulatory Institution) in which several radiological laboratories participated [8]. Analyses were carried out in duplicate. The results of the analysed samples are given in Table 2. The values are presented with one sigma overall uncertainty, including uncertainties coming from all the variables used in calculating the ^{90}Sr concentration.

This proposed chemical procedure combined with Cerenkov counting notably improves the traditionally used methods. The time for sample preparation is considerably reduced, and the chemical recovery is greater than 80% in all cases. The results were in good agreement with the known values (confidence intervals quoted in Table 2).

Table 2 *Results obtained for the reference material and intercomparison samples.*

Sample	Weight (g)	Chemical recovery (%)	Activity concentration (Bq. kg^{-1})	
			Obtained value	Confidence interval
SOIL-6	10	83 ± 5	29.4 ± 3.2	26.7 - 34.0
	10	85 ± 5	27.6 ± 3.4	
IAEA-375	10	82 ± 5	120.6 ± 14.5	101.5 - 114.5
	10	89 ± 5	111.3 ± 16.2	
LMR-1	30	87 ± 5	7.2 ± 0.9	6.7 - 7.8
	30	90 ± 6	6.3 ± 0.7	

According to the Currie [9] criteria, and using the conditions established in the proposed method for a common soil sample, the analysis of a 30 g dried soil sample gave a chemical recovery of 80%, the minimum detectable activity for ^{90}Sr being 0.42 Bq. kg^{-1}. These values make the application of this method possible as a routine procedure in environmental monitoring, and it can be used to investigate the activity concentrations due to fallout in soils.

4 CONCLUSION

The analysis of several soil samples for ^{90}Sr by Cerenkov counting in a LSC QUANTULUS 1 220 indicated that this method is both simple and highly reliable, with acceptable detection limits for environmental monitoring. The problems associated with counting solid samples, such as self-absorption, non-uniformity of the sample mounting, and variation in counting efficiency, are avoided. This method would be preferably used for the determination of ^{90}Sr in soil and filter samples for routine radioactive monitoring purposes. In emergency cases, the method could also be used, but the possible degree of equilibrium between ^{90}Sr and its daughter ^{90}Y should be taken into consideration.

Acknowledgements Work supported by Ministerio de Ciencia y Tecnología (Project No. BFM2000-0811-C02-01). D. Pérez Sánchez acknowledge the Guest Research Position TEM01/0009 by Consejería de Educación, Ciencia y Tecnología (Junta de Extremadura).

References

1 R. D. Wilken and E. I. Joshi, *Radioact. & Radiochem.*, 1991, **2**,14.
2 H.E. Bjørnstad, H.N. Lien, Yu-Fu Yu and B. Salbu, *J. Radioanal. Nucl. Chem.*, 1992, **156**, 165.
3 G. Majón, F. Vaca and M. García-León, *Appl. Radiat. Isot.*, 1996, **47**,1097.
4 J. Suomela, L. Walberg and J. Mellin, *Methods for Determination of Strontium-90 in Food and Environmental Samples by Cerenkov Counting*, SSI-rapport 91-11,1993.
5 H.H. Ross, *Anal. Chem.*, 1969, **41**, 1260.
6 F. Vaca, G. Majón and M. García-León, *Nucl. Inst. and Methods A*, 1998, **406**, 267.
7 J.M. Torres, J. F. García, M. Llauradó and G. Rauret., *Analyst*, 1996, **121**, 1737.
8 *Informe de Evaluación de la intercomparación analítica de radionucleidos en muestras ambientales, campaña 2000*, Consejo de Seguridad Nuclear, Spain, 2000.
9 L.A. Currie, *Anal. Chem.*, 1968, **40**, 568.

INVESTIGATION OF INORGANIC CONTAMINANTS IN PACKAGING MADE FROM RECYCLED PAPER AND BOARD

S J Parry & D S J Aston

Department of Environmental Science & Technology
Imperial College of Science, Technology & Medicine
Silwood Park, Buckhurst Road, Ascot, Berkshire, SL5 7PY

1. ABSTRACT

Neutron activation analysis (NAA) has been used to compare the elemental composition of 20 samples of virgin paper and board with 20 samples of recycled paper and board, and to analyse samples of newspaper and magazine. In general the concentrations of all the elements in retail recycled paper and board are higher than those found in virgin paper and board, and closely reflect the values obtained from the samples of recycled paper and board provided by a paper mill. The concentration factor for the recycled samples appears to be around tenfold. However, the physical nature of the paper and board also affects the concentration of elements regardless of their origin, indicating that substances used in the manufacturing process, such as aluminium sulfate, calcium carbonate and titanium dioxide, also have a role to play. Such common elements include sodium, magnesium, aluminium, chlorine, potassium, calcium, scandium, titanium, vanadium, manganese and iron. Fifteen other elements were detected, including chromium, copper, nickel, copper, zinc, gallium, arsenic, selenium, molybdenum, cadmium, antimony, barium, mercury, thorium and uranium. These elements were generally present at extremely low concentrations (<1 mg/kg). The values for chromium, zinc and barium were higher than 1 mg/kg, with concentrations in the range of 4-15, 6-77 and 18-36 mg/kg, respectively. Another 34 elements were not determined because their concentrations fell below their detection limits (ranging from 1 μg/kg to 1 mg/kg).

2. INTRODUCTION

In Europe the use of recycled fibres in the production of paper and board is increasing and in some EU member states use is made of up to 75% of recycled fibres. The proportion of recycled paper and board used in indirect and direct food contact applications is not known but an investigation of retail samples in the United Kingdom has indicated that an appreciable proportion of the paper and board packaging used for foodstuffs comprises recycled fibres. Of this, a smaller proportion is in direct contact with food. Although special cleaning treatments are incorporated into the recycling process, there is a potential for contamination from printing inks, adhesives, varnishes and lacquers.

During a period between 1993 and 1996, a feasibility study was carried out at

Imperial College to test the suitability of Neutron activation analysis (NAA) for the determination of trace elements in paper and board packaging[1, 2]. The project was successfully concluded, and over forty elements were monitored at one time, including the key toxic metals[3]. Common elements, such as Na, Mg, Al, Cl, K, Ca, Sc, Ti, V, Mn and Fe were detected. These are used to enhance the quality of the paper, these include retention aids (aluminium sulphate, sodium aluminate, polyaluminium chloride), biocides (chlorine, sodium hypochlorite, hypo-bromous acid), and coatings (calcium carbonate, calcium sulphoaluminium, clay, titanium dioxide). The data exhibited a wide range of concentrations within and between sample types, with fast food boxes showing the highest concentrations for Na, Al, Ca and Ti. Positive values were also obtained for Cr, Co, Ni, Cu, Zn, Ga, Sb and Ba, in at least one of the samples analysed, and detection limits were obtained for those samples where the element was not determined. These data confirmed the findings of Castle *et al.*[4]. Finally, there were a large number of less important elements measured, which were below the detection limit of around 1 mg/kg or less in all of the samples. These were Br, Rb, Ag, In, Cs, La, Ce, Nd, Sm, Eu, Gd, Ho, Er, Tb, Tm, Yb, Lu, W, Re, Hf, Ta, Au, Th, U, Rh, Pd, Ru, Os, Ir, Pt.

This preliminary work demonstrated that NAA was a suitable method for the determination of trace elements in paper and board. Detection limits were sufficiently low to provide adequate assurance of the quality of the packaging for all the toxic elements determined. The preliminary study highlighted elevated concentrations of Cr in baking parchment (92 mg/kg), greaseproof paper (51 mg/kg) and take-away food packaging (5 mg/kg); high concentrations of Cu in green and blue napkins (300 and 233 mg/kg, respectively) and in orange kitchen roll (30 mg/kg); Sb in cake cases (2.5 mg/kg) and Ba in a pizza base (69 mg/kg).

As a result of these preliminary studies[1-5], the Food Standards Agency funded a project to compare the concentrations of elements of potential interest in recycled paper and boards with those in virgin paper and boards. This paper describes the results of this work.

2.1 Experimental

The method described below is based on an earlier feasibility study to test the viability of using neutron activation for the analysis of paper and board[3]. The method is capable of determining a total of over 60 elements, including the common elements: Na, Mg, Al, Cl, K, Ca, Sc, Ti, V, Mn and Fe; 15 elements of potential interest: Cr, Co, Ni, Cu, Zn, Ga, As, Se, Mo, Cd, Sb, Ba, Hg, Th and U; 23 other trace elements: Ge, Br, Rb, Sr, Zr, Nb, Ru, Rh, Pd, Ag, In, Sn, Te, I, Cs, Hf, Ta, W, Os, Re, Ir, Pt, Au and 14 of the lanthanides: La, Ce, Pr, Nd, Sm, Eu, Gd, Tb, Dy, Ho, Er, Tm, Yb and Lu.

The accuracy and precision of the analytical method is usually checked using certified reference materials with known concentrations of the elements of interest. In this case there are no reference materials that are certified for the entire range of elements studied in this project. The method was initially validated for a limited number of elements using IAEA Cotton Cellulose V9 in the previous study[3] and subsequently checked in this work for a larger number of elements using the certified reference material NIST SRM Orchard leaves 1571. It was also important to demonstrate that samples of paper and board could be analysed in a representative manner. Three samples of cardboard were analysed to demonstrate homogeneity within different examples of similar recycled board. Samples were selected from the same source but they contained three different varieties of the same product.

2.2 Sample preparation

Samples of virgin and recycled paper and board packaging that are used in contact with dry food were sourced from manufacturers of recycled materials, retailers and packing companies. In addition, material typically used in recycling was analysed. Newspaper and magazine samples were typical examples of retail material, plus high quality and standard office paper which would form the bulk of office paper waste sent for recycling. Where necessary, layers of printed paper were removed and in some cases inner liners were analysed separately.

Non-powdered latex gloves were used for handling the paper and board samples. Polyethylene sample containers were cleaned in hexane, 10% nitric acid, deionised water and acetone in sequence, and allowed to dry. The food contact paper samples weighing 0.4-0.8 g were torn to size to avoid the use of metal scissors.

2.3 Standards

Standards were prepared from single element stock solutions, pipetted onto filter papers, dried and analysed to provide a database for the determination of each of the elements of interest, where the radionuclide was short-lived. This database constitutes the routine analytical procedure for calibration under the laboratory quality assurance protocol and has been validated thoroughly through the use of many hundreds of reference materials, proficiency testing samples and 'in house' quality control monitors. For elements where the radionuclide has a long half-life, multi-elemental standards were produced containing the range of elements determined on a 'long irradiation'. Appropriate volumes of multi-elemental standard were pipetted onto cellulose powder and dried, to reproduce the geometry of the paper and board samples.

2.4 Neutron Activation Analysis (NAA)

All irradiations were carried out in the Imperial College Consort reactor, with a thermal neutron flux of 10^{16} m^{-2} s^{-1} and gamma ray spectrometry measurements were made with a high purity germanium semiconductor detector (EG&G Ortec, efficiency 20% at 1.33 MeV, resolution 1.87 keV FWHM at 1.33 MeV). Samples including unknowns, standards and reference materials were irradiated in sequence for 10 min in the automated large volume irradiation system (ALVIS) for the measurement of short-lived radionuclides. After irradiation the samples were allowed to decay for 5 min and then counted for 10 min, then allowed to decay for a further 5 min, followed by a second 10 min count. The samples were then re-irradiated in the core tube irradiation site for two days (15 h), counted for 25 min after a decay of 16 h and for 1.4 h after a further 10 days. In the case of the long-lived radionuclides the samples were transferred into clean containers before gamma ray spectrometry. Over 60 elements were identified and quantified using gamma ray spectrometry.

3. RESULTS

3.1 Method validation

Duplicate sets of results were obtained for the NIST certified reference material SRM 1571

Orchard Leaves. The data are shown in table 1. The values obtained by NAA are not statistically significantly different from the recommended values. The results obtained for the elements of potential interest also agreed well with consensus values, except for Hg and U where the results are close to, or below, detection limits. Although determination of the remaining elements could not be validated using SRM 1571, the work of the Analytical Services Group is subject to a quality system that ensures that all analyses are carried out using standards that are constantly used to check a wide range of reference materials.

The results, in table 2, show that there is no significant difference in the elemental composition of the three homogeneity test samples, with the single exception of Ba that is higher in sample 3. The values for each sample are not significantly different to the overall mean, demonstrating that the board is very homogeneous and therefore it is valid to take samples weighing only 0.4-0.8 g to get a representative analysis of the sample of packaging.

Table 1 Concentrations of elements in NIST SRM 1571 Orchard Leaves: Concentrations in mg/kg except K% and Ca% (2 standard deviation uncertainty in parentheses)

Element	Sample 1	Sample 2	Consensus value (mg/kg)
Na	88 (10)	95 (11)	89 (15)
Mg	5581 (235)	5613 (245)	6050 (380)
Al	379 (16)	375 (15)	323 (112)
Cl	646 (26)	631 (31)	730 (40)
K%	1.38 (0.05)	1.48 (0.12)	1.44 (0.07)
Ca%	1.91 (0.08)	1.84 (0.08)	2.04 (0.12)
Sc	0.064 (0.003)	0.068 (0.003)	0.063 (0.014)
Ti	27 (10)	<33	20 (7)
V	0.63 (0.15)	0.72 (0.10)	0.50 (0.11)
Mn	87 (3)	93 (3)	89 (5)
Fe	307 (29)	305 (26)	286 (28)
Cr	3.0 (0.2)	2.4 (0.2)	2.6 (0.6)
Co	0.21 (0.02)	0.15 (0.02)	0.16 (0.08)
Cu	15 (4)	17 (5)	12.0 (2.8)
Zn	26.1 (1.4)	24.3 (1.2)	25 (4)
As	12.3 (0.5)	11.7 (0.4)	10.7 (2.6)
Sb	3.9 (0.1)	2.9 (0.2)	2.9 (0.6)
Ba	37 (4)	46 (5)	43 (8)
Hg	<0.12	0.084 (0.034)	0.155 (0.028)
Th	0.068 (0.012)	0.068 (0.010)	0.058 (0.024)
U	<0.075	0.042 (0.016)	0.029 (0.006)

Table 2 Analyses of three cardboard drums from the same source: *Concentrations in mg/kg (2 standard deviation uncertainty in parentheses)*

Element	Sample 1	Sample 2	Sample 3	Mean
Na	600 (47)	737 (47)	631 (48)	656 (72)
Mg	2910 (270)	2570 (270)	2900 (280)	2790 (190)
Al	9440 (360)	10,360 (390)	10,050 (380)	9,950 (470)
Cl	203 (25)	231 (29)	267 (33)	234 (32)
K	490 (110)	430 (77)	448 (96)	456 (31)
Ca	13,670 (700)	14,330 (750)	14,490 (750)	14,160 (440)
Sc	0.51 (0.02)	0.61 (0.02)	0.59 (0.02)	0.57 (0.05)
Ti	462 (39)	502 (46)	500 (43)	488 (23)
V	3.8 (1.1)	4.7 (0.9)	5.0 (1.2)	4.5 (0.6)
Cr	9.0 (0.4)	8.8 (0.4)	9.3 (0.5)	9.0 (0.3)
Mn	29.8 (1.1)	26.3 (1.0)	28.3 (1.0)	28.1 (1.8)
Fe	710 (40)	737 (47)	736 (35)	728 (15)
Co	1.03 (0.05)	1.02 (0.05)	1.17 (0.07)	1.07 (0.08)
Ni	13 (4)	<12	<13	<13
Cu	<28	<24	<21	<24
Zn	34 (4)	33 (5)	37 (5)	35 (2)
Ga	1.66 (0.22)	1.71 (0.26)	1.84 (0.25)	1.74 (0.09)
Ge	<10	<20	<14	<15
As	0.382 (0.030)	0.359 (0.032)	0.378 (0.033)	0.373 (0.012)
Se	<0.55	<1.0	<0.66	<0.74
Br	5.8 (0.5)	3.5 (0.4)	5.6 (0.4)	5.0 (1.3)
Rb	5.8 (1.2)	5.9 (1.4)	5.8 (1.5)	5.83 (0.06)
Sr	24 (6)	24 (6)	27 (8)	25 (2)
Mo	1.20 (0.20)	1.15 (0.21)	1.24 (0.21)	1.20 (0.05)
Sb	0.32 (0.01)	0.35 (0.02)	0.29 (0.01)	0.32 (0.03)
Cs	0.55 (0.08)	0.58 (0.08)	0.64 (0.08)	0.59 (0.04)
Ba	35 (4)	33 (4)	106 (6)	54 (42)
La	2.04 (0.06)	2.19 (0.07)	2.23 (0.07)	2.15 (0.10)
Ce	4.9 (0.2)	5.0 (0.2)	5.1 (0.2)	5.0 (0.1)
Nd	1.8 (0.4)	1.8 (0.4)	2.0 (0.4)	1.8 (0.2)
Sm	0.24 (0.02)	0.23 (0.02)	0.25 (0.02)	0.24 (0.01)
Eu	0.066 (0.009)	0.074 (0.011)	0.071 (0.010)	0.071 (0.003)
Tb	0.051 (0.009)	0.056 (0.010)	0.060 (0.011)	0.056 (0.005)
Dy	0.228 (0.011)	0.209 (0.011)	0.228 (0.011)	0.222 (0.011)
Tm	0.067 (0.023)	0.071 (0.023)	0.083 (0.024)	0.074 (0.008)
Yb	0.085 (0.015)	0.088 (0.012)	0.084 (0.012)	0.086 (0.002)
Lu	0.012 (0.002)	0.012 (0.002)	0.014 (0.002)	0.0123 (0.001)
Hf	0.46 (0.04)	0.54 (0.04)	0.52 (0.04)	0.51 (0.05)
Ta	0.145 (0.014)	0.173 (0.016)	0.159 (0.015)	0.159 (0.014)
W	0.65 (0.11)	0.66 (0.10)	0.83 (0.14)	0.71 (0.10)
Au	0.0031 (0.0006)	0.0023 (0.0006)	0.0030 (0.0005)	0.0028 (0.0005)
Hg	<0.16	<0.17	<0.18	<0.17
Th	0.70 (0.03)	0.75 (0.03)	0.72 (0.03)	0.72 (0.03)
U	0.231 (0.015)	0.250 (0.018)	0.253 (0.018)	0.245 (0.012)

3.2 Common elements in paper and board

Both virgin and recycled paper and board contained a wide range of elements, some at high concentrations. These elements may result from chemicals used in paper production, including fillers, coatings and inks. Common elements including Na, Mg, Al, Cl, K, Ca, Sc, Ti, V, Mn, and Fe were all detected, some at percentage levels. The concentrations of common elements in virgin paper and board are given in table 3. There is a wide variation in composition due to the nature of the substances used in the manufacture of paper and board such as aluminium sulfate, calcium carbonate and titanium dioxide. The Mg, Al, Ca and Ti concentrations in the different items of packaging show a wide range of concentrations covering several orders of magnitude, presumably depending on the processing agents used. Smaller variations are found in Na, K, Cl, Sc, Ti, Mn and Fe. Mn shows a particularly small range of concentrations (<0.2-14.3) with the exception of one of the rice boxes and an oats box that contain 27 and 30 mg/kg respectively.

Table 3 Common elements in virgin paper and board packaging: *Concentrations in mg/kg (2 standard deviation uncertainty in parentheses)*

TG0 code	Package	Na	Mg	Al	Cl	K
419	Dried peas box	233 (10)	1863 (83)	1976 (79)	129 (7)	<24
424	Rice box	785 (30)	4137 (190)	10468 (397)	1229 (49)	683 (155)
503	Raisin box	436 (17)	291 (106)	4382 (169)	352 (15)	126 (39)
505	Cereal bar card	781 (30)	394 (94)	6140 (234)	197 (11)	348 (110)
506	Sugar bag	226 (9)	44 (12)	462 (18)	318 (13)	<21
507	Sandwich card	241 (10)	<224	4111 (300)	156 (9)	112 (22)
509	Rice box	284 (11)	443 (113)	6532 (249)	172 (7)	246 (34)
511	Frozen potato waffle box	<14	<70	<17	<1.6	<58
513	Flour bag	244 (9)	48 (14)	783 (30)	319 (13)	<47
514	Icing sugar bag	495 (19)	3212 (133)	516 (20)	114 (5)	<54
515	Oats box	215 (9)	202 (66)	1305 (49)	44 (4)	155 (48)
517	Microwaveable chips box	410 (16)	135 (14)	218 (9)	519 (20)	<52
518	Frozen fish fingers box	148 (6)	1282 (77)	1735 (66)	128 (6)	<40
520	Chocolate biscuit wrapper	862 (33)	1929 (96)	929 (37)	378 (16)	<147
521	Pancake card	1011 (38)	156 (53)	3593 (140)	796 (31)	704 (168)
523	Savoury pie box	433 (17)	765 (67)	2377 (92)	397 (16)	<87
526	Muffin bag	610 (16)	347 (94)	896 (21)	505 (15)	<102
527	Cake card	546 (13)	163 (54)	441 (11)	442 (13)	90 (33)
529	Chocolate bar card	409 (16)	2826 (179)	5548 (212)	394 (16)	<102
530	Buns card	241 (9)	269 (55)	2872 (109)	273 (11)	<62

Table 3 continued:

TG0 code	Package	Ca	Sc	Ti	V	Mn	Fe
419	Dried peas box	3184 (181)	0.077 (0.004)	<53	<0.6	5.6 (0.2)	163 (16)
424	Rice box	32847 (1440)	0.44 (0.02)	491 (47)	3.7 (1.1)	27.2 (1.0)	466 (24)
503	Raisin box	7183 (368)	0.31 (0.01)	221 (20)	2.8 (0.9)	3.6 (0.2)	182 (20)
505	Cereal bar card	25500 (1114)	0.22 (0.01)	76 (18)	1.3 (0.3)	5.4 (0.2)	253 (15)
506	Sugar bag	360 (31)	0.004 (0.001)	324 (13)	<0.09	0.67 (0.04)	<14
507	Sandwich card	7303 (381)	0.035 (0.002)	<37	<0.9	3.2 (0.2)	95 (18)
509	Rice box	3081 (168)	0.048 (0.003)	<34	<0.7	14.3 (0.5)	139 (16)
511	Frozen potato waffle box	<16	0.083 (0.004)	63 (18)	<0.08	<0.2	211 (25)
513	Flour bag	266 (47)	0.006 (0.001)	396 (16)	<0.09	0.43 (0.02)	<17
514	Icing sugar bag	188 (21)	0.042 (0.002)	<8	0.30 (0.09)	1.58 (0.07)	303 (22)
515	Oats box	3234 (157)	0.009 (0.001)	<9	<0.1	30.2 (1.1)	35 (8)
517	Microwaveable chips box	339 (30)	<0.003	<6	<0.08	6.0 (0.2)	<47
518	Frozen fish fingers box	3111 (155)	0.075 (0.004)	24 (8)	0.5 (0.2)	5.5 (0.2)	170 (20)
520	Chocolate biscuit wrapper	949 (87)	0.013 (0.001)	<24	<0.3	1.35 (0.07)	88 (13)
521	Pancake card	6752 (341)	0.022 (0.001)	<33	<0.8	2.19 (0.08)	53 (9)
523	Savoury pie box	6644 (317)	0.019 (0.001)	<28	<0.5	12.3 (0.5)	64 (10)
526	Muffin bag	827 (81)	0.13 (0.01)	126 (20)	<0.1	3.3 (0.1)	110 (10)
527	Cake card	6725 (266)	0.021 (0.001)	<5	<0.06	2.4 (0.1)	56 (13)
529	Chocolate bar card	18825 (815)	0.31 (0.01)	85 (10)	2.0 (0.2)	5.8 (0.2)	389 (22)
530	Buns card	3660 (178)	0.016 (0.001)	<15	<0.3	1.7 (0.1)	48 (7)

The concentrations of common elements in recycled paper and board are given in table 4. It appears that the sample of coloured magazine contains higher concentrations of most elements when compared to the sample of newspaper or office paper. This may be due to the presence of inks and the substances used to produce a glossy paper finish. The concentrations of the common elements in recycled paper and board represent a much more consistent dataset than those for virgin samples. The values for Mg, Al, Ca and Ti vary to a smaller extent than they do in virgin packaging and the concentrations of all these elements tend to be higher. There is a clear difference between the two types of packaging with respect to trace element concentrations where, with the exception of kitchen roll, the ranges of Sc (0.3 –0.9 mg/kg), V (1.0-6.0 mg/kg), Mn (21-69 mg/kg) and Fe (247-1164 mg/kg) are much higher than the values generally found in virgin paper and board.

The inner liner of the loose tea packet sample contains concentrations of common elements found in virgin paper and board samples. It is possible that the lower values found in the kitchen roll and egg box samples are due to the fact that they are not composed of 'end of use' recycled paper.

Table 4 Common elements in recycled paper and board packaging: *Concentrations in mg/kg (2 standard deviation uncertainty in parentheses)*

TG0 code	Package	Na	Mg	Al	Cl	K
534	High quality office paper	300 (20)	485 (55)	329 (19)	231 (20)	127 (40)
535	Standard office paper	3961 (155)	1326 (74)	408 (21)	1279 (53)	<134
310	Magazine	1642 (61)	<1779	52857 (2002)	416 (21)	4606 (364)
311	Newspaper	474 (18)	924 (141)	7010 (270)	100 (9)	533 (71)
312	Mill 186 gsm	499 (18)	1861 (202)	11598 (440)	354 (17)	740 (86)
313	Mill 140 gsm	451 (17)	1757 (203)	12125 (461)	391 (17)	447 (33)
314	Mill Fine flute	560 (21)	1429 (194)	8481 (326)	379 (19)	524 (78)
315	Mill Fine chip	659 (24)	1849 (239)	9510 (366)	395 (19)	602 (85)
106	Sea salt box	935 (59)	1450 (270)	9370 (360)	753 (42)	694 (99)
108	Rice box	1403 (66)	790 (230)	11340 (430)	137 (26)	<82
202	Lentil box	538 (20)	1490 (290)	20010 (760)	197 (10)	1500 (110)
203	Table salt drum	569 (22)	1310 (180)	8790 (340)	351 (15)	616 (80)
303	Gravy drum inner liner	816 (30)	503 (68)	3253 (123)	147 (8)	228 (59)
403	Oat packet	616 (24)	7291 (306)	19584 (745)	301 (13)	1153 (173)
406	Mushroom bag	370 (16)	4139 (182)	9943 (376)	142 (7)	626 (103)
408	Apple pie case	408 (17)	6308 (263)	16212 (616)	172 (13)	939 (126)
410	French fries	459 (18)	1782 (280)	15132 (575)	140 (8)	522 (108)
412	Loose tea drum card	695 (48)	9217 (475)	23115 (876)	1559 (73)	1654 (242)
412	Loose tea drum inner liner	149 (7)	1153 (70)	3480 (134)	176 (8)	386 (84)
417	Cannelloni box	190 (10)	1744 (259)	13982 (532)	161 (10)	656 (63)
422	Lasagne box	524 (21)	3982 (182)	9971 (382)	763 933)	583 (67)
427	Pizza base	627 (24)	1969 (267)	9130 (349)	327 (16)	563 (80)
428	Custard powder drum	420 (18)	5140 (219)	13969 (529)	347 (17)	1043 (403)
532	Fruit tray	1052 (40)	1088 (162)	10605 (406)	690 (28)	780 (114)
301	Kitchen roll	582 (22)	317 (44)	1964 (75)	601 (25)	162 (64)
302	Egg box	254 (10)	402 (53)	1066 (42)	151 (9)	195 (65)

Table 4 continued:

TG0 code	Package	Ca	Sc	Ti	V	Mn	Fe
534	High quality office paper	69736 (2958)	0.108 (0.005)	<29	<0.3	42.8 (1.6)	197 (13)
535	Standard office paper	18370 (826)	0.046 (0.002)	<33	0.72 (0.10)	93 (3)	106 (12)
310	Magazine	4784 (314)	0.92 (0.13)	177 (64)	<3	46.1 (1.7)	1841 (329)
311	Newspaper	3396 (233)	0.18 (0.03)	<85	<1	36.3 (1.3)	<412
312	Mill 186 gsm	26458 (1148)	0.50 (0.02)	492 (27)	3.5 (0.8)	32.5 (1.2)	785 (33)
313	Mill 140 gsm	27383 (1185)	0.48 (0.02)	506 (28)	3.3 (0.7)	32.1 (1.2)	792 (33)
314	Mill Fine flute	15789 (769)	0.34 (0.01)	509 (44)	2.9 (1.1)	28.1 (1.1)	646 (28)
315	Mill Fine chip	18522 (914)	0.40 (0.01)	574 (54)	3.7 (0.7)	36.9 (1.4)	692 (29)
106	Sea salt box	19530 (960)	0.38 (0.01)	592 (43)	3.3 (1.1)	33.6 (1.2)	649 (34)
108	Rice box	10220 (570)	0.68 (0.03)	1810 (83)	6.0 (1.1)	15.2 (0.3)	269 (19)
202	Lentil box	40600 (1700)	0.67 (0.02)	1634 (72)	4.7 (1.3)	29.9 (1.1)	847 (34)
203	Table salt drum	32600 (1400)	0.31 (0.01)	339 (31)	2.6 (0.8)	25.6 (0.9)	324 (24)
303	Gravy drum inner liner	5514 (251)	0.13 (0.01)	185 (15)	1.2 (0.2)	22.4 (0.8)	247 (14)
403	Oat packet	28763 (1262)	0.79 (0.03)	884 (50)	5.8 (1.4)	32.4 (1.2)	979 (40)
406	Mushroom bag	16499 (749)	0.34 (0.01)	490 (37)	2.8 (0.8)	28.8 (1.1)	638 (27)
408	Apple pie case	24591 (1075)	0.66 (0.03)	2663 (131)	<2.1	28.5 (1.1)	668 (28)
410	French fries	19692 (859)	0.63 (0.02)	1192 (70)	4.4 (1.4)	26.0 (1.0)	557 (29)
412	Loose tea drum card	22086 (1232)	0.93 (0.04)	558 (72)	5.6 (1.9)	31.3 (1.2)	1164 (46)
413	Loose tea drum inner liner	210 (45)	0.063 (0.003)	<46	<0.6	24.2 (0.9)	152 (29)
417	Cannelloni box	40400 (1713)	0.67 (0.03)	566 (63)	5.6 (1.7)	33.3 (1.2)	675 (32)
422	Lasagne box	36835 (1603)	0.39 (0.02)	426 (44)	<1.5	31.7 (1.2)	461 (23)
427	Pizza base	20186 (957)	0.39 (0.02)	556 (54)	3.9 (1.4)	36.1 (1.3)	713 (36)
428	Custard powder drum	23450 (1056)	0.50 (0.02)	572 (53)	3.7 (0.6)	33.3 (1.2)	888 (38)
532	Fruit tray	24254 (1065)	0.39 (0.02)	277 (22)	2.8 (0.5)	20.6 (0.8)	490 (37)
301	Kitchen roll	11629 (524)	0.063 (0.003)	65 (9)	<0.2	5.1 (0.2)	111 (12)
302	Egg box	28096 (1202)	0.056 (0.002)	<28	1.0 (0.3)	69 (3)	130 (13)

3.3 Elements of potential interest in paper and board

The concentrations of elements in virgin paper and board are given in table 5. In general the values are low, that is below 1 mg/kg, for Cr, Co, Ga, As, Mo, Cd, Th and U, and below 0.1 mg/kg for Sb and Hg. The detection limit for Ni was around 5 mg/kg and from 3 to 50 mg/kg for Cu which is badly affected by high gamma background immediately after irradiation. The actual detection limits for many elements are dependent on the concentrations of the common elements in the sample and are given in table 5 where appropriate.

Exceptionally high concentrations of certain elements were measured in some items, such as 1.2–241 mg/kg Cr, 5-45 mg/kg Ni, 25-74 mg/kg Cu, 1.6-3.2 mg/kg Ga, 27 mg/kg Sb, 6-101 mg/kg Ba. Certain items had exceptional concentrations of more than one element, such as the microwaveable chip box with 241 mg/kg Cr and 27 mg/kg Sb and the muffin bag with 38 mg/kg of Cu and 101 mg/kg of Ba. Zn concentrations in the samples were remarkably consistent, with all but one falling within the range 3.0-6.9 mg/kg. The exception was the rice box, which contained 17 mg/kg Zn.

Table 5 Elements of potential interest in virgin paper and board packaging: *Concentrations in mg/kg (2 standard deviation uncertainty in parentheses)*

TG0 code	Package	Cr	Co	Ni	Cu	Zn	Ga	As
419	Dried peas box	3.8 (0.2)	0.35 (0.02)	5.7 (1.1)	<24	3.1 (0.5)	0.5 (0.1)	<0.08
424	Rice box	5.2 (0.3)	0.92 (0.05)	<5	<51	17 (1)	3 (1)	0.3 (0.1)
503	Raisin box	1.8 (0.2)	0.39 (0.03)	<4	<25	3.8 (0.7)	1.6 (0.2)	<0.05
505	Cereal bar card	0.98 (0.15)	0.18 (0.02)	<4	<30	4.1 (0.6)	2.4 (0.4)	0.23 (0.03)
506	Sugar bag	0.20 (0.07)	<0.03	<1	74 (4)	2.7 (0.3)	<0.3	<0.03
507	Sandwich card	0.89 (0.12)	0.09 (0.02)	<2	<20	2.8 (0.4)	0.6 (0.2)	0.18 (0.02)
509	Rice box	0.73 (0.12)	0.09 (0.01)	<2	<16	5.1 (0.6)	1.5 (0.2)	0.18 (0.03)
511	Frozen potato waffle box	0.87 (0.15)	<0.05	<3	<4	4.8 (0.6)	1.0 (0.3)	<0.08
513	Flour bag	0.22 (0.08)	<0.04	<2	25 (2)	4.5 (0.5)	<0.4	<0.05
514	Icing sugar bag	11.6 (0.5)	0.54 (0.04)	14 (3)	<15	6.9 (0.7)	<0.4	0.20 (0.04)
515	Oats box	0.33 (0.08)	<0.04	<2	<5	6.0 (0.7)	<0.4	0.19 (0.03)
517	Microwaveable chips box	241 (8)	<0.1	<4	<4	3.9 (1.0)	<0.5	<0.07
518	Frozen fish fingers box	3.5 (0.2)	0.30 (0.02)	5.7 (1.2)	<8	3.3 (0.6)	<0.4	0.06 (0.02)
520	Chocolate biscuit wrapper	0.47 (0.13)	0.10 (0.04)	<3	<9	3.3 (0.7)	<1	<0.2
521	Pancake card	0.47 (0.08)	0.05 (0.01)	<2	<26	3.4 (0.5)	<1	0.13 (0.05)
523	Savoury pie box	0.64 (0.08)	<0.03	<1	<16	3.0 (0.5)	<0.7	<0.08
526	Muffin bag	1.2 (0.1)	<0.05	<3	38 (12)	4.7 (0.8)	<2	0.14 (0.04)

527	Cake card	0.53 (0.08)	0.05 (0.02)	45 (15)	<3	3.4 (0.5)	<0.9	<0.09
529	Chocolate bar card	7.6 (0.3)	0.67 (0.04)	11 (2)	<19	4.3 (0.7)	3.2 (0.9)	0.13 (0.4)
530	Buns card	0.56 (0.07)	<0.03	<2	<12	3.3 (0.4)	<1	0.10 (0.04)

Table 5 continued:

TG0 code	Package	Se	Mo	Cd	Sb	Ba	Hg	Th	U
419	Dried peas box	<0.5	<0.6	<0.9	<0.02	<4	<0.05	0.076 (0.007)	<0.05
424	Rice box	<0.7	1.2 (0.5)	<1.7	0.12 (0.03)	34 (4)	<0.10	0.77 (0.03)	0.30 (0.03)
503	Raisin box	<0.6	1.4 (0.2)	<0.8	<0.06	<5	<0.11	0.26 (0.2)	0.16 (0.03)
505	Cereal bar card	<0.6	<0.4	<0.8	0.09 (0.02)	9 (2)	<0.12	0.56 (0.03)	0.79 (0.05)
506	Sugar bag	<0.3	0.4 (0.1)	<0.6	0.03 (0.01)	53 (3)	0.064 (0.024)	<0.01	<0.05
507	Sandwich card	<0.5	<0.4	<0.7	0.05 (0.01)	9 (2)	0.10 (0.03)	0.14 (0.01)	0.19 (0.03)
509	Rice box	<0.5	0.6 (0.2)	<1.0	0.07 (0.02)	13 (2)	<0.08	0.22 (0.01)	0.31 (0.04)
511	Frozen potato waffle box	<0.6	<0.5	<1.2	<0.06	8 (3)	<0.10	0.27 (0.02)	<0.1
513	Flour bag	<0.6	0.6 (0.2)	<1.0	0.59 (0.03)	2.1 (0.7)	<0.06	<0.02	<0.04
514	Icing sugar bag	<0.6	<0.7	<1.3	<0.03	<1.9	<0.06	<0.018	<0.05
515	Oats box	<0.4	<0.5	<0.9	<0.03	6 (1)	<0.06	0.076 (0.008)	0.05 (0.02)
517	Microwaveable chips box	<2.3	<0.8	<1.8	27 (1)	2.8 (1.1)	<0.19	<0.05	<0.06
518	Frozen fish fingers box	<0.5	<0.4	<0.8	<0.03	3.0 (0.9)	<0.06	0.089 (0.009)	0.04 (0.01)
520	Chocolate biscuit wrapper	<0.9	<1	<2.1	<0.04	<4	<0.10	0.04 (0.01)	<0.08
521	Pancake card	<0.4	<0.8	<1.6	0.05 (0.01)	6.0 (1.5)	<0.06	0.100 (0.009)	0.13 (0.03)
523	Savoury pie box	<0.5	<0.5	<1.1	<0.04	<4	<0.05	0.026 (0.006)	0.06 (0.02)
526	Muffin bag	<0.7	<0.6	<1.1	<0.06	101 (4)	<0.09	1.78 (0.07)	0.41 (0.02)
527	Cake card	<0.3	<0.5	<0.9	<0.04	6.1 (1.9)	<0.05	0.063 (0.008)	0.06 (0.01)
529	Chocolate bar card	<0.7	<0.5	<1	<0.03	4.8 (1.0)	<0.08	0.39 (0.02)	0.08 (0.01)
530	Buns card	<0.4	<0.5	<0.98	<0.03	6.3 (0.8)	<0.05	0.075 (0.007)	0.05 (0.01)

The results for samples of recycled paper and board presented a much more consistent dataset, as shown in table 6, and the magazine has the highest concentrations within the group of sources for recycling, including Cr, Cu, Zn, Ga, As, Mo, Sb and Ba. In general the concentrations of elements in recycled paper and board were a factor of ten higher than those found in virgin packaging. The exceptions were the kitchen roll and egg box, which were not representative of the group. The concentrations of elements determined in the packaging reflected the values in the mill samples. In particular, Cr, Zn and Ba were consistently high at 4.4-15.2 mg/kg, 6-77 mg/kg and 18-86 mg/kg, respectively, compared to values in virgin paper and board of 0.2-11.6 mg/kg, 3.0-6.9 mg/kg and 2.1-9 mg/kg, respectively. The range of concentrations for the other elements included Co (0.27-1.2 mg/kg), Ni (up to 11 mg/kg), Ga (1.0-7.6 mg/kg), As (0.16-1.1 mg/kg), Mo (<0.7-2.7 mg/kg) and Sb (0.05-0.66 mg/kg). Hg was always below 0.2 mg/kg. Other elements such as Se and Cd were always below the detection limit, with the exception of the lasagne box which contained 5.5 mg/kg of Cd and also had high concentrations of Ag (16 mg/kg) and Cu (55 mg/kg).

Generally there were far fewer exceptional concentrations than had been the case for the virgin paper and board. However, there were items with Cr in the range of 12-15 mg/kg and several with Cu over 50 mg/kg. The oat packet contained 15 mg/kg Cr, 74 mg/kg Zn and 86 mg/kg Ba; and the tea drum contained 15 mg/kg Cr and 822 mg/kg Ba. The tea drum was separated into the main body of recycled board, the printed outer layer, and the inner white liner in contact with the tea, which appeared to be made of virgin paper. The inner liner was cleaner than the recycled board, except that it contained 51 mg/kg of Sb. It appeared that Cr, Zn and Ba were more generally present at consistent concentrations in recycled packaging, and at levels that could potentially be of interest.

Table 6 Elements of potential interest in recycled paper and board packaging: *Concentrations in mg/kg (2 standard deviation uncertainty in parentheses)*

TG0 code	Package	Cr	Co	Ni	Cu	Zn	Ga	As
534	High quality office paper	0.5 (0.1)	0.28 (0.02)	<5	<12	22 (1)	<1	<0.09
535	Standard office paper	2.4 (0.1)	0.09 (0.01)	<3	17 (5)	9.2 (0.2)	<2	<0.1
310	Magazine	4.8 (1.3)	<0.9	<45	207 (38)	36 (13)	14 (1)	3.8 (0.2)
311	Newspaper	4.0 (1.2)	<0.8	<41	<42	<30	1.7 (0.3)	0.16 (0.04)
312	Mill 186 gsm	5.7 (0.2)	0.89 (0.04)	<3.7	41 (9)	37 (1)	2.9 (0.3)	<1.2
313	Mill 140 gsm	5.5 (0.2)	0.89 (0.04)	<3.6	48 (9)	33 (1)	3.0 (0.3)	<1.2
314	Fine flute	4.4 (0.2)	0.65 (0.03)	<3.5	<49	33 (1)	2.0 (0.3)	<1.2
315	Fine chip	5.8 (0.2)	0.88 (0.04)	<3.7	<60	48 (2)	2.3 (0.4)	<1.5
106	Sea salt box	6.9 (0.3)	3.5 (0.1)	<11	<25	41 (4)	3.4 (0.5)	0.56 (0.04)
108	Rice box	5.3 (0.3)	0.64 (0.05)	<10	<25	<12	3.0 (0.7)	0.10 (0.04)
202	Lentil box	6.5 (0.3)	1.94 (0.07)	<6	<41	22 (3)	3.8 (0.4)	0.76 (0.03)
203	Table salt drum	5.0 (0.3)	1.20 (0.05)	<6	<74	30 (3)	1.7 (0.4)	0.37 (0.03)
301	Kitchen roll	0.93 (0.1)	0.29 (0.02)	<2.2	<11	5.8 (0.7)	<0.9	<0.1
302	Egg box	0.52 (0.1)	0.36 (0.02)	<2.6	<11	10.2 (0.8)	<0.6	<0.1
303	Gravy drum inner liner	5.5 (1.8)	0.27 (0.02)	<2.6	18 (6)	22 (1)	1.0 (0.3)	0.16 (0.05)

403	Oat packet	15.1 (0.6)	1.14 (0.05)	8.1 (2.4)	<78	74 (3)	4.8 (0.6)	0.73 (0.06)
406	Mushroom bag	9.9 (0.5)	0.87 (0.05)	<6.2	<36	38 (2)	2.5 (0.4)	0.51 (0.05)
408	Apple pie case	10.0 (0.5)	0.96 (0.07)	7.2 (2.7)	<139	35 (2)	5.0 (0.6)	0.70 (0.06)
410	French fries	8.2 (0.4)	1.21 (0.06)	<6.4	<61	19 (1)	4.2 (0.5)	0.37 (0.05)
412	Loose tea drum card	15.2 (0.7)	1.16 (0.6)	6.7 (2.5)	<93	47 (2)	7.6 (1.3)	1.09 (0.08)
413	Loose tea drum inner liner	<1.2	2.8 (0.10	<5.6	<30	5 (1)	1.2 (0.2)	0.21 (0.04)
417	Cannelloni box	9.3 (0.4)	1.01 (0.05)	<5.4	<71	37 (2)	3.1 (0.2)	0.41 (0.03)
422	Lasagne box	5.5 (0.3)	0.75 (0.04)	<4.3	55 (18)	22 (1)	2.2 (0.2)	0.33 (0.05)
427	Pizza base	6.0 (0.3)	0.88 (0.07)	<5.7	59 (22)	47 (2)	2.5 (0.3)	0.46 (0.07)
428	Custard powder drum	12.2 (0.5)	1.08 (0.05)	8.6 (2.2)	<60	77 (3)	3.8 (0.3)	0.60 (0.07)
532	Fruit tray (green)	5.3 (0.3)	0.82 (0.05)	<7.1	43 (14)	24 (2)	3.9 (1.4)	0.59 (0.08)

Table 6 continued:

TG0 code	Package	Se	Mo	Cd	Sb	Ba	Hg	Th	U
534	High quality office paper	<0.6	<0.5	<1.0	<0.07	<12	<0.08	0.06 (0.01)	0.04 (0.01)
535	Standard office paper	<0.4	<0.7	<1.3	0.044 (0.008)	<8	<0.05	0.030 (0.006)	0.25 (0.02)
310	Magazine	nd	4.0 (0.5)	<2.2	0.72 (0.04)	60 (6)	nd	nd	nd
311	Newspaper	nd	<0.7	<1.5	0.07 (0.02)	8 (2)	nd	nd	nd
312	Mill 186 gsm	<0.6	1.1 (0.4)	<2.7	0.37 (0.04)	44 (4)	nd	nd	nd
313	Mill 140 gsm	<0.7	1.2 (0.4)	<2.6	0.42 (0.04)	43 (4)	nd	nd	nd
314	Fine flute	<0.6	1.0 (0.4)	<2.5	0.38 (0.03)	43 (3)	nd	nd	nd
315	Fine chip	<0.7	<1.1	<2.9	0.54 (0.04)	51 (3)	nd	nd	nd
106	Sea salt box	<0.5	1.5 (0.2)	<1.4	0.41 (0.02)	39 (6)	<0.18	0.87 (0.03)	0.43 (0.02)
108	Rice box	<0.9	2.1 (0.2)	<1.4	0.05 (0.02)	<13	<0.18	0.77 (0.03)	0.17 (0.01)
202	Lentil box	<0.5	1.3 (0.1)	<1.1	0.21 (0.02)	27 (3)	<0.17	1.93 (0.06)	0.87 (0.03)
203	Table salt drum	<0.4	1.1 (0.1)	<1.3	0.24 (0.02)	21 (4)	<0.17	0.76 (0.03)	0.43 (0.02)
301	Kitchen roll	<0.7	<1.0	<2.1	<0.04	9.3 (2.1)	nd	nd	nd
302	Egg box	<0.6	<0.8	<1.7	0.05 (0.01)	<9	nd	nd	nd
303	Gravy drum inner liner	<0.5	<0.8	<1.7	0.17 (0.02)	18 (3)	nd	nd	nd
403	Oat packet	<1.2	2.4 90.3)	<1.1	0.32 (0.02)	86 (6)	<0.13	2.0 (0.1)	0.70 (0.03)

406	Mushroom bag	<1.1	1.0 (0.2)	<1.2	0.35 (0.02)	33 (3)	<0.12	0.78 (0.03)	0.36 (0.03)
408	Apple pie case	<1.4	1.5 (0.3)	<1.5	0.26 (0.02)	41 (5)	<0.14	1.9 (0.1)	0.66 (0.04)
410	French fries	<1.0	<0.7	<1.4	0.17 (0.03)	25 (3)	<0.12	1.9 (0.1)	0.38 (0.03)
412	Loose tea drum	<1.5	2.7 (0.4)	<1.8	0.60 (0.03)	822 (38)	<0.15	3.3 (0.1)	0.84 (0.04)
413	Loose tea drum inner liner	<3.5	<1.1	<2.7	51 (2)	7 (2)	<0.27	0.32 (0.03)	0.17 (0.03)
417	Cannelloni box	<0.8	1.4 (0.4)	<1.6	0.29 (0.03)	50 (4)	0.14 (0.04)	1.16 (0.04)	0.56 (0.04)
422	Lasagne box	<0.7	0.9 (0.3)	5.5 (1.4)	0.14 (0.05)	33 (3)	<0.09	0.75 (0.03)	0.47 (0.05)
427	Pizza base	<1.0	1.3 (0.5)	<2.3	0.49 (0.05)	66 (7)	<0.13	0.87 (0.04)	0.36 (0.05)
428	Custard powder drum	<0.9	1.8 (0.5)	<2.7	0.66 (0.05)	57 (7)	<0.12	1.36 (0.05)	1.2 (0.1)
532	Fruit tray (green)	<1.1	1.4 (0.4)	<2.0	0.20 (0.06)	49 (6)	<0.14	1.00 (0.04)	0.37 (0.03)

3.4 Other elements of interest

Over 60 elements have been determined in the packaging using NAA. Many of the elements are of limited interest, since they occur at trace concentrations that have no potential impact on health. In many cases they are below a detection limit of 0.1 mg/kg and therefore do not represent a potential hazard, even if 100% of the element migrated into the food. Typical detection limits are given as mg/kg in parentheses: Ge (7), Br (0.8), Rb (0.7), Zr (30), Sr (6), Nb (90), Ag (0.2), In (0.005), Sn (5), Te ((1.7), I (0.2), Cs (0.03), Ru (0.1), Rh (0.1), Pd (2), La (0.05), Ce (0.1), Pr (2), Nd (0.3), Sm (0.001), Eu (0.01), Gd (0.4), Tb (0.01), Ho (0.02), Tm (0.03), Yb (0.03), Lu (0.003), Hf (0.01), Ta (0.02), W (0.1), Os (0.03), Ir (0.0003) and Pt (0.5).

4. DISCUSSION

In general the concentrations of all the elements determined in recycled paper and board are higher than those found in virgin paper and board. Both sets of samples contained a range of items, including boxes and bags, representing a variety of grades of packaging. The elemental composition was influenced more by its nature (i.e. virgin or recycled) than by the grade of material, indicating that it was the recycling process that concentrated the elements. In general terms the concentration factor appears to be around tenfold, as demonstrated in table 4 for the common elements. However, individual items demonstrate elevated levels of certain elements, regardless of the origin of the paper and board, suggesting that additives have a role to play. For example, those items with shiny surfaces display higher Cr values, possibly due to the use of greaseproofing agents containing Cr, confirming earlier findings. In general Cr, Zn and Ba are consistently high in recycled paper and board, as demonstrated in table 6. However, it is extremely unlikely that even these high levels would present a hazard to health

In conclusion, NAA provides a sensitive method for the determination of elements in paper and board. The results of a comparison between virgin and recycled paper and board demonstrate that the elements are concentrated to a level of about tenfold during the recycling process.

ACKNOWLEDGEMENTS

This work was carried out for the Food Standards Agency under project number A03026. A copy of the full report is available from the Food Standards Agency library[6].

References

1. Z. F. Ibrahimi, Determination of trace elements in paper and board food contact materials by instrumental neutron activation analysis. MSc Thesis, Imperial College of Science, Technology & Medicine, 1994.
2. S.S. Kim, The UK paper industry and its environmental impact. MSc Thesis, Imperial College of Science, Technology & Medicine, 1996.
3. S.J. Parry, *J. Radioanal. Nucl. Chem.*, 2001, **248**, 143-147.
4. L. Castle, C. P. Offen, M. J. Baxter, and J. Gilbert, *Food Additives & Contaminants*, 1992, **14** (1), 35-44.
5. S. J. Parry and D.S.J. Aston, In: *International Conference on Nuclear Science & Technology, Washington, USA..*, 2000, **83**, 485-486.
6. Project A03026, Investigation of the migration of inorganic contaminants into dry food from packaging made from recycled paper and board. The Library, Food Standards Agency, Aviation House, 125 Kingsway, Holborn, London, WC2B 6NH

EQUILIBRIUM FACTOR F AND EFFECTIVE DOSE EQUIVALENT WITH SOLID STATE NUCLEAR TRACK DETECTORS

A. Chávez, M. Balcázar and M. E. Camacho.

Instituto Nacional de Investigaciones Nucleares, Apdo Postal 18-1027, México D.F. 11801, México

1 INTRODUCTION

The most important natural radioactivity source in the air is the ^{222}Rn and its decay products. ^{222}Rn is produced by the decay of ^{226}Ra, which is present in soil and building materials. It is known that radon and its decay product concentrations are higher in closed environments than in the opened ones, because the high concentration due to radon and its decay products slowly builds up in atmospheres and in poorly ventilated areas. These high-activity air concentrations produced by radon and its descendants have been studied as sources of radioactive pollution for their possible effects on the health of the population[1].

The activity concentration in the air due to ^{222}Rn and its α emitting decay-products, such as ^{218}Po and ^{214}Po, can be determined using solid-state nuclear track detectors (SSNTD). These detectors are used, with adaptations, to quantify the activity due to radon and that due to its decay products, since experience has shown their capability for dosimeter purposes[2].

The ^{222}Rn in the atmosphere is not necessarily in equilibrium with its short life progeny, so it becomes necessary to know a relative equilibrium relationship between those and the existing radon concentration by introducing the equilibrium factor F[3].

In a previous work[4], we reported the presence of radon decay products attached to cloths of several workers at the Tandem accelerator facility, as determined by a whole body counter. It was found that the activity of attached radon decay products depends on the type of cloth material and indoor radon. The purpose of this work is to obtain the effective dose equivalent of several working areas at the Nuclear Centre.

2 METHODS AND RESULTS

Each sampler consisted of two LR115 type plastic detectors, one of which was placed on the inner bottom of a cylindrical can 12.5 cm in length and 10 cm in diameter. The opened extreme of the can was covered with a thin polyethylene film 5μm in thickness, which stops the entrance of aerosols, permits a quick diffusion of radon inside and maintains a radon equilibrium atmosphere in the detection volume. The second detector was placed on the external part of the can to expose it directly to the environment without any aerosol protection, with the aim of calculating the F factor and the H_E effective dose equivalent.

Each sampler was placed at 1.5 m high from the floor, in the facilities where the studies took place and remained there for a period of 60 days; at the end of which, the detectors where chemically etched with the usual chemical methods, and the number of tracks counted automatically by means of a Spark Counter[5].

The activity of Radon and its decay products where evaluated during two bimonthly periods at ten selected places and two facilities.

The ten selected places were mainly study cubicles and laboratories. Table 1 shows the results for the spring monitoring period; the second and third columns display the specific activity for the internal D_0 and external D detectors. Forth and fifth columns show the F factor and the calculated effective dose equivalent H_E. Table 2 displays the same parameters for the summer period.

The two facilities are the TRIGA MARK III research reactor displayed in Figure 1, which is divided in four areas: three warehouses (1,2,3), and hallway (4). The Tandem Accelerator, Figure 2, had five areas: accelerator room (1), bombarding room (2), control room (3), warehouse (4) and chief room (5). Table 3 and Table 4 show the results obtained in these areas, with the same parameters described above, for spring period only.

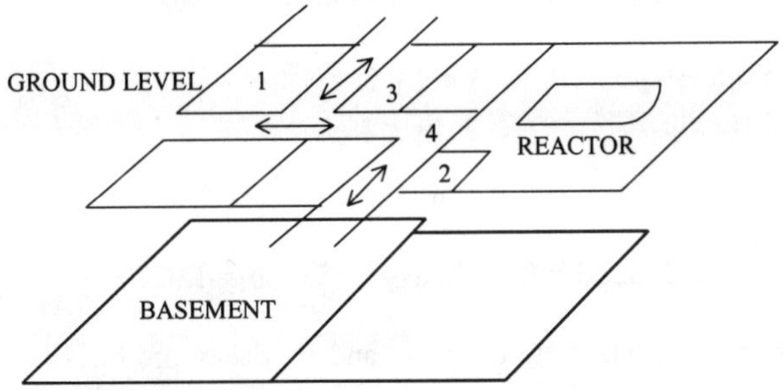

Figure 1 *Reactor building sketch*

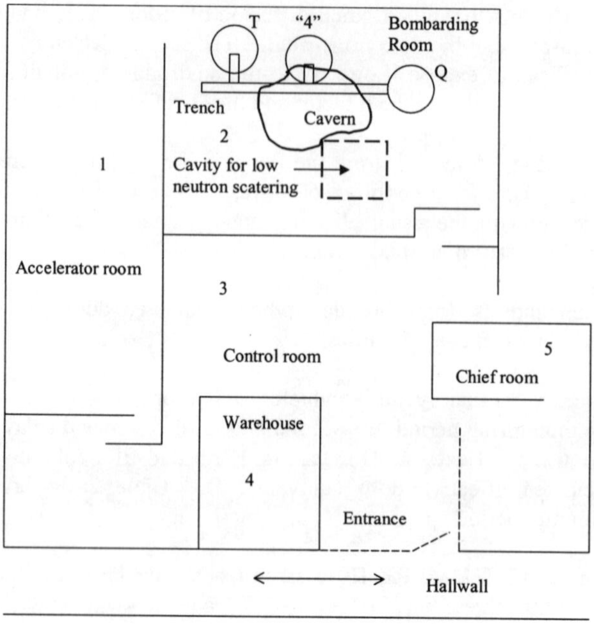

Figure 2 *Accelerator building sketch*

Planinic and Faj proposed[2] two equations, which relate the observed α-track density on two simultaneously exposed SSNTD; one facing directly to the environment (D), and the other one protected by a membrane, permeable to radon (D_0)

$$1)\ F_1(x) = a_1 \exp(b_1 x) - d_1 \qquad 1 < x < 2$$

$$2)\ F_2(x) = a_2 x - b_2 \qquad 2 \leq x \leq 3$$

Where $x = D/D_0$

And a_1=0.0041; b_1= 2.2252; d_1= 0.0352; a_2= 0.6757; b_2= 1.0270.

The effective dose equivalent H_E of radon and its decay products is calculated with the following equation[3]:

$$H_E = c_0(d_0 + d_E F)$$

Where d_0=0.33 $\mu Svy^{-1}Bq^{-1}m^3$, d_E=80$\mu Svy^{-1}Bq^{-1}m^3$ and c_0=D_0/k being k= 3.127x10^{-3}m, k is a factor related to the detector's sensitivity to α particle registration and c_0 is the real radon concentration in the air.

The results of the activity due to radon and its decay products obtained for the detectors D and D_0 from the ten mentioned sampling places and the two facilities are shown in the Tables 1, 2, 3 and 4.

Table 1 *Ten selected monitoring places, spring period.*

SAMPLER	D (Bq/m^3)	D$_o$ (Bq/m^3)	F Factor	H$_E$ (mSv/year)
1	156	145	0.0097	0.14
2	227	178	0.0135	0.46
3	348	267	0.0393	0.78
4	352	241	0.0705	1.22
5	380	221	0.1529	2.78
6	194	107	0.1960	1.72
7	322	173	0.2227	3.14
8	697	358	0.2768	6.79
9	220	104	0.4023	3.38
10	649	270	0.5971	11.0

Table 2 *Ten selected monitoring places, summer period.*

SAMPLER	D (Bq/m^3)	D$_o$ (Bq/m^3)	F Factor	H$_E$ (mSv/year)
1	124	106	0.0201	0.02
2	169	81	0.3827	2.51
3	269	173	0.0902	1.32
4	339	271	0.0311	0.76
5	278	144	0.2657	3.11
6	157	77	0.3507	2.19
7	85	49	0.1594	0.64
8	260	63	0.7616	8.90
9	107	73	0.0717	0.44
10	439	106	0.7714	15.06

Table 3 *Monitoring places at the Reactor facility. Springtime.*

SAMPLER	D (Bq/m^3)	D$_o$ (Bq/m^3)	F Factor	H$_E$ (mSv/year)
1(warehouse)	3027	636	2.1889	94.18
2(warehouse)	3354	3122	0.0095	2.89
3(warehouse)	2847	639	1.9835	85.75
4(hallway)	364	326	0.0139	0.40

Table 4 *Monitoring places at the Accelerator facility. Springtime.*

SAMPLER	D (Bq/m^3)	D$_o$ (Bq/m^3)	F Factor	H$_E$ (mSv/year)
1(accelerator room)	2350	1575	0.0782	8.76
2(bombarding room)	3226	2334	0.0536	9.10
3(control room)	3134	3038	0.0063	2.53
4(warehouse)	3004	1569	0.2552	27.47
5(chief room)	2424	660	1.4546	65.00

3 DISCUSSIONS AND CONCLUSION

For the ten selected places (Table 1 and Table 2), radon concentration was on average nearly double in the spring (206 ± 80 Bqm^{-3}) than during summer (114 ± 66 Bqm^{-3}). The opposite occurs for the specific activity of the attached decay products with values of 345 ± 184 Bqm^{-3} and 222 ± 113 Bqm^{-3} respectively. The reason is the dry and rather warm environment during springtime, which favours radon exhalation from the pores of the emanating material. During the raining summer season, wet and rather cool weather increases the humidity, filling in the pores and partially screening radon emanation. The increase in humidity, produce an increment of the average $D/D_0=1.69 \pm 0.41$ for the dried season to $D/D_0=2.15 \pm 1.08$ for the wet season. Therefore, although the radon concentration decreases during the wet summer season, the increase of D/D_0 makes an increase of F= 0.198 ± 0.188 to F= $0.290 \pm 0.0.281$. This effect maintains an average effective dose equivalent of around 3 $mSvy^{-1}$ for both seasons.

Sampling point 2 in Table 1 is another example of the above statement, although there is a reduction in radon concentration, by a factor of 2.2, and a reduction in the activity measured by internal detector, by a factor of 1.3, the F factor increases from 0.035 to 0.3327, and H_E by a factor of 5.4. For the same sampling period and similar radon concentrations (see sampling points 3 and 10 for spring period), the increase in aerosol concentration facilitate the attachment of decay products, increasing the activity by a factor of 1.8, but with an increase in H_E by a factor of 4.

This effect is particularly seen for the accelerator facility in the chief room, which has the lowest radon concentration for that facility, but the highest effective dose equivalent (65 $mSvy^{-1}$. A remedial action consisted in removing the thick carpet, which had no vacuum cleaning, and sealing small fractures at the ceiling to reduce the high humidity in the room. Those actions reduced dramatically the H_E value down to 2.8 $mSvy^{-1}$ for an F factor of 0.013 and $D/D_0 = 1.28$.

The highest radon concentration is found in the control room (3038 Bqm^{-3}) followed by the bombarding room (2334 Bqm^{-3}) and the accelerator room (1575 Bqm^{-3}). The most probable source of radon is from a cave near below the floor of the bombarding room and a low-background neutron-scattering cavity. Both, the bombarding room and the control room are maintained in clean conditions and hence have a low F factor. This is not the case for the warehouses.

The three mentioned rooms are connected, usually the control room is kept at warm temperature by a local heater followed by the accelerator room, where the equipment installed heats the environment; then, the air diffuses from the bombarding room (the coolest room of the three of them) to the other two rooms. The proposed remedial action is to install an extracting air system in the bombarding room to see the resulting radon effect in the other two.

The warehouses in the accelerator and the reactor facilities are dusty places with little or no ventilation, and are usually closed. Therefore, they have an associated high F factor; no personnel are permanently working in those areas.

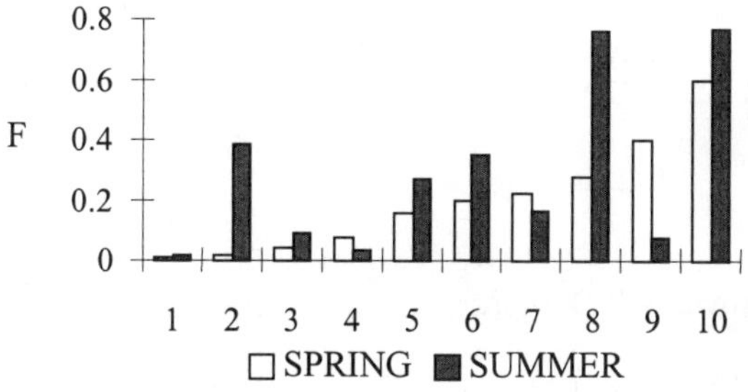

Figure 3 *The F factor in spring and summer*

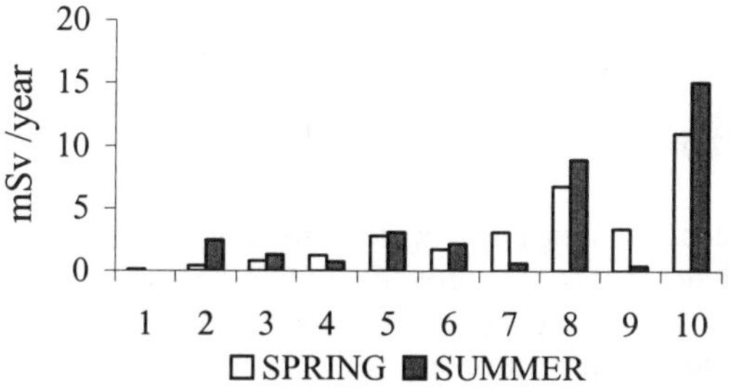

Figure 4 *The effective dose equivalent H_E in the spring and the summer*

References

1. R. Crameri and W. Burkart, *Radiat. Phys. Chem*, 1989, **34**, 251.
2. J. Planinic and Z. Faj, *Nuclear Instruments and Methods in Physics Research*, 1989, **A278**, 550.
3. J. Planinic and Z. Faj, *Radiation Protection Dosimetry*, 1991, **35**, 265.
4. M. Balcázar, A. Chávez, G. Piña-Villalpando, M. Alfaro and D, Mendoza, *Radiation Measurements,* 1999, **31**, 337.
5. M. Balcázar and A, Chávez, *Nuclear tracks and Radiation Measurements*, 1984, **8**, 617.

BIOLOGICAL MONITORING OF RADIOACTIVITY AND METAL POLLUTION IN EDIBLE EGGS OF *Liometopum apiculatum* (ANTS) FROM A RADIOACTIVE WASTE SITE IN CENTRAL MEXICO

M.I. Gaso[1*]., N. Segovia N. [1], O. Morton O.[2] and M.A. Armienta[2]

[1]ININ, Ap. Post 18-1027, 11801 Mexico D.F.,
[2]IGFUNAM, Ciudad Universitaria, 04510 Mexico D.F., Mexico.

1 INTRODUCTION

Many ancient Mexican cultures have made insects a special ingredient in their diets providing an excellent source of protein. Immature stages of reproductives from *Liometopum apiculatum,* belonging of the Order Hymenopterae locally known as "escamoles", are considered in Mexico to be a delicacy. Nowadays, "escamoles" are sold in local native markets and served in restaurants mainly in Mexico City and surrounding States. Insects considered to be "social", or those that develop in groups such as bees, ants and tree worms, are the most often eaten. In Mexico, "escamoles" cost ten times more than meat, and the white agave worms, fourteen times more. Today insects are exported to major cities around the world and sold at incredible prices in gourmet food shops.[1] It is worth mentioning that a *Liometopum* specie (*Liometopum microcephalum*), found at the Alpes region, was probably also included in the diet of ancient European cultures. This particular specie build their nests at a warmer arboreo stratum, instead of underground cavities.[2]

Insects such as *Liometopum apiculatum* are an extremely rich source of protein (67%), minerals (5%) and fat (12%). This characteristic makes them a particularly attractive food source for developing nations where protein is not as readily available as it is in more industrialized countries. The "management" of the nests by the peasants has permitted their exploitation for many years. One mature nest can yield as much as 3 kg of "escamoles" and each "escamol" can measure 1.5 mm.[1,2]

Because of their ability to adapt, insects have conquered practically all existing habitats worldwide, even such unlikely environments as petroleum contaminated fields, mines and, as in our case, storage centres for radioactive waste. In order to observe, under natural field conditions, the effect of some trace elements and pollution from uranium ore tailings and cesium, samples of "escamoles", nest trabecula and soil from the Mexican Storage Centre for Radioactive Waste (SCRW), were analysed. The samples were taken from the SCRW and from a non-polluted area located in a semi-arid region from the border of Mexico and Hidalgo States.[3,4] Ants express contamination as a whole through ingestion of polluted food such as live plants, micro organisms, soil, water but also through cutaneous contact. In this paper, transfer factors soil-"escamoles", soil-nest trabecula and soil-local vegetation have been calculated for different trace stable and radioactive elements.

2 METHOD AND RESULTS

2.1 Sampling and analysis methodologies

The SCRW site (19° 47' 39" N; 98° 50' 04" W) is located in a semi-arid region at an altitude of 2475 m in the middle part of the Neovolcanic Mexican belt. At this site, solid and liquid radioactive waste were stored for two decades and uranium ore tailing piles also stood for some time before being buried in specific containers. The presence of the ore contaminated the soil with radium. A local contamination in the soil with ^{137}Cs also occurred at the site as a consequence of a broken industrial source.[3] The climate at the zone is temperate sub humid and the mean annual precipitation is 638.5 mm, mainly during the rainy season, from June to October. *Liometopum apiculatum* feeds from local vegetation such as *Opuntia megacantha, Agave atrovirens* and *Schinus molle*. The ants construct theirs nests 60-80 cm depth under the stones located near the above mentioned plant roots. The nest is constructed during the rainy season through transport of several biological materials and little branches available in the soil of the near-by region (contaminated in this case). The anastomosed nest trabecula has a hemispherical structure of 35 cm diameter. Over the nest trabecula, like a hard net or sponge, the "escamoles" are grown by the ants. In March-April, the "escamoles" must be recollected, just before the nuptial flight of the ants, occurring between April and May, the day after a strong rain, when environmental temperature is around 26 °C.[2]

During 2000 and 2001, "escamoles", nest trabecula and the soil near the nest were sampled at the SCRW. Reference "escamoles" samples were purchased at local native markets 15 km around. The samples were dried at 110 °C, disaggregated, homogenized and ground following a previously reported methodology.[3] Major elements were measured with an atomic absorption spectrophotometer Perkin Elmer 2380. For the trace metal elements determination, 1 g samples were digested with a mixture of concentrated acids (5 ml HCl, 5 ml HNO_3 and 10 ml $HClO_4$). The solution was evaporated to dryness and the residue was subsequently dissolved and diluted to 100 ml with HNO_3 2%. The quantification was carried out with an ICP-MS VG Elemental model PQ3. Matrix effects and instrumental drift were eliminated by the use of an ^{115}In (10 ppb) internal standard.[5] The validity of the analytical procedure has been assessed by comparison with Standard Reference Materials (IAEA SOIL 7; SRM-2586 (NIST) and 1573a (NIST)).

The samples were also analysed for ^{137}Cs, ^{226}Ra and ^{40}K activity concentration, by low background spectrometry with an HPGe detector, Princeton Gamma Tech., Model N-IGC 29. The geometry used was a 500 ml Marinelli beaker. The measurement time was from 1000 to 60 000 s. Counting errors for the measurements were usually lower than 10 %.[2] The transfer factors (TF) soil-"escamoles", soil-nest trabecula and soil-plants for ^{226}Ra, ^{137}Cs and other trace elements were determined.

2.2 Results and discussion

Trace elements at the SCRW samples were compared with reference samples. The results together with TFs soil-"escamoles" and soil-nest trabecula are shown in Table 1. The higher values for all the elements except Se, were obtained at the SCRW "escamoles" samples. Elements such as Mn, Co, Mo, Tl, Pb, Rb, Sr and Ba are mainly related to the unprotected pile of waste uranium ores and other wastes that generated the contamination of soil and surface waters.[3] As mentioned before, the SCRW has experienced, along the time, several localized soil contaminations and also partial soil decontamination was performed between 1993 and 2001, by removing the surface soil layer and confining it into containers.

In the case of barium, being an alkaline earth metal like calcium, when the ants extract calcium from the environment, they inevitably take its chemical analogues (Sr, Ba, Ra). Barium can enter biological systems in ionic form due to its chemical similarity to calcium; the ionic radii and coordination chemistry are very similar. Given the potential toxicological and ecologic effects of barium, understanding its transport and localization in ecosystems is a matter of interest.

Table 1 *Trace elements content (mg kg^{-1} (dry wt.)) in the "escamoles" from the SCRW and the reference site (REF.); trace elements content in the soil and in the nest trabecula from SCRW (2001-2002); soil-"escamoles" and soil-nest trabecula transfer factors (TF's).*

ELEMENT	TOXIC LEVEL	ESCAMOLES			SOIL SCRW	TF Soil-"escamol"	NEST SCRW	TF Soil-nest
		SCRW	REF.	SCRW/REF.				
V	10	15.23	7.28	2.09	41.41	0.37	72.27	1.75
Cr	1	8.60	2.59	3.32	42.20	0.20	42.06	1.00
Mn	100	54.26	4.88	11.12	270.61	0.20	450.23	1.66
Co	3	1.11	0.03	37.17	7.73	0.14	7.54	0.98
Ni	2	2.48	1.24	2.00	16.44	0.15	21.45	1.31
Cu	250	19.49	13.68	1.42	69.25	0.28	20.66	0.30
Zn	500	134.03	88.49	1.51	82.40	1.63	104.63	1.27
As	1	6.09	4.03	1.51	2.34	2.60	6.97	2.98
Se	1	1.65	1.89	0.87	0.78	2.12	3.78	4.86
Mo		4.96	1.18	4.20	1.86	2.66	4.02	2.16
Cd	0.2	0.14	0.06	2.22	0.01	23.04	0.38	62.53
Sb		0.87	0.50	1.74	0.11	7.82	1.56	14.02
Tl	0.09	0.28	0.04	7.00	0.20	1.42	0.31	1.57
Pb	3	1.24	0.35	3.54	5.23	0.24	25.64	4.90
Rb		7.00	1.00	7.00	23.95	0.29	26.88	1.12
Sr		29.75	1.20	24.79	239.62	0.12	428.91	1.79
Ba	500	58.34	1.36	42.90	449.65	0.13	460.98	1.03
Cs		0.29	0.20	1.45	1.49	0.19	1.21	0.81

For the biodegradation of lignocelluloses the manganese peroxidase lignin-degrading enzyme plays a predominant role. The ants efficiently mineralize lignin molecules and accelerate the biodegradation processes of lignin if larger amounts of manganese are present at the soil. A relatively high concentration of manganese (450.23 mg kg^{-1}) was measured in the nest trabecula. The enhanced Mn content can influence the uptake of micronutrients (Cu, Zn) and toxic elements (Cd, Pb; Ni) investigated.[6]

Thallium is naturally found in soil at levels from 0.3 to 0.7 mg kg^{-1}.[7] Since Tl is an element generated from the decay of ^{232}Th, its enhancement in the present results is probably due to the local U tailings and radioactive sources management at the site. It is noticeable that Tl is reported 14 times higher in the SCRW soil as compared with to the reference,[3] and 7 times higher at the site "escamoles" as compared with the reference. Increased Tl levels in plant tissues are highly toxic to both plants and animals. The cation Tl^{+} is highly associated with K and Rb, it tends to bind with sulphide compounds and can interfere with pyruvate in the carbohydrate metabolism. Exposure to thallium occurs mainly from eating food. Significant routes of exposure near hazardous waste sites are through swallowing thallium contaminated soil or dust, drinking contaminated water and skin contact with contaminated soil. The US Environmental Protection Agency (EPA) has determined a water quality criteria level of 0.013 mg kg^{-1} of Tl in water to protect humans from harmful effects of drinking water containing thallium.[8]

Nickel was twice at the SCRW "escamoles" compared with the reference; Ni, is an element classified by the International Agency for Research on Cancer[9] in the first priority group due its high toxicity. For the reference man (70 kg), the oral reference dose is 1.4 mg d^{-1}.[10] The workers involved in mining and milling ores are exposed to hazardous agents such as radioactive (uranium and thorium) and stable (nickel, thallium) elements, that can cause a health detriment.

The soil-"escamoles" TFs higher than 1 at the SCRW were found for Cd, Sb, As, Mo, Se, Zn and Tl, the highest being for Cd. This element has a large mobility and its uptake exceeds the values of Pb by two orders of magnitude. Cd binds to sulfhydryl groups and, as Cd^{2+}, can substitute Zn^{2+} in enzymes. Zinc influences the permeability of membranes, stabilizes cellular components and is believed to stimulate the resistance of many organisms to dry and hot weather.[11]

Soil samples obtained near the nest, together with the nest trabecula were also analysed for some major element contents. For Ca and Mg, the nest trabecula had 3.4 and 1.6 times higher values as compared with the soil, indicating that an enrichment for these elements occurs through enzymatic processes realized by the ants.

Determinations of ^{226}Ra, ^{137}Cs and ^{40}K have been performed in the surface soil of the SCRW and the reference sites from 1991-2001.[3] The average activity concentration and the values range obtained at the SCRW, were 3281 (71-6384) Bq kg^{-1} (dry wt.) for ^{226}Ra and 290 (4-4374) Bq kg^{-1} (dry wt.) for ^{137}Cs. At the reference sites, the average activity concentration, and the range in the soil samples were 37 (6-90) Bq kg^{-1} (dry wt.) for ^{226}Ra and 4 (1.8-19) Bq kg^{-1} (dry wt.) for ^{137}Cs. For ^{40}K the average values were similar (290 and 339 Bq kg^{-1} (dry wt.)) at the SCRW and reference sites respectively.

The soil samples taken near the ant nest (70 cm depth) had 137, 3.7 and 313 Bq kg^{-1} (dry wt.) for ^{226}Ra, ^{137}Cs and ^{40}K respectively. Only the nest trabecula had detectable radioactivity values: 60 Bq kg^{-1} (dry wt.) for ^{226}Ra, 13 Bq kg^{-1} (dry wt.) for ^{137}Cs and 473 Bq kg^{-1} (dry wt.) for ^{40}K. "Escamoles" samples from the SCRW and the reference sites had below the limit of detection (<0.6 and < 0.3 (Bq kg^{-1}, dry wt.) ^{226}Ra and ^{137}Cs values. ^{40}K average values in these samples was 287 and 195 (Bq kg^{-1}, dry wt.) at SCRW and reference respectively. The transfer factors soil-nest trabecula from SCRW were 0.44 and 3.5 for ^{226}Ra and ^{137}Cs respectively.

The impact of natural radioactivity due to uranium and thorium decay series has become an increasingly important environmental question. ^{226}Ra, is one of the more radiotoxic radionuclides, being of concern particularly in connection with uranium milling. ^{226}Ra, together with ^{40}K are the main naturally occurring radionuclides which enter vegetable foods, thereby becoming a source of internal radiation to man and animals.

The *Agave atrovirens* samples were obtained near the ant nest region. Soil samples from the vicinity of the plant had 83 and 3.7 Bq kg^{-1} (dry wt.) for ^{226}Ra and ^{137}Cs respectively. This agave specie (known as "maguey" in Mexico) has been used traditionally to obtain an alcoholic beverage called "pulque" that appears in pre-Hispanic archeological evidences playing an important social, economic and religious role. The sap of the plant, called "aguamiel" or honey water, becomes "pulque" through a natural fermentation process which usually occurs within the center of the plant (the sampled part in the present work). The activity concentrations for ^{226}Ra, ^{137}Cs, and ^{40}K for the agave sample were: 160, 1 and

463 (Bq kg $^{-1}$, dry wt.) respectively. A sample of the same plant specie taken at a reference site had <11, <0.6 and 534 (Bq kg $^{-1}$, dry wt.) for ^{226}Ra, ^{137}Cs, and ^{40}K.

For *Opuntia megacantha,* similar values were obtained (162, 5 and 507 (Bq kg $^{-1}$, dry wt.) for ^{226}Ra, ^{137}Cs, and ^{40}K respectively, while at the reference site activity concentration values of <10, <0.7 and 958 (Bq kg $^{-1}$, dry wt.) for ^{226}Ra, ^{137}Cs, and ^{40}K respectively were measured.

It should be emphasized that consumption of the above-mentioned species as a delicacy is a common practice. In a few countries, on several occasions, official publications have indeed warned people against the possibility of poisoning caused by heavy metals in wild edible foods from semi natural ecosystems.[12] Cause and effect relationships are not evident for physicians in poisoning cases but some equivocal cases of poisoning may result from repetitive consumption of metal polluted wild foods. The presence of toxic elements such as V, Cr, Ni, Tl in the dried "escamoles" (dry matter ≅ 70%) from the SCRW samples, were higher than the toxic level accepted for these elements (10, 1, 2, 0.09 mg kg^{-1} (dry wt.) respectively).[13]

3 CONCLUSION

This study provides a basis for estimating uptake of metal and radioactive elements by social insects and the possible application as bio indicators. The evaluation of transfer coefficients, shows the bioavailability of toxic metals and of radium and caesium uptake by ants native species growing at a waste site.

The impact of contamination of the terrestrial environment by radionuclides and metallic trace elements is difficult to evaluate because of the complexity of soil ecosystems and high soil variability. This work points to the need for further laboratory and field studies on metal and radioactive accumulation by ants. To predict the environmental impact of chemicals on soil ecosystems, some authors[14] propose the use of a laboratory microcosm with insects in standard conditions, as "sentinels" and metal pollution bio indicators, for the assessment of the bioavailability of anthropogenic elements in their environment.

Acknowledgements

The authors acknowledge E. Quintero, E. Hernandez, G. Valentin, R. Benitez, V. Rojas and F. Montes for technical assistance.

References

1 J. Ramos-Elorduy. *Creepy Crawly Cuisine. The gourmet guide to edible insects,* Park Street Press, Rochester, Vermont, 1998, 150.
2 J. Ramos Elorduy, B. Delage-Darchen, N. Galindo and J. M. Pino. *Anales Inst. Biol.* UNAM, 1987, **58**, 341.
3 M.I. Gaso, N. Segovia, L. Cervantes and S. Salazar, *Environmental Radiochemical Analysis,* ed. G.W.A. Newton, The Royal Society of Chemistry, Cambridge, 1999, 50.
4 M.I. Gaso, N. Segovia and O. Morton, *Radioprot. Colloques*, 2002, **37**, 865.
5 O. Morton-Bermea, E. Hernández, I. Gaso and N. Segovia, *Bull. Environ. Contam. Toxicol,* 2002, **68**, 383.
6 L. Rácz, L. Papp, V. Oldal and Zs. Kovács, *Microchem. Journal,* 1998, **59**, 181.

7 A. Kabata-Pendias and H. Pendias, *Trace elements in soil and plants*, CRC Press Inc., Boca Raton, 1985, 315.
8 ATSDR (Agency for Toxic Substances and Disease Registry), *Toxicological profile for thallium*, Atlanta, 1992, 8737.
9 IARC (International Agency for Research on Cancer), *Monographs on the evaluation of carcinogenic risk to humans. Chromium, nickel and welding,* Lyon, 1990.
10 IRIS (Integrated Risk Information System), *Data base. EPA-U.S. Environmental Protection Agency,* Washington D.C., 1993.
11 D. Aruguete, J. Aldstadt and G. Mueller, *Sci. Total Environ.*, 1995, **173**, 41.
12 P. Kalac and L. Svoboda, *Food Chem.*, 2000, **69**, 273.
13 M. Awadalah, H. Amrallah and F. Grass, *Environmental Radiochemical Analysis,* ed. G.W.A. Newton, The Royal Society of Chemistry, Cambridge, 1999, 330.
14 A. Gomot de Vaufleury, *Ecotoxicol. Environ. Saf.,* 2000, **46,** 41.

PREPARATION OF ENVIRONMENTAL SAMPLES FOR RADIOANALYTICAL ASSAY: AN OVERVIEW*

S. M. Cogan and K. S. Leonard

Centre for Environment, Fisheries and Aquaculture Science, Lowestoft Laboratory, Pakefield Road, Lowestoft, Suffolk NR33 0HT, UK

1 INTRODUCTION

The key issues for any quality analytical determinations are accuracy, sensitivity and reproducibility. Factors that contribute to these issues include aspects such as the underlying science, chemical procedures, techniques and instrumentation. The latter are frequently discussed and reported thoroughly to demonstrate that the quality of a result is maintained or improved. Analytical accreditation (from an appropriate authority) has become a significant part of radioanalytical assay, and a retrospective appraisal of the advantages and disadvantages of accreditation has been reported recently.[1]

The preliminary stages before analysis, such as sampling and initial preparations (excluding chemical procedures), are not often discussed in any significant detail. Indeed, for many, sample preparation is the tedious part of the analytical procedure, hence the reason for the lack of interest on a wider scale. It is not intended to discuss the sampling aspects for chemical assay, as this is a subject of it own, and it has received widespread comment in certain environmental disciplines. For example, for soil sciences, it was recently concluded that without the harmonisation of sampling protocols comparability of any (biological or chemical) data remain questionable, and the best quality of analysis is useless when doubts are thrown on the representativeness of the analysed samples.[2] Furthermore, it is also worth noting that, as far as quality assurance is concerned, sampling in the field cannot be conducted under the same controlled conditions and to the same extent, as for the preparation of samples in the laboratory.

The aim of this paper is to provide an overview of the objectives, rationales and practices of a typical environmental preparation service. By highlighting the contribution to the radioanalytical assay, it is anticipated that the information given here will provide an awareness of the requirements to operate a preparation group. Samples are prepared in accordance with standard operating procedures (SOPs), providing specific instructions for the particular activities, but the intention here is to inform laboratory personnel of some of the important issues that are required to be considered in assessing the reliability of their procedures.

2 PREPARATION OBJECTIVES

The overall objective for preparation is to provide the "analyst" with a suitable aliquot of the collected environmental sample for further chemical separation, purification and chemical determination. For radiometric assay, the sample matrices extend over a wide range of samples including soils and sediments, waters and effluents, plant and animal tissues and food products. Nevertheless, for all types of matrices considered, it is essential that the following standards are adhered to, for all samples, to ensure quality of service.

- Fully identifiable
- Provided with the appropriate supporting data (e.g. dry/wet ratios)
- Representative of the original environmental matrix
- Free from contamination (including from other samples)
- Provided in the required form (e.g. edible or other fraction)
- Provided in the required physical state and/or geometry
- Provided to the required customer time-scale

3 SAMPLE IDENTIFICATION

To ensure that a sample is fully identifiable, a robust system is required to track the sample from initial receipt to the laboratory and thereafter through the various preparation schemes. The system must also allow for correct identification during radiochemical analysis, storage (and/or further transport) and allow for the communication of the result.

Sample identification is best maintained by using a unique tracking number on receipt of the sample to the designated area. Prior to assigning a unique identifier, it is necessary to check the integrity of the sample for breakage and/or leakage of the sample package or sample container. Any evidence of the latter should necessitate a corrective action to rectify the observation (e.g. request for replacement sample) or provide notification of the observation. A visual inspection of the sample is undertaken to ensure that:

a) the sample is in reasonable condition; and
b) the associated sample details (e.g. sample location etc.) are correct and appropriate within the expected criteria of the sampling schedule (or similar).

Non compliance with these criteria will also necessitate a corrective action. A unique tracking number is assigned to the containment of the sample by either the use of a paper record system and/or computer database, once any discrepancies have been resolved. Having carried out the receipt, the sample can be processed in accordance with preparation procedures (including short-term storage) and laboratory tracking systems. Sample tracking is a process by which the location and status of a sample can be identified throughout. In its simplest form, this may involve transferring the identification number from one sample container to another as part of a particular preparation process (e.g. grinding to produce homogeneity). Sample chain of custody is an important process ensuring the maintenance and integrity of the sample. This requires that documentation of transfer is kept and authorised by both parties (relinquisher and the receiver), and is particularly important for samples that are physically transferred to and from preparation and analytical functions.

4 PROVISION OF SUPPORTING DATA

Most laboratories now have access to computer systems (such as a LIMS database) and these provide the facility to record supporting information at the same time as the sample is assigned its unique sample number. As well as the sample number, information to be documented might include sampling location details, date and time samples were collected and received, requested analyses, type of sample matrix, and any special instructions or observations. As a result of certain preparation processes, further data or information may be generated which can be recorded throughout the procedures and used later to generate an analytical result. One example is the measurement of dry/wet weight ratios, for which both the wet and dry weights of the sample are determined (by removal of the water content by controlled drying) to obtain the ratio.

5 REPRESENTATIVE OF THE ORIGINAL SAMPLE

It is necessary that the sample prepared is representative of the original sample. As a result of the preparation procedures there is the possibility that material may be lost from the sample during certain preparation procedures which could adversely affect the determination. The most likely imprecision can occur from losses through volatilisation of the radionuclide to be analysed. Indeed, the moisture content, and hence the chemical composition of a solid may be altered during simple grinding and milling procedures.[3] Other losses that could also provide potential uncertainties are a) reactions between the radionuclides and the container walls; and b) losses of dust and small particles.

It is important, wherever possible, to demonstrate that procedures are appropriate by testing the existing preparation methodologies. In the example given, here, a sample of crab was collected (May 2002) in the vicinity of Sellafield and the edible fraction was extracted. The sample was thoroughly homogenised to provide three replicate sub-samples of approximately 200g (wet weight). One of the samples was initially dried at a low temperature ($100 \pm 10\ ^{o}$ C), and then dry ashed, prior to the addition of yield recovery tracers to the sample. To test the validity of this current preparation procedure, the two remaining samples were prepared by an alternative, more labour intensive/less cost effective, method. For both these crab samples, yield recovery tracer was added to the wet sample and the samples were wet ashed. The three prepared samples were then provided to the radioanalyst for the determination of plutonium radionuclides ($^{239+230}$Pu and ^{238}Pu) by alpha counting, according to standard methods routinely carried out at CEFAS Lowestoft.[4] The results of this exercise are given in Table 1. Both plutonium determinations produced consistent results (within experimental error) for each sub-sample. The results demonstrate that the original sample was prepared homogeneously, and furthermore, that there was no loss of activity during the processing.

Table 1. *Reproducibility of plutonium analysis in a crab sample using different preparation procedures.*

Sub sample	Wet weight taken g	239+240 - Pu mBq kg $^{-1}$ (wet)	238 - Pu mBq kg $^{-1}$ (wet)	239+240 - Pu/238 - Pu
Sellafield 1 #	206.60	196 ± 3.56%	39.5 ± 6.84%	4.96 ± 7.17%
Sellafield 2 #	201.90	203 ± 3.69%	46.7 ± 6.68%	4.36 ± 7.03%
Sellafield 3 *	222.05	190 ± 3.28%	41.2 ± 6.07%	4.61 ± 6.37%

Errors quoted are ± 1σ percentage counting propagated errors only

Tracer added to wet sample, wet ashed and analysed
* Dried and ashed, and tracer added prior to analysis

Losses due to interactions between sample and container walls occur due to adsorption processes. However these losses can be minimised. For example, for water samples it may be necessary to acidify the samples, whilst being stored prior to processing, to minimise adsorption effects. During dry ashing of solid samples, platinum crucibles should be left uncovered as this minimises the reduction of samples to base metals which may form alloys with platinum. Also, increasing the amount of sample for dry ashing increases the amount of ash, thus permitting the radionuclide to react with the ash instead of the container.

A number of processes may also be responsible for the loss of small particulates. Dry ashing forms a fine ash residue that is susceptible to air-flow over the sample. For example, an air-flow is generated by opening the furnace door when it is hot. Losses are minimised by ashing samples at as low a temperature as possible, gradually increasing the temperature when the sample has stopped combusting at the pre-set temperature. This artefact is not restricted to environmental samples, for example, we have observed that the resin routinely used to concentrate ^{99}Tc from seawater is particularly vulnerable to suspension.

The addition of isotopic yield recovery tracers, at the earliest stage of the radioanalytical procedure is most desirable. Recovery tracers, as well as quantifying losses such as in the dry ashing examples described earlier, also provide a quantitative measurement of losses due to volatilisation from heating. However, in some circumstances it may be necessary to demonstrate the use of a non-isotopic yield monitor. Martinez-Lobo *et al.*[5] considered that none of the twenty or so other nuclides of technetium could constitute a completely ideal yield recovery tracer for the analysis of ^{99}Tc. Harvey *et al.*[6] investigated the use of stable rhenium as non-isotopic yield monitor for ^{99}Tc analysis. A number of dry ashing experiments were performed on a variety of biological matrices for the development of the procedure, to evaluate the losses of ^{97m}Tc and $KReO_4$, at different ashing temperatures. Selected data are reproduced here in Table 2, to indicate the optimum conditions under which both technetium and rhenium are retained quantitatively whilst carbonaceous material is destroyed as completely as possible.

Table 2. *Recovery (%) of technetium and rhenium from biological samples under different ashing temperatures.* [#]

Ashing temperature ^{0}C	Recovery of Tc ^{97m} (%)	Recovery of $KReO_4$ (%)
400	100.9 $\pm$ 1.2	100.2 $\pm$ 1.7
450	99.6 $\pm$ 1.5	100.0 $\pm$ 1.9
500	98.3 $\pm$ 1.8	98.6 $\pm$ 2.0
550	87.5 $\pm$ 5.4	82.2 $\pm$ 1.2
600	41.0 $\pm$ 2.2	48.8 $\pm$ 2.0

[#] Selected data from Harvey *et al.* [6]

6 CONTAMINATION

Contamination may arise from a number of sources which will inevitably bias the result of the sample, either systematically or non-systematically, dependent upon the source of contamination. Possible sources may include equipment/glassware, airborne particles and reagents.

To minimise contamination from equipment/glassware it is essential that well established cleaning protocols are adhered to and are well documented in the SOP's. For the selection of grinders, it is useful to consider the methods required for cleaning between samples; ideally they should be easy to dismantle and contain as few pieces as possible (that are in contact with the sample). Airborne contamination can occur most readily from the grinding of samples. During this procedure, some of the particles produced are 10 μm or less in size and are sufficiently small to be suspended in the air. These small particles will eventually settle out contaminating any surface that they fall on. Therefore, grinding procedures should be carried out in a fume cupboard (or similar) to prevent dispersion or deposition of airborne contamination. A further precaution is to ensure that other samples are removed from the immediate area, during these types of processes, or covered to eliminate the possibility of cross contamination of other samples.

Contamination from radiochemical impurities in reagents can be a serious problem, particularly for low level determinations. A comprehensive report is available which reviews radiochemical contamination of analytical reagents.[7] Care can be taken to minimise contamination using appropriate cleaning procedures. Also, by carrying out blank analytical determinations, it is possible to readily identify any contamination from chemical reagents. It is worth, however, considering carrying out duplicate analyses on a given sample to demonstrate preparation procedures to reduce contamination are working effectively.

7 CONTINUOUS ASSESSMENT

A most useful method of assessing radio-assay procedures is to participate in inter-comparison exercises or proficiency testing schemes. The primary purpose of these schemes is to improve the overall quality of the procedures by comparison of the obtained analytical results with expected results (target values) and results obtained by other participating laboratories. Whilst the inter-comparison sample may be viewed differently from routine samples by analysts (because of its status), the sample should be processed like any other sample. Hence, preparation procedures should be included in these exercises (although not always requested by the proficiency scheme organisers) and carried out as appropriate to current methodologies.

Table 3. *Recent inter-comparison exercises for ^{137}Cs determinations in solid materials.*

Inter-comparison Exercise	Date	Material	CEFAS Result Bq kg^{-1} ± 1σ	Target Value Bq kg^{-1} ± 1σ	All Reported Results (Range) Bq kg^{-1}
FSA	2000	Dried Lobster	95.7 ± 3.1	101.5 ± 15.3	82.0 - 118.9
		Dried Mussel	407 ± 12	405.4 ± 60.8	305 - 516
		Dried Liver	1085 ± 32	1165 ± 175	930 - 1310
IAEA-414	2001	Dried Fish	5.26 ± 0.28	5.16 ± 0.30	4.3 - 6.3
IAEA Almera Proficiency	2001	Spiked Soil	34.09 ± 0.76	34.9 ± 1.0	NA
AIWG ICL01/03	2002	Dried Grass	363 ± 6	370 ± 14	326 - 404
		Dried Cabbage	18.9 ± 0.5	19.3 ± 0.7	17.9 - 21.6

The information given in Table 3 provides selected radiometric results, for ^{137}Cs, of a number of different sample matrices undertaken for recent inter-comparison exercises. The results compare favourably with target values and the range of reported results by other participating laboratories. Although samples were received in a dried state, samples were re-dried and processed, following receipt and prior to analysis, in accordance with our laboratory SOPs. By incorporating the preparation aspects of the analysis, this rationale provides a more comprehensive validation of the complete procedure.

8 REQUIRED FORM

The information below (in Table 4) summarises the main processes to facilitate sample preparation.

Table 4. *Key processes necessary for the preparation of a sample for radiochemical and radiometric assay.*

Processes (for samples)	**Typical Function**
Wet processing	Obtaining a specific fraction for analysis e.g. edible fraction
Dry processing	Removing water content by controlled drying
Homogeneity processing	Providing a representative sample by grinding or milling
Ash processing	Removal of organic material by use of a muffle furnace
Geometry processing	Providing an appropriate counting source e.g. disc or tub

Wet Processing can be considered as a function that excludes part of the sample that is not identified as part of the matrix required for analysis and is carried out before more specific preparations (as given in Table 4). Exclusion of material could constitute a number of environmental materials such as stones, leaves etc. and guidance for removing unwanted material should be documented in standard operating procedures (SOPs) or similar documentation. It is usual to identify, weigh and document any material that is removed from the sample. Another common practice of wet processing, which is more specific to biota and food samples, is to exclude certain tissue from the sample material or to prepare the sample in a manner as would be before being consumed. An example of these could include washing, depurating and cooking a sample of shellfish to obtain a "flesh" fraction for analysis. Jackson & Rickard [8] demonstrated that evacuation of sediment from the gut of winkles resulted in the loss of 40% to 79% of total caesium and 25% to 67% of total americium and plutonium (at 48 hours).

The techniques for dry processing include traditional methods such as air drying, freeze drying and the use of vacuum desiccators and ovens at elevated temperatures. These techniques are to remove moisture or evaporate liquids (to provide a result of dry weight) and to breakdown organic material for subsequent analysis and/or ashing. Heating is by far the most common practice and the important criteria to be considered are the composition of the container material, the increase in heating rates, the maximum temperature and the time available. More recently microwave ovens have been utilised during sample preparation and their use has been recently reviewed.[9]

As mentioned earlier, the use of grinding or blenders to produce a homogeneous sample is a potential source of contamination. However, as well as ensuring that the whole mass of the sample is represented for analysis, the reduction of particle size also facilitates the effect of the subsequent addition of chemical reagents. Dry grinding is most often a better option than wet grinding because it is simpler and quicker. Furthermore, by the use of sieves (of the required mesh size) grinding efficiency of the sample can be well characterised. Samples may also require geometry processing such as for large-planchet total beta or tubs or discs for gamma spectrometry. For the latter the operations may involve filling and weighing (tubs) and the use of hydraulic press and sample moulding die (discs). For constant geometry determinations (gamma spectrometry) there is occasionally insufficient sample to fill tubs. In such circumstances, it a common practice to include an additive such as cellulose powder, however the final sample must be homogeneous for reasons described earlier.

The objective of dry ashing (as well as determining the ash weight) is; a) to combust all of the organic material, and b) to concentrate the activity into a smaller sample size, prior to subsequent treatment using wet ashing or fusion techniques. This procedure usually involves combustion in a muffle furnace to a controlled temperature and the sample is contained in a crucible or open container.

9 SUMMARY

This paper has discussed some of the issues associated with the preparation of environmental samples for radiochemical and radiometric assay, but has not exhausted the subject.

Broadly, the generic requirements for effective preparation as a precursor for radio-assay can be described as:

- Personnel
- Technical Equipment
- Procedures

In this paper, we have outlined some of the important aspects of the working procedures. In our view, the most important requirement is that the laboratory has an effective and appropriate quality assurance system, which includes testing and re-evaluating the procedures.

Acknowledgements

The Food Standards Agency funded this work as part of their food surveillance programme (Contract RA103). The authors would also like to express their sincere thanks to our colleagues in the CEFAS Radioanalytical Services (RAS) Team, for their contributions to the preparation and radioanalysis of samples.

References

1. P. C. Hodson and A. N. Dell, in *Environmental Radiochemical Analysis*, ed. G. W. A. Newton, Royal Society of Chemistry, Cambridge, 1999, p. 312-317.
2. P. Quevauviller, *Sci. Total Environ.*, 2001, **264**, 1.
3. J. A. Dean, in *Analytical Chemistry Handbook*, McGraw-Hill, Inc., New York, Chapter 3, 1995.

4. M. B. Lovett, S.J. Boggis and P. Blowers. The determination of α-emitting nuclides of plutonium, americium and curium in environmental materials: Part 1. Seawater. Aquatic Environmental Protection, Analytical Methods, Ministry of Agriculture, Fisheries and Food, Directorate of Fisheries Research, Lowestoft, 1990. p.1-36.
5. A. Martinez-Lobo, M. Garcia-Leon, G. Madurga, *Int. J. Appl. Radiation Isotope*, 1987, **37**, 438.
6. B. R. Harvey, K. J. Williams, M. B. Lovett and R. D. Ibbett, *J. Radioanal. Nucl. Chem., Articles*, 1992, **158**, 417.
7. J. R. Devoe, *Radioactive contamination of materials used in scientific research*, Publication 895, NAS-NRC, 1961.
8. D. Jackson and A. Rickard, *Rad. Prot. Dosimetry*, 1998, **75**, 155.
9. P. S. Walter, S. Chalk and H. Kingston, in *Overview of Microwave-assisted Sample Preparation*, eds. H. Kingston and S. Haswell, American Chemical Society, Washngton, DC, Chapter 2, 1997.

COMPARATIVE INVESTIGATIONS ON HYDROPHILICITY AND MOBILITY OF HUMIC SUBSTANCES USING RADIOLABELLING TECHNIQUES

H. Lippold, D. Roessler and H. Kupsch

Institut fuer Interdisziplinaere Isotopenforschung, Permoserstrasse 15, 04318 Leipzig, Germany

1 INTRODUCTION

Humic and fulvic acids (HA, FA), natural organic colloids comprising the soluble fraction of humic substances (HS), are known to associate with organic and inorganic contaminants by hydrophobic binding and complexation, respectively.[1,2] Since HS in aquifer systems are subject to partition processes between the aqueous phase and the sediment, migration of colloid-borne pollutants can be enhanced or confined depending on the present geochemical conditions. Due to the complex nature of humic colloids, liquid-solid partitioning is essentially influenced by the variable properties of the solute itself, aside from the surface characteristics of the solid matrix. The investigations presented here are focused on the hydrophilicity of humic materials under varying solution conditions, excluding adsorption equilibria, which are mainly governed by the pH-dependent surface charge of minerals. In order to characterise HS in this respect, octanol-water partitioning of HA and FA from various sources was studied depending on pH and ionic strength. Radiolabelling was applied as a prerequisite for reliable measurements at low concentrations relevant to geochemical systems. Additionally, mobility was evaluated by gel permeation chromatography in a sequential elution procedure.

Several labelling techniques for HS have been proposed in literature, e.g. coupling with pre-labelled organic entities (preferably ^{14}C compounds) or direct radiohalogenation with ^{131}I / ^{125}I by electrophilic substitution at the hydrocarbon skeleton.[3-5] Chloramine-T or hydrogen peroxide in the presence of peroxidase enzymes were used as oxidising agents for the iodide. A problem of this procedure consists in the necessary removal of the auxiliary reagents, as they are likely to be entangled in the colloidal network. To avoid this, we utilised the iodogen method, which was introduced as a gentle option for radioiodination of proteins.[6] Iodogen is nearly insoluble in water, i. e. the reaction is performed in a heterogeneous system.

Alternatively, complexation with $^{111}In^{3+}$ was employed as a labelling reaction, even though the stability of metal-humate complexes is rather low compared to covalent bonds. In the case of trivalent metals, however, complexation is in part irreversible[7,8] and hence comparable to covalent labelling, provided that the loosely bound fraction is separated carefully.

2 EXPERIMENTAL

2.1 Materials

Natural aquatic and soil HS were sampled from the peat bog "Kleiner Kranichsee" in the upper Ore Mountains nearby Johanngeorgenstadt, Germany. Isolation of HA and FA was accomplished following the procedures adopted by the IHSS.[9,10] A commercially available HA from Sigma-Aldrich (Steinheim, Germany) was purified according to the method proposed by Kim and Buckau.[11] The synthetic HA M42 was supplied from Forschungszentrum Rossendorf, Germany.[12] [^{111}In]$InCl_3$ and [^{131}I]NaI were purchased from Amersham (Braunschweig, Germany).

2.2 Methods

2.2.1 Octanol-Water Partitioning. Since HS are prone to be adsorbed at the walls of the containing vessel, both phases have to be analysed directly. In the case of strongly asymmetric partition equilibria, the sensitivity of usual detection methods is not sufficient. Radiolabelling with ^{111}In or ^{131}I was applied to yield a signal intensity that allows reliable measurements even at low concentrations as encountered in nature. Partition experiments were performed in polyethylene test tubes. After shaking for 30 min, aliquots of both phases were analysed with a gamma counter 1480 Wizard (Wallac, Finland).

2.2.2 Sequential Chromatographic Analysis. The method was implemented as a specific application of gel permeation chromatography, providing comprehensive information on stability and selectivity of radiolabelling in the special case of HS.[13] Furthermore, assessments can be conducted with regard to the overall mobility of HS based on the pH-dependent recovery. A Knauer GPC system with UV detector was additionally equipped with a NaI(Tl) scintillation detector. Subsequent to the size fractionation of the mobile HS constituents under selected solution conditions, the immobilised species are eluted in two rinsing steps with 0.1 M NaOH (2nd sequence) and 0.1 M HCl (3rd sequence). Separations were carried out with Fractogel TSK-HW 50(F).

2.2.3 Radiolabelling. For complexation, [^{111}In]$InCl_3$ was allowed to react with the HS stock solution for 30 minutes. Radiohalogenation with [^{131}I]NaI was performed in 500 µl glass tubes with 150 µg iodogen coating at pH 8 (borate buffer), shaking permanently for 10 minutes. Input activities were suited to yield specific activities of about 10 MBq per mg HS after purification by ultrafiltration (3 kD).

3 RESULTS AND DISCUSSION

Figure 1 shows sequential chromatograms of two HA labelled with ^{111}In and ^{131}I, respectively. As regards ^{111}In, the label appears to be persistent at moderate pH values, which is evident from the absence of any bleeding in the 1st sequence of the radiochromatogram. The immobilised In-humate fraction is, however, attacked by the treatment with NaOH in the 2nd sequence. A certain amount of ^{111}In is deposited as hydroxide and resolubilised with HCl in the 3rd sequence.

In contrast, the radiochromatogram of the ^{131}I-labelled HA contains an additional peak within the 1st sequence. The peak area increases significantly if the sample is exposed to daylight over a longer period of time, indicating the photosensitivity of C-I bonds. Consequently, experiments must be conducted in the dark to avoid loss of label.

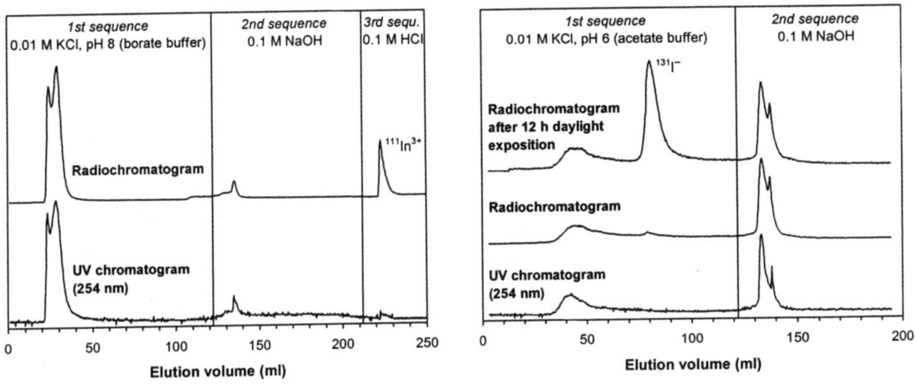

Figure 1 *Sequential chromatographic analysis of ^{111}In-labelled bog soil HA (left) and ^{131}I-labelled Aldrich HA (right)*

In Figure 2, partition ratios of different HS are represented as negative logarithms to indicate the degree of hydrophilicity. According to their nature as polyelectrolytes, HS exhibit a strong preference for the aqueous phase, thereby reflecting characteristic structural features. HA extracted from solid material (Aldrich HA, bog soil HA) show a lower affinity for the aqueous phase than does the aquatic bog HA. The associated FA is even more hydrophilic, which is due to the higher proportion of carboxylic groups. The synthetic HA M42, a polycondensate of xylose and glutamic acid,[12] behaves similarly.

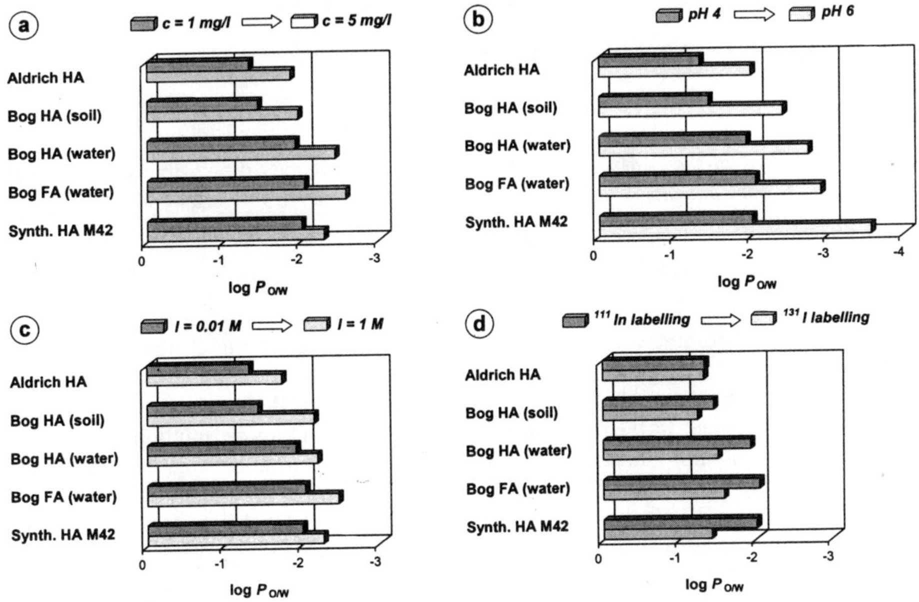

Figure 2 *Octanol-water partition ratios $P_{O/W} = c^{oct} / c^{aq}$ of radiolabelled HS depending on HS concentration (a), pH value (b), ionic strength (c) and labelling (d). The upper parts of the bar pairs refer to 1 mg/l [^{111}In]HS in 0.01 M $NaClO_4$ (pH 4), the lower parts represent the effect of the respective parameter variation.*

Partition ratios proved to be dependent on the concentration (a). Since partitioning of HS displays the sum of numberless interdependent equilibria, Nernst's law does not apply in this case. Raising the pH value from 4 to 6 enhances the hydrophilicity (b). This finding is explained by the increased deprotonation of the acidic groups. An increase in the ionic strength has a similar effect (c), which may be attributed to structural changes induced by the background electrolyte, leading to an augmented exposure of hydrophilic units within the colloids.

Variations in the partition ratios depending on the kind of label (d) are indicative of different selectivities of the labelling reactions. Whereas the target structures of radioiodination are evenly distributed over the HS constituents, the functional groups susceptible to complexation with ^{111}In are concentrated in smaller colloids, rendering them more hydrophilic than the bulk of the polydisperse system.

The pH-dependent distributions of mobile and immobile HS constituents as obtained from the sequential chromatographic analysis are compiled in Figure 3. In general, the pattern resembles the specific partitioning in the octanol-water system, indicating that the overall hydrophilicity of HS is likewise relevant to their partitioning in solid-liquid systems.

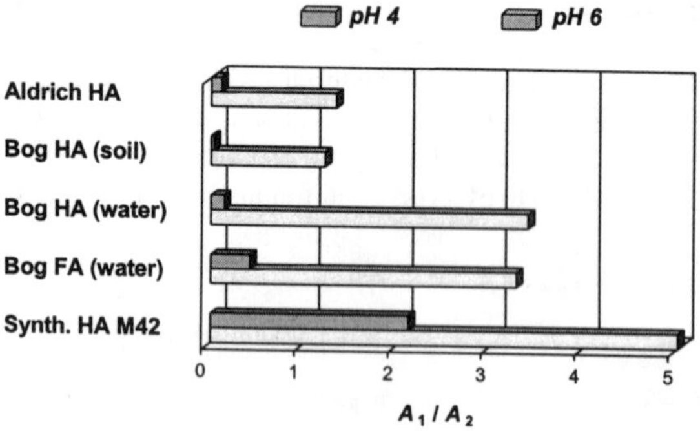

Figure 3 *Ratios of the peak areas A of 1st and 2nd sequence taken from the sequential chromatograms (UV) of different HS at pH 4 and pH 6 (I = 0.01 M)*

4 CONCLUSIONS

Labelling with radioactive isotopes provides analytical access to direct measurements of octanol-water partition ratios for humic materials at environmentally relevant concentrations, which was found to be an appropriate method to characterise these substances with respect to their overall hydrophilicity depending on solution parameters. Both complexation with ^{111}In and halogenation with ^{131}I proved to be suitable methods for introducing a radiolabel to HS, taking into account that ^{111}In-humate complexes are attacked under strongly alkaline conditions, whereas the ^{131}I label is photosensitive. The measurements revealed that humic materials differ in hydrophilicity, depending on their origin. Furthermore, partitioning is influenced by pH, ionic strength and HS concentration and is related to the distribution of mobile and immobile species as determined by sequential gel chromatography. Albeit many other factors must be considered to evaluate the mobility of colloid-borne contaminants in real geosystems, the observed sensitivity toward solution conditions is of relevance with regard to risk assessments of waste management strategies.

ACKNOWLEDGEMENTS

Financial support by the Federal Ministry of Economics and Technology (project ref. no. 02 E 9329) is gratefully acknowledged. The synthetic humic acid M42 was kindly supplied by Dr. S. Sachs (Forschungszentrum Rossendorf).

References

1 C.T. Chiou, R.L. Malcolm, T.I. Brinton and D.E. Kile, *Environ. Sci. Technol.*, 1986, **20**, 502.
2 J. Buffle, *Complexation Reactions in Aquatic Systems: An Analytical Approach*, Ellis Horwood, Chichester, UK, 1988.
3 J.V. Christiansen and L. Carlsen, *Radiochim. Acta*, 1991, **52/53**, 327.
4 L. Carlsen, P. Lassen, J.V. Christiansen, P. Warwick, A. Hall and A. Randall, *Radiochim. Acta*, 1992, **58/59**, 371.
5 P. Warwick, L. Carlsen, A. Randall, R. Zhao and P. Lassen, *Chemistry and Ecology*, 1993, **8**, 65.
6 P.J. Fraker and J.C. Speck, *Biochem. Biophys. Res. Commun.*, 1978, **80**, 849.
7 E. Tombácz and J.A. Rice, in *Understanding Humic Substances: Advanced Methods, Properties and Applications*, eds. E.A. Ghabbour and G. Davies, Royal Society of Chemistry, Cambridge, UK, 1999, p. 69.
8 H. Geckeis, T. Rabung, T. Ngo Manh, J.I. Kim and H.P. Beck, *Environ. Sci. Technol.*, 2002, **36**, 2946.
9 G.R. Aiken, in *Humic Substances in Soil, Sediment and Water: Geochemistry and Isolation*, eds. G.R. Aiken, D.M. MacKnight, R.L. Wershaw and P. MacCarthy, Wiley-Interscience, New York, NY, 1985, p. 363.
10 R.S. Swift, in *Methods of Soil Analysis. Part 3: Chemical Methods*, ed. D.L. Sparks, Soil Sci. Soc. Am., Madison, WI, 1996, p. 1018.
11 J.I. Kim and G. Buckau, *Characterization of Reference and Site Specific Humic Acids*, RCM report 02188, TU Munich, 1988.
12 S. Pompe, A. Brachmann, M. Bubner, G. Geipel, K.H. Heise, G. Bernhard and H. Nitsche, *Radiochim. Acta*, 1998, **82**, 89.
13 D. Roessler, K. Franke, R. Suess, E. Becker and H. Kupsch, *Radiochim. Acta*, 2000, **88**, 95.

DETERMINATION OF GROSS ALPHA IN DRINKING WATER AND NUCLEAR WASTE WATER USING ACTINIDE RESIN™ AND LSC

P.J.M. Kwakman and M. Witte

National Institute of Public Health and the Environment, Laboratory for Radiation Research, 3720 BA Bilthoven, the Netherlands.

1. INTRODUCTION

The European Commission has given two parameters for radioactivity monitoring in the Drinking Water Directive (98/83/EC). These are a tritium concentration of 100 Bq/l, and a total indicative dose of 0.1 mSv per year[1]. For practical monitoring purposes, screening values are set to 0.1 Bq/l for gross alpha activity and 1.0 Bq/l for gross beta activity[1]. If the measured activities are below the screening values, the drinking water is considered to be in compliance with the parametric values for the indicative dose of 0.1 mSv per year[2,3].
In the Netherlands, the laboratories of drinking water producing companies have limited experience with the determination of gross alpha activity using gas flow counting. Eichrom has described a promising method using Actinide resin material, which has a strong affinity for trivalent and tetravalent ions in acidified water[4]. In principle, adsorption on a resin removes (part of) the interferences from the matrix and is therefore preferred above evaporation techniques[5], which may result in higher concentrations of interferences. Detection of gross alpha activity takes place with a Liquid Scintillation Counter (LSC) with alpha-beta discrimination. In this study we investigated the practical usefulness of the Eichrom method for drinking water laboratories. The method should be simple, straightforward and easy to introduce in a laboratory with limited experience in the field of radiochemistry.
Next to drinking water, the analysis of nuclear waste water needs investigation. The traditionally used technique of evaporating a sample on a plate combined with gas flow counting often suffers from bad results, caused by non-horizontal plates or irreproducible crystallization processes on the plate. Both aspects may have a large and unknown effect on the counting efficiency. Theoretically these problems are solved by using Actinide resin, which takes away the effect of the sample matrix, and LSC as detection method, which has an almost 100 % counting efficiency for alpha's. Therefore, following the experiments on drinking water, the Eichrom method is also tried on waste water from Urenco, a uranium enrichment company in Almelo (NL).

2. EXPERIMENTAL

2.1 Adsorption procedure

The Eichrom ACW11 method[4], using Actinide resin, was adapted to practical conditions in our laboratory. Drinking water (untreated water and tap water) was analyzed without filtering. Nuclear waste water was filtered with a paper filter before analysis. To 100 ml water sample, acidified to pH 2 with HCl, 300 mg Actinide resin were added, and the solution was stirred for at least 4 hours. The resin was filtered off using a 55 mm S&S glassfibre filter (black ribbon). The filter and resin were put in an LSC vial. Remaining resin particles on the glassware were flushed into the LSC vial using 5 ml 0.1 M HCl.
The recovery of the adsorption was established by adding weighed amounts of a ^{241}Am standard to the 100 ml water sample, and by comparing the result with the ^{241}Am-activity which is weighed directly into an LSC vial.

2.2 LSC counting conditions

Fifteen ml Ultima Gold LLT (Packard, Groningen, The Netherlands) were added to the LSC vial and the samples were counted for four hours using a Packard 2700 TR LSC counter. Alpha/beta-discrimination was performed at a Pulse Decay Discriminator (PDD) setting of 124. The PDD value was optimized using ^{241}Am and ^{36}Cl as model alpha- and beta-emitters, respectively. In the LSC cocktail the filter becomes transparant leading to reproducible counting conditions (Fig. 1). Organic constituents in the cocktail elute the DIPEX adsorber from the resin and the alpha emitter is homogeneously distributed in the vial.

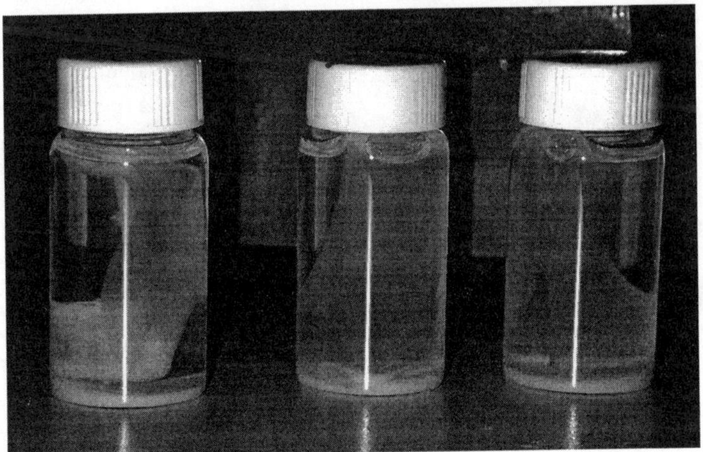

Figure 1 *LSC vials containing resin and filter*

3. RESULTS AND DISCUSSION

3.1 Drinking water

The Eichrom ACW11 method[4] suggests 500 mg of resin material in 100 ml of water sample. As drinking water usually has a low salt content we investigated the adsorption of ^{241}Am on resin quantities of 100 - 200 - 300 mg in 100 ml water sample. In order to be sure that all resin particles are counted the resin is filtered off on a 55 mm S&S filter. Both filter and resin were placed in the LSC vial. Using 100 or 200 mg of resin material resulted in recoveries of about 40 – 60 %, with 300 mg resin we found quantitative adsorption. Competing ions such as Ca^{2+} or Fe^{3+} may have a negative influence on the adsorption efficiency. In the Netherlands, drinking water may contain at most 200 ppm of calcium. There was no measurable effect of calcium in the range of 50 to 220 ppm on the adsorption of ^{241}Am. If concentration exceeding 50 mg.l^{-1} of Fe^{3+} are present in the sample, reduction to Fe^{2+} with ascorbic acid is recommended as the divalent ion is much less adsorbed on the resin[4]. In Dutch drinking water the limit for Fe^{3+} is 0.2 mg.l^{-1}, so no problems are expected from this ion. In Table 1 the most important analytical data are summarized. The data are obtained in series of ten samples 'raw' untreated water and ten samples tap water. In November 2001 RIVM participated in the French OPRI drinking water intercomparison (68 SH 300) using this method. The value for the gross alpha activity concentration which was reported by RIVM (0.068 ± 0.012) Bq.l^{-1} (2s), agrees satisfactorily with the target value of OPRI (0.077 ± 0.016) Bq.l^{-1} (2s).
It took only two days to optimize and introduce this method in the drinking water laboratory of Amsterdam (GWA, Leiduin, NL) using different equipment and glassware. This illustrates the simplicity of the method.

Table 1 *Data obtained in untreated water and tap water*

Analytical data	reported value
recovery for ^{241}Am	(100.7 ± 1.3) % (n=20)
alpha counting efficiency	98.6 %
optimum resin quantity*	300 mg
detection limit	0.03 Bq.l^{-1}
LSC counting time	240 min

* provided that the salt content is within the limits described in the text.

3.2 Nuclear waste water

The use of alpha/beta-discrimination in the determination of gross alpha activity in nuclear waste water will only give reliable results if the effect of varying quench is fully understood. The optimum PDD is highly dependent on the amount of quench in the LSC vial. A too low PDD will cause a spill of beta´s into the alpha window. This is unfavourable because in waste water the activity concentration of beta´s is generally much higher than that of alpha´s. A too high PDD, however, will cause a spill of alpha´s into the beta window. This is unfavourable too, but may be tolerated as long as it is known how many alpha´s are lost in the beta window. Usually ^{241}Am and ^{36}Cl are chosen as model nuclides for optimization of the PDD. In order to be as close as possible to nuclear waste water from Urenco in Almelo (NL), a uranium enrichment company, we chose $^{234/238}$U for studying the dependency of the PDD on quenching. In the LSC spectra below (Fig 2) it is clear that, at a moderate quench level, it is difficult to fully separate the soft uranium alpha's from the hard ^{234m}Pa beta's.

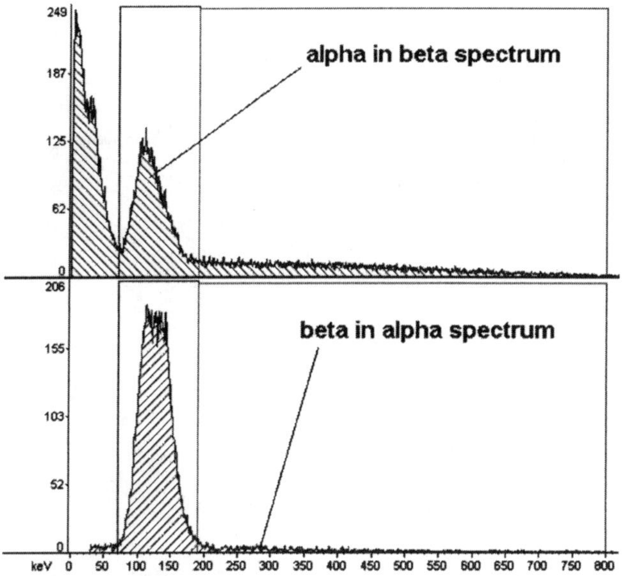

Figure 2 *LSC spectra of uranium (PDD 110; tSIE 346)*

Shifting the PDD to lower values than 110 will result in more beta's in the alpha spectrum, and a higher PPD value than 110 will result in more alpha's in the beta spectrum. As $^{234/238}$U is a weaker alpha emitter than ^{241}Am, and its daughter ^{234m}Pa is a stronger beta emitter than ^{36}Cl (Table 2) we compared the optimum PDD obtained for $^{234/238}$U with the PDD obtained for ^{241}Am and ^{36}Cl. For an automated PDD measurement pure alpha and beta standards are necessary. As the chemical yield in the purification of ^{234}Th/^{234m}Pa has large uncertainties, we calculated the PDD from LSC spectra at quench levels ranging from tSIE 350 - 240 (Fig 3). From Figure 3 we can see that at a PDD up to 80 all beta's are found in the alpha spectrum. At a PDD of 130 and higher a very small percentage of beta's are found in the alpha spectrum while at the same time alpha's are spilled into the beta spectrum depending strongly on the quenching in the sample. These values found for $^{234/238}$U are different from the model nuclides ^{241}Am and ^{36}Cl. For ^{241}Am, at a quench value of tSIE = 350 and a PDD of 130, a spill of only 2-3 % of alpha's into the beta spectrum is found. For ^{36}Cl a spill of 6-8 % is found at the same quench level and a PDD of 110. This

Table 2 *Properties of ^{238}U and daughters, ^{241}Am and ^{36}Cl*

nuclide	Half life	alpha energy MeV	beta energy keV
^{238}U	$4.47{\cdot}10^9$ y	4.2	
^{234}Th	24.1 d		188
^{234m}Pa	1.2 min		2280
^{234}U	$2.44{\cdot}10^5$ y	4.7	
^{241}Am	432 y	5.5	
^{36}Cl	$3.01{\cdot}10^5$ y		709

shows that the determination of gross alpha in nuclear waste water has large uncertainties. These uncertainties can only be minimized by using a PDD setting which is optimized with nuclides of approximately the same alpha and beta energy as the nuclides that are expected in the samples.

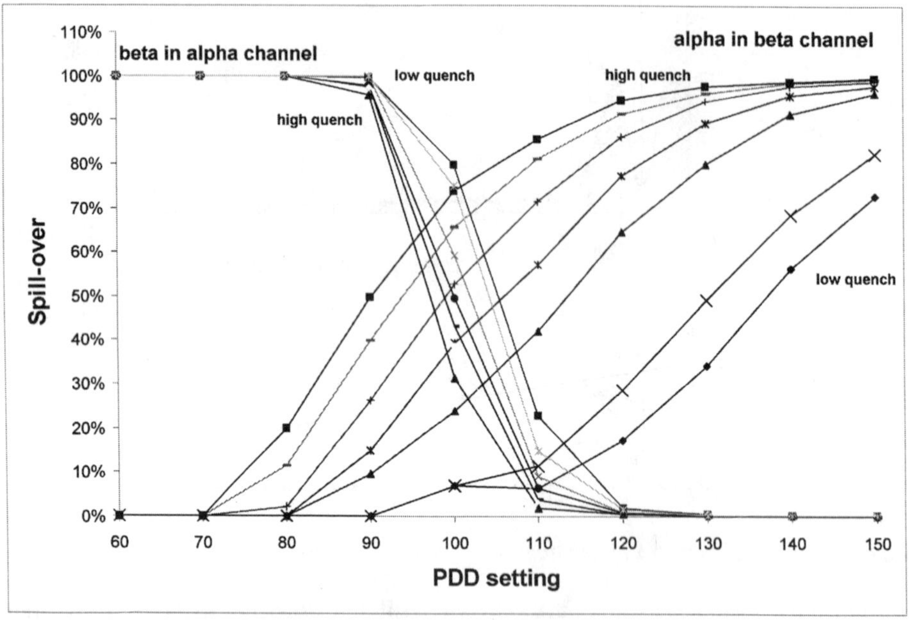

Fig 3 *α/β-discrimination for uranium at various quench levels (low quench: tSIE = 350; high quench: tSIE = 240).*

In order to deal with these uncertainties in practice we analyzed a number of Urenco waste water samples at a PDD of 130 and, depending on the quench value in the sample, we corrected the gross alpha activity for the loss of alpha's. This method works well provided that the quench value is not out of the range which is shown in Fig. 3. A tSIE value higher than 350, meaning lower quenching, is unlikely as these values are similar to those obtained in drinking water. A tSIE value lower than 240 in highly quenched samples, can be improved by changing the filtration prior to the adsorption procedure.

4. CONCLUSIONS

The principle of adsorbing actinides in drinking water and nuclear waste water on resin material and counting with LSC works well. Using ^{241}Am as a model nuclide chemical recoveries of 100 % were found and the counting efficiency was 98.5 %. Transfer of the resin material into the LSC vial including the glassfibre filter which was used for filtration is a simple way of bringing over the resin quantitatively. The disadvantage is, however, that interfering constituents in waste water samples are concentrated which may lead to highly quenched samples. As the PDD setting is usually optimized at a fixed quench level, the reported gross alpha activity may differ largely from the 'true value' in samples with higher or lower quench levels. Samples should preferably contain the least amount of quenching and the quench level should be as close as possible to the quench level at which the PDD setting was optimized. In this study we showed that not only quenching but also the energy of both the alpha and beta emitter has a large influence on alpha/beta-discrimination in the LSC measurement. The alpha energies of $^{234/238}$U (4.2 and 4.7 MeV) are considerably lower than that of ^{241}Am (5.5 MeV) leading to a much higher spill over of uranium alpha's in the beta window as would be expected from the PDD optimized with ^{241}Am. For beta's the opposite goes. The strong ^{234m}Pa beta's are much more likely to spill into the alpha window than would be expected from the optimization with ^{36}Cl. Nevertheless, the described method works well for uranium in waste water, provided that the quench value in the sample is within a range in which the alpha/beta-discrimination has been investigated. In practice, the beta spill over into the alpha channel can be minimized by chosing a PDD of 130, and the loss of alpha's can be corrected for.

References

1 Draft amendment in the Directive on the quality of water intended for human consumption, Directive 98/83/EC.

2 L. Salonen, "Natural radionuclides in drinking water – regulations and monitoring", paper presented at the International Conference on Advances in Liquid Scintillation Spectrometry. Karlsruhe, Germany, May 7-11, 2001.

3 C. Kralik, "Performance of LSC cocktails in gross-beta analysis of drinking water", paper presented at the International Conference on Advances in Liquid Scintillation Spectrometry. Karlsruhe, Germany, May 7-11, 2001.

4 Eichrom Industries, Analytical procedures ACW11, "Gross alpha radioactivity in water", July 27, 2001. See www.eichrom.com/methods .

5 J.A. Sanchez-Cabeza and U. Pujol, "A rapid method for the simultaneous determination of gross alpha and beta activities in water samples using a low background liquid scintillation counter", Health Physics vol. 68 (1995), 674-682.

CHROMIUM SPECIATION IN ACIDIC SOLUTIONS USING RADIOCHROMATOGRAPHIC METHODS

Sérgio H. Pezzin[1], José F. Lugo Rivera[2], Kenneth E. Collins[3], and Carol H. Collins[3]

[1]Centro de Ciências Tecnológicas, Universidade do Estado de Santa Catarina, 89223-100 Joinville-SC, Brazil;
[2] Centro Regional de Estudios Nucleares, Universidad Autónoma de Zacatecas, 98000 Zacatecas, ZAC, Mexico;
[3]Instituto de Química, Universidade Estadual de Campinas, Cx. Postal 6154, 13083-970 Campinas-SP, Brazil

1 INTRODUCTION

Chromium speciation is an important analytical operation for laboratories that determine chromium in natural water, in drinking water or in other consumables.[1-5] Trivalent chromium is essential in animals, playing a role in glucose and lipid metabolism.[6] On the other hand, hexavalent chromium is known to be toxic and carcinogenic even at very low levels.[7] Although chromium can be present in aqueous solutions as Cr(VI) or as any of various kinetically stable forms of Cr(III), whose distribution depends on the chemical history of the aqueous sample,[8] most chromium speciation procedures in the literature consider only Cr(VI) and Cr(III) (as the hexaaquo species).

The procedure for preserving Cr in water samples requires filtering (0.45 μm) into plastic bottles followed by acidification with HNO_3 to a pH < 2.[9] If samples are not acidified, precipitation of Fe compounds could take place, with subsequent adsorption of Cr(VI). In addition, the American Public Health Association advises refrigeration of water samples when Cr(VI) is to be determined, and analysis of samples within 48 h, presumably because of the potential for reduction of Cr(VI). Nevertheless, it was verified that the acidification of the samples resulted in loss of Cr(VI) from solution.[9-12]

The present work gives examples of characterizations involving Cr(VI) and several typical forms of Cr(III) using both open column (BioRad AG50W-X8, Na^+ form) and HPLC (Partisil SCX, 125 x 4.6 mm, 10 μm) cation exchange procedures.

2 EXPERIMENTAL

2.1 Reagents

Recrystalyzed $K_2Cr_2O_7$, dried for 2 hours at 160°C, was used to prepare stock solutions of Cr(VI). All solutions were prepared with triply-distilled deionized water (Nanopure, Barnstead, or Milli-Q, Millipore) and analytical grade reagents. The eluents for HPLC were passed through 0.45 μm HA type membrane filters (Millipore) and degassed by ultrasound and vacuum.

2.2 Radioactive Material

Some $^{51}Cr(VI)$ was prepared from aqueous $^{51}CrCl_3$ (CNEN-IPEN, Brazil) by oxidation with peroxide in basic medium.[13] Other portions were obtained as aqueous solutions of $Na_2{}^{51}CrO_4$ (Amersham, England). Specific activities ranged from 6,000 to 31,000 Mbq mg^{-1}. All samples were submitted to radionuclidic and radiochemical analysis[14] to show that ^{51}Cr was the only radionuclide present and that at least 98 % of the ^{51}Cr was initially present as $^{51}Cr(VI)$.

2.3 Radioactivity Measurements

The 0.320 MeV gamma rays from the decay of ^{51}Cr in the collected fractions were determined using a well type NaI(Tl) detector in a modular single channel gamma analyser (EG&G Ortec and Hewlett Packard modules).

2.4 Studies of the Stability of Cr(VI) in Acidic Solutions

2.4.1 Stability of Cr(VI) in Nitric Acid Solutions. A set of acidic solutions of Cr(VI) with concentrations from 10^{-7} to 10^{-3} mol/L were prepared in nitric acid whose concentration varied from 2 to 10^{-5} mol/L. The 10^{-7} to 10^{-5} mol/L Cr(VI) solutions were prepared directly from the radioactive solution, whereas those of 10^{-4} mol/L and 10^{-3} mol/L Cr(VI) were prepared from a stock solution of $K_2Cr_2O_7$ and spiked with $^{51}Cr(VI)$. All of the solutions were stored in glass flasks in the absence of light. After appropriate time periods, aliquots of these solutions were analyzed as described in the 'Chromatographic Analysis' section.

2.4.2 Behavior of Trace Cr(VI) in Acidic Solutions. To a 5 mL screw-capped Erlenmeyer flask were added 1.0 mL of concentrated acid (HCl, HF, H_2SO_4, $HClO_4$ or HNO_3) and 100 µL of high specific activity $^{51}Cr(VI)$. The system was homogeneized and maintained at room temperature (19.6-20.6°C). After appropriate time periods, 10µL aliquots were taken and immediately analyzed or diluted in water for later analysis ("aquation solutions"). For the open tube chromatographic analyses, aliquots from concentrated acids were placed into the column reservoir containing 4 mL of water above the resin level. For cation-HPLC analyses, these aliquots were diluted in 2 mL of deionized water and 613 µL were rapidly injected.

2.5 Chromatographic Analysis

2.5.1 Chromium Species Separation by Open Column Cation Exchange Chromatography. Cation exchange columns were prepared by placing 0.5 mL of pretreated[14] resin (Bio-Rad AG50W-X8, 100-200 or 200-400 mesh, Na^+ form) in a glass column tube (40-80 x 5 mm) having a small (5 mL) open reservoir at the top and a porous polyethylene support disk at the exit. Flow was controlled by a small PTFE stopcock. Just prior to use, the resin bed (20 mm high) was treated with 0.5 mL of a 0.02 mol L^{-1} $Na_2Cr_2O_7$ solution (to eliminate possible reducing impurities) and washed with 6 mL of 0.01 mol L^{-1} $HClO_4$ and 10 mL of deionized water. Then, 0.01 to 1 mL of sample solution containing a few micromoles of the ionic species was carefully placed on the top of the column and was allowed to begin passage through the resin. When the solution to be analyzed had an acid concentration or ionic strength greater than 0.01 mol L^{-1}, the sample solution was added to an appropriate quantity of deionized water contained in the open reservoir. After the liquid level reached the top of the column, a small portion of the first eluent was used to carefully rinse the walls. Elution was carried out using the sequence of eluents: 0.01, 0.1 and 1 mol L^{-1} $HClO_4$; 0.25, 0.5 and 1 mol L^{-1} $Ca(ClO_4)_2$ at pH 2, adjusted with $HClO_4$.[15] Two mL fractions (ISCO 328 fraction collector) of the eluate were collected for radioactivity measurements. The resin was also counted to estimate non-eluted species. This procedure permitted construction of elution profiles whose peaks were identified by comparison with authentic samples.

2.5.2 Chromium Species Separation by Open Column Anion Exchange Chromatography. Anion exchange columns of 0.5 mL of Bio Rad AG1-X8 resin (200-400 mesh, Cl^- form) were prepared in similar glass columns. After extensive conditioning with water and 0.01 mol L^{-1} $HClO_4$, the sample (0.01-1 mL) was placed on the column. Elution was carried out with 8 mL of 0.01 mol L^{-1} $HClO_4$, followed by 12 mL of deionized water and the resin was also counted after elution.

2.5.3 Estimation of Relative Quantities of Cr(VI) and Cr(III). The aliquots from the solutions described in section 2.4.1 were passed through the respective cationic and anionic resins and the columns were eluted with 8 mL of 0.01 mol/L $HClO_4$. The relative quantities of Cr(VI) and Cr(III), the latter a reduction product in the acid solution, are estimated by comparing the total radioactivity eluted from each column with the summed eluted radioactivities.

2.5.4 Chromium Species Separation by Cation-HPLC. Most of the work has been carried out with either a Milton Roy LDC 396 pump with pulse dampener or a Waters 510 pump, Rheodyne 5012 solvent selection valve and Rheodyne 7010 injector valves. Two mL fractions were collected for "off-line" radiochemical detection. A 120 x 4.0 mm stainless steel column was packed at 300 bar with Partisil SCX (10μm) (Alltech, USA). The suspension solvent was carbon tetrachloride and the pressurizing solvent was methanol. The elution sequence was similar to that of the open column chromatography,[15] while the flow rate was maintained at 1.2 mL min^{-1}.

3 RESULTS AND DISCUSSION

3.1 Stability of Cr(VI) in Nitric Acid Solutions

The results show that low concentrations (up to 10^{-3} mol L^{-1}) of Cr(VI) are not stable in an acid solution. Detailed studies of the reduction of low concentrations of Cr(VI) in nitric acid solutions have shown that the rate of reduction increases as the concentration of the acid increases (Figure 1) or as the concentration of the Cr(VI) decreases (Figure 2). It is noteworthy the difference between the reaction rates of 10^{-5} mol/L Cr(VI) in 1 and 0.1 mol/L HNO_3, showing the importance of H^+ concentration in the acid reduction process. On the other hand, solutions with an initial Cr(VI) concentration of 10^{-3} mol/L are relatively stable for 3 months at pH 1.

When the initial Cr(VI) concentration was 10^{-7} mol/L the solutions were very instable, being totally reduced in less than 7 days in 1 mol/L HNO_3. Table 1 shows the products from the reduction of 10^{-6} mol/L Cr(VI) in nitric acid solutions. For 2 and 1 mol/L HNO_3 the reduction is nearly complete in 6 and 34 days, respectively. As shown in Table 1, both the 1+ and 2+ species, presumably $[Cr(H_2O)_4(NO_3)_2]^+$ and $[Cr(H_2O)_5NO_3]^{2+}$, respectively, are relatively stable, as a significant quantity of each remains even after prolongated storage.

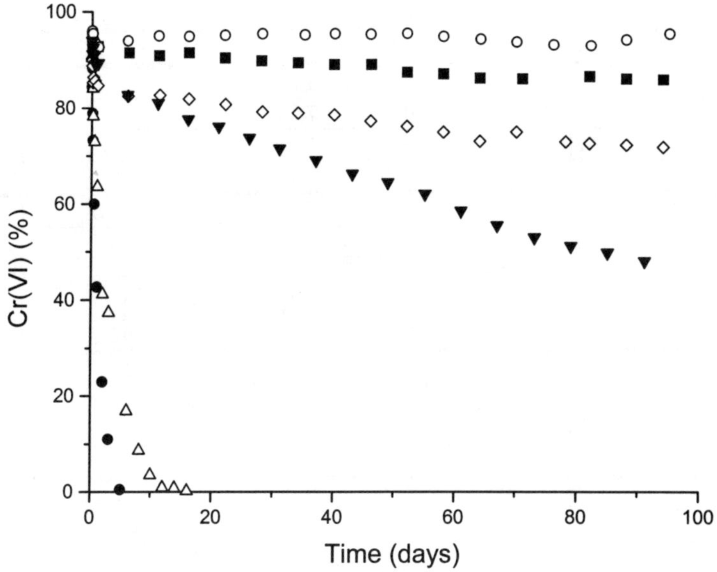

Figure 1 *Behavior of* $^{51}Cr(VI)$ 10^{-5} *mol/L in* HNO_3 *at different concentrations (mol/L):* ● *2;* △ *1;* ▼ 10^{-1}*;* ◇ 10^{-2}*;* ■ 10^{-3}*;* ○ 10^{-4}

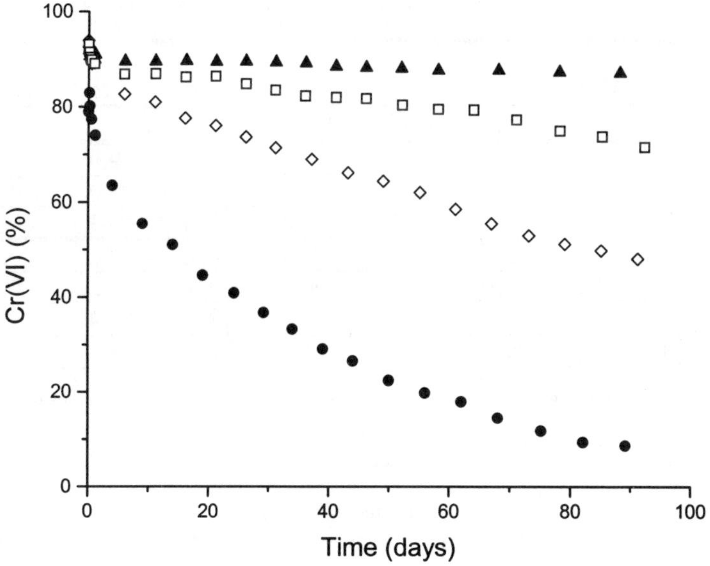

Figure 2 *Behavior of* $^{51}Cr(VI)$ *in 0.1 mol/L* HNO_3 *at different Cr(VI) concentrations (mol/L):* ● 10^{-6}*;* ◇ 10^{-5}*;* □ 10^{-4}*;* ▲ 10^{-3}

Table 1 *Products (%) from the reduction of 10^{-6} mol L^{-1} [Cr(VI)] in 2 to 10^{-5} mol/L HNO_3.*

$[HNO_3]$	Time (h)	Cr(VI)	neutral	1+	2+	3+	Dimer	Resin
	1	25.4	1.5	5.6	3.5	63.1	0.3	0.7
2 mol L^{-1}	6	1.7	0.8	7.0	1.2	88.9	0.1	0.4
	42	0	0.2	1.2	0.9	96.7	0.3	0.7
	58	0	0.4	1.1	1.5	96.2	0.5	0.4
	1	79.9	1.2	2.0	1.3	15.0	0.3	0.3
	5	47.9	1.3	3.1	1.0	46.4	0.1	0.3
1 mol L^{-1}	16	2.8	0.2	5.7	0.9	89.8	0.3	0.4
	34	0.7	0.2	5.1	0.7	92.7	0.3	0.4
	63	0.6	0.6	4.2	1.0	92.7	0.6	0.5
	1	84.8	0.8	1.2	1.0	10.5	1.0	0.7
0.1 mol L^{-1}	7	69.2	0.9	2.0	1.0	25.3	0.9	0.8
	27	39.7	1.0	2.8	1.9	52.9	0.9	0.7
	63	2.6	4.5	4.1	3.3	83.2	1.2	1.3
	1	78.3	0.8	2.0	1.2	16.4	0.7	0.6
10^{-2} mol L^{-1}	26	46.2	1.1	2.6	1.4	47.2	1.2	0.4
	66	26.5	1.0	3.3	1.5	64.8	1.7	1.2
	90	23.1	0.8	3.5	1.5	68.6	1.7	0.8
	1	91.1	0.6	0.9	1.5	4.9	0.5	0.4
10^{-3} mol L^{-1}	60	58.0	1.2	4.4	3.9	31.1	0.4	1.1
	90	49	1.4	5.6	3.6	38.7	0.5	1.2
10^{-4} mol L^{-1}	1	93.7	1.2	1.0	0.9	2.7	0.2	0.2
10^{-5} mol L^{-1}	1	93.4	1.3	1.1	1.0	2.5	0.4	0.3

Table 2 *Products(%) from the reduction of 10^{-4} mol L^{-1} [Cr(VI)] in 2 to 10^{-2} mol/L HNO_3.*

$[HNO_3]$	Time (h)	Cr(VI)	neutral	1+	2+	3+	dimer	resin
	1	54.6	5.0	0.5	1.9	29.0	1.4	7.4
2 mol L^{-1}	28	0	0	0.1	0.7	98.0	0.2	1.1
	61	0	0.1	0.0	0.7	98.3	0.1	0.9
	1	85.0	4.2	0.4	0.3	8.6	0.3	1.2
1 mol L^{-1}	29	0,1	0.1	0.3	0.4	98.3	0.1	0.8
	60	0	0.1	0.3	0.5	98.4	0.1	0.7
0.1 mol L^{-1}	1	89.1	0.6	0.3	0.7	5.3	1.7	2.3
	31	83.6	0.8	0.3	0.4	11.6	0.9	2.4
	1	91.6	2.3	0.1	0.3	1.9	0.9	2.8
10^{-2} mol L^{-1}	31	91.3	1.9	0.4	0.4	3.2	1.0	1.8

These results show that low concentrations of Cr(VI) are not stable in acid solutions, an observation which has implications with respect to the storage of Cr(VI) solutions in studies of environmental speciation.

3.2 Behavior of Trace Cr(VI) in Acidic Solutions

The results show that trace quantities of Cr(VI), monitored by means of radiochromium (^{51}Cr), are reduced in the presence of mineral acids such as hydrochloric, sulphuric, perchloric and nitric acids, even in the absence of conventional reducing agents, producing several different Cr(III) species which can be separated by cation exchange chromatography.

Figure 3 shows a typical chromatogram obtained by open column chromatography, while Figure 4 shows an HPLC analysis for the same sample.

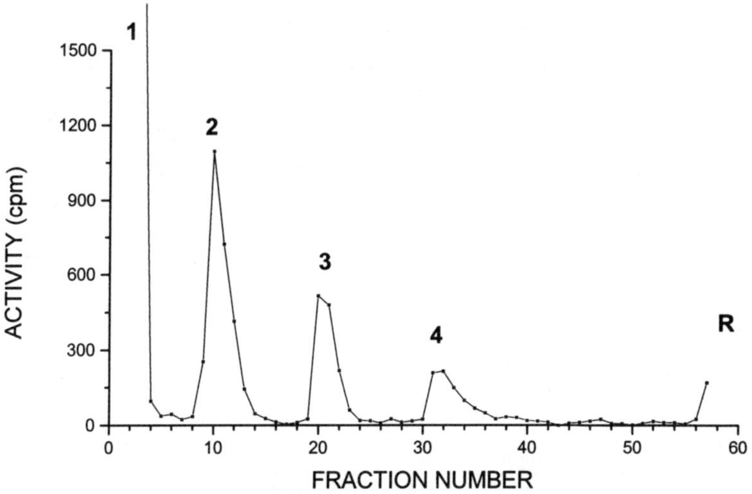

Figure 3 *Open column cation exchange chromatogram of the* $^{51}Cr(III)$ *products from the reaction of* $^{51}Cr(VI)$ *in 40% HF (1 h). Peak identification: 1-* $Cr(H_2O)_3F_3$*; 2 –* $[Cr(H_2O)_4F_2]^+$*; 3 -* $[Cr(H_2O)_5F]^{2+}$*; 4 -* $[Cr(H_2O)_6]^{3+}$*; R –Cr(III) polymers*

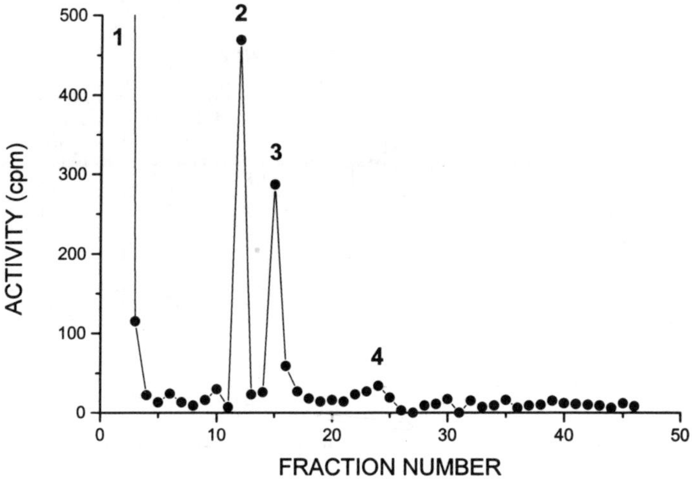

Figure 4 *High performance cation exchange chromatogram of the* $^{51}Cr(III)$ *products from the reaction of* $^{51}Cr(VI)$ *in 40% HF (1 h). Peak identification: 1-* $Cr(H_2O)_3F_3$*; 2 –* $[Cr(H_2O)_4F_2]^+$*; 3 -* $[Cr(H_2O)_5F]^{2+}$*; 4 -* $[Cr(H_2O)_6]^{3+}$*; R –Cr(III) polymers*

When trace $^{51}Cr(VI)$ reacts with HCl, HF or H_2SO_4, complexed species, $Cr(H_2O)_3Cl_3$, $Cr(H_2O)_3F_3$ and $Cr(H_2O)_2(SO_4)_2^-$, are, respectively, the main initial products (Figures 3 and 4 and Table 3).

Table 3 *Summary of products from trace $^{51}Cr(VI)$ reacting with HCl and H_2SO_4 observed after different reaction times, t(conc), and after specified aquation times, t(dil)*

Acid	t (conc)	t (dil)	Anionic/ Neutral	Charge 1+	Charge 2+	Charge 3+	Resin
HCl	1 h	0	85.4	9.7	3.8	0.8	0.3
HCl	1 h	2 h	53.2	10.6	35.2	0.8	0.3
HCl	1 h	24 h	6.1	1.5	88.0	3.9	0.5
HCl	1 h	48 h	1.0	0.4	89.7	8.3	0.7
HCl	1 h	72 h	0.4	0.3	85.6	13.0	0.7
HCl	1 h	168 h	0.2	0.3	71.0	27.6	0.9
HCl	840 h	0	80.9	11.8	7.3	---	---
H_2SO_4	1 h	0	99.5	0.5	---	---	---
H_2SO_4	1 h	24 h	60.0	37.0	---	3.0	---
H_2SO_4	1 h	96 h	67.0	20.0	---	13.0	---
H_2SO_4	1 h	168 h	54.0	23.0	---	23.0	---
H_2SO_4	1 h	360 h	39.0	20.0	---	41.0	---
H_2SO_4	1 h	504 h	29.0	22.0	---	49.0	---
H_2SO_4	1 h	2200 h	4.0	20.0	---	76.0	
H_2SO_4	24 h	0	100	---	---	---	---

By contrast, the major product when trace concentrations of $^{51}Cr(VI)$ reacts with concentrated HNO_3 and $HClO_4$ is the 3+ species, $Cr(H_2O)_6^{3+}$, and the smaller amounts of 1+ and 2+ species are rapidly converted to the 3+ species on aquation (Table 4).

Table 4 *Summary of products from trace $^{51}Cr(VI)$ reacting with perchloric and nitric acids observed after different reaction times, t(conc), and after specified aquation times, t(dil)*

Acid	t (conc)	t (dil)	Anionic/ Neutral	Charge 1+	Charge 2+	Charge 3+	Resin
$HClO_4$	5 min	0	3.5	2.7	23.7	68.0	2.0
$HClO_4$*	5 min	0	5.1	---	20.4	71.7	2.8
$HClO_4$	5 min	2 h	2.6	2.5	21.5	72.2	1.3
$HClO_4$	5 min	96 h	1.8	4.1	16.1	75.2	2.7
$HClO_4$	25 min	0	0.6	2.9	14.0	82.5	---
$HClO_4$	1 h	0	5.9	---	9.5	82.2	2.3
HNO_3	1 h	0	0.3	0.3	59.0	33.8	6.6
HNO_3	1 h	2 h	0.5	0.2	28.9	65.4	5.0
HNO_3	1 h	48 h	0.2	---	1.4	98.0	0.4

*by cation HPLC

However, even the small amounts of 2+ and 1+ species seen with the "non-complexing" anions suggest that the mechanism of the reduction reaction probably involves a ligand capture step to form complexes.[16]

4 CONCLUSION

The results reported here on the reduction of Cr(VI) in solutions of oxidizing acids, not involving a conventional reducing species such as chloride, suggest that "acid", i.e., the proton, may be a defining reagent in the reduction pathway.

Thus, the role of the different anions present, which depend on the acid used, should contribute importantly, as they form complexes whose stability may influence the overall kinetics of the acid-reduction process and the resulting product distributions.

These results are apparently not consistent with the well-known acidification procedure for the determination of Cr(VI) in natural water samples, and the use of acidic Cr(VI) solutions as titrimetric and spectrophotometric standards.[17]

References

1 J. G. Farmer, R. P. Thomas, M. C. Graham, J. S. Geelhoed, D. G. Lumsdon and E. Paterson, *J. Environ. Monitoring*, 2002, **4**, 235.
2 Y. L. Chang and S. J. Jiang, *J. Anal. Atom. Spectrom.*, 2001, **16**, 858.
3 H. Gurleyuk and D. Wallschlager, *J. Anal. Atom. Spectrom.*, 2001, **16**, 926.
4 Y. J. Li and H. B. Xue, *Anal. Chim. Acta*, 2001, **448**, 121.
5 B. Wen, X. Q. Shan and J.Lian, *Talanta*, 2002, **56**, 681.
6 S. A. Katz, *Environ. Health Perspect.*, 1991, **92**, 13.
7 P. O'Brien and A. Kortenkamp, *Transition Met. Chem.*, 1995, **20**, 636.
8 J. W. Ball and D. K. Nordstrom, *J. Chem. Eng. Data*, 1998, **43**, 895.
9 K. G. Stollenwerk and D. B. Grove, *J. Environ. Qual.*, 1985, **14**, 396.
10 J. Pavel, J. Kliment, S. Stoerk and O. Suter, *Fresenius Z. Anal. Chem.*, 1985, **321**, 587.
11 C. Archundia, P. S. Bonato, J. F. Lugo Rivera, L. C. Mascioli, K. E. Collins and C. H. Collins, *Sci. Total Environ.*, 1993, **130/131**, 231.
12 C. Archundia, J. F. Lugo Rivera, C. H. Collins and K. E. Collins, *J. Radioanal. Nucl. Chem., Art.*, 1995, **195**, 363.
13 K. E. Collins and C. Archundia, *Int. J. Appl. Radiat. Isot.*, 1984, **35**, 910.
14 K. E. Collins, P.S. Bonato, C. Archundia, M. E. L. R. De Queiroz and C. H. Collins, *Chromatographia*, 1988, **26**, 160.
15 C. H. Collins, S. H. Pezzin, J. F. Lugo Rivera, P. S. Bonato, C. C. Windmöller, C. Archundia and K. E. Collins, *J. Chromatogr. A*, 1997, **789**, 469.
16 S. H. Pezzin, J. F. Lugo Rivera, C. H. Collins and K. E. Collins, *J. Brazilian Chem. Soc.*, to be published.
17 M. Gil, D. Escolar, N. Iza, J. L. Montero, *Appl. Spectrosc.*, 1986, **40**, 1156.

INFLUENCE OF GEOCHEMICAL PARAMETERS ON THE MOBILITY OF METAL-HUMATE COMPLEXES

A. Mansel and H. Kupsch

Institut für Interdisziplinäre Isotopenforschung, Permoserstr. 15, D-04318 Leipzig, Germany

1 INTRODUCTION

In case of a release from subterranian waste repositories (SWR), the spreading of radionuclides and heavy metals may be affected by the presence of humic substances in the surrounding aquifer.[1,2,3,4] Depending on geochemical parameters (pH, ionic strength, surrounding geomatrix, redox kinetics), migration can also be influenced by colloid sorption and flocculation processes.[5,6,7,8] In order to investigate metal complexation and sorption at pico- and nanomolar concentrations under near field conditions, radionuclides are necessary. ^{203}Pb (as a tracer for divalent heavy metals) and ^{64}Cu (as a tracer for transition metals) were complexed with ^{131}I radiolabelled humic acids to investigate metal complexation and sorption.

2 METHOD AND RESULTS

2.1 Experimental

^{64}Cu ($T_{1/2}$ = 12.7 h) was produced by irradiation of copper(II)nitrate in the TRIGA-reactor Mainz (Institut für Kernchemie, Germany) at a flux of 4.2 * 10^{12} n cm^{-2} s^{-1} for 6 h, yielding an absolute activity of 34.6 MBq / mg Cu and was used directly for the preparation of the analyte solutions. ^{203}Pb ($T_{1/2}$ = 51.9 h) was synthesized by the $^{203}Tl(p,n)$ ^{203}Pb-reaction at the cyclotron in Hannover (Medizinische Hochschule, Germany) at a current of ~ 6 μA h^{-1}, yielding an absolute activity of ~ 37 MBq / 1.7 g Tl. Radiolead was separated from thallium by dissolving the metallic target with nitric acid, followed by coprecipitation three times with $Fe(OH)_3$ / ammonia and by anion exchange using a DOWEX 1 X 8-column in 6 M hydrochloric acid.[9] For higher molar lead concentrations, inactive Pb^{2+} was added as carrier. ^{131}I was purchased from Amersham. The commercially obtainable Aldrich HA was purified by repeated precipitation.[10] A synthetic HA (type M42) was provided by the Institut für Radiochemie (Forschungszentrum Rossendorf, Germany).[11] The geomatrices sandy soil (Merck), Phyllite, Granite and Diabase (Harz Mountains, Germany) were ground and sieved to a defined grain size. All solutions had an ionic strength of 0.1 M $NaClO_4$ and were buffered with morpholinoethanesulfonic acid. The complexation studies were performed by anionic exchange with Sephadex DEAE-25 (indirect speciation method).[12] Batch experiments were carried out for investigations on sorption data (10 ml /

2.5 g geomatrix). The HA were radiolabelled with ^{131}I by the IODOGEN-method.[13,14] The radiolabelled HA were characterized with the sequential chromatographic analysis (SCA).[15]

2.2 Results and Discussion

The adsorption isotherms of Cu^{2+} on four SWR-relevant geomatrices are shown in Figure 1. The sandy soil adsorbs less than the other three geomatrices, because of a lower pH at the equilibrium of sorption.

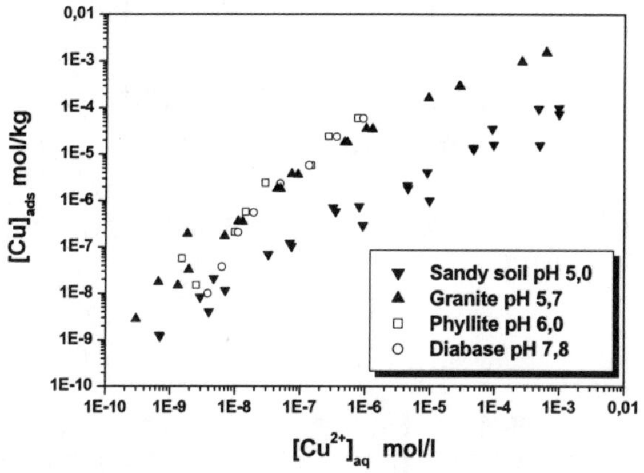

Figure 1 *Adsorption isotherms of Cu^{2+} on four different geomatrices; 0.1 M $NaClO_4$.*

The complexation constants for the metal-humates were calculated using the metal ion charge neutralization model [2] and are shown in Figure 2 for Cu^{2+}, Pb^{2+}, Aldrich- and M42 HA at pH 5 and 6. The synthetic M42 HA forms nearly the same strong complexes as the Aldrich HA. Cu^{2+} (full symbols) shows an increase in log β_{LC} of ~ 1 with decreasing metal ion concentration in contrast to Pb^{2+} (open symbols). In Figure 3 the distribution of Cu^{2+}-species in the ternary system Cu^{2+} / HA / sandy soil is illustrated. The sorbed Cu^{2+}-species $[Cu]_{ads}$ decreases with increasing Cu^{2+} concentration because of the saturation of the small surface of the sandy soil. The dissolved Cu-humate also decreases as the maximum loading capacity of the HA in solution was achieved. The characterization of Pb^{2+}-humate by SCA is shown in Figure 4. In the first sequence (size exclusion chromatography) nearly 90 % of the lead is unbound to HA, but the iodine- and the lead-signals course parallel, so lead is bound to all mobile HA molecular sizes. The same is true for the second sequence (immobile HA). Signals of the radioactivity (^{131}I) and the UV-detector (HA) are in good agreement showing that the radiolabel is stable.

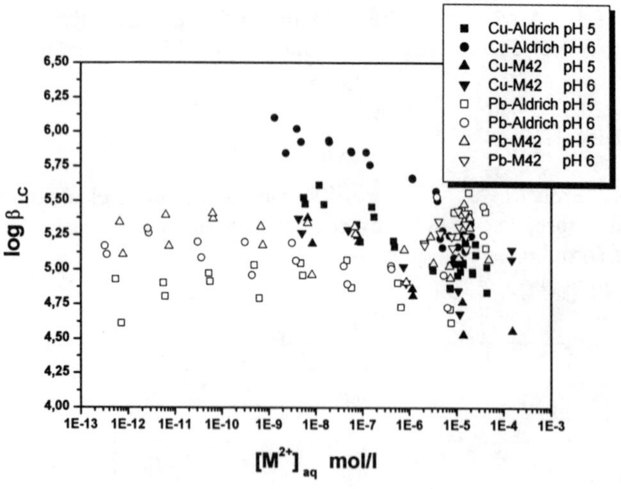

Figure 2 *Complexation constants for Cu^{2+}, Pb^{2+}, Aldrich and M42 HA; 10 mg / l HA; 0.1 M $NaClO_4$; LC = loading capacity.*[2]

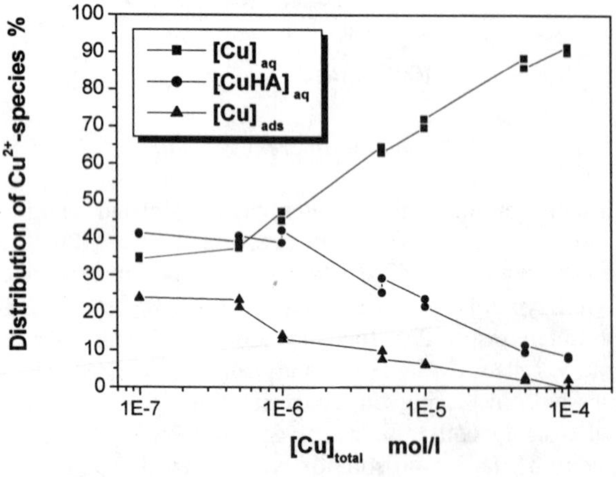

Figure 3 *Distribution of Cu^{2+}-species in the System Cu^{2+} / HA / sandy soil; pH 5; 10 mg / l Aldrich HA.*

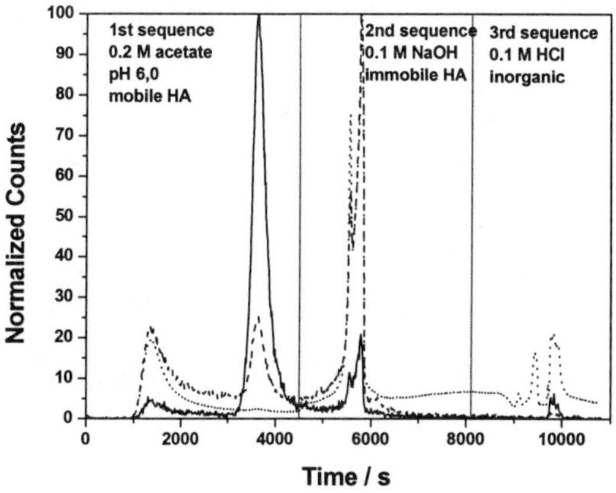

Figure 4 *SCA of a radioactive labelled Aldrich HA as well as labelled with ^{131}I (dashed) and ^{203}Pb (solid), respectively. The UV-signal (254 nm) is given in dotted line.*

3 CONCLUSION

Indirect speciation using anionic exchange is an excellent method to determine humate-complexation constants for heavy metals and the distribution of heavy metals in the ternary system heavy metal ion / humic substances / geomatrix. The SCA can characterize the bound heavy metal to the mobile and immobile HA in combination with size exclusion chromatography.

Acknowledgement

The work performed in this paper is supported by funding from the German Federal Ministry for Education and Research (BMBF) under the Contract No. 02 C 0709.

References

1 G. R. Choppin, *Radiochim. Acta*, 1988, **44/45**, 23.
2 J. I. Kim and K. R. Czerwinski, *Radiochim. Acta*, 1996, **73**, 5.
3 P. Benes and J. Mizera, *Radiochim. Acta*, 1996, **74**, 185.
4 M. Filella and R. M. Town, *Fresenius J. Anal. Chem.*, 2001, **370**, 413.
5 J. I. Kim, *Radiochim. Acta*, 1991, **52/53**, 71.
6 C. Degueldre, H.-R. Pfeiffer, W. Alexander, B. Wernli, R. Bruetsch, R. Grauer, A. Laube, A. Oess and H. Silby, *Appl. Geochem.*, 1996, **11**, 677 and 697.
7 P. Schmitt, A. Kettrup, D. Freitag and A. G. Garrison, *Fresenius J. Anal. Chem.*, 1996, **354**, 915.
8 H. Geckeis, Th. Rabung, T. Ngo Manh, J. I. Kim and H. P. Beck, *Environ. Sci. Technol.*, 2002, **36**, 2946.
9 S. M. Qaim, R. Weinreich and H. Ollig, *Int. J. Appl. Radiat. Isotopes*, 1979, **30**, 85.

10 J. I. Kim and G. Buckau, Report RCM 02188, 1988, Institut für Radiochemie der Technischen Universität München.
11 S. Pompe, A. Brachmann, M. Bubner, G. Geipel, K. H. Heise, G. Bernhard and H. Nitsche, *Radiochim. Acta*, 1998, **82**, 89.
12 G. Montavon, A. Mansel, A. Seibert, H. Keller, J. V. Kratz and N. Trautmann, *Radiochim. Acta*, 2000, **88**, 17.
13 P. J. Fraker, J. C. Speck, *Biochem. Biophys. Res. Commun.*, 1978, **80**, 849.
14 K. Franke, D. Rößler and H. Kupsch, Proceedings of 5th International Conference on Nuclear and Radiochemistry, NRC 5, 3 - 8 September 2000, Pontresina (Switzerland) Vol. **2**, 466.
15 D. Rößler, K. Franke, R. Süß, E. Becker and H. Kupsch, *Radiochim. Acta*, 2000, **88**, 95.

ULTRA SENSITIVE MEASUREMENTS OF GAMMA-RAY EMITTING RADIONUCLIDES USING HPGE-DETECTORS IN THE UNDERGROUND LABORATORY HADES

M. Hult[1], J. Gasparro[1], L. Johansson[1], P. N. Johnston[2] and R. Vasselli[1]

[1]European Commission, Joint Research Centre, Institute for Reference Materials and Measurements (EC-JRC-IRMM), Retieseweg, B-2440 Geel, Belgium
[2]Department of Applied Physics, Royal Melbourne Institute of Technology, GPO Box 2476V, Melbourne 3001, Australia

1 INTRODUCTION

The field of underground radioactivity measurements is a growing field. One reason for this is that since the late 1990s, high quality commercial ultra low-background HPGe-detectors are readily available. Since 1992 the Institute for Reference Materials and Measurements (IRMM) has performed Ultra Low-level Gamma-ray Spectrometry (ULGS) in the underground laboratory HADES.[1] Initially, measurements were directed towards an improved understanding of sources of background radiation and how to reduce the effect of those sources.[2] As this understanding increased, new and better detector systems were installed. This paper describes the special features of the fourth HPGe-detector that was installed in HADES. Increasing the measurement capacity means that new fields of science can exploit the use of ULGS. In 1999 to 2002 some 20 different projects have been carried out using the ULGS facility in HADES.[3] ULGS is particularly useful in environmental studies. Man-made radioactivity in the environment has been steadily decreasing since the 1960s and new sensitive measurement techniques would greatly assist many investigations. In some cases it is possible to replace complicated and expensive radiochemistry work by ULGS. The main advantage of gamma-ray spectrometry is that sample preparation is very simple. Although an ULGS measurement may require many days of counting time, it can proceed unattended and requires little manpower. In certain cases ULGS measurements in combination with radiochemistry offer possibilities of measuring radionuclides with unprecedented sensitivity. This is the case for measurement of Zn impurities in GaAs,[4] where neutron activation in combination with radiochemical separation and ULGS has proven to be more than a factor of 100 more sensitive than other techniques, *e.g.* glow discharge mass spectrometry. Finally, in other cases ULGS complements radiochemistry allowing investigations of a multitude of radionuclides as in the case of measurements of radioactivity in a bone ash reference material.[5] In many underground laboratories, tasks that are related to the environmental field dominate. At IRMM, recent measurements have included evaluating the effect of the JCO accident[6] in Tokai-mura in 1999 and measuring radionuclides in samples downstream from a U-mine.[7]

A key part of the mission of IRMM is to perform special reference measurements in various fields. The ULGS measurements therefore require multipurpose detectors rather than detectors specially designed for one particular type of sample. As a consequence, the new detector described here has high efficiency both at low and high energy. This is achieved by having a large Ge-crystal with a thin deadlayer at the top contact.

1.1 Applications

At regular intervals it is necessary to assist other groups at IRMM with high sensitivity measurements of radioactive impurities in various samples. In the past two years there were two interesting cases where radioactive impurities in radioactive solutions needed to be determined. The radioactive solutions in question contained the radionuclide ^{204}Tl, which decays by 97.4% β^--decay and 2.6% electron capture, and the pure electron capture radionuclide, ^{41}Ca. These two radionuclides have no known γ-ray emission and so allow very high sensitivity measurements at energies above the beta decay energy (for ^{204}Tl) and above the internal bremsstrahlung energy (for ^{41}Ca).

1.1.1 Impurities in a solution containing ^{204}Tl. During the recent international key-comparison of the activity of a ^{204}Tl-solution, there was a surprisingly large variation in the measured specific activities.[8] The trend was that methods based on Liquid Scintillation Counting (LSC) resulted in a narrow spread and gave a systematically higher specific activity than methods based on measurements with proportional counters. Johansson could show that the methods based on proportional counters generally gave the wrong results due to problems in correcting for the self-absorption of β-rays.[9] As part of the study to find out reasons for discrepancies, highly sensitive measurements of γ-ray emitting impurities were carried out.

1.1.2 Impurities in a solution containing ^{41}Ca. OSTEODIET is a consortium of laboratories grouped in an EU Measurement & Testing research project, which is developing a novel method to study osteoporosis. The method is based on using ^{41}Ca, which does not occur naturally. The radionuclide will be given to patients and it will allow the study of Ca bone metabolism with unprecedented accuracy. It is important to have full control of all contaminants in the solution including those emitting γ-rays. To obtain the highest sensitivity for γ-ray emitting radionuclides, the original solution was measured in HADES.

2 MATERIALS AND METHODS

2.1 Detector and Shielding

The measurements presented here were carried out using a newly installed HPGe-detector (Ge-4). It is a 105% relative efficiency, coaxial detector with a carbonepoxy window of thickness 0.5 mm. The measured resolution, FWHM (Full Width Half Maximum) when installed in HADES is 2.1 keV at 1332 keV and 1.5 keV at 122 keV. The active volume of the Ge-crystal is 407 cm^3 and the crystal diameter is 80 mm. The detector is installed in the underground research facility HADES at the SCK•CEN site in Mol, Belgium. HADES is located 223 m below ground-level (500 m water equivalent). The shield (Fig. 1) is relatively similar to a previously installed shield for another HPGe-detector (Ge-3).[1] In

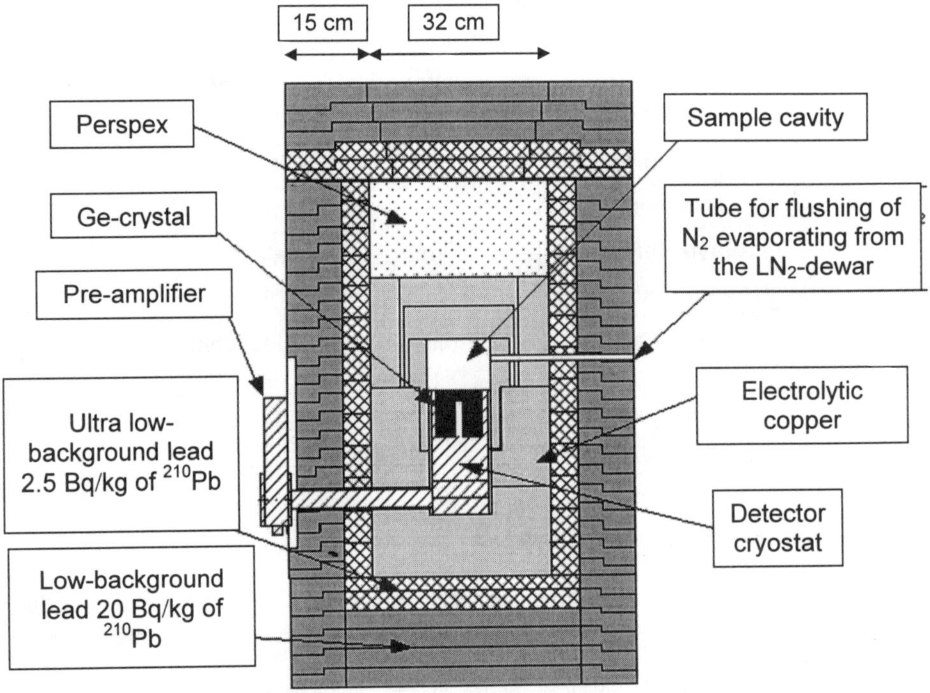

Figure 1 *Schematic drawing of Ge-4 and the Pb/Cu shield.*

order to minimise gaps in the shield for radon to penetrate, the top five lead discs (the lid) of the shield rest on the lead rings that constitute the side of the shield. The lid can be lifted using a pneumatic hand-pump and slid over the LN_2–dewar. The contribution to the background spectrum from the radon daughters is further minimised by filling all empty volumes in the shield and making it as airtight as possible. The sample volume is also flushed by nitrogen evaporating from the LN_2 dewar. The lead is shaped as 3 cm thick (vertically) rings, which are stacked to form a shield without gaps through which gas can penetrate. The innermost 5 cm of lead has a ^{210}Pb activity concentration of 2.5 Bq kg^{-1} and the outermost 10 cm of 20 Bq kg^{-1} The inner shield is made from freshly produced electrolytic copper (taken underground immediately after leaving the copper refinery in order to minimise the cosmogenic production of radionuclides) and it is 11 cm thick near the Ge-crystal. It effectively attenuates bremsstrahlung produced by the β^--decay of ^{210}Bi in the lead as well as γ-rays from other radionuclides in the lead. The modular Cu-shield is designed to accommodate different types of sample containers, including Marinelli beakers. Above the copper is a piece of Perspex in order to fill the shield volume beneath the sliding lid. Perspex is light and easy to handle as well as very radiopure. A shield of the type described here is not suitable for a detector above ground[10] since cosmic ray interactions with the shield would induce too much radiation and give high background.[11]

Table 1 *Some characteristics of two detectors installed in HADES.*

	Ge-3	Ge-4
Relative efficiency	60%	105%
Ge-volume	251cm^3	412 cm^3
Ge-mass	1.34 kg	2.20 kg
Deadlayer thickness	1.0 mm	0.0006 mm
Crystal-diameter	6.4 cm	8.0 cm
Cryostat	Aluminium KryAl	Electrolytic copper
Window	0.7 mm KryAl	0.5 mm Carbonepoxy

2.2 Modelling of the Detector Efficiency

As is commonly the case with ultra low-level gamma-ray spectrometry, standards for detector calibration were not available for the varied sample geometries and radionuclides of these essentially one-off measurements, so the efficiencies for the samples had to be estimated by modelling. A physical model of the detector was developed based on specifications from the manufacturer and horizontal orthogonal projections from radiography. The model was implemented in the Monte Carlo code EGS4,[12] which was used to model the full-energy peak efficiency at the energies of the γ-rays. The model parameters were determined as follows:

(i) The manufacturer's specification provides details about the dimensions of the detector cryostat, crystal size, window thickness and the composition of the materials of which the detector constructed.

(ii) The radiographs showed that the crystal was not skew, tilted or misaligned. The radiographs also showed that the distance from the Ge detector crystal to the carbonepoxy window was 3 mm at the centre and 6 mm at the rim. This provides the fundamental parameters regarding the position of the crystal when at liquid nitrogen temperatures.

(iii) The remaining parameters relate to the deadlayers of the detector crystal which act as absorbers. These deadlayers on the front, sides and rear of the detector can only be estimated by comparison between experimental calibration and the model results.

The procedure for adjusting the parameters in the EGS4 model were to firstly calibrate Ge-4 experimentally by measuring a set of standardised point sources from CERCA (prepared by drop evaporation at LMRI) at various distances. The modelled efficiency values were compared with the experimental calibration. The contact at the front of the crystal (near to the window) is ion implanted and known to be very thin in order to allow for high detection efficiencies for low energy γ-rays. Adjusting the model to the experimental calibration data, resulted in a front deadlayer thickness of 0.6 μm. This detector has a conventional contact (diffused lithium) on the outer surface. Such contacts are known to be between 0.5 and 1.5 mm thick but it is important to know the deadlayer with better accuracy. The outer deadlayer affects all γ-ray energies as it effectively determines the solid angle without attenuating the γ-rays, although when using Marinelli beakers, the deadlayer on the side also affects the efficiency through attenuation of γ-rays. Using the value 1.1 mm for the outer deadlayer gave the best fit with experimental data. The deadlayer inside the re-entrant is also somewhat important but for this detector it is very thin and therefore need not be known as accurately as the other two deadlayers. It is important to stress that the values for the deadlayer thicknesses given here should not be interpreted as the actual physical dimensions of these parameters, *e.g.* the range of implanted or diffused ions. The values given here simply optimize the fit between experimental data and the model that was used in the EGS4 simulations. The parameters are also not independent of each other and other parameters of the model.

Note that the calibration approach mentioned above cannot be reliable unless a radiograph from two angles is made. It is usually possible to find a set of detector-parameters that allows the model to reproduce the results of the experimental efficiency calibration, even if the detector crystal is skew, tilted or misaligned and the crystal in the model is not. Manufacturers may position the crystal accurately when it is assembled but the situation when the crystal is cooled to LN_2 temperature may change considerably.

2.3 Samples for Impurity Measurements

The ^{204}Tl was in a radioactive solution that contained 0.26 mg g^{-1} of TlCl in 0.1 M HCl. The solution was in a glass vial, the mass was 3.6 g and the total activity was 71 kBq.

There was 26 ml of the liquid solution with the radionuclide ^{41}Ca ($CaCl_2$ in diluted HCl, 2 M). It had been transferred to a Teflon bottle and the activity was 5.2 MBq. During measurement, both samples were placed on a 2 mm thick Teflon plate protecting the detector window.

3 RESULTS

3.1 Background Measurements

The background spectrum from Ge-4 in Figure 2, was collected over a period of 6 months starting 16 months after the detector was taken to HADES. The detector could thus be considered to be relatively "cool", since there was time for certain radionuclides produced above ground by cosmic rays to decay. In order to judge the performance of the detector it was compared with Ge-3, which is also a coaxial HPGe-detector (although with a thick deadlayer at the front), but from a different company (Table 1). Furthermore, Ge-3 is housed in a similar type of shield. It is evident from Figure 2 that the two detectors have very similar background characteristics. At energies above 2 MeV the normalised background spectra coincide. The quantitative analysis of the peak count rates is given in Table 2. Ge-3 displays two main peaks at 46 keV and 662 keV. We believe that the 46 keV γ-ray from ^{210}Pb originates from trace impurities inside the Ge-3 detector, whilst the 662 keV γ-ray has its origin both from the Cu-shield[1] and trace impurities in the detector.

The background spectrum of Ge-4 includes one dominating peak, which is the 1460 keV γ-ray from ^{40}K. The other peaks that are relatively large come from the two ^{222}Rn daughters ^{214}Pb and ^{214}Bi. Initially it was thought that the ^{40}K peak was due to environmental radioactivity from the surroundings (clay and construction materials in the

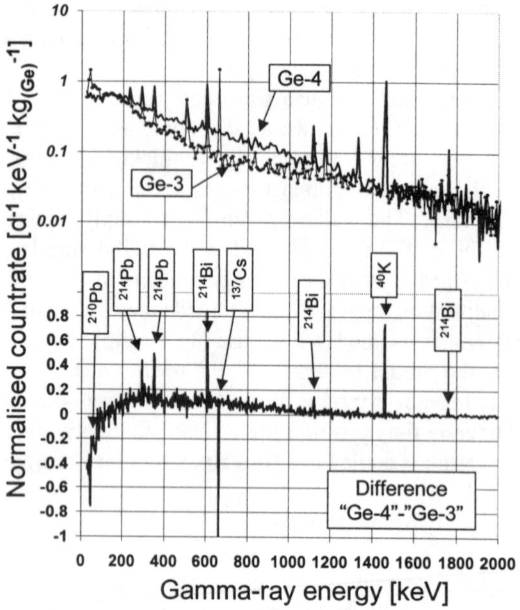

Figure 2 *The top figure shows the background spectra of Ge-3 and Ge-4 normalised to the Ge-crystal mass. The lower figure shows the difference between the spectra in the upper figure.*

gallery), but after comparing the results of attenuation and efficiency calculations (1460 keV γ-rays are attenuated by a factor of $5 \cdot 10^5$ when penetrating the Pb/Cu shield) the suspicion was directed towards the detector. The detector is built using state of the art techniques and only from very radiopure materials. In the three first detectors placed in HADES the window was made from pure Al (KryAl) or electrolytic Cu so the carbonepoxy window of Ge-4 was investigated. The detector manufacturer supplied some carbonepoxy from a recent batch and it was measured in HADES. The result was that there was 3.2(4) Bq kg^{-1} of ^{40}K in the carbonepoxy. Running the computer model with the same ^{40}K concentration in the window resulted in a countrate of 7.4 counts per day, which is comparable with the measured countrate of 10.8 counts per day in the background (Table 2). Therefore it is likely that the major part of the countrate in the 1460 keV peak comes from the detector window. Analysis of the carbonepoxy also revealed the presence of ^{214}Pb and ^{214}Bi which are short lived daughters of ^{226}Ra which is believed to be present in the detector window. Using the measured activities of ^{214}Pb and ^{214}Bi in the detector window, the peak count rate in the 186 keV peak due to ^{226}Ra has been calculated to be 0.5 counts per day which is below the detection limit of 1.8 d^{-1} (Table 2). So the lack of a 186 keV peak is consistent with the calculated amount of ^{226}Ra in the window which admittedly relies on a crude assumptions regarding Rn-emanation. Due to the flushing of N_2 into the sample volume it is not likely that the presence of ^{222}Rn daughter nuclides, ^{214}Pb (295 and 351 keV) and ^{214}Bi (609, 1120 and 1764 keV) comes from radon in the laboratory air. There are also γ-ray peaks from ^{208}Tl and ^{212}Pb in the spectrum, which indicate traces of ^{232}Th in the detector.

Table 2 *List of background γ-peaks for two of the detectors in HADES. The uncertainty is the combined standard uncertainty and the detection limits are given according to ISO.*[13]

Eγ (keV)	Nuclide	Ge-3 60%-coaxial (counts d^{-1})	Ge-4 105%-coaxial (counts d^{-1})
46	^{210}Pb	3.7(4)	1.18 (7)
63	^{234}Th	< 1.1	< 1.7
93	^{234}Th	< 1.0	< 1.6
186	^{226}Ra/^{235}U	< 0.9	< 1.8
238	^{212}Pb	1.3(3)	2.8 (2)
242	^{214}Pb	< 0.8	0.55 (4)
295	^{214}Pb	0.6(2)	2.4 (2)
351	^{214}Pb	1.1(2)	5.4 (3)
511	Annihil.	3.1(3)	4.0 (2)
583	^{208}Tl	< 0.5	< 1.3

Eγ (keV)	Nuclide	Ge-3 60%-coaxial (counts d^{-1})	Ge-4 105%-coaxial (counts d^{-1})
609	^{214}Bi	1.2(2)	4.8 (3)
661	^{137}Cs	6.2(4)	< 1.2
911	^{228}Ac	< 0.4	0.60 (5)
1120	^{214}Bi	0.43(12)	1.3 (2)
1332	^{60}Co	0.60(13)	0.93 (10)
1460	^{40}K	2.0(22)	10.8 (7)
1764	^{214}Bi	0.3(1)	1.2 (2)
2614	^{208}Tl	0.29(9)	1.0 (2)
40 – 2700		394(2)	782 (4)
Measuring time		84 days	55 days

3.2 Impurities in a Solution with ^{204}Tl

Figure 3a shows the spectrum from the analysis of the ^{204}Tl vial. This vial was sitting in a plastic container. The empty plastic container was measured separately afterwards. Bremsstrahlung from the β-particles and Hg X-rays dominate the spectrum up to the β-endpoint energy of 763 keV. Above that energy only background is expected, which allows detection with high sensitivity. Table 3 shows the quantitative results of the impurities that were detected in the underground measurement.

Table 3 *Radionuclides found in the ^{204}Tl solution. The combined standard uncertainty is given.*

Radionuclide	Activity concentration (mBq g^{-1})	Relative activity compared to the ^{204}Tl-solution (%)
^{137}Cs	0.7 (3)	1×10^{-6}
^{60}Co	190 (60)	3×10^{-4}
^{134}Cs	9 (3)	1×10^{-5}
^{40}K	5100 (1500)	7×10^{-3}
^{228}Th	14 (6)	2×10^{-5}
^{226}Ra	40 (15)	6×10^{-5}

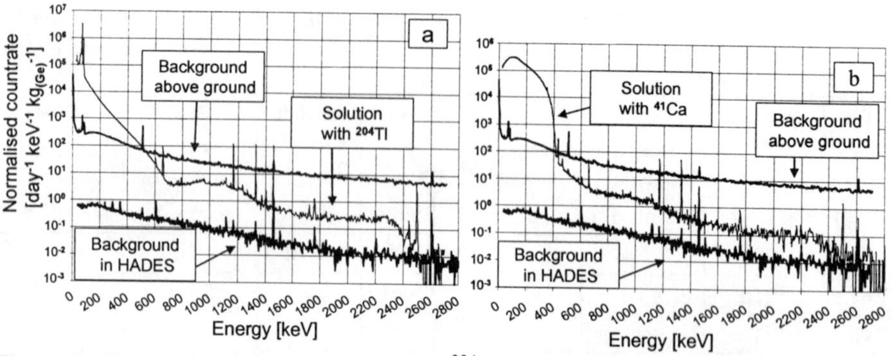

Figure 3. *Spectra from measurements of the ^{204}Tl solution (a) and the ^{41}Ca solution (b). The background above ground and in HADES are shown in both spectra as comparison.*

3.3 Impurities in a Solution with ^{41}Ca

Figure 3b shows the spectrum from the analysis of the ^{41}Ca solution. The internal bremsstrahlung dominates the spectrum in the absence of γ-rays, external bremsstrahlung and X-rays. Above the threshold energy at 413 keV very low detection limits can be obtained. The solution has been measured previously for impurities using γ-ray spectrometry[14] but never underground. In the previous measurement only ^{133}Ba was detected in the ^{41}Ca solution. In this study four more radionuclides were detected. Table 4 shows the quantitative results of the impurities that were found.

Table 4 *Activity concentration of radionuclides in the ^{41}Ca-solution. The combined standard uncertainty is given*

Radionuclide	Activity concentration (mBq g^{-1})	Mass concentration (g g^{-1})
^{60}Co	3.06(16)	72(4)·10^{-18}
^{40}K	1.7(6)	1.1(4)·10^{-9}
^{133}Ba	183(18)	19.4(19)·10^{-15}
^{137}Cs	1.83(16)	0.94(8)·10^{-15}
^{138}La	0.102(18)	0.113(20)·10^{-6}

4 DISCUSSION

4.1 Background Analysis

Deep underground measurements make it possible to detect very low levels of activity in the detector and surrounding materials. It is important to carefully analyse the background spectrum in order to identify impurity sources in the detector and the near shield materials. If one subtracts the contribution of ^{40}K, ^{214}Bi and ^{214}Pb to the Ge-4 background spectrum, the normalised background spectra of Ge-3 and Ge-4 are almost identical. Considering the high efficiency at both high and low energies of Ge-4 and its large surface area, one can conclude that Ge-4 is a very good detector for ultra low-level counting in close geometry. It may be possible to replace the window in the future with one containing less ^{40}K.

4.2 Impurities in Radioactive Solutions with ^{204}Tl and ^{41}Ca

Any radioactive impurity created in the production of a radionuclide or due to contamination may destroy the high accuracy and precision required for a primary standardisation. The activity of sources for primary standardisation is normally in the order of a kBq and impurities above hundreds of mBq have to be determined. This could be done before opening the ampoule containing the solution for standardisation, as described in this paper, or could be performed on an individual source.

The ^{204}Tl is produced by neutron activation in a reactor. It is most likely that the fission and activation products found in this specific vial are contamination in the ampoule and not in the solution itself. The values given in Table 3 can thus be interpreted as upper limits for the contamination. Furthermore, the activities of the radionuclide impurities detected and determined by ULGS were insignificant for the standardisation. However, if the level of impurities had been shown to be significant, individual sources, containing only a few microgram of the solution or less in the case of dilutions, could benefit from the low detection limit available at the HADES facility. For example, the presence of the activation product ^{134}Cs in the solution would cause interference in the case that it was used as a tracer for β-efficiency tracing in 4πβ-γ coincidence counting for standardisation of ^{204}Tl. Usually, several methods of detection are applied in the standardisation and an activity of a few hundred mBq of ^{134}Cs in a source would already disturb the 4πβ-γ coincidence measurements and would require a correction. Low impurity concentrations are needed in other types of investigations as *e.g.* when determining the double internal bremsstrahlung of ^{204}Tl.[15]

Thallium-204 decays by 97.4% via β^- decay to the ground state of ^{204}Pb and by 2.6% via electron capture (EC) to the ground state of ^{204}Hg. These transitions are first forbidden and the β^- decay end point is 763.0 keV. The data collected here would allow the setting of an upper limit of 10^{-7} for the γ-ray emission probability for a γ-ray from a possible competing decay branch with higher Q-value or higher excited states of ^{204}Tl. The theoretical probability for such a transition is close to zero.[16]

^{41}Ca is produced in the ^{40}Ca(n,γ)^{41}Ca nuclear reaction. It decays only by EC directly to the ground state of ^{41}K. As this is the only branch no γ-rays are emitted. For other possible decay modes the same reasoning as for ^{204}Tl is valid. The γ-ray emitting impurities were found to be acceptable[17] for the osteoporosis study. The values obtained here will also serve as benchmark for possible future purification actions.

Acknowledgements

The HADES crew of SCK•CEN is gratefully acknowledged for their work. The authors thank Mikael Berglund and Marcus Osterman of IRMM for initiating part of this work and José das Neves for design work. Gerd Heusser, MPI für Kernphysik, Heidelberg, Germany, has contributed with fruitful discussions and helpful advice.

References

1 M. Hult, M.J. Martinez, M. Köhler, J. Das Neves and P.N. Johnston, *Appl. Radiat. Isot.*, 2000, **53**, 225.
2 D. Mouchel and R. Wordel, *Appl. Radiat. Isot.*, 1996, **47**, 1033.
3 M. Hult, *Information on IRMM Measurements Carried out in the Underground Laboratory HADES 1999-2002*, Internal report at IRMM, 2002, GE/R/RN/03/02.
4 M. Köhler, A.V. Harms and D. Albert, *Appl. Radiat. Isot.*, 2000, **53**, 197.
5 R. Pilviö, J.J. LaRosa, D. Mouchel, R. Wordel, M. Bickel and T. Altzitzoglou, *J. Environ. Radioactivity,* 1999, **43**, 343.
6 M. Hult, M.J. Martinez, P.N. Johnston and K. Komura, *J. Environ. Radioactivity,* 2002, **60**, 307.
7 M. Köhler, *Determination of radioactive disequilibria in environmental samples*, Internal report at IRMM, 1999, GE/R/RN/02/99.
8 G. Ratel and P. Cassette. Private communication.
9 L. Johansson, G. Sibbens, T. Altzitzoglou and B. Denecke, *Appl. Radiat. Isot.*, 2002, **56**, 199.
10 J. Verplancke, *Nucl. Instr. And Meth. A*, 1992, **312**, 174.
11 P. Voytola and P.P. Povinec, *Appl. Radiat. Isot.*, 2000, **53**, 185.
12 W.R. Nelson, H. Hirayama and D.W.O. Rogers, *The EGS4 code system*, 1985, SLAC Report, 265.
13 ISO, *Determination of the detection limit and decision threshold for ionizing radiation measurements Part 3,* ISO. 2000, ISO 11929-3:2000.
14 M. Paul, I. Ahmad and W. Kutschera, *Z. Phys. A,* 1991, **340**, 249.
15 A. Bond and H. Lancman, *Phys Rev. C*, 1972, **6,** 2231.
16 D. De Frenne, Private communication.
17 J. W. Haggith, *Int. J. Nucl. Med. Biol.*, 1983, **10**, 85.

SEDIMENTATION RATES AND METALS IN SEDIMENTS FROM THE RESERVOIR RIO GRANDE – SÃO PAULO/BRAZIL

S.R.D. Moreira,[1] D.I.T. Fávaro,[2] F. Campagnoli, [3] and B.P. Mazzilli [1]

[1]Departamento de Radiometria Ambiental/NA-Instituto de Pesquisas Energéticas e Nucleares/IPEN-Travessa R, 400 – Cid. Universitária, Butantã - São Paulo, Brazil - 05508-900

[2]Laboratório de Análise por Ativação Neutrônica/LAN–Instituto de Pesquisas Energéticas e Nucleares/IPEN

[3]Divisão de Geologia/DIGEO-Instituto de Pesquisas Tecnológicas do Estado de São Paulo/IPT–Av. Prof. Almeida Prado, 532 – Cidade Universitária – São Paulo, Brasil – CEP: 01064-970.

1. INTRODUCTION

The Metropolitan Area of São Paulo - MASP is the greatest and most populous urban agglomerate of South America and one of the largest industrial complexes of the world. The problems of high demography density and lack of water resources are typical of great urban centres that have to live with environmental impacts, such as disposal of residues, insufficient water availability for public and industry, air and soil pollution, etc.

These problems can easily be observed in reservoirs that supply water for cities because they receive an expressive amount of sediments, causing silting and flooding with considerable losses for the public administration.[1]

In order to evaluate, identify and characterize areas of sediment production and the corresponding silt deposition, a project was established in one of the most important reservoirs of São Paulo city, Rio Grande reservoir, located in the Southeast portion of the MASP. This reservoir is responsible for the water supply of four counties (São Bernardo do Campo, São Caetano do Sul, Santo André and Diadema) and has been seriously affected by the urban expansion of the MASP, due to the chaotic urban occupation, with irregular use of the land. This region presents extensive degraded areas caused not only by erosion, but also by pollution of diffuse loads, as recent sedimentary deposits.[2]

One of the objectives of this project is to determine sedimentation rates in different parts of the reservoir and also to verify whether the sediments contain an historical registration of antropic activity. The sedimentation rates were calculated by ^{210}Pb method and neutron activation analysis was used to determine elements As, Ba, Br, Co, Cr, Cs, Fe, Na, Rb, Sb, Sc, Se, Ta, Th, U, Zn and rare earths Ce, Eu, La, Lu, Nd, Sm, Tb and Yb in two sediment cores collected in the reservoir

2. METHODS

Two sediment cores, point 26, 80cm long and point 29, 54cm long, were extracted in January 1998, using a Piston Corer sediment sampler, at a water-depth of 9.1 m and 10.2 m, respectively.

Grain-size analysis of sediment samples showed that the grain sizes are dominantly silt and clay, indicating the prevalence of settling processes. Sediment organic matter contents in the first centimetres of the two cores are very high and decrease with depth.

The measurement of the radionuclides ^{226}Ra and ^{210}Pb were used to determine the dates and sedimentation rate. These radionuclides were determined in each slice of the core. The samples of sediments were dried at 60^0C and sifted in sieves of 0.065 mm. The samples were then dissolved in mineral acids, HNO_3conc. and HF 40%, and H_2O_2 30%, in a microwave digestor and submitted to the radiochemical procedure for the determination of Ra and Pb.

This procedure consists of an initial precipitation of Ra and Pb with 3M H_2SO_4, dissolution of the precipitate with nitrilo-tri-acetic acid at basic pH, precipitation of $Ba(^{226}Ra)SO_4$ with ammonium sulphate and precipitation of 30% $^{210}PbCrO_4$ with sodium chromate. The ^{226}Ra concentration was determined by gross alpha counting of the $Ba(^{226}Ra)SO_4$ precipitate[3] and the ^{210}Pb concentration through its decay product, ^{210}Bi, by measuring the gross beta activity of the $^{210}PbCrO_4$ precipitate[4]. The chemical yields for the radionuclides, ^{226}Ra and ^{210}Pb, were determined by gravimetric analysis. Both radionuclides were determined in a low background gas flow proportional detector. The dates were calculated by the Constant Rate of Supply (CRS) model.[5]

For the multielemental analysis of samples, approximately 200 mg of sediment (duplicate samples), about 150 mg of reference materials and synthetic standards were accurately weighed and sealed in pre-cleaned double polyethylene bags, for irradiation. Single and multi-element synthetic standards were prepared by pipetting convenient aliquots of standard solutions (SPEX CERTIPREP) onto small sheets of Whatman nº 41 filter paper.

Sediment samples, reference materials and synthetic standards were irradiated for 16 hours, under a thermal neutron flux of 10^{12} cm^{-2} s^{-1} in the IEA-R1m nuclear reactor at IPEN. Two series of counting were made: the first after one-week decay and the second after 15-20 days. The counting time was 2 hours for each sample and reference material and half an hour for each synthetic standard. Gamma spectrometry was performed using a Canberra gamma X HPGe detector and associated electronics, with a resolution of 0.88keV and 1.90keV for ^{57}Co and ^{60}Co, respectively. The analysis of the data was made by VISPECT program to identify the gamma-ray peaks[6] and by ESPECTRO program to calculate the concentrations. Both programs were developed at the Radiochemistry Division, IPEN. The elements analysed using this methodology were As, Ba, Br, Co, Cr, Cs, Fe, Hf, Na, Rb, Sb, Sc, Se, Ta, Th, U, Zn, Zr and the rare earths Ce, Eu, La, Lu, Nd, Sm, Tb and Yb. The precision and accuracy of the method were verified by measuring the IAEA reference materials Buffalo River Sediment (NIST SRM 2704) and Soil 7 (IAEA).

3. RESULTS AND DISCUSSION

Isotopic ages are presented in Figure 1. For point 26 the ^{210}Pb profile depicts two well-defined linear trends: from level 0 cm (year 1998) to level 24 cm (year 1950) indicating an average accumulation rate of 0.50 $cmy^{-1}$and from level 36cm (year 1948) to level 54cm (year 1910) indicating an average accumulation rate of 0.47 cmy^{-1}. The lower dated level (54 cm) yields a sedimentation rate of 0.61 cmy^{-1}. For point 29 the ^{210}Pb profile depicts a well-defined linear trend, from level 0 cm (year 1998) to level 72 cm (year 1968) indicating an average accumulation rate of 2.4 cmy^{-1}. The lower dated level (80.5 cm) yields a sedimentation rate of 1.22 cmy^{-1}. Lower rates were related to the period prior to the water dam, when the loading of the sediments was stabilised. Higher rates were related to rainy seasons and the urban expansion in the surroundings of the reservoir.[7]

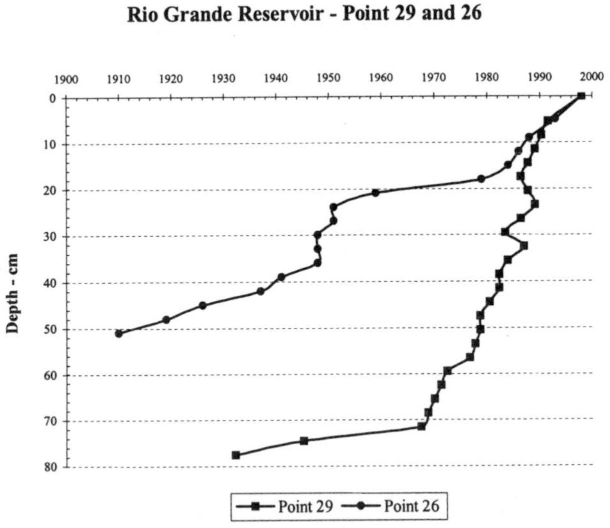

Figure 1 *Age-depth relationship for the two cores*

The results obtained for Buffalo River Sediment and Soil 7 reference materials, as well as certified and information values are presented in Table 1. It can be seen that the results for the analysed elements are, in general, in good agreement with the certified or information values. Figures 2 and 3 show the calculation of the standardised difference or z-value that was made according to Bode[8]. If $/z/ < 3$, it means that the individual result of the control sample (reference material) should be in the 99% confidence interval of the target value.[9-11]

Table 2 and 3 show the results obtained for multielemental analysis in the sediment samples by INAA for the cores point 26 and point 29. The precision, based on duplicate sample analysis, was better than 10%. For point 26, Cr, Fe, Hg, Zn, As, Br, Co, Cs and Th concentrations increase with depth and for the other elements no variation was observed. For point 29, As, Br, Cr, Fe, Na, Sb, Sc and Zn concentrations decrease with depth and Cs and Hf concentration increase with depth. Comparing the results obtained for the two cores with shale reference values, most of the analysed elements are enriched in Rio Grande reservoir, indicating an antropic contamination.

Table 1 *Results obtained for the reference materials SOIL 7 (IAEA) and Buffalo River Sediment (NIST SRM 2704) by INAA*

		SOIL-7		NIST SRM 2704	
Element	Concentration unity	Measured values	Literature	Measured values	Literature
As	μg	13.5 ± 0.7	13.4 ± 0.84	23.2 ± 0.5	23.4 ± 0.8
Ba	μg	151 ± 12	(159)	417 ± 33	414 ± 12
Br	μg	8.7 ± 0.5	(7)	7.0 ± 0.7	(7)
Ce	μg	52.3 ± 0.7	61 ± 7	58.5 ± 0.7	(72)
Co	μg	8.21 ± 0.01	8.9 ± 0.9	12.4 ± 0.3	14.0 ± 0.6
Cr	μg	64 ± 1	60 ± 13	132 ± 2	135 ± 5
Cs	μg	5.0 ± 0.1	5.4 ± 0.7	5.8 ± 0.1	(6)
Eu	μg	0.89 ± 0.04	1.0 ± 0.2	1.20 ± 0.04	(1.3)
Fe	(%)	2.53 ± 0.08	(2.57)	4.0 ± 0.1	4.1 ± 0.1
La	μg	27.3 ± 0.4	28 ± 1	30.0 ± 0.7	(29)
Lu	μg	0.27 ± 0.02	(0.3)	0.52 ± 0.06	(0.6)
Na	μg	2303 ± 102	(2400)	5860 ± 247	5470 ± 140
Nd	μg	25 ± 3	30 ± 6		
Rb	μg	48 ± 3	51 ± 5	108 ± 5	(100)
Sb	μg	1.5 ± 0.1	1.7 ± 0.2	4.23 ± 0.09	3.79 ± 0.15
Sc	μg	8.09 ± 0.01	8.3 ± 0.1	11.8 ± 0.1	(12)
Sm	μg	4.86 ± 0.02	5.1 ± 0.4	6.3 ± 0.2	(6.7)
Tb	μg	0.70 ± 0.06	0.6 ± 0.2	8.9 ± 0.1	(9.2)
Th	μg	7.6 ± 0.3	8.2 ± 1.1		
Yb	μg	2.10 ± 0.07	2.4 ± 0.4	2.8 ± 0.1	(2.8)
Zn	μg	99 ± 3	104 ± 6	422 ± 8	438 ± 12

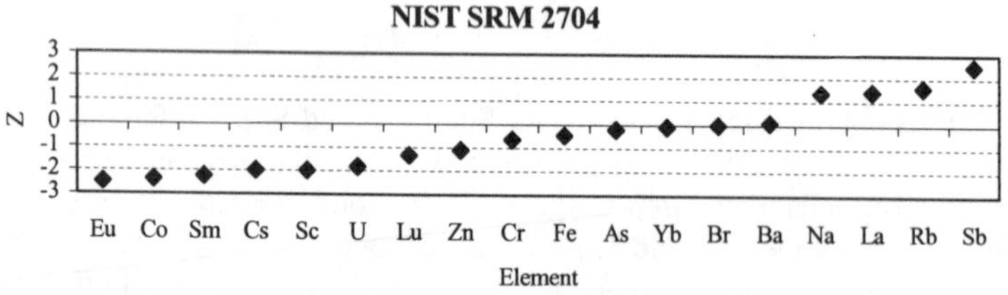

Figure 2 *Control chart (z-values) for inspection of the normalised concentrations of some elements in the NIST SRM 2704 reference material sample*

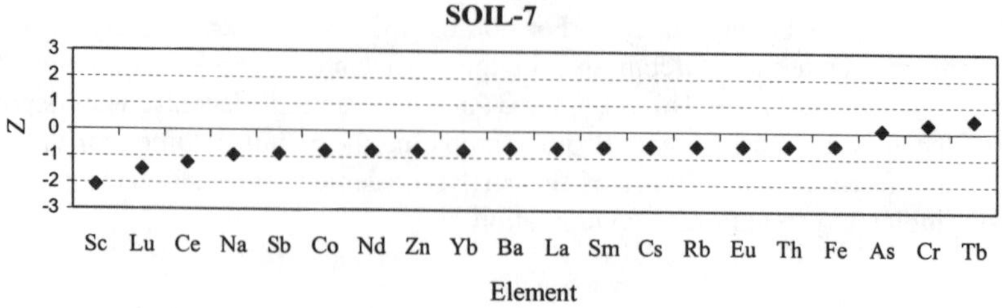

Figure 3 *Control chart (z-values) for inspection of the normalised concentrations of some elements in the SOIL-7 reference material sample*

Table 2 *Concentration of the elements (µg g^{-1}) determined by INAA in the different depths of the sediment core for point 26 (Mean of two individual determinations)*

Depth (cm)	As	Ba	Br	Ce	Co	Cr	Cs	Eu	Fe%	Hf	Ho	La	Lu	Na	Nd	U	Th	Rb	Sb	Sc	Sm	Ta	Tb	Yb	Zn	Zr
0-5	9.6	526	9.8	111	8.3	124	7.5	1.3	5.3	7.2	2.2	45	0.6	1148	40	6.1	21.3	62	1.1	20	7.2	2.5	1.1	4.0	252	328
5-9	6.7	489	6.5	82	8.5	78	10.3	1.2	3.4	7.2	1.6	43	0.5	839	37	4.9	15.5	67	0.8	18	7.5	2.4	0.9	3.1	156	258
9-12	4.5	613	2.4	66	5.9	60	13.0	1.1	1.7	12.7	1.7	42	0.6	1034	50	7.1	12.0	95	0.1	17	6.1	3.7	1.1	4.1	78	287
12-15	2.9	651	2.6	66	5.8	66	13.2	1.2	1.7	13.5	1.7	42	0.7	1195	50	7.2	13.1	101	0.1	19	6.3	4.1	1.0	4.2	65	306
15-18	3.8	717	2.8	69	6.0	65	13.1	1.2	1.7	13.6	1.6	42	0.7	1183	41	6.8	13.3	91	0.5	19	6.7	3.5	0.7	4.2	102	405
18-21	4.0	695	3.6	67	6.0	67	13.0	1.2	1.7	13.8	1.4	41	0.8	1150	36	6.2	13.2	98	0.5	19	6.3	3.6	0.75	4.4	112	366
21-24	4.3	805	2.9	74	6.5	67	13.0	1.2	1.9	12.0	1.6	42	0.6	1011	43	8.1	13.9	99	0.7	19	6.9	3.8	1.1	4.1	74	235
24-27	4.1	785	4.1	76	6.5	68	13.2	1.3	2.0	12.4	1.9	44	0.7	1152	48	7.1	14.1	104	0.7	19	7.3	3.9	1.1	4.4	75	301
27-30	6.0	673	3.6	70	6.2	70	12.5	1.2	2.0	13.0	2.0	43	0.7	1074	34	6.7	14.2	102	0.7	18	6.7	3.5	1.0	4.6	93	289
30-33	6.5	608	4.4	66	6.1	65	12.1	1.1	2.0	13.0	2.1	41	0.7	881	32	6.7	14.0	100	0.2	18	6.3	3.5	1.1	4.4	90	270
33-36	5.8	649	5.8	71	6.5	70	13.3	1.2	2.3	13.9	1.7	43	0.8	1003	38	7.8	15.2	109	0.3	19	6.5	3.9	1.0	4.6	96	305
36-39	5.9	660	5.3	69	6.6	69	13.1	1.3	2.1	12.6	2.0	43	0.8	978	35	9.2	14.7	105	0.6	19	6.6	4.0	1.0	4.5	87	290
39-42	5.2	600	5.2	66	6.7	65	13.0	1.2	2.3	10.8	2.2	49	0.7	1152	41	6.5	13.7	109	0.6	18	6.9	3.6	1.1	4.0	85	231
42-45	5.9	567	4.7	68	6.6	63	13.2	1.2	2.4	10.4	2.2	49	0.7	1060	43	7.0	13.8	107	0.7	18	7.2	3.5	1.0	3.8	87	269
45-48	6.5	583	4.9	68	6.6	68	13.5	1.3	2.7	10.0	1.4	45	0.7	1010	37	6.8	13.6	104	0.2	19	5.9	3.3	0.9	3.9	89	248
48-51	5.0	585	4.4	71	6.6	67	13.0	1.3	2.3	9.6	1.8	47	0.7	971	39	7.2	13.8	104	0.6	19	6.8	3.4	0.7	4.0	87	263
51-54	3.9	659	3.6	77	6.9	71	14.7	1.3	1.9	11.3	1.9	48	0.7	1001	27	6.4	15.3	110	0.6	20	6.6	3.6	1.3	4.3	98	283
54-57	4.1	698	3.6	87	6.5	77	14.0	1.5	1.7	11.6	2.1	52	0.8	1006	44	7.4	15.6	101	0.7	20	7.6	3.2	1.4	4.4	108	309
Shale	13	580	-	80	19	90	5	1.0	4.7%	2.8	1.0	92	0.7	*8200 - 11000	24	3.7	12	140	1.5	13	6.4	1.8	1.0	2.6	95	*200-210

Table 3 *Concentration of the elements (μg g^{-1}) determined by INAA in the different depths of the sediment core for point 29. (Mean of two individual determinations)*

Depth-cm	As	Ba	Br	Ce	Co	Cr	Cs	Eu	Fe%	Hf	La	Lu	Na	Nd	Rb	Sb	Sc	Se	Sm	Ta	Tb	Th	U	Yb	Zr	Zn
0-5.5	10.2	513	5.2	156	10.4	77	5.4	1.3	5.8	17.3	51.91	1.2	1676	50	96	0.8	21.2	2.2	6.6	4.1	1.7	28.4	8.7	7.7	573	270
5.5-8.5	8.6	477	6.6	144	10.2	81	5.7	1.3	6.0	11.5	48.95	0.9	1496	47	88	0.8	21.7	2.4	7.7	3.6	1.2	27.0	7.9	5.8	314	201
8.5-11.5	8.9	559	6.0	120	9.9	95	5.1	1.1	6.4	7.2	36.28	0.6	1454	29	85	0.8	21.2	1.5	6.6	2.6	0.8	22.7	6.2	3.7	290	267
11.5-14.5	9.4	531	5.8	116	9.2	98	4.4	1.0	7.5	7.1	33.63	0.6	1362	24	81	0.8	21.7	1.4	2.9	2.7	0.8	22.4	5.9	3.9	209	270
14.5-17.5	11.5	562	11.6	139	12.7	107	5.6	1.5	6.6	4.8	48.32	0.6	1303	58	70	1.4	21.8	3.6	8.7	2.9	1.2	28.2	7.8	3.4	210	384
17.5-20.5	11.1	493	6.1	127	8.7	120	6.0	2.9	6.9	4.4	39.84	0.5	1844	52	84	0.8	23.5	3.5	7.4	2.9	1.1	27.0	8.0	3.2	218	244
20.5-23.5	11.0	462	4.7	140	9.6	103	5.3	1.1	7.3	8.8	42.83	0.8	1840	36	82	0.7	24.0	2.1	12.5	2.7	0.9	27.7	8.0	4.6	248	284
23.5-26.5	9.8	428	3.1	140	11.7	90	4.1	1.2	7.1	18.1	34.14	1.6	1262	43	92	0.7	25.3	1.7	9.7	2.6	1.4	28.4	7.7	8.3	444	218
26.5-29.5	11.9	520	4.4	155	8.9	83	6.1	1.4	7.2	8.3	49.50	0.9	1587	53	91	0.7	22.7	1.6	9.3	2.9	1.2	28.8	8.9	5.6	350	210
29.5-32.5	12.7	495	4.5	138	8.9	105	3.8	1.2	7.1	18.1	38.36	1.5	1578	57	67	0.6	23.9	1.6	8.4	2.2	1.5	28.3	9.4	8.9	706	184
32.5-35.5	12.6	586	4.6	121	8.6	91	4.5	1.1	7.6	10.7	30.48	0.7	1424	27	91	0.7	26.0	2.1	8.5	2.4	0.5	26.2	7.3	4.3	310	200
35.5-38.5	11.2	550	5.4	132	8.5	89	5.1	1.1	7.6	8.1	36.96	0.7	1369	29	94	0.6	24.7	2.2	10.2	2.8	0.6	27.0	7.6	4.2	297	253
38.5-41.5	9.7	594	3.3	121	8.7	88	4.5	1.1	7.0	11.8	33.27	1.2	1620	39	86	1.0	24.1	1.8	6.9	2.6	1.0	25.4	8.0	6.0	368	217
41.5-44.5	11.9	568	4.8	129	9.7	99	5.1	1.1	7.8	6.7	36.70	0.6	1448	41	87	0.7	24.8	1.8	6.8	2.7	0.4	27.3	7.9	3.8	369	270
44.5-47.5	12.2	582	5.0	118	10.2	99	4.7	1.3	8.3	5.7	32.55	0.5	1352	36	87	0.3	26.5	1.7	5.9	2.9	0.8	26.9	7.4	3.3	121	345
47.5-50.5	11.8	573	6.1	124	9.2	96	4.4	1.2	8.7	5.6	33.90	0.5	1269	37	83	0.6	26.9	1.4	6.0	2.8	0.7	27.1	7.5	3.2	264	172
50.5-53.5	11.5	396	8.1	112	8.5	90	3.6	1.0	7.9	3.5	28.94	0.4	1199	25	73	0.6	25.2	1.4	9.7	2.9	0.5	26.0	6.4	2.7	133	119
53.5-56.5	10.2	441	4.7	108	8.5	82	3.8	1.0	7.4	5.9	26.89	0.6	1340	28	73	0.5	23.8	1.3	10.0	2.8	0.5	23.6	7.0	3.4	161	141
56.5-59.5	8.8	685	3.0	125	8.7	79	4.2	1.1	6.4	12.6	35.04	0.9	1999	41	85	0.6	22.0	1.6	7.0	2.5	1.0	23.5	8.9	5.9	328	176
59.5-62.5	10.3	605	3.4	135	9.0	87	4.5	1.2	7.6	7.8	35.23	0.6	1816	38	83	0.6	23.7	2.0	7.0	2.7	0.8	25.3	8.9	4.1	317	157
62.5-65.5	10.7	393	8.6	137	7.8	98	5.5	1.3	6.8	3.1	44.27	0.4	1126	42	65	0.9	22.2	1.4	7.6	2.6	0.7	26.8	6.8	2.9	149	136
65.5-68.5	9.2	333	16.2	96	7.9	85	7.2	1.2	4.3	3.00	37.8	0.4	708	40	48	1.2	18.1	1.4	6.6	2.2	0.7	19.3	5.0	2.3	131	151
68.5-71.5	4.8	479	7.0	112	8.8	60	8.5	1.4	2.1	22.3	56.7	1.2	1017	53	66	0.6	16.2	1.1	10.0	3.0	1.2	19.9	9.4	7.4	610	95
71.5-74.5	3.5	562	5.6	110	7.8	56	8.3	1.4	1.6	26.4	56.6	1.4	1119	54	70	0.4	16.7	1.3	10.0	3.2	1.2	19.6	9.6	8.2	848	74
74.5-77.5	2.6	521	4.1	87	8.0	59	11	1.3	1.6	17.9	49.6	0.9	1121	46	90	0.4	17.6	1.5	8.3	3.6	1.1	18.6	8.3	6.5	542	56
77.5-80.5	2.1	507	3.4	112	8.1	59	12	1.5	1.7	23.6	60.3	1.0	1120	58	108	0.4	17.6	1.5	10.8	3.7	1.4	23.3	9.5	7.8	708	86
shale	13	580	-	80	19	90	5	1.0	4.7%	2.8	92	0.7	8200-11000	24	140	1.5	13	0.6	6.4	1.8	1.0	12	3.70	2.6	200-210	95

In order to investigate the association among the elements in sediments, R-mode factor analysis was performed. Through the Varimax method, five factors were extracted. Results can be interpreted as follows:

- Point 26: **F1**: (45.7% of the total variance) – load >0.70: points to the association of As, Br, Ce, Co, Cr, Fe, Th and Zn; load > -0.70, for Cs and Rb.
  F2: (15.4% of the total variance) and F3 (10.4% of the total variance) have high loading on Lu, U, and Yb, and Eu, La, Sm, and Sc, respectively.
  F4: (8,4% of the total variance) has high loading on Ho and Zr.
  F5: (5.3% of the total variance) has high loading on Na and Nd.

- Point 29: **F1**: (36.4% of the total variance) - load > 0.70: points to the association of Hf, Lu, Nd, Tb, U, Yb and Zr.
  F2: (20.1% of the total variance) has high loading on As, Ce, Co, Fe, Th, Zn.
  F3: (13.8% of the total variance) - load > 0.70: has high loading on Ba, K, Na and Rb; load > -0.70: has high loading on Br and Sb.
  F4: (7,3% of the total variance) has high loading on Cs, La, Ta.
  F5: (4.9% of the total variance) has high loading on Eu and Se.

These results, in combination with future mineralogical studies, will provide an improved and more in-depth analysis of source identification.

4. CONCLUSIONS

The sedimentation rates obtained for the two cores studied show that the reservoir is affected by the rainy season and urban expansion in its surroundings. The latter being the main factor that causes silting in rivers or reservoirs. The evolution of the sediment deposits is usually related to the history of land use. These rates show the tolerance of erosive process in a basin and are characteristic of geological substratum, geomorphology, types of soils and land use. Furthermore, these rates can be taken as reliable environment geo-indicators to measure the effectiveness of preventive and corrective actions adopted in the hydrographical basins to mitigate soil degradation.

Neutron activation analysis of the sediment samples showed increased concentrations for many metals when compared with shale[12] values, indicating that the reservoir has an antropic contamination.

References

1 F. Campagnoli – III Encontro Nacional de Engenharia de Sedimentos - ENES - Anais da ABRH Associação Brasileira de Recursos Hídricos, 1998, 135.
2 S. R. D. Moreira, F. Campagnoli and B. P. Mazzilli. V Congresso de Geoquímica dos Países de Língua Portuguesa - VII Congresso Brasileiro de Geoquímica, Porto Seguro, Bahia, Brazil, 1999.
3 J. Oliveira. Instituto de Pesquisas Energéticas e Nucleares, dissertação de mestrado, 1993.
4 S. R. D. Moreira. Instituto de Pesquisas Energéticas e Nucleares, dissertação de mestrado, 1993.
5 I. Crespi, N. Genova, L. Tositti, O. Tubertini, G. Berttolli, M. Odonne, S. Meloni, A. Buzeno. *J. Radioanalytical and Nuclear Chemistry*, 1993 **168**(1),107.

6 J.S. Noller – *Quaternary Geochronology: Methods and Applications*. J. S. Noller, J. M. Sowers & W. Lettis, Editors. American Geophysical Union, Washington, DC, 2000, p. 115-120.
7 A. I. Bulnayev. *Analyst*, 1995, **120**, 445.
8 Departamento de Águas e Energia Elétrica do Estado de São Paulo/(DAEE)- Fundação Centro Ecnológico de Hidráulica/FCTH. Banco de Dados Pluviométricos do Estado de São Paulo – CD-ROM, 1998.
9 P. BODE. Instrumental and Organizational Aspects of a Neutron Activation Analysis Laboratory, Interfaculty Reactor Institut, Delft University of Technology, Delft, The Netherlands, 1996, 147.
10 I. C.Dinescu, O. Duliu, M. Badea, N.G. Mihăilescu, I. Vanghelie. *J. Radioanalytical and Nuclear Chemistry*, 1998, **238**(1-2), 75.
11 J. Al-Jundi, K. Randle. International Symposium of Nuclear and Related Techniques in Agriculture, Industry, Health and Environment, Havana (Cuba), 1997, p. 28-30.
12 S. R. Taylor and S. M. McLennan. *The continental crust: its composition and evolution*, Blackwell Scientific, Palo Alto, Ca., 1985, p. 25-50.

OPERATIONAL SPECIATION OF ^{226}Ra AND U-ISOTOPES IN SEDIMENTS AFFECTED BY NON-NUCLEAR INDUSTRY WASTES

J. L. Aguado[1], J. P. Bolívar[1], E. G. San-Miguel[1] and R. García-Tenorio[2]

[1]Departamento Física Aplicada, Universidad de Huelva. EPS La Rábida, 21819 Huelva, Spain.
[2]Departamento Física Aplicada II, Universidad de Sevilla. ETS Arquitectura, Avda Reina Mercedes 2, 41012 Sevilla, Spain.

1 INTRODUCTION

The Odiel River (SW Spain) has been historically perturbed by anthropogenic activities such us mining[1] and phosphoric acid industries located at its estuary[2]. Particularly, the Odiel estuary has received natural radioactivity impact due to waste releases to the river from the phosphoric industries[2, 3].

In these industries, the phosphoric acid is formed from dissolution of phosphate rock with sulphuric acid; the waste called phosphogyspum (PG) is mainly calcium sulphate[4]. During the 1968-1997 period, 20% of the PG produced by the factories were directly released to the Odiel river, while the rest were stored in open-air piles (gyp-stacks) with the waters used for PG transportation to the piles directly released to the estuary. Nowadays, the PG produced by the industries (3 million metric tones per year) is totally stored in the piles while the waters used for its transportation are chemically treated, purified and re-used.

Since these wastes are enriched in several radionuclides of the Uranium series, specially ^{226}Ra [4], it is not surprising to find, even after cessation of the releases, that riverbed sediments in the estuary are enriched in radionuclides such as $^{234,238}U$ and ^{226}Ra [3]. Particularly, the contamination levels found in the Odiel river sediments one year after the change in waste management policy will be reported in this paper. Additionally, these estuarine sediments, with their associated radionuclide contamination, are of radiological concern. They could act as delayed sources of the mentioned radionuclides to the waters following desorption and/or remobilization processes, with consequent implications in relation to their availability for biota and the redistribution of the radioactive impact.

Direct speciation studies of radionuclides in sediments from this estuary have been performed. Speciation techniques are useful tools to study the behaviour of natural and artificial radionuclides in aquatic systems. The main objective of this work is to determine and compare the direct operational speciation of uranium isotopes and ^{226}Ra in superficial sediments collected from the estuary.

2 MATERIAL AND METHODS

A sampling campaign was performed along the Odiel river estuary (December 1998). Superficial sediment samples were collected from the Odiel channel, where phosphate rock industries are located, (A - G samples), and also from the Ria de Huelva channel (H - K samples) which connects the estuary with the Atlantic ocean and is formed after the confluence of the Odiel and Tinto river mouths (Figure 1).

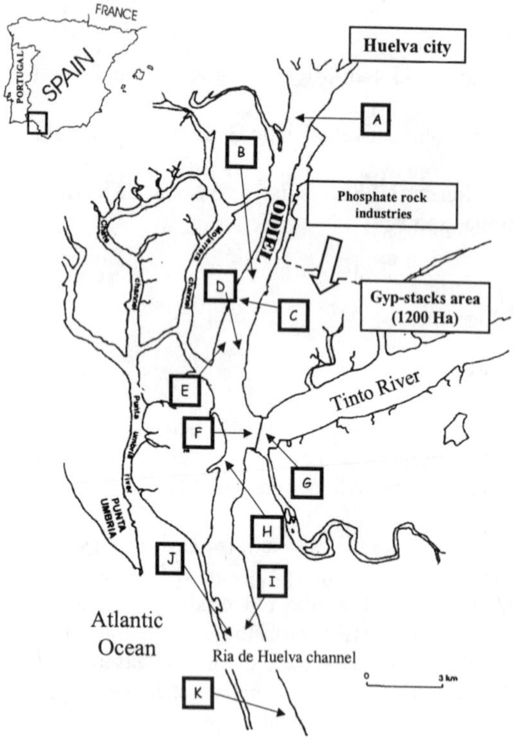

Figure 1 *Map of Odiel river estuary (SW of Spain). Sampling points are plotted.*

2.1 Direct speciation scheme: selective leaching

All samples were dried at 60ºC for 24 hours and homogenized; two 0.5 g aliquots were taken for analysis. A sequential radiochemical procedure (discussed below) was applied to the first aliquot in order to determine total ^{226}Ra and $^{234, 238}$U concentrations of the sediments (bulk samples), while a selective leaching scheme[5] (Table 1) was performed on the second aliquot for operational speciation of ^{226}Ra and $^{234, 238}$U.

Table 1. *Sequential extraction scheme for superficial sediment samples (0.5 g sample)*

Fraction	Procedure
F1 (exchangeable)	1 M $MgCl_2$ (8 ml), pH 7, 1 h ca 20°C, continuous agitation (end-over-end, 40 rpm).
F2 (acidic)	1 M NaOAc (8 ml), pH 5, 5 h ca 20°C, continuous agitation (end-over-end, 40 rpm).
F3 (reducible)	0.4 M $NH_2OH \cdot HCl$ in 25% acetic acid (20 ml), 6 h, 96°C, manual shaking every 30 min.
F4 (oxidizable)	0.02 M HNO_3 (3 ml) + 30% H_2O_2 (5 ml), pH 2, 2 h, 85°C, manual shaking every 30 min; further 30% H_2O_2 (3 ml), pH 2, 3 h, 85°C, manual shaking every 30 min; then 3.2 M NH_4OAc in 20% HNO_3 (5 ml), 30 min, ca 20°C, continuous agitation (end-over-end, 40 rpm).
F5 (residual)	Mixture 5/7.5/1 of $HF/HNO_3/HClO_4$ (13.5 ml), 5 h, 170°C

Between each successive extraction, separation was carried out by centrifugation at 10000 rpm for 10 min. The supernatant was removed by using Pasteur pipette and stored at 4°C in sealed polyethylene vessels for radiochemical analysis. To condition the remaining residue for the following extraction, the residue was washed with 8 ml of water, centrifuged for 10 min, and the supernatant discarded. The selective leaching procedure is particularly suitable to the characteristics of the sediments of this area, especially enriched in iron oxides[5].

2.2 Radiochemical procedure

Aliquots taken for the determination of total radium and uranium isotopes concentrations were dissolved by acid digestion. Afterwards, the bulk acidic solution and each selective leachate resulting from the speciation procedure were spiked with ^{225}Ra (taken from a standard solution where ^{225}Ra is in secular equilibrium with ^{229}Th) and ^{232}U. Then, they were evaporated and the residues redissolved with 0.1 M HNO_3. Radium was then coprecipitated with Pb as sulphate while uranium remained in the supernatant, which was evaporated to dryness and its residue redissolved with 8 M HNO_3. Two solvent extractions with tributylphosphate (TBP) were performed to isolate U from Po and Th[6]. Once uranium was isolated in aqueous solution, it was electrodeposited onto stainless steel discs[7].

The lead sulphate precipitate containing the radium was dissolved by EDTA 0.1 M. Radium was chemically isolated by using two chromatographic columns[8]. First, thorium was separated from the solution through an ionic exchange column and, afterwards, radioactive traces of uranium, polonium and chemically (barium) interfering elements were removed from the solution by using a cation-exchange column. Radium was eluted from this last column with nitric acid 6 M. The acidic radium solution was evaporated to dryness and the obtained residue redissolved 0.1 M. nitric acid. Finally, a radioactive radium source was obtained by electrodeposition onto stainless steel discs in organic media[9].

Chemical recoveries for uranium and radium were greater than 50% for both bulk samples and selective extracts. Discs were counted by alpha-particle spectrometry, for 48-

72 hours, using 450 mm^2 low background silicon ion-implanted detectors with 25% counting efficiency. A minimum detectable activity (MDA) lower than 1 mBq either for ^{226}Ra and uranium isotopes was obtained. This value was suitable for the proper determination of radium and uranium isotopes in the selective leachates.

The radiochemical method was checked for ^{226}Ra and U-isotopes measurements. Validation exercises for ^{226}Ra (by comparing ^{226}Ra-concentrations in reference materials) and the result of an international intercomparison exercise are shown. (Table 2). Uranium concentrations in sediments obtained using alpha-spectrometry and using ICP-MS are shown in Figure 2.

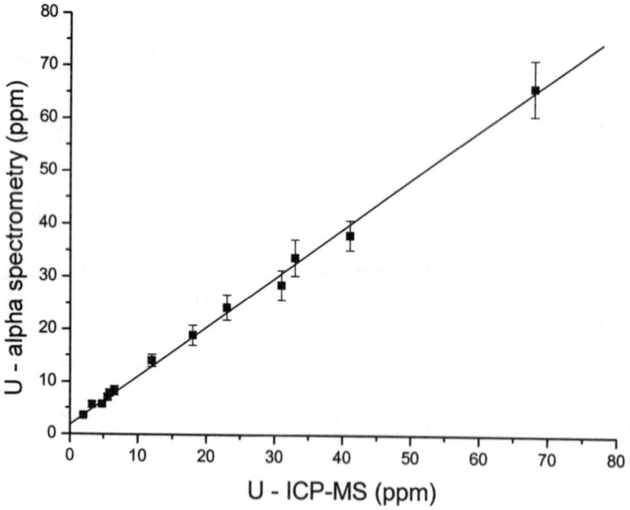

Figure 2 *Comparison between the uranium concentrations determined in sediments by alpha-particle spectrometry (^{238}U) and ICP-MS. Slope of the line: 0.94 ± 0.04, linear regression coefficient: 0.9942.*

The good agreement obtained (slope compatible with 1 and regression coefficient greater than 0.99) provide satisfactory confidence for the determination of ^{226}Ra and U-isotopes concentrations in the extracts obtained from the operational speciation procedure.

Table 2 *Results of validation exercises for ^{226}Ra determination in environmental samples by using the radiochemical method described in this paper and alpha-particle spectrometry.*

Sample	Reference value	Alpha spectrometry
IAEA Soil	69.6 – 93.4 (Bq/kg)	87.8 ± 14.0 (Bq/kg)
^{226}Ra solution	34.74 ± 0.32 (Bq/l)	35.6 ± 2.4 (Bq/l)
Aqueous waste (intercomparison exercise)	2.00 ± 0.19 (Bq/l)	2.00 ± 0.15 (Bq/l)

The absence of contamination after the application of the speciation procedure was checked by comparing the sum of ^{226}Ra and U isotopes activities in all fractions of the sequential extraction process to the bulk total activity obtained for the other aliquot of the same sample. Figure 3 shows, as an example, the linear fitting obtained for ^{238}U measurements. Slopes and regression coefficients for ^{226}Ra and ^{234}U were 0.99 ± 0.02, 0.9931 and 0.98 ± 0.02, 0.9939, respectively. These results give evidence of the agreement found in the comparison.

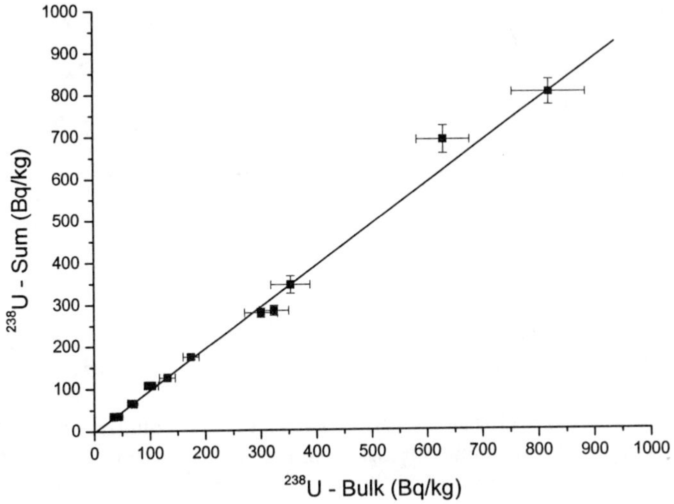

Figure 3 *Checking of the speciation procedure for ^{238}U in superficial sediments. Slope of the line is 0.98 ± 0.02 and regression coefficient is 0.9932.*

3 RESULTS AND DISCUSSION

Total activities of U-isotopes and ^{226}Ra obtained in samples collected from the Odiel river estuary are shown in Figure 4. Although PG releases ceased in 1997, the sediment samples show, in general, enhanced levels of U-isotopes and ^{226}Ra. Samples collected from the Ria de Huelva channel (H to K), far away from the PG-discharge points from the factories (which are placed in the vicinity of sampling points B to E), are also contaminated. Previous studies have shown that, after the PG releases, contamination "bags" travelled along the estuary with the tides[10], the radionuclides being incorporated to sediments after changes in water conditions (pH, redox potential) in zones where river waters from Odiel and Tinto mix with marine waters from the Atlantic Ocean during tidal periods.

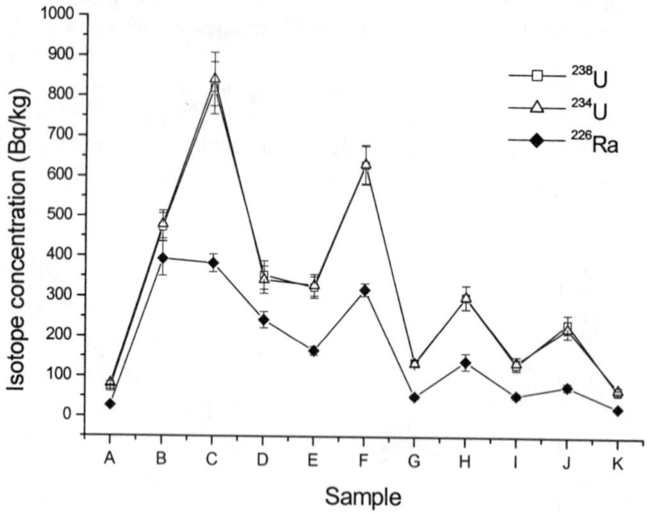

Figure 4 *Plot of ^{226}Ra and U-isotopes total concentrations (Bq/kg) in superficial sediments collected from the Odiel river estuary.*

It can be also observed that U-isotopes are in secular equilibrium in the sediments analysed. This result indicates that releases from the phosphate industries are the source of the uranium contamination observed in the sediments. The ^{226}Ra/^{238}U ratios found in these sediments are clearly less than one, although this ratio in PG-wastes ranges from 3 to 4 [4]. This ^{226}Ra/^{238}U ratio comparison gives clear evidence that the main impact of the PG released to the estuary is not the direct deposition of gypsum itself. Sediment radioactive impact is related to: a) the different leaching properties of the radionuclides from the wastes[2], and, b) the different nuclide-specific processes that determine their incorporation from the waters to the sedimentary environment. These processes could be investigated through speciation studies.

Table 3 summarizes the results obtained for the operational speciation of the sediments collected in the estuary. First we defined a background level (*B*), given by the speciation results of several estuarine samples with a total radionuclide activity close to typical values for uncontaminated sediments of this area reported by the literature[11]. Then, we calculated the mean value for isotope concentrations in the five operational forms defined by the speciation scheme among the collected Odiel river estuarine sediments (*M*).

Finally, we determined the deviation between each mean concentration value and the corresponding background level (*D*).

Table 3 *Results (Bq/kg) obtained after direct operational speciation of superficial sediments. Errors = 1 standard deviation. B: background level, M: mean value, D: deviation of mean value from background level*

	^{226}Ra			^{238}U		
	B	*M*	*D*	*B*	*M*	*D*
F1	2.4 ± 0.5	6.4 ± 1.7	4.0 ± 1.8	4.4 ± 0.5	7.9 ± 1.9	3.5 ± 2.0
F2	1.8 ± 0.3	8.0 ± 3.0	6.2 ± 3.0	6.0 ± 0.7	56 ± 10	50 ± 10
F3	5.2 ± 0.7	92 ± 23	86 ± 23	4.8 ± 0.6	131 ± 38	126 ± 38
F4	1.7 ± 0.4	47 ± 15	45 ± 15	3.7 ± 0.5	68 ± 19	64 ± 19
F5	14.6 ± 0.3	28 ± 7	14 ± 7	15 ± 1	39 ± 11	23 ± 11

3.1 ^{226}Ra results

Speciation results for ^{226}Ra show that its incorporation into fractions F1 or F2 is insignificant, indicating that its potential for remobilisation under ordinary environmental conditions is low. Greater than 90% of the total ^{226}Ra activity above the typical ^{226}Ra background is associated with refracting forms (F3+F4+F5), the majority being found in the reducible fraction F3.

Thus 60% of the total activity above the background level is incorporated to the sediment by coprecipitation with other metals such as Fe. This result is related to the fact that more than 90% of radium in estuarine waters is in the dissolved phase[13]. On the other hand, the oxidizable fraction (F4) shows less activity of radium than F3, with 30% bound to this fraction.

Plotting the F3- and F4-^{226}Ra concentrations versus the bulk activity of the sediment (Figure 5), it is possible to deduce that: a) levels in the oxidizable fraction are quite uniform in the 50-250 Bq/kg total activity range, while F3 increases in activity close to linearity; b) when bulk radium concentration is higher than 250 Bq/kg, F3 seems to be radium-saturated and F4-^{226}Ra activity increases remarkably. This behaviour suggests that incorporation of radium to the F3 and F4 phases of the sediments may not be independent, and could be related to the redox conditions in the sediments. However, we need to take in account that fraction F4 is operationally obtained after F3. Although the speciation scheme was designed for iron oxide rich sediments[5], under high total radium concentration conditions we can not neglect the possibility that the solid:liquid ratio of the extraction procedure may be insufficient to leach all the radium incorporated to the reducible fraction, consequently enriching the oxidizable fraction (F4). Additional research is being performed with the more contaminated samples to clarify this point; some helpful information was obtained in the U speciation results discussed below.

Finally, an incomplete leaching of the ^{226}Ra associated with the previous operational fractions could explain the enrichment of residual fraction (F5) observed in several samples with total activity >100 Bq/kg. These radium enriched samples (C, D, E) are located near the points where PG was released to the river. It is possible that some unprocessed mineral particles from the phosphate rocks used in the factories may be incorporated to sediments near the sites of discharge. The PG wastes released to the Odiel river may contain these refractory particles, which are well known to be enriched in radionuclides from the uranium series close to secular equilibrium[4].

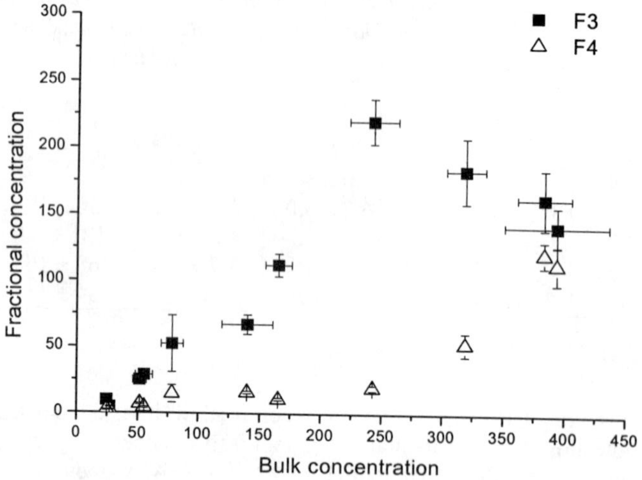

Figure 5 *^{226}Ra concentration in fractions F3 and F4 (Bq/kg) versus the total radium isotope activity of the sediment (Bq/kg).*

3.2 U-isotopes results

We observed that $^{234}U/^{238}U$ ratio is close to unity in the selective extracts. Only the low uranium enriched sediments show an isotopic ratio close to 0.9 for exchangeable (F1) and residual (F5) fractions, which could be related with alpha-recoil processes[14].

A significant difference between radium and uranium speciation results is that an important fraction of uranium is weakly bound to the sediment because of the $^{234,\ 238}U$ content of the acidic phase (F2). In this fraction, 20% of the total U activity above background was found, suggesting that U is mainly present in the suspended matter of the estuarine waters[2]. As a consequence, uranium has a greater potential for mobilization from the sediments than radium.

Nevertheless, a major percentage of the U incorporated to the sediments is associated with the reducible phase (F3), as it occurs with ^{226}Ra. The percentage of anthropogenic contamination associated with (co)precipitated forms (F2+F3) is around 60% for both uranium and ^{226}Ra.

In the case of U, no saturation of the F3 phase was observed when comparing U activity in this fraction with the bulk activity of the sediment. In fact, F3- and F4- $^{234,238}U$ activities increase linearly with uranium bulk activity, as shown in Figure 6 for the oxidizable fraction. These results are remarkably different from those for ^{226}Ra; this can be considered as an indication that no fractionation artefacts affect the results obtained for Ra in these operational fractions.

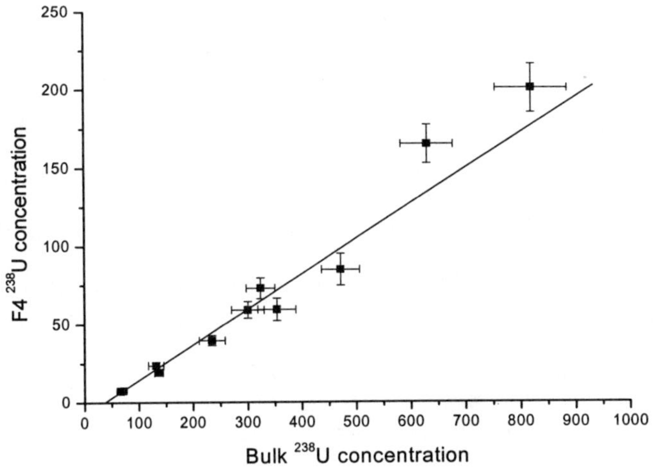

Figure 6 *^{238}U concentration (Bq/kg) in the oxidizable fraction (F4) versus total activity of ^{238}U in the sediment (Bq/kg). Parameters of the regression line: $a = -8.4 \pm 0.9$ Bq/kg, $b = 0.23 \pm 0.01$, $r = 0.9852$.*

Finally, 9% of the total uranium present in the sediments above the background level was found in the residual phase (F5); this percentage is higher in some contaminated samples collected in the vicinity of the PG release points, similar to results obtained for Ra. The origin of the anthropogenic inputs to this fraction was discussed previously for ^{226}Ra .

4 CONCLUSIONS

An operational speciation procedure by selective leaching was applied to superficial sediments collected from the Odiel river. ^{226}Ra and U-isotopes concentrations in the selective extracts were obtained by a suitable and validated radiochemical method for radium and uranium isolation and electrodeposition followed by alpha-particle spectrometry. Results revealed that radium and uranium are preferably bound to (co)precipitated forms (F2+F3). Nevertheless, radium content in the acidic fraction (F2) is lower than uranium concentrations in this phase. As this phase could be altered by pH changes, this suggests that the potential remobilization of U-isotopes contamination from sediments to the aqueous phase is greater than for radium under typical environmental conditions of this estuary.

References

1 J. E. Martín, R. García-Tenorio, M. A. Raspaldiza, J. P. Bolívar and M. F. Da Silva, *Nucl. Instrum. Methods Phys. Res. B,* 1998, **136-138**, 1000.
2 J. P. Bolívar, R. García-Tenorio and F. Vaca, *Water Resources,* 2000, **34**, 2941.
3 J. P. Bolívar, R. García-Tenorio, J. L. Mas and F. Vaca, *Environment International,* 2002, **27**, 639.

4 J. P. Bolívar, R. García-Tenorio and M. García-León, *J. Radioanal. Nucl. Chem.*, 1996, **214**, 77.
5 J. L. Gómez-Ariza, I. Giráldez, D. Sánchez Rodas and E. Morales, *Intern. J. Environ. Anal. Chem.*, 1999, **75**, 3.
6 E. Holm and R. Fukai, *Talanta*, 1977, **24**, 659.
7 L. Hallstadius, *Nucl. Instrum. Methods,* 1984, **223**, 226.
8 J. L. Aguado, J. P. Bolívar and R. García-Tenorio, *Czech. J. Phys.*, 1999, **49**, 439.
9 G. Hancock and P. Martin, *Appl. Radiat. Isotop.*, 1991, **42**, 63
10 R. Periañez, J. M. Abril and M. García-León, *J. Environ. Radioactivity*, 1996, **31**, 253.
11 E. G. San-Miguel, PhD Thesis, University of Seville, 2001.
12 H. R. Von Gunten and P. Benes, *Radiochimica Acta*, 1995, **69**, 1.
13 R. Periañez, J. M. Abril and M. García-León, *J. Radioanal. Nucl. Chem.*, 1994, **183**, 395.
14 J. K. Osmond and M. Ivanovich, in *Uranium Series Disequilibrium: Applications to earth, marine and environmental sciences*, eds., M. Ivanovich and R. S. Harmon, Clarendon Press, Oxford, 1992, Chapter 8, pp 262-264.

RADIOCHRONOLOGY OF SEDIMENT CORES COLLECTED IN AN ESTUARY STRONGLY AFFECTED BY FERTILIZER PLANTS RELEASES

E.G. San Miguel[1], J.L. Aguado[1], J.P. Bolívar[1] and R. García-Tenorio[2]

[1]Departamento de Física Aplicada, E.P.S. La Rábida, Universidad de Huelva, Huelva, 21819-SPAIN
[2]Departamento de Física Aplicada II, E.T.S. Arquitectura, Universidad de Sevilla, Sevilla, 41012-SPAIN

1 INTRODUCTION

Tinto and Odiel rivers flow through Huelva province (southwest of Spain) from north to south, forming in their mouth, the Huelva estuary (Figure 1). In this province, important mining activities have been carried out around the basin of these rivers for more than twenty centuries. These activities were heavily increased from 1875 to 1926, but later on the production fell down steeply until 10% of the maximum reached. Historically, the pH of the waters in the Tinto and Odiel rivers has been quite low (2-3) due to these mining activities, allowing heavy metals and other pollutants to travel in dissolution up to the estuary, where the abrupt change in pH caused a heavy metal precipitation[1].

In the Huelva estuary is also located a large industrial complex including, among others, two factories to produce phosphate fertilizers. The wastes of these factories have been released either directly into the Odiel river waters or stored in phosphogypsum piles since 1968, originating a high input of radionuclides from the Uranium series (U-isotopes, ^{230}Th, ^{226}Ra, ^{210}Pb,...) and a clear radioecological impact in the estuary[2]. However, specifically the U input in the estuary can also be caused by the mining activities due to mobilisation of this radionuclide under acid conditions. This would allow U to be transported up to the mouth of Odiel and Tinto rivers and to precipitate there.

We are interested in establishing a reliable chronology in sediment cores from this estuary to relate the temporal evolution of several pollutants (heavy metals, U-isotopes...) with the mining and industrial activities. Nevertheless in the time-span we are interested (150-200 years) the anthropogenic inputs of ^{210}Pb during the last 30-35 years do not allow to apply in full extent the ^{210}Pb dating method, which is a technique commonly used for that purpose. However, in a previous pilot study performed in one sediment core[3] we have shown that by means of the analysis of the ^{230}Th/^{232}Th vertical profiles it is possible to assign the date of the beginning of the fertilizer plants releases –1968- to the depth in the core where there is an observed increase in the values of this activity ratio (due to the inputs of ^{230}Th).

In this work, we will confirm the utility of the ^{230}Th/^{232}Th activity ratios profiles along the estuary to establish chronological marks in the sediment cores and additionally we will show that ^{226}Ra/^{228}Ra activity ratios are also a useful tool for this purpose. In this way we are able to estimate an average sedimentation rate for the upper layers of sediment cores and to delimitate the thickness of them with anthropogenic inputs of ^{210}Pb. Then, discarding the contaminated zone of the cores, the ^{210}Pb dating technique could be applied to the rest (deeper layers) of the sedimentary column being possible to establish a confident chronology in each sediment core covering the last 100-150 years.

The usefulness of Th and Ra isotopic ratios as markers of the industrial contamination has been validated in this work through its intercomparison with an independent dating technique -the ^{137}Cs dating method- comprising practically the same time interval. Additionally, and as an example, it is shown how the chronology assigned to one of the sediment cores analysed allows to explain the evolution during the last century of the vertical profile of ^{238}U activities.

2 MATERIALS AND METHODS

Four sediment cores were collected in different zones of the estuary (Figure 1), using a cylindrical home-made corer, and immediately frozen. In the laboratory, these frozen cores were cut into sections 1-2 cm thick, removing the outer layers of every slice in contact with the tube. All the sections were then weighed, dried at 60° C and afterwards reweighed in order to determine their dry weight and their porosity values (volume percentage of water in the sediment). The porosity values obtained were used to determine the bulk density ρ (grams of dry sediment per cm^{-3} of wet sediment) and the sediment mass per surface area

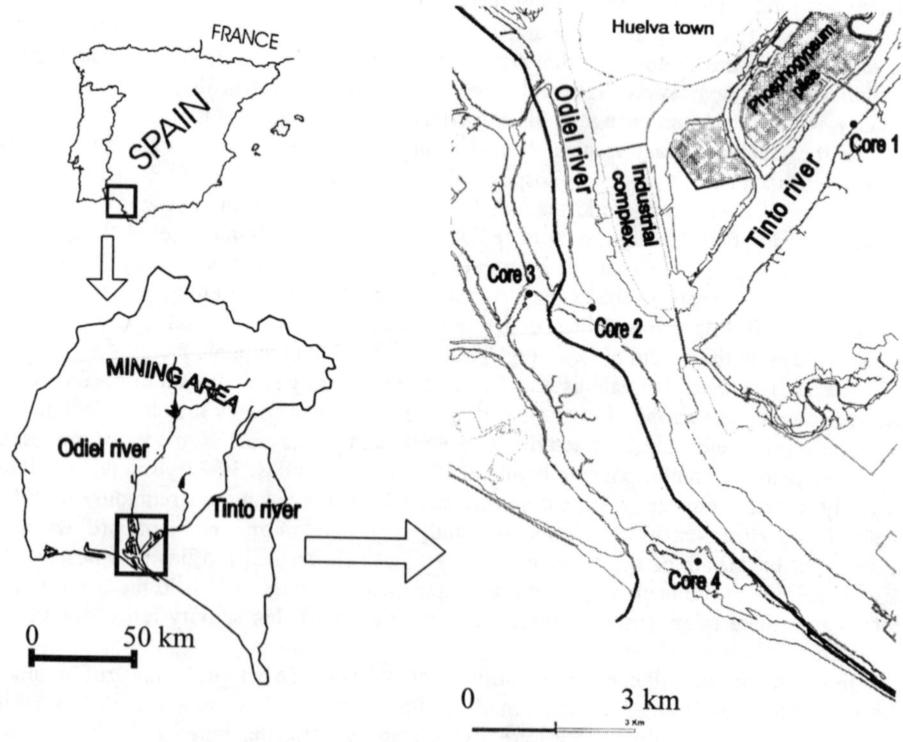

Figure 1 *Study area with the location of collection points*

(gcm^{-2}) in each layer of the sediment cores.

Aliquots of the sediment core sections were taken for radionuclide determination after its homogenisation and grounding.

2.1 Gamma spectrometry

Gamma measurements were performed using an XtRa coaxial Ge detector (Canberra), with 38% relative efficiency and FWHM of 0.95 keV at 122 keV and 1.9 keV at 1333 keV. The detector was coupled to a conventional electronic chain, including a multi-channel analyser, and was shielded with Fe 15 cm thick.

For radionuclide determination we have developed an original efficiency calibration in the 150-1800 keV energy range[4]. Then, ^{137}Cs activities were measured by its 662 keV gamma photon, ^{228}Ra determinations were carried out by the emission of 911 keV of ^{228}Ac and ^{226}Ra activities were estimated via the 352 keV emission of ^{214}Pb (the secular equilibrium between ^{214}Pb and ^{226}Ra can be assured because samples were sealed and stored at least during one month prior to the measurements). On the other hand, ^{210}Pb activities have been also determined from the same measurement by using an independent and original efficiency calibration developed specifically for the photon of 46.5 keV emitted by this radionuclide[5].

2.2 Alpha spectrometry

The Th-isotopes (^{230}Th, ^{232}Th) activity concentrations were quantified directly by alpha-particle spectrometry in the different layers of each sediment core. In core 3, ^{210}Pb concentrations were determined through the measurement of its grand-daughter ^{210}Po, also by the previously reported technique, assuming secular equilibrium. For Th-isotopes, ^{238}U and ^{210}Po isolation and determination we have applied a sequential radiochemical method[6,7] that can be found elsewhere[3].

3 RESULTS AND DISCUSSION

3.1 Delimiting the influence of fertilizer plants releases

Figure 2 shows the vertical profiles of Thorium isotope ratios ($^{230}Th/^{232}Th$) in the sediment cores. As can be seen, the values of Th-isotopic ratios in the deeper layers are about unity, typical for non-contaminated sediments world-wide[8], while the values clearly higher than one in the upper layers, with the exception of core 4, indicate the influence of releases from fertilizer plants. Therefore, two zones can be distinguished in each core: a zone contaminated by fertilizer plants releases whose Th-isotopic ratios are clearly higher than unity, and a second and deeper zone with Thorium isotope ratios in the range for uncontaminated sediments.

Then, we are able to distinguish in sediment cores from Huelva Estuary between thicknesses influenced or not by the releases of fertilizer plants. Only in the case of core 4 all the layers show Th-isotopic ratios typical of non-contaminated sediments, indicating that this sediment core was not affected by the fertilizer releases because there is not direct contact between the waters from the most polluted zone of the Huelva Estuary -near the factories and piles- and the waters from the collection point (Figure 1).

The Th-isotopic signature also allows to obtain a mean sedimentation rate for the contaminated zone of the sediment cores[3]. In fact, from Figure 2 it is possible to observe

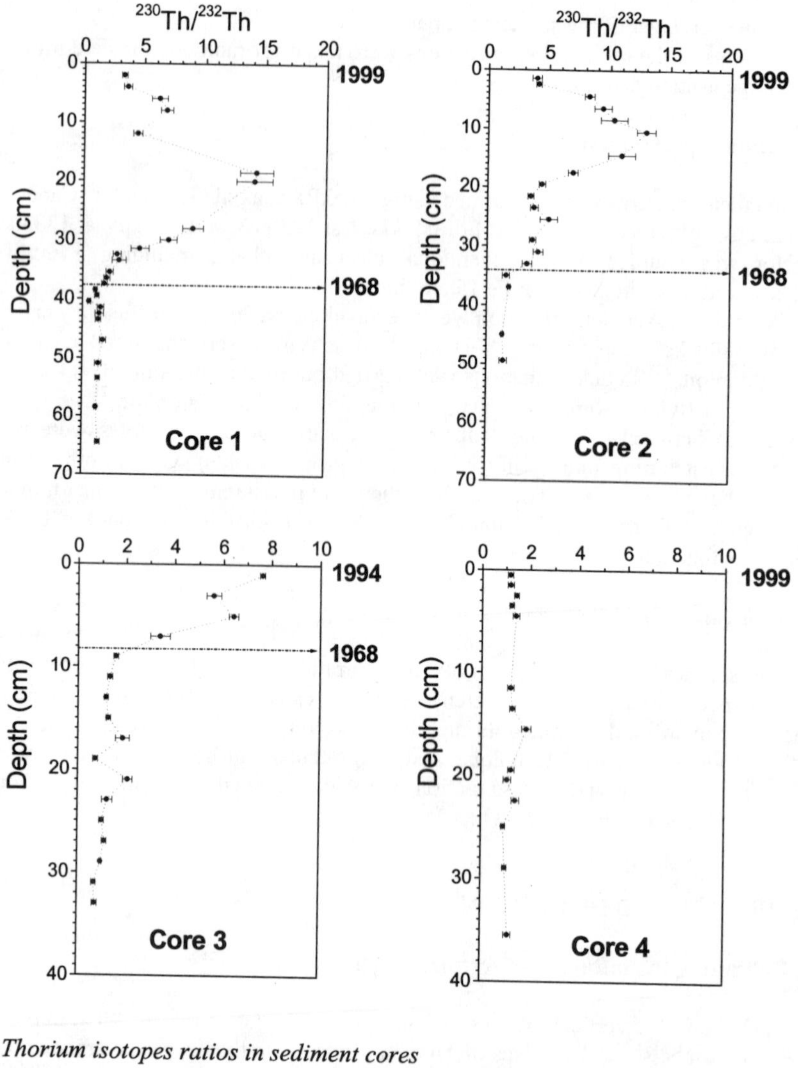

Figure 2 *Thorium isotopes ratios in sediment cores*

that the first non-contaminated layer for cores 1, 2 and 3 are located at 37-38 cm, 34-36 cm and 8-10 cm depth respectively.

Then, by assigning the date when the releases started to these layers –1968-, a mean sedimentation rate can be obtained by dividing the thickness/cumulative dry mass in the contaminated zone of the cores by the time elapsed between 1968 and the collection date. In this way, the average sedimentation rate for the contaminated zone of the core 1 ranges between 1.19-1.23 $cmyr^{-1}$, while in the case of sediment cores 2 and 3 the values obtained are about 1.10-1.16 and 0.31-0.38 $cmyr^{-1}$ respectively. The application of this approach is not possible in core 4 because it was not affected by the releases.

On the other hand, and looking at the Figure 3 the Ra-isotopic ratios from cores 1 and 2 (as examples), it is also possible to conclude that the ^{226}Ra/^{228}Ra and the ^{232}Th/^{230}Th

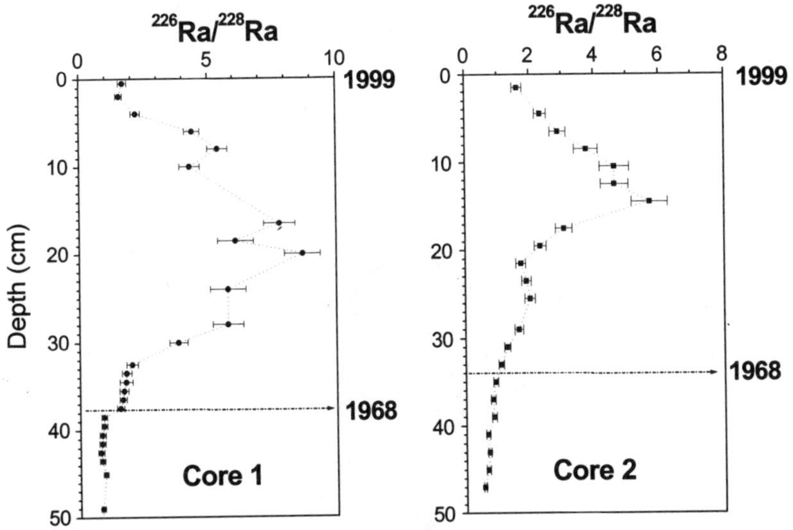

Figure 3 *Radium isotopes ratios in sediment cores 1 and 2*

activity ratio profiles provide the same information. Indeed, as can be seen from Figure 3, the layer of the core in which the Ra-isotopic ratios begin to increase agrees with that for the Th-isotopic ratios. Then, taking into account that these Ra isotopes can be measured by gamma-spectrometry, the use of the Ra-isotopic ratios instead of the Th-isotopic ratios provides the advantages of gamma spectrometry against alpha spectrometry: it is a non-destructive, more simple and less time-consuming technique.

Additionally, in Figure 4 we have plotted simultaneously the vertical profiles of Ra-isotopic ratios and ^{137}Cs activities in core 1. As can be seen from this Figure, there is a well defined peak in activities of ^{137}Cs in layer 39-40 cm, that can be assigned to the date of 1963, year characterised for a clear maximum in the concentrations of ^{137}Cs in the atmosphere [9].

Therefore if we take into account that through the Th and Ra-isotopic ratios we have assigned the year 1968 to the layer 37-38 cm, we can affirm that the ^{137}Cs dating technique and the Th and Ra-isotopic ratios provide a quite similar information (specially if we take into account that the sedimentation rate before 1968 was clearly lower as we will show later). Then the usefulness of Th and Ra-isotopic ratios as markers of the influence of fertilizer plants releases in sediment cores is validated through an independent technique which comprises the same time-interval.

3.2 Dating of the non contaminated zones of sediment cores

Furthermore, as we have determined the layers of sediment cores affected by the releases, and consequently with anthropogenic inputs of ^{210}Pb, we can discard them and apply the ^{210}Pb dating method in the rest of the sedimentary columns to establish chronologies covering the last 150-200 years.

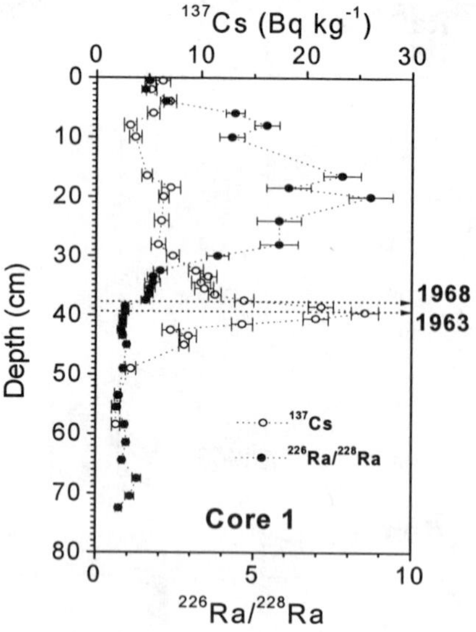

Figure 4 *Radium isotopes ratios and ^{137}Cs activities in core 1*

In order to apply the ^{210}Pb dating method in the non-contaminated zones of the sediment cores, the supported fraction of ^{210}Pb was taken in each layer as the ^{226}Ra activity, while the residual inventories of unsupported ^{210}Pb for each layer *n*, $A_\Sigma - A_n$,were determined through the equation:

$$A_\Sigma - A_n = \sum_{i=n+1}^{\infty} m_i C_i \tag{1}$$

where A_Σ represents the total ^{210}Pb unsupported inventory in the sediment core, and m_i and C_i are, respectively, the mass thickness (gcm^{-2}) and the specific activity of ^{210}Pb unsupported ($Bqkg^{-1}$) in layer *i* (in equation (1) the sum is really extended up to the last layer we find ^{210}Pb unsupported).

In Figure 5 we have plotted the vertical profiles of the residual inventories of ^{210}Pb unsupported versus cumulative dry mass for the non-contaminated zones of these sediment cores. As can be seen, these profiles show a linear trend in the case of sediment cores 1, 2 and 4, while this relationship in the core 3 seems to indicate a quadratic function. Therefore, we can apply the CRS/MV ^{210}Pb dating model[10] by accomplishing a least square weighted fitting of the residual inventory against cumulative dry mass, taking into account that the ^{210}Pb unsupported inventory in the non-contaminated zone of the sediment cores is given by:

$$A_\Sigma - A_C = \sum_{i=n_0}^{\infty} m_i C_i \tag{2}$$

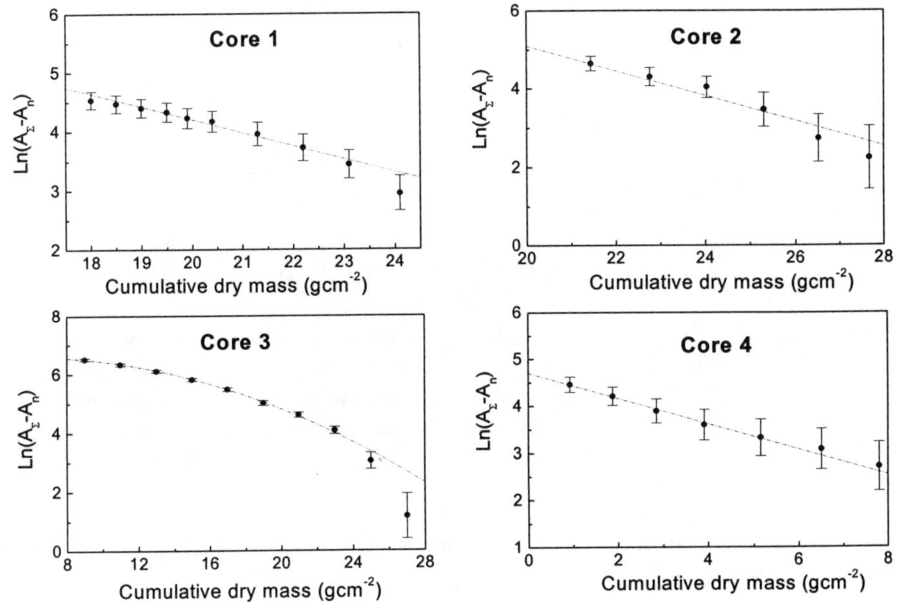

Figure 5 *Logarithmic profiles of residual inventories of ^{210}Pb versus cumulative dry mass*

where A_C is the inventory of ^{210}Pb unsupported in the contaminated zone of the core, and n_0 is the index used to identify the first layer not affected by the fertilizer plants releases. Then, if we name t_0 to the time elapsed between 1968 (beginning of industrial activities) and the collection date of a given sediment core, and t_n to the age of a given layer n of the not contaminated zone of the sediment core, the residual inventory of ^{210}Pb unsupported in this layer (n) can be expressed as follows:

$$A_\Sigma - A_n = \left(A_\Sigma - A_C\right)e^{-\lambda\Delta t} \tag{3}$$

where $\Delta t = t_n - t_0$ and λ is the ^{210}Pb radioactive decay constant (0.031 yr^{-1}). The age-cumulative dry mass relationship can be obtained by determining a function $t_n = t_0 + f(m)$ through a least square weighted fitting of the residual inventory against depth or cumulative dry mass to obtain the function $f(m)$.

Table 1 shows the mean sedimentation rates (in $gcm^{-2}yr^{-1}$ and in $cmyr^{-1}$) for the contaminated (layers deposited after 1968) and non-contaminated zones (deposited before 1968) of the analysed sediment cores. The mean sedimentation rates in the non-contaminated zones were determined after the application of the ^{210}Pb dating method (CRS/MV model) through the equation:

$$\omega = \frac{\Delta m}{\Delta t}\ (\text{g cm}^{-2}\ \text{yr}^{-1}) \quad \text{or} \quad \omega' = \frac{\Delta z}{\Delta t}\ (\text{cm yr}^{-1}) \tag{4}$$

Table 1 *Sedimentation rates obtained in sediment cores before 1968 (through ^{210}Pb dating method) and after 1968 (through Th-isotopic ratios).*

	cm yr^{-1}		*g cm^{-2} yr^{-1}*	
Core	After 1968	Before 1968	After 1968	Before 1968
1	1.21(1)	0.28(5)	0.56(1)	0.14(2)
2	1.13(2)	0.16(4)	0.67(2)	0.10(3)
3	0.34(3)	0.19(2)	0.31(3)	0.17(2)
4	0.10(2)	0.10(2)	0.12(2)	0.12(2)

where Δt (year) is the time elapsed between the formation of two layers whose cumulative dry mass (or depth) differs in Δm, gcm^{-2}, (or in Δz, cm), while the mean sedimentation rates in the contaminated zones were determined through the Th and Ra isotopic profiles. From the obtained data, it can be seen that the sedimentation rate has experienced a high increase since the beginning of the industrial releases and the simultaneous construction of several dikes in the estuary.

3.3 Application: vertical distribution of concentrations of ^{238}U during the last century in a dated sediment core

By means of the chronology established in these sediment cores we can study different environmental problems. For example it is possible to evaluate the historical impact of anthropogenic activities in the sediments of this estuary.

In Figure 6 we have shown as an example, the vertical distribution of the ^{238}U activities versus depth in core 3. The chronology established in this sediment core has been also represented marking some relevant dates.

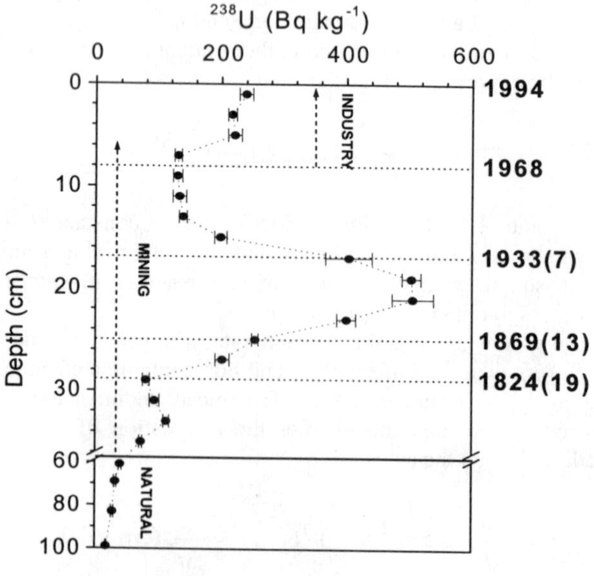

Figure 6 *Vertical profile of ^{238}U activities in core 3 versus depth*

As can be deduced from this Figure, the vertical profile of ^{238}U clearly reflects the main anthropogenic activities affecting the estuary. Indeed, the ^{238}U activities are about 30-40 Bq kg^{-1} (typical of non-contaminated sediments) in the deeper layers of the core deposited at least two hundred years ago, while in the depth-interval 18-24 cm there is a big enhancement in the ^{238}U activities in agreement with the period of the maximum mining activities (1875-1926). This fact indicates a strong impact of mining activities on the ^{238}U levels in this sediment core. Finally, and after a steep decrease in the U activities in the 8-18 cm depth interval (in agreement with the decrease of the mining activities after 1926) it is possible to see an increase just after 1968 as a direct consequence of the influence of the fertilizer plants releases.

4 CONCLUSIONS

We have established a reliable chronology in sediment cores from an estuary strongly contaminated by U-series radionuclides by using Th and Ra isotopic ratio profiles and by applying the ^{210}Pb dating method. Indeed, the Th and Ra isotopic ratios have allowed to delimitate the influence of fertilizer plants releases in the sediment cores from this estuary in order to apply the ^{210}Pb dating method exclusively in the non-contaminated fraction of the sediment cores. The chronologies established allow to analyse the temporal evolution of the concentrations of different pollutants in the estuary, and to relate it with the known mining and industrial activities that have affected historically the estuary.

References

1 J.E. Martín, R. García-Tenorio, M.A. Respaldiza, J.P. Bolívar and M.F. Da Silva, *Nucl. Instr. and Meth. B*, 1998, **136-138**, 1000.
2 J.P. Bolívar, R. García-Tenorio, J.L. Más and F. Vaca, *Environ. Internat.*, 2002, **27**, 639.
3 E.G. San Miguel, J.P. Bolívar, R. García-Tenorio and J.E. Martin, *Environ. Pollut.*, 2001, **112**, 361.
4 J.P. Pérez-Moreno, E.G. San Miguel, J.P. Bolívar and J.L. Aguado, *Nucl. Instr. And Meth. A*, 2002, **491**, 152.
5 E.G. San Miguel, J.P. Pérez-Moreno, J.P. Bolívar, R. García-Tenorio and J.E. Martín, *Nucl. Instr. And Meth. A*, 2002, **493**, 111.
6 E. Holm and R. Fukai, *Talanta*, 1977, **24**, 659.
7 J.P. Bolívar, R. García-Tenorio and M. García-León, *Anales de Física*, 1992, **B 92**, 101 (In spanish).
8 M. Ivanovich and R.S. Harmon, in *Uranium series disequilibrium: applications to environmental problems*. Clarendom Press, Oxford, 1982.
9 C.I. Sánchez, R. García-Tenorio, R., M. García-León, J.M. Abril and F. El-Daoushy, *Nucl. Geophys.*, 1992, **6**, 395.
10 W.R. Schell and M.J. Tobin, In *Low-Level measurements of radioactivity in the environment: Techniques and Applications*, eds. M. García-León and R. García-Tenorio, World Scientific, Singapur, 1994, p. 355.

CHEMICAL FRACTIONATION OF IODINE-129 AND CESIUM-137 IN CHERNOBYL CONTAMINATED SOIL AND IRISH SEA SEDIMENT

X. L. Hou [1], C.L. Fogh [1], J. Kucera [2], K.G. Andersson[1], H. Dahlgaard [1], S.P. Nielsen [1]

[1] Risø National Laboratory, NUK-202, DK-4000 Roskilde, Denmark
[2] Nuclear Physics Institute, CZ-250 68 Rez near Prague, Czech Republic

1 INTRODUCTION

The Chernobyl accident release is one of the main anthropogenic sources of ^{129}I in the environment, especially in the areas which were most affected by the Chernobyl accident. Due to the long half-life and continuous production and release of ^{129}I from nuclear fuel reprocessing plants, the environmental fate and geochemical cycle of ^{129}I are becoming of increasing concern (Katagiri et al., 1997). In this context, the chemical speciation of ^{129}I in environmental samples is a key factor. Many parameters influence the chemical speciation of iodine. It has been observed that the soil to pasture transfer factor for ^{129}I is quite different from that for natural ^{127}I (Schmitz & Aumann, 1995). This is related to the different sources and different chemical speciation of these two isotopes in the soil. Wilkins (1989) and Schmitz & Aumann (1995) have investigated the association of ^{129}I in different components of the soil from the vicinity of nuclear reprocessing plants. However, no investigations of ^{129}I association in Chernobyl-contaminated soil have been reported. In this study, a strongly Chernobyl-contaminated soil was sampled, and the association of ^{129}I and ^{137}Cs with different soil components is investigated by a sequential extraction combined with neutron activation analysis and gamma spectrometry. In addition, for comparison, a sediment sample collected from the Irish Sea is also investigated for the association of ^{129}I.

2 MATERIAL AND METHODS

2.1 Soil Sampling

Soil cores in Belarus, 10 km far from Chernobyl nuclear power plant were collected using UPONYL polyethylene tubes with an inner diameter of 81.0 mm and a length of 25-50 cm. The collected soil core samples were wrapped up in polyvinyl-chloride foil for transport and storage. Before slicing, 200 ml of distilled H_2O was added to the tubes, which were then frozen at -20°C. The soil cores were then sliced in 10 or 15 mm sections with a diamond saw. The slices were dried at 60°C for 2 days on plastic plates. After measuring the ^{137}Cs activity, the samples were ground and sieved to <1.5 mm. Sample was then thoroughly mixed and subsamples of 30 g were taken for the sequential extraction procedure.

One sediment sample (0-20 cm) was collected from the Irish Sea (15 km north of Sellafield). After freeze drying, the sediment sample was ground, sieved to < 1.5 mm and thoroughly mixed. A subsample of 30 grams was taken for the sequential extraction procedure.

2.2 Sequential Extraction Procedure

The prepared meadow soil (MaB-3, the layer corresponding to a depth of 25-35 mm) from the Chernobyl area (Massany, 10 km northwest of the Chernobyl power plant) and a sediment sample from the Irish Sea were weighed (30 g) and extracted sequentially. The extraction procedure was carried out as follows:

F1- Water-soluble fraction :

300 ml of H_2O was added to 30 g of the soil (or sediment), and the suspension was shaken for 24h at room temperature. The leachate was separated by centrifugation at 24000 rpm for 30 min and the residue was subjected to the second extraction step.

F2- Exchangeable fraction :

300 ml of 1.0 mol/L NH_4OAc (pH 8.0) was added to the residue and the suspension was shaken for 24h at room temperature. The leachate was separated by centrifugation, and the residue was washed with 30 ml of H_2O and the wash was combined with the leachate. The residue was subjected to the third step.

F3- Carbonate fraction

300 ml of 1.0 mol/L NH_4OAc (pH5.0) was added to the residue, then shaken for 24h at room temperature. The leachate was subsequently separated by centrifugation, and the residue was washed with 30 ml of water. The leachate and wash were combined, while the residue was subjected to the fourth step.

F4- Metal oxides fraction

300 ml of 0.04 mol/L $NH_2OH{\cdot}HCl$ in 25%(v/v) HOAc (pH 2.0) were added to the residue and the suspension was shaken for 24h at 95±5°C. After separation of the leachate by centrifugation, the residue was washed with 30 ml of water and the wash was combined with the leachate. The residue was dried at 95±5°C for 30h. After grinding and homogenisation, half of the residue was saved for determination of ^{129}I, while the other half was used for the next extraction step.

F5- Organic matter fraction:

150 ml of 30% H_2O_2-HNO_3 (pH 2.0) was added to half of the residue, and the suspension was shaken for 2h at 95±5°C. After separation of the leachate by centrifugation, another 75 ml of 30% H_2O_2-HNO_3 (pH 2.0) was added to the residue, and leached for 2h at 95±5°C. The residue was separated by centrifugation, and 75 ml of 3.2 mol/L NH_4OAc-20%HNO_3 (pH 2.0) was added. The suspension was then shaken for 1h at room temperature before separation by centrifugation. The leachates were combined and the residue was dried at 95°C for 30h.

F6- Residue.

2.3 Separation and Determination of ^{129}I

2.3.1 Leachate. ^{129}I was separated from the leachates containing H_2O, NH_4OAc and NH_2OH-HOAc. First, 1.0 mg of ^{127}I (KI) carrier, 1 kBq of ^{125}I tracer and 0.5 ml of a 0.3 mol/L $KHSO_3$ solution were added to the leachate. Iodine was then extracted by CCl_4 after acidification of the leachate to pH <2 with HNO_3 and addition of 0.2 ml of a 5% $NaNO_2$ solution. The extraction was repeated and the CCl_4 phases were combined. Iodine in the CCl_4 phases was back-extracted with a H_2SO_3 solution. This extraction and back-extraction was repeated to purify the iodine fraction. The final back-extracted phase was used to prepare the ^{129}I target for neutron activation analysis (Hou et al. 1999).

2.3.2 Soil/sediment and their extraction residues. ^{129}I was separated from the original soil and sediment and the residues before and after leaching by the H_2O_2-HNO_3 solution, according to Hou et al. (1999). Three g of the original soil or sediment sample, or 12-15 g of the extraction residues were taken to a crucible, and 6 g (30 g for the residue) of KOH, 1.0 mg of ^{127}I (KI), and 1 kBq of ^{125}I tracer were added. The sample was completely mixed, and the mixture was dried at 80°C, and fused at 650°C for 4h. After cooling, iodine was leached with hot water. Then 2 ml of 0.3 mol/L $KHSO_3$ were added, and the leachate was acidified to pH<2 with HNO_3. Iodine was separated by the extraction and back-extraction cycle described above.

2.3.3 Preparation of ^{129}I sample and standard for neutron activation analysis. A 0.3% LiOH solution was added to the back-extracted iodine sample until pH>9 was reached. The solution was carefully evaporated to about 0.5 ml on a hotplate and then transferred to a quartz ampoule and dried at 70-80°C. The ^{129}I standard was prepared from a ^{129}I standard solution (NIST-SRM-4949c) and from stable ^{127}I (KI). It contained 1.76×10^{-10} g of ^{129}I and 4.05 mg of ^{127}I. The method has been described in detail elsewhere (Hou et al. 1999).

2.3.4 Neutron activation analysis for ^{129}I. Sealed samples and standard ampoules were irradiated at a thermal neutron fluence rate of 4×10^{13} cm^{-2} s^{-1} for 10 h in the Danish DR-3 reactor, or 8×10^{13} cm^{-2} s^{-1} for 6.0 h in the Czech LWR-15 reactor. After a 10-15 h decay, iodine in the irradiated samples was further purified by extraction and precipitation of PdI_2 (Hou et al. 1999). The PdI_2 precipitate on the filter was sealed in a polyvial for measurement. The 536 keV γ-line of ^{130}I produced by neutron activation of ^{129}I was counted using an HPGe detector. The ^{129}I content was calculated by comparison with the standard. Activity of ^{125}I was also measured to determine the chemical yield.

2.4 Separation and determination of ^{137}Cs

2.4.1 Preparation of $Cu_2Fe(CN)_6$ AG1×4 exchange resin. A slurry of AG1×4 resin in Cl^- form was poured in to a column (1.0 cm in diameter, 30 cm in length) and the column was eluted with a 2 mol/L KOH solution to transfer the resin to OH^- form. Then the column was washed with water to pH <8. Subsequently, 30 ml of a 0.5 mol/L $K_4Fe(CN)_6$ solution was added to the column to transfer the resin to $Fe(CN)_6^{4-}$ form. The column was washed with water to remove excessive $K_4Fe(CN)_6$ from the column. Then 20 ml of a 0.5 mol/L $Cu(NO_3)_2$ solution was added to form $Cu_2Fe(CN)_6$ on the resin. The column was then washed with H_2O to remove excess $Cu(NO_3)_2$.

2.4.2 Separation of ^{137}Cs from the leachate. Two ml of the prepared $Cu_2Fe(CN)_6^{2-}$ AG1×4 exchange resin was poured to a column. The leachate of H_2O, NH_4OAc, $NH_2OH{\cdot}HCl$ or H_2O_2-HNO_3 (after separation of ^{129}I) was then added to the column, and the effluent was discarded. The column was washed with 30 ml of H_2O. The resin was then transferred to a counting vial for the measurement of ^{137}Cs.

2.4.3 Measurement of ^{137}Cs. Two g of original soil, extraction residues or resin with adsorbed ^{137}Cs from leachate was sealed in a polyvial. The ^{137}Cs activity was measured using an HPGe detector system.

3 RESULTS

The distributions of ^{129}I in different fractions of the Chernobyl (Massany) soil and Irish Sea sediment are shown in Table 1 and Fig.1. A similar distribution of ^{129}I in soil and sediment is observed. ^{129}I is mainly bound to organic matter and oxides, which account for 70 % and 85% of the total ^{129}I in soil and sediment, respectively. Less ^{129}I is bound to carbonates or exists in exchangeable form. Fig. 2 compares the distribution of ^{129}I and ^{137}Cs in the Chernobyl soil. A significantly different association is observed. A large part of the ^{137}Cs (73%) exists in the residue after sequential extraction, while only 6.8% of ^{129}I is contained in this fraction. This implies different bioavailablities and mobilities of these two radionuclides in the soil.

Table 1 *Distribution of ^{129}I in different fractions of Chernobyl soil and Irish Sea sediment*

Fraction	Soil (MaB-3)		Sediment	
	Concentration, 10^{-12} g/g	%	Concentration, 10^{-12} g/g	%
Water soluble	1.10	12.74	6.70	6.55
Exchangeable	0.67	7.79	3.27	3.20
Carbonate	0.37	4.33	1.61	1.57
Fe-Mn oxides	2.66	30.81	38.44	37.54
Organic bound	3.38	39.20	48.55	47.42
Residues	0.59	6.80	4.13	4.04
Sum	8.77	101.68	102.70	100.30
Whole*	8.63	100.0	102.39	100.0

* the correspondence between the sum of concentrations in the different fractions and the concentration in the whole sample can be seen as a verification of the procedure

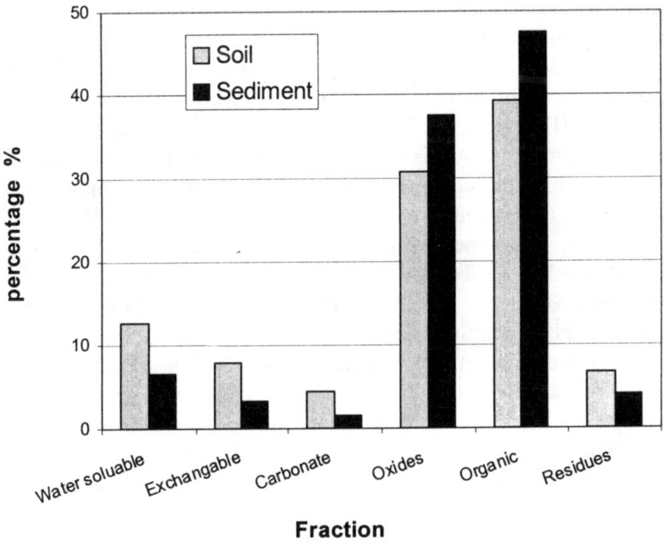

Figure 1 *Distribution of ^{129}I in different fractions of Chernobyl soil (MaB-3) and Irish Sea sediment*

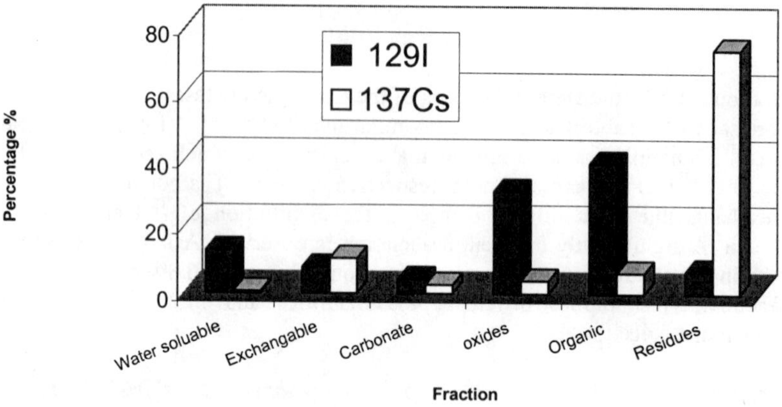

Figure 2 *Distribution of ^{129}I and ^{137}Cs in different fractions of soil from Chernobyl region (MaB-3)*

4 DISCUSSION

4.1 Fractionation of ^{129}I in Soil and Sediment

Sequential extraction procedures have been widely used for the investigation of the association of radionuclides and trace element with soil and sediment components (Tessier et al. 1979); Fawaris & Johanson, 1995; Krouglov et al. 1998; Andersson & Roed, 1994; Oughton et al. 1992; Schmitz & Aumann, 1995; Wilkins, 1989). Of these, the procedure proposed by Tessier et al. (1979) has found wide application because the reagents were chosen to simulate changes in pH and E_h that could occur naturally. Wilkins (1989) modified this procedure for the investigation of ^{129}I associated with soil and sediment. In the modified procedure, the dissolution of organic material was carried out with a mixture of hydroxylamine and sodium carbonate solution to avoid the loss of ^{129}I in the leaching process by the H_2O_2-HNO_3 mixture used in the Tessier procedure. This modified procedure was also used by Schmitz & Aumann (1995) to assess ^{129}I in soil. However, there is no evidence that the basic hydroxylamine can completely destroy the organic matter in the soil and release all iodine to the solution as inorganic ions. Consequently, results obtained by this method can, especially when considering other radionuclides, not easily be compared with results obtained by the Tessier procedure. Therefore, a mixture of 30% H_2O_2 and HNO_3 was still used to leach ^{129}I bound to the organic matter in this work. Leached ^{129}I is unstable in H_2O_2+HNO_3 solution since it easily converts to I_2, which is lost from the leachate. Therefore, the ^{129}I bound to organic matter is calculated as the difference of the ^{129}I contents between the sample before and after H_2O+HNO_3 leaching. In this step, the sample was leached twice, each leaching having a duration of 2 h. The reason for this is that the H_2O_2 in HNO_3 solution is not stable and can easily be decomposed. Washing with 3.2 mol/L NH_4OAc-20% HNO_3 followed the two leachings to prevent re-adsorption of extracted ^{129}I to the residue.

4.2 Association of ^{129}I in the Chernobyl Soil and Irish Sea Sediment

Similar distributions of ^{129}I in the different fractions are observed in the Chernobyl soil and Irish Sea sediment. In these two samples, more than 70% of ^{129}I is bound to oxides and organic matter. About 20% of the ^{129}I in the Chernobyl soil and 10% of the ^{129}I in the Irish Sea sediment was found to be associated with the readily-available fractions (i.e. water-soluble and exchangeable components). Wilkins (1989) reported that in the Irish Sea sediment, 11% of the ^{129}I exists in readily available form and 80% of the ^{129}I is associated with oxides and organic matter. This is in agreement with our analytical result for sediment. However, for the soil sample from the vicinity of the Sellafield reprocessing plant, Wilkins (1989) reported that more than 80% of the ^{129}I is bound to oxides, while 1.5-10% ^{129}I exists in readily available form.

Schmitz & Aumann (1995) analysed some soil samples from the vicinity of Karlsruhe nuclear fuel reprocessing plant in Germany. Their results show that only 12-23% of the ^{129}I is bound to oxides and organic phases, while 45-60% of the ^{129}I exists in readily available form in this soil. The differences in distribution of ^{129}I in the Chernobyl soil and the soil from the vicinity of the reprocessing plants in Germany and UK may result from different soil composition and properties, as well as from differences in sources and chemical speciation of ^{129}I deposited on the ground. Ibrahimi et al (2000) reported that ^{129}I was released from the Sellafield reprocessing plant to the atmosphere as both gaseous inorganic (I_2, HI) and gaseous organic iodine (CH_3I, etc.), in which more than half is organic iodine. In April-May, 1986, Nedveckaite & Filistowicz (1993) measured the chemical speciation in the atmosphere of ^{129}I and ^{131}I originating from the Chernobyl accident. They observed that 24-50% of the ^{129}I was bound to aerosol, whereas 10-20% was gaseous inorganic iodine and 40-55% was gaseous organic iodine. It should be noted that this relationship is different from that at the release from a reprocessing plant. For instance, photochemical processes may result in changes in the chemical form of atmospheric iodine (Jenkin et al., 1985).

It is not clear how iodine can be bound to oxides, but it has been demonstrated that Fe_2O_3 and Al_2O_3 can adsorb iodide (Whitehead, 1974). Parfitt (1978) proposed that protonation of M-OH groups on the surface of oxide phases may generate positively charged sites and hence lead to attraction and sorption of anions, i.e. I^-, IO_3^-

The final state of organic matter in soil and sediment is humic substance, such as humic and fulvic acids. Schnitzer (1978) formulated a "model humic acid" containing 24.0% of alophatic-, 20.3% of phenolic- and 32% of benzene carboxylic groups, respectively. It has been reported that humic substances typically contain thiol groups (Choudry, 1981). It is well known that iodine reacts easily with polyphenols leading to iodine-substitued species. The thiol group in proteins can react with iodine to form iodinated protein (Jirovsek & Pritchard, 1971). Hou, et al. (2000b) have observed that iodine is mainly bound to protein and polyphenol in seaweed. Therefore, ^{129}I in the organic fraction in the soil and sediment may be mainly bound to phenolic and thiol groups in the humic substance. The experiment also confirmed that microorganisms and oxygen participate in the reaction and combination of iodine to the humic substance (Christiansen, 1990). On this background, it can be concluded that ^{129}I in Chernobyl soil mainly exists in the organic phase, in which it is probably bound to phenolic and thiol groups in the humic substance.

4.3 Association of ^{137}Cs in the Chernobyl Soil

The association of ^{137}Cs with the Chernobyl soil components is different from that of ^{129}I (Fig.2). As much as 73% of the ^{137}Cs is so strongly bound that it remains in the sequential extraction residue, while only 6.8% of ^{129}I remains in this fraction. The fractions of ^{137}Cs

bound to oxides and organic matter are only 3.9% and 6.2%, respectively, and only slightly more (10.4%) is in the exchangeable phase. Thus, the speciation of ^{137}Cs is very different from that of iodine and the mobility and availability of ^{129}I is higher than that of ^{137}Cs, as would be expected, since ^{137}Cs is selectively retained at certain sites in micaceous soil substances, from where it is virtually inexchangeable (Andersson & Roed, 1994). This would be expected to result in a deeper migration of ^{129}I compared with ^{137}Cs and a higher ratio of $^{129}I/^{137}Cs$ in the deeper layers of the soil, which has also been observed.

Andersson & Roed (1994) and Oughton et al.(1992) have investigated the association of ^{137}Cs in the Chernobyl soil using a similar sequential extraction procedure and found that more than 70% of the ^{137}Cs is bound in the residue and leachate of 7 mol/L HNO_3, while about 15 % of ^{137}Cs exists in exchangeable form. This is in good agreement with our results.

5 CONCLUSION

The relative fractions of ^{129}I removed by the various sequential extractions performed on Chernobyl soil and on Irish Sea sediment are similar: more than 70% of the ^{129}I is bound to organic matter and oxide phases and 10-20% of the ^{129}I exists in readily available phases (water soluble and exchangeable fractions). In comparison, most of the ^{137}Cs in the soil (>70%) remains in the soil residue after the sequential extractions. This indicates that the mobility of ^{129}I in the Chernobyl soil is much higher than that of ^{137}Cs.

References

1 K. G. Andersson, & J Roed, 1994. The Behaviour of Chernobyl ^{137}Cs, ^{134}Cs and ^{106}Ru in undisturbed soil: Implications for external radiation. *Journal of Environmental Radioactivity*, **22**, 183-196.

2 G.G. Choudry, 1981. Humic substances, Part I: structural aspects. *Toxicological and environmental chemistry*, **4**, 209-260.

3 J.C. Christiansen, 1990. The behaviour of iodine in the terrestrial environment. *Risø-M-2851*, Risø National Laboratory, Denmark.

4 B. H. Fawaris and K. J. Johanson, 1995. Fractionation of caesium (^{137}Cs) in coniferous forest soil in central Sweden. *The Science of the Total Environment*, **170**, 221-228.

5 X. L. Hou, H. Dahlgaard, B. Rietz, U. Jacobsen, S. P. Nielsen, and A. Aarkrog, 1999. Determination of ^{129}I in seawater and some environmental materials by neutron activation analysis. *Analyst*, **124**,1109-1114.

6 X. L. Hou, X. J Yan and C. F. Chai, 2000. Chemical species of iodine in some seaweed II: iodine-bound biological macromolecules. *Journal of Radioanalytical and Nuclear Chemistry*, **245**, 461-467.

7 Z. F. Ibrahimi, M. J. Fulker, S. J. Parry and D. Jackson, 2000. The effect of a change in ^{129}I aerial source at BNFL Sellafield on the behaviour of ^{129}I through the air-grass-cow-milk pathway in west Cumbra in 1997, *Extended abstract of 5th International Conference on Nuclear and Radiochemistry*. Pontresina, Switzerland, Sept. 3-8, 2000. 497-498.

8 M. E. Jenkin, R. A. Cox and D. E. Candeland, 1985. Photochemical aspects of tropospheric iodine behaviour. *Journal of Atmospheric Chemistry*, **2**, 359-375.

9 L. Jirosek and E. F. Pritchand, 1971. On the chemical iodine of tyrosine with protein sulfenyl iodine and sulfenyl periodide derivatives: the behaviour of thiol protein-iodine system. *Biochem. Biophys. Acta*, 243, 230-238.

10 K. Katagiri, T. Shimizue, Y. Akatsu and H. Ishiguro, 1997. Study on the behaviour of ^{129}I in the terrestrial environment. *Journal of Radioanalytical and Nuclear Chemistry*, **226** (1-2), 23-27.

11 S.V. Krouglov, A. D. Kurinov and R. M. Alexakhin, 1998. Chemical fractionation of ^{90}Sr, ^{106}Ru, ^{137}Cs and ^{144}Ce in Chernobyl-contaminated soil: an evolution in the course of time. *Journal of Environmental Radioactivity*, **38**, 59-76.

12 T. Nedveckkaite and W. Filistowicz, 1993. Determination of gaseous and particulate ^{129}I in atmospheric air by neutron activation analysis. *Journal of Radioanalytical and Nuclear Chemistry, Articles*, **174**, 43-47.

13 D. H. Oughton, B. Salbu, G. Riise, H. Lien, G. Østby and A. Nøren, 1992. Radionuclide mobility and Bioavailablility in Norwegian and Soviet Soil. *Analyst,* **117**, 481-486.

14 R. L. Parfitt, 1978. Anion adsorption by soil and soil materials, *Adv. Agron.*, **30**, 1-50.

15 K. Schmidtz and D. C. Aumann, 1995. A study on the association of two iodine isotopes, of natural ^{127}I and of the fission product ^{129}I, with soil components using a sequential extraction procedure. *Journal of Radioanalytical and Nuclear Chemistry, Articles*, **198**, 229-236.

16 M. Schnitzer, 1978. Some observations on chemistry of humic substances. *Agrochimica*, **22** (3-4), 216-225.

17 A. Tessier, P. G. C. Campbell and M. Bisson, 1979. Sequential extraction procedure for the speciation of particulate trace metals. *Analytical Chemistry,* **51**, 844-851.

18 C. D. Whitehead, 1974. The sorption of iodide by soil components. *J.Sci.Food Agric.,* **25**, 73-79.

19 B. T. Wilkins, 1989. Investigation of iodine-129 in the natural environment: Results and implications, NRPB-R225, National Radiological Protection Board, UK.

Isotope Index

Subject Index